HUMAN BEHAVIOR GENETICS

HUMAN BEHAVIOR GENETICS

Compiled and Edited by

ARNOLD R. KAPLAN

Director, Laboratory of Medical Genetics
Ohio Department of Mental Health and Mental Retardation Research Center
Cleveland, Ohio
Visiting Professor, Department of Preventive Medicine and Public Health
Creighton University School of Medicine
Omaha, Nebraska

With Technical Assistance of Wilma E. Powell

QH457
K36
1976

CHARLES C THOMAS · PUBLISHER
Springfield Illinois · U.S.A.

Published and Distributed Throughout the World by
CHARLES C THOMAS · PUBLISHER
Bannerstone House
301-327 East Lawrence Avenue, Springfield, Illinois, U.S.A.

This book is protected by copyright. No part of it
may be reproduced in any manner without written
permission from the publisher.

© 1976, by CHARLES C THOMAS · PUBLISHER
ISBN 0-398-03378-1
Library of Congress Catalog Card Number: 74-30141

With THOMAS BOOKS careful attention is given to all details of manufacturing and design. It is the Publisher's desire to present books that are satisfactory as to their physical qualities and artistic possibilities and appropriate for their particular use. THOMAS BOOKS will be true to those laws of quality that assure a good name and good will.

Library of Congress Cataloging in Publication Data

Kaplan, Arnold R
   Human behavior genetics.

   Includes indexes.
   1. Behavior genetics. 2. Human genetics. I. Title.
[DNLM: 1. Genetics, Behavioral. BF701 K17h]
QH457.K36   155.7   74-30141
ISBN 0-398-03378-1

Printed in the United States of America

An ancillary benefit from editing a book involves the pleasure of writing its dedication. This volume is dedicated, with gratitude and affection, to Edward N. Hinko, M.D., Director, Ohio Mental Health and Mental Retardation Research Center, under whose leadership I had the good fortune to work for more than a decade. His high ethical standards, ideals, humanistic dedication, social concern, and abiding faith have characterized his professional and personal life, with little attention to personal honors or acquisition. Dr. Hinko has devoted his career to working for improved mental health programs: to develop better programs and facilities for patient care, to select and train clinicians and researchers, to facilitate basic research for increasing our understanding of the nature of the disorders and their control, and to integrate these programs at state psychiatric institutes, providing academic and community leadership for progress rather than mere custodial care. Despite setbacks associated with political expediencies, progress will continue—because of the dedication and value systems of people like Dr. Edward N. Hinko.

—A.R.K.

CONTRIBUTORS

KAZUHIKO ABE, M.D.: Associate Professor, Department of Psychiatry, Osaka City University Medical School, Osaka, Japan.

RICHARD ALLON, Ph.D.: Staff Psychologist, Division of Clinical Psychology; Associate, Research Division, Department of Psychiatry, Toronto General Hospital; Assistant Professor, Departments of Behavioural Science and Psychiatry, Faculty of Medicine, University of Toronto, Toronto, Ontario, Canada.

THOMAS J. BOUCHARD, JR., Ph.D.: Professor, Psychology Department, University of Minnesota, Minneapolis, Minnesota.

ROBERT CANCRO, M.D.: Professor, Department of Psychiatry, University of Connecticut Health Center, Farmington, Connecticut.

H. J. EYSENCK, Ph.D., D.Sc.: Professor, Department of Psychology, Institute of Psychiatry, Maudsley Hospital, London, England.

ROBIN FOX, Ph.D.: Professor of Anthropology, The Graduate School, Rutgers University, New Brunswick, New Jersey.

EDWARD N. HINKO, M.D.: Director, Ohio Department of Mental Health and Mental Retardation Research Center, Cleveland, Ohio.

ARNOLD R. KAPLAN, Ph.D.: Director, Laboratory of Medical Genetics, Ohio Department of Mental Health and Mental Retardation Research Center, Cleveland, Ohio; Visiting Professor, Department of Preventive Medicine and Public Health, Creighton University School of Medicine, Omaha, Nebraska.

HENRY T. LYNCH, M.D.: Professor and Chairman, Department of Preventive Medicine and Public Health, Creighton University School of Medicine, Omaha, Nebraska.

AUSTIN E. MOORHOUSE, B.A.: Administrative Specialist, Ohio Department of Mental Health and Mental Retardation Research Center, Cleveland, Ohio.

ARNO G. MOTULSKY, M.D.: Professor of Medicine and Genetics; Director, Center of Inherited Diseases, University of Washington, Seattle, Washington.

GILBERT S. OMENN, M.D., Ph.D.: Associate Professor of Medicine, Division of Medical Genetics, University of Washington RG-20, Seattle, Washington.

WILLIAM POLLIN, M.D.: Coordinator of Research Programs, Division of Narcotic Addiction and Drug Abuse, National Institute of Mental Health, Rockville, Maryland; Chief, Section of Twin and Sibling Studies, Adult Psychiatry Branch, National Institute of Mental Health, Bethesda, Maryland.

WILMA E. POWELL, B.A.: Administrative Specialist, Ohio Department of Mental Health and Mental Retardation Research Center, Cleveland, Ohio.

JOHN D. RAINER, M.D.: Chief, Psychiatric Research (Medical Genetics), New York State Psychiatric Institute, New York, New York; Professor of Clinical Psychiatry in Human Genetics and Development, Columbia University, New York, New York.

DIANE SANK, Ph.D.: Department of Anthropology, City College of the City University of New York, New York, New York, and Research Center, Rockland State Hospital, Orangeburg, New York.

PROFESSOR J. P. SCOTT: Director, Center for Research on Social Behavior, Bowling Green State University, Bowling Green, Ohio.

CHARLES R. SHAW, Ph.D., M.D.: Chief, Section of Medical Genetics, Department of Biology, The University of Texas at Houston, M.D. Anderson Hospital and Tumor Institute, Houston, Texas.

ALEXANDER THOMAS, M.D.: Professor of Psychiatry, New York University Medical Center, New York, New York.

STEVEN G. VANDENBERG, Ph.D.: Department of Psychology and Institute for Behavioral Genetics, University of Colorado, Boulder, Colorado.

PROFESSOR DETLEV V. ZERSSEN, M.D.: Max-Planck-Institut für Psychiatrie, München, West Germany.

FOREWORD

Research support in the United States has undergone dramatic changes in recent years. We have witnessed the end of an era marked by extensive support for basic biological and behavioral research. That was the context in which human behavior genetics began to flourish. Government support for research has been markedly diminished, and the available research funds now tend to be channeled into projects labeled "applied" or "clinical." The State of Ohio has followed the examples of the U.S.P.H.S. National Institutes of Health and is currently phasing out its support for biological research in mental health and mental retardation. The elimination of a luxury such as basic research will save the taxpayers relatively little money. But, it means the termination of much relevant research in behavior, mental disorder, and mental retardation; and an end to enrichment programs that once held promise of ultimately attracting, for competitive selection, high-caliber professionals and trainees. Instead of continuing to develop academically-oriented and progressive institutional settings, there is reversion to a largely custodial orientation. The eagerness to obliterate programs that took years to build up, i.e. as part of the plan to upgrade state hospital programs and staffs and to facilitate research and interest in areas where basic knowledge is nearly as deficient today as it was two decades ago, has been rationalized as required for budgetary expediency. This philosophy evidently crosses political party lines and involves liberal as well as conservative leaders.

Little contribution can come from despondency over "what might have been." Many investigators have found it necessary and expedient to radically alter their professional activities. For those of us who had organized our professional lives within the context of basic science research, the changes have certainly been discouraging. Nevertheless, much was accomplished before the changes occurred, and we can be happy with those achievements.

Living beings and their institutions are dynamic, and the future includes promise for change and improvement.

PREFACE

The human species is remarkable for its intraspecies variation. Each human population manifests extensive ranges of individual differences within that population. These are effects of interactions between the experiential variables and constitutionally polymorphic and extensively heterozygous groups of individuals. Even human isolate populations cannot be equated with highly inbred and homozygous strains of laboratory animals. Behavioral extrapolations to human populations, from strains of laboratory and/or farm animals, de-emphasize the characteristically human species quality of enormous intrapopulation differences. Such extrapolations also evade the characteristically human involvement of culture. Determination of valid generalities in human behavior requires recognition of the characteristically human individuality within all human populations and the behavioral uniqueness of each member of the species.

Dangers of overgeneralizing and of confounding facts with theories are not unique to human behavior genetics. Actual and potential social and political aspects of the discipline of human behavior genetics, however, structure the need for exceptionally careful controls and conservative judgments. In such an area, it is particularly appropriate to emphasize that even an immaculately controlled scientific study can, at most, determine the character of an association between two variables. That is the limit of the scientifically determined facts. For interpretations of the facts, for their meaning, we are at the mercy of the interpretors: scientifically determined evidence are *facts,* but interpretations of evidence are *theories.*[1] Interpretations and theories, while they may be based on objective facts, are themselves prone to be influenced by subjective bias. Sometimes the fire that the smoke betrays is within, and interpretors' explanations of facts may sometimes tell us more about the interpretors than about the facts.[2,3]

REFERENCES

1. Montagu, A.: Introduction. In Montagu, A. (Ed.): *Man and Aggression.* New York, Oxford U Pr, 1968.
2. Montagu, A.: "Original sin" redivivus. *J Historical Studies, Spring*:132,1969.
3. Taylor, G. G.: *Sex in History.* New York, Vanguard, 1954.

ACKNOWLEDGMENTS

This book would not have been conceived or organized without the help, support, and inspiration of numerous colleagues and friends. My principal thanks are due to each of the contributing authors for their scholarly contributions, cooperation, patience, and tolerance. Dr. Edward N. Hinko developed the professional setting which inspired the idea for this volume, and his interest and help facilitated its completion even after the editor's diversion to another professional context. The technical help of Wilma Powell has been invaluable. The technical assistance of Austin Moorhouse is also warmly appreciated. Special thanks are due to my dear friends, Carol and Leonard and Lisa Ullman, to Betty and Larry Thurman, and to the principal women in my life, Anita and Pamela and Joanne Kaplan, for their moral support and patience and understanding. Finally, on the occasion of the eighty-fourth birthday of my father, William Kaplan, I am delighted to express enthusiastic gratitude to my parents for their most significant contributions to my behavior patterns and value systems, as well as my genotype.

—A.R.K.

CONTENTS

Foreword ix
Preface xi

Chapter
1. INTRODUCTION 3
2. BASIC HUMAN GENETICS 5
3. THE FUTURE OF BEHAVIOR GENETICS 33
4. ON THE GENETICS OF BEING HUMAN 49
5. MEASURING HUMAN BEHAVIOR 63
6. TWIN STUDIES 90
7. BEHAVIORAL INDIVIDUALITY IN CHILDHOOD 151
8. GENETIC FACTORS IN INTELLIGENCE 164
9. GENETIC FACTORS IN PERSONALITY DEVELOPMENT 198
10. PHYSIQUE AND PERSONALITY 230
11. GENETIC AND ENVIRONMENTAL DETERMINANTS OF NEUROSIS 279
12. GENETICS AND HOMOSEXUALITY 301
13. GENETIC AND ENVIRONMENTAL VARIABLES IN SCHIZOPHRENIA ..317
14. GENETICS, CYTOGENETICS, DERMATOGLYPHICS, CLINICAL HISTORIES, AND SCHIZOPHRENIA ETIOLOGY 330
15. PSYCHOPHARMACOGENETICS, PHYSIOLOGICAL GENETICS, AND SLEEP BEHAVIORS 347
16. PSYCHOPHARMACOGENETICS 363
17. HUMAN BIOCHEMICAL VARIATION 385
18. TASTE SENSITIVITY AND HUMAN VARIATION 401

Glossary 424
Subject Index 451
Reference Author Index 459

HUMAN BEHAVIOR GENETICS

Chapter 1

INTRODUCTION

Arnold R. Kaplan

All creature characteristics that enter into human actions are involved in our symbolic systems of cultural interpretations.[1] An individual's actions and reactions are influenced by the cultural and social systems within which the individual is operating and has previously operated.[2] The acts of pointing out and exploring cultural involvements do not deny biological aspects of the physiological elements. Reciprocally, the acts of pointing out and exploring biological involvements do not deny cultural elements. Different combinations of different biological and cultural variables affect individual differences in behavior. A living system consists of a coordinated system of genes plus an adaptive system which is subject both to the genetic system and to the environment.[3] Populations are intrinsically polymorphic, and this property is reflected in morphology, biochemistry, and behavior.[4,5] The relevance to behavior of many different kinds of variables, biological and cultural, may not be regarded as separable contributions. Rather, for each individual, all of the variables are moments of the total being.[6]

The difficulties of distinguishing between genetically-determined and environmentally-determined aspects of differences are particularly great for traits which show continuous variation. Sociocultural and psychogenic influences may facilitate transmission of a behavioral trait as consistently as genetic factors. Particular genetic factors may be manifested only by appropriate genotype-environment combinations. Genes interact with other genes and with the environment to influence the physiological or functional expression of a trait.[7]

We can study the behavior of *an* organism, the genetics of *a* population, and individual differences in the expression of some behavior by members of *that* population.[8] Problems of genetical analysis and behavioral analysis may be separated for behavior-genetic analyses only in the context of understanding the system of genotype-environment interactions responsible for phenotypes.[8,9]

The human species is characterized by a remarkable degree of intraspecies variation, morphologically and biochemically, as well as socioculturally. This species characteristic and the primary role of learning and cultural experience in a human being's "humanization" minimize the validity of extrapolating behavior-genetic phenomena from other species to humans. The difficulties and impossibilities of structuring adequate controls for behavioral and biological variables in human studies may induce investigators and theoreticians to rely upon analogies extrapolated from other species. Such activities may structure stimulating controversies, but they do not contribute to elucidation of human behavior-genetic phenomena. The human beings' primary dependence upon culture, combined with the species' remarkable polymorphism, make human behavior genetics a unique discipline.

REFERENCES

1. Goldschmidt, W.: *Comparative Functionalism.* Berkeley, U of Cal Pr, 1966.
2. Kaplan, A.R.: Concluding remarks: Genetics and schizophrenia. In Kaplan, A.R. (Ed.): *Genetic Factors in "Schizophrenia."* Springfield, Thomas, 1972.
3. Thoday, J.M.: Components of fitness. *Symp Soc Exp Biol, 7*:96, 1953.
4. Kaplan, A.R.: Introduction. In Kaplan, A.R. (Ed.): *Genetic Factors in "Schizophrenia."* Springfield, Thomas, 1972.
5. Hirsch, J.: Individual differences in behavior and their genetic basis. In Bliss, E.L. (Ed.): *Roots of Behavior.* New York, Har-Row, 1962.
6. Merleau-Ponty, M.: *Phenomenology of Perception* (Translated by Smith, C.). London, Routledge and Kegan Paul, 1962.
7. Montagu, A.: Social interest and aggression as potentialities. *J Individ Psychol, 26*:17, 1970.
8. Hirsch, J.: Behavior-genetic analysis and the study of man. In Mead, M., *et al.* (Eds.): *Science and the Concept of Race.* New York, Columbia U Pr, 1968.
9. Hirsch, J.: Behavior-genetic analysis and its biosocial consequences. *Seminars in Psychiatry, 2*:89, 1970.

Chapter 2

BASIC HUMAN GENETICS

Arnold R. Kaplan

INTRODUCTION

Individual differences in predisposition to acquire a particular trait, characteristic, or disorder in any specific environment are the consequences of individual differences effected by interacting genetic and nongenetic factors. The etiology of a physiological or pathological characteristic may be evaluated in terms of the interactions of genetically transmitted diatheses and exposures to precipitating factors. Any particular property in an organism derives from a complex series of reactions, interactions, and developmental processes. Interference with these at any of many different levels may affect the final end-product.

The complexity of metabolic activities within each cell is such that vast numbers of genes influence almost any activity. A particular gene or group of genes determines an indefinite but limited assortment of effects, the different effects being associated with differences in environment and/or with differences in other aspects of the total genetic makeup or genotype. One's genetic material, derived from one's parents, is a set of potentialities and not a set of already-formed or predetermined characteristics.

Controversy over the relevance of genetic differences to the etiology of any category of characteristics or traits may be instigated by concern for the social consequences of such relevance. The involvement of gene-transmitted constitutional differences in different diatheses does not necessarily imply a poor prognosis or attenuate the value of compensatory therapy. The occurrence of even a primary relevance of genetic differences regarding different predispositions to a category of traits does not mean that the traits are any less amenable to environmental or therapeutic modification than they would be if all the principle etiological variables for the same traits were environmental.

A trait or characteristic is an effect of, and may be affected by, genetic and/or environmental alterations. Theoretically, no characteristic is necessarily unpreventable or unchangeable; but the environmental/therapeutic changes which must be introduced in order to affect prevention, control, cure, or modification may be unknown. Identification of the gene-transmitted differences which are relevant to a trait's etiology may facilitate that trait's prevention or amelioration by purely environmental methods. The more it becomes possible to precisely characterize an individual's genotype and the potential range of phenotypic (i.e. trait) variation resulting from interactions of that genotype with different environmental variables, the more likely are we to learn how to select and modify environments according to the different individuals' needs. Identification of genotypes which manifest certain diatheses, in particular environmental milieux, provides a basic step in the direction of effective control. Determination of the environmental differences between the genetically predisposed individuals who develop an undesirable trait and those who do not will characterize the environmental modifications necessary to prevent and control the undesirable trait. Determination of the constitutional differences associated with the specific genes related to different diatheses for an undesirable trait will contribute to knowledge of the trait's etiology and the environmental or therapeutic modifications necessary to prevent or modify its manifestation.

PHENOTYPES, GENOTYPES, PHENOCOPIES, AND GENOCOPIES

The term *phenotype* refers to an individual's observable properties—physiological and pathological, structural and functional—which are effects of the interactions between the individual's genotype and his particular environment. The sum total of an individual's genetic material, the total genetic constitution, is his *genotype*. The characteristics of an individual's development and growth, from conception to birth to maturity to death, are effects of the interactions of his genotype with his environment.

A trait or disorder which is usually associated with a particular gene or group of genes may, in response to particular environmental variables, be manifested without that gene or group of genes. Nonhereditary phenotypic modifications, which are caused by special environmental conditions and which mimic similar phenotypes characteristically associated with particular genes, have been termed *phenocopies*. Specific effects of genic action can be imitated by certain environmental influences interacting with genotypes whose norms of reaction or phenotypic flexibilities include the potentials for such manifestations, even though the genotypes do not include the particular gene(s) which is/are characteristically associated with these effects. Series of grades of phenocopies may occur to resemble effects characteristically associated with various combinations of multiple genes. A trait or disorder which is characteristically associated with a particular gene or group of genes may also occur in the absence of such gene(s) effected by other (i.e. different) genes. The effects of such mimetic genes have been termed *genocopies*.

Certain phenocopies are only inducible when a particular environmental influence is applied during a specific, or within a limited, period of development. That is, for some effects, there is a limited sensitive or critical period during which the effects may be determined. This suggests that during such critical stages in development there exist alternative paths which lead to different effects, and the path which is to be followed may be determined by genetic and/or environmental factors. The existence of a particularly sensitive period in development, during which a specific disorder is most easily initiated, means that induction of a phenocopy may depend upon the stage of development during which the environmental influence occurs, as well as upon the interacting genotype (i.e. the potential responsiveness or norm of reaction of that genotype) and the nature, intensity, and duration of the environmental influence. Only a part of the population segment consisting of individuals who are genetically predisposed to a particular disorder actually do develop that disorder.

Many criteria have been studied for indications of the significance of genetic factors in etiologies of various disorders, including elevated morbidity risks in relatives of index cases compared with the general population, increasing morbidity risks associated with increasing degrees of genetic kinship to index cases, greater concordance in monozygotic than in dizygotic pairs of twins, and variation in frequencies of the disorders in different populations. These observations alone are not rigorous and conclusive evidence of genetic etiology. Environmental factors common to relatives can simulate genetic determinants. It is not possible to conclusively prove that a trait is genetic except by showing that it could be due to one of several modes of inheritance. The genetic method consists of attempting to extract from a possibly heterogeneous trait one or more genetic entities due, in increasing order of refinement, to a single inheritance pattern, locus, or allele. Sometimes genetic traits depend on the presence of several genes (i.e. complementary epistasis). If the effect of each gene is recognizable, each component may be studied separately. Otherwise, the analysis may be beyond the limits of available methods in human genetics. The only unequivocally reliable approach to a nonexperimental system is to defer final determination of mode of inheritance until the relevant genes are individually recognizable. Traits determined by many genes, not individually recognizable, are unfavorable for genetic analysis.

The occurrence of multiple etiologies and associated multiple morbidity risks may be masked by a diagnostic system which is based upon the clinical symptoms or effects of a disease rather than its etiology. Thus, multiple morbidity risks involving clinically heterogeneous populations may be combined to indicate a composite morbidity risk figure which is invalid and misleading.[1]

Relatives of index cases can be divided grossly into categories according to their memberships in high-risk versus low-risk families if such a division of particular data can be statistically supported by segregation analysis.[2] Such a dichotomy could be based, for example, upon familial versus nonfamilial or sporadic index cases, or upon concordant versus discordant pairs of monozygotic twins. Even if the traits or disorders in most or many of the affected individuals were associated with simple modes of genetic transmission, some cases could be sporadic due to occurrence of phenocopies, genocopies, diagnostic errors, etc. The assumption of the occurrence of sporadic cases may lead to recognition of heterogeneity in data previously regarded as homogeneous and may even show that the risk in some high-risk families is great enough to suggest a simple genetic hypothesis.[3] Genetic ratios within sibships may be utilized to test for Mendelian modes of genetic transmission based upon observations of individuals in only a single generation. Pedigree analysis, however, which utilizes available data regarding individuals in various generations of the family tree, is far more informative and is a principal method of genetic study in humans.

PEDIGREE ANALYSIS

Essentially, pedigree analysis involves utilization of basic Mendelian principles for extrapolation of genotypes from information on phenotypes and genetic relationships.

The term *family* is usually applied to a pair of parents and their children. The term may also be used to refer to a more extensive association of relatives, although a larger group of individuals who are related to each other genetically and through marriage is usually called a *kindred*. Females are usually symbolized by circles and males by squares, but sometimes other symbols are used, e.g. the symbol for Mars to indicate a male, the one for Venus to indicate a female. (See Fig. 2-1.) The symbols of a pair of parents are joined by a horizontal mating line, and their offsprings' symbols are located in a horizontal row below that line. Double horizontal mating lines indicate a consanguinous combination (e.g. in Fig. 2-1, III 2 and 3 is a consanguinous mating). The offsprings' symbols are connected by vertical lines to a horizontal line above them which is connected further above by a single vertical line to the horizontal mating line. The children of a parental pair form a *sibship;* and the children in a sibship, regardless of sex, are *siblings* or *sibs* of each other. Thus, in Figure 2-1, there are six sibships: II, 1 and 2; II, 3, 4, and 5; II, 6 and 7; III, 1 and 2; III, 3, 4, 5, and 6; and IV, 1, 2, and 3. Possession of the trait being examined in the pedigree is designated by shading the symbol (e.g. circle or square) of the individual involved, and an unshaded symbol

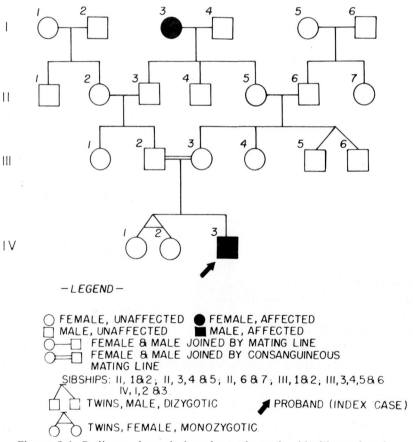

Figure 2-1. Pedigree chart designed to schematize kinship and trait associations in a kindred.

denotes absence of the character or disease. Twins are designated by lines extending from the two symbols involved to a common point on the horizontal sibship line. In Figure 2-1, two pairs of twins are shown: III, 5 and 6; and IV, 1 and 2. Monozygotic (MZ) twinship is designated by insertion of a horizontal line below this common point (e.g. IV, 1 and 2, in Fig. 2-1). The affected individual with whom a particular pedigree study was initiated may be identified as the *index case* or *proband* or *propositus,* and this is shown by an arrow in the schematic. Generations are customarily identified with Roman numerals, ranging from the earliest at the top of a pedigree schematic to the most recent at the bottom. Within each generation, the individuals are numbered from left to right with Arabic numerals. Thus, each individual in a particular pedigree may be identified by the combination of a Roman numeral and an Arabic numeral.

Pedigree patterns characteristically provide information on the Mendelian principles of segregation and independent assortment, and they may also provide information on allelism and linkage.[4] A specific pedigree pattern depends upon whether effects of the gene involved are manifested in single dosage or only in double dosage and upon whether the gene is located on one of the autosomes or on a sex chromosome. An individual who possesses an identical pair of genes on a particular locus of a pair of homologous chromosomes is *homozygous* for the locus, and an individual whose pair of genes on a particular locus are different from each other is *heterozygous* for the locus. In a heterozygous individual, if one of the pair of genes is recognizably manifested while the other one is suppressed, the former gene is termed *dominant* and the latter one is *recessive.* Genetic dominance is a manifestation of a gene's ability to be expressed in the phenotype of a genetically heterozygous individual. The failure of a gene to be expressed in the phenotype, when it occurs in a genetically heterozygous genotype, is genetic recessivity. A recessive gene may be carried and transmitted through innumerable generations without being manifested. A dominant gene, however, is characteristically manifested in the phenotypes of the individuals who possess either one (i.e. in the heterozygous individual) or a pair (i.e. in the homozygous individual) of them.

The manifestation of a trait transmitted by a single autosomal dominant gene with complete penetrance is indicated in a pedigree in which the trait does not "skip" generations. An individual with the trait has a parent with the trait, and approximately one-half the offspring of couples in which one of the parents is affected are similarly affected. A trait which is transmitted by a pair of autosomal recessive alleles, by contrast, may occur without any previous family history of the trait in either the maternal or the paternal side. Children born subsequent to an index case, establishing both parents as heterozygous carriers of the recessive gene for the trait, will include approximately one fourth affected, and approximately two thirds of the unaffected siblings will be heterozygous "carriers" of the recessive gene. In the case of a very rare trait (in the population) transmitted by a pair of autosomal recessive alleles, consanguinity may be expected to occur more frequently in affected families than in the general population.

Genes which are located on the X chromosome, like those on autosomes, may be dominant or recessive. A female, with two X chromosomes, may be either homozygous or heterozygous for a particular locus on the X chromosome. A male, with only one X chromosome, is hemizygous for any X-linked gene. Such a gene may be dominant or recessive in the female. In the hemizygous male, with no other gene at the locus, an X-linked gene is always expressed. Thus, X-linked genetic transmission is characterized by the absence of male-to-male transmission because a son necessarily inherits a Y chromosome from his (XY) father and an X chromosome from his (XX) mother. The X chromosome of a male is characteristically transmitted to none of his sons but to all of his daughters.

One's sex may affect the expression of a gene, and a trait transmitted by a gene which is manifested differently in the two sexes is *sex-influenced*. That is, a sex-influenced trait is one in which phenotypic manifestations are different in the two sexes. If the gene is completely suppressed in one of the two sexes, then it is *sex-limited*. A sex-influenced or sex-limited gene (i.e. regarding its manifestation in the phenotype) may occur on an autosome or on an X chromosome.

Genetic *linkage* involves the occurrence of different genetic loci on the same chromosome. If two loci occur on separate nonhomologous (i.e. not pairing at meiosis) chromosomes, then independent assortment occurs. If two loci occur on the same chromosome, and particularly if they occur relatively close to each other, then the genes occurring on the two loci would usually tend to be transmitted in the same combination (i.e. without independent assortment) from generation to generation. Close genetic linkage of different loci may be difficult to distinguish from allelism, i.e. occurrence of alternate gene forms at a single locus on a particular chromosome. The term, *X-linked*, refers to the location of a gene on the X chromosome. A gene whose locus occurs on the Y chromosome is termed *Y-linked* or *holandric*. Such a gene would characteristically affect only males and would be transmitted only from father to sons.

PENETRANCE, EXPRESSIVITY, AND HERITABILITY

A trait or disorder may be regarded as the dependent variable which has resulted from interactions of independent variables (e.g. environmental factors) and the organism with its intervening variables (e.g. morphological and biochemical characteristics). The intervening variables do not remain constant, but are themselves modified by the independent variables which affect the organism. The *penetrance* of a gene or group of genes and the nature of its/their *expressivity*, observed as phenotypic manifestations, differ in different *milieux*, environmental and/or genetic. Thus, there are differences in penetrance of the gene(s) associated with predisposition to a disease which are associated with different environmental milieux and different genetic milieux (i.e. the complete genotypes). When the gene(s) is/are penetrant, these differences affect variability in expressivity. Complete penetrance is shown by a genotype which is always associated with a particular phenotype. Even when a genotype associated with predisposition to a particular pathology is penetrant, there may be clinical differences between different affected individuals. Thus, even when penetrant, a genotype may show variable expressivity. Environmental, as well as ancillary genetic, variables interpose numerous influences between primary gene products and the final mechanisms by which genetic manifestations are actuated. Environmental variables may affect reductions in correlations between genotypes and observable phenotypic effects, causing variable expressivity and/or incomplete penetrance.

Heritability determinations are estimates of the proportion of the total phenotypic variance (i.e. individual differences) shown by a trait that can be attributed to genetic variation in some particular population at a single generation under one set of conditions. The heritability of a particular pathology may be defined as the extent to which the variation in individual risk of acquiring the pathology is due to genetic differences. A trait will show a greater-than-zero heritability if two or more segregating alleles, which manifest different effects upon the trait, occur on at least one genetic locus. A relevant genetic locus, which is associated with phenotypic variation in one population because it is represented by two or more different segregating alleles, might show no variation in another population because the same locus involves only a single allele (i.e. identical genes on homologous loci) in that other population. One environment may activate a particular gene, while another may not. Thus, a trait may show a particular greater-than-zero heritability in one allele-segregating population but other heritabilities in other populations which involve population-genetic differences and/or environmental differences. The different heritabilities shown for a particular trait in different populations do not indicate that any particular gene manifests different degrees of heredity in the different populations involved. The ontogeny of an individual's phenotype (i.e. observable outcome of development) has a norm or range of reaction which is not predictable in advance. Even in the most favorable investigations, only an approximate estimate can be obtained for the norm of reaction, when, as in plants and some animals, an individual genotype can be replicated many times and its development studied over a range of environmental conditions.[5] The more varied the conditions, the more diverse might be the phenotypes developed from any one genotype. Different genotypes do not have the same norm of reaction. Therefore, the limits set by a particular gene or group of genes cannot be entirely specified. These limits are plastic within each individual and differ between individuals.[5] Extreme environmentalists have been wrong to hope that one law or set of laws might describe universal features of modifiability, and extreme hereditarians have been wrong to ignore the *norm of reaction*.[5] In its strictest sense, a heritability measure provides an estimate of the proportion of the variance which a specific population shows in trait (i.e phenotype) expression and which is correlated with segregation of independently acting alleles of a relevant genetic locus or of relevant genetic loci. Heritability is a property of a specific population and not of a trait (i.e. in all populations or in all *milieux*). In any particular population, heritability coefficients are based on observed correlations among individuals with different degrees of genetic kinship. Many published twin, family, twin-family, and extended-pedigree studies of disease incidence have essentially been studies of heritability, and they have not indicated a clearly-defined mode of heredity (i.e. a specific mode of genetic transmission). Heritabilities may vary between different groups because of different environmental contexts, as well as because of numerous secondary genetic (i.e. in addition to the genetic factors predominantly associated with the liability to the pathology) differences. Thus, the findings from different studies of heritability for any category of disorder which does not involve a clear and simple mode of genetic transmission with complete penetrance may be expected to vary from each other. The demonstration of significant heritability for a pathology in one particular group indicates the major relevance of genetic differences for different predispositions to the pathology. The demonstration of lower heritability for the same pathology in another group may indicate that the effects of these genetic factors (i.e. within the genetically defined norm or range of reaction) may be profoundly modulated by other genetic factors and/or by nongenetic influences.[6] Different combinations of different genetic and nongenetic factors may manifest similar phenotypes. The same genetic or nongenetic factor affecting different individuals, who differ with regard to other genetic factors and/or environmental influences, may show different manifestations. Thus, different genes may be associated with the same phenotype, and different phenotypes may be associated with the same gene(s), depending upon the total interacting genetic and environmental contexts. If the incidence of a particular pathology varies from one population to another and the difference is not due solely to diagnostic inconsistencies, the variance can be ascribed to differences of mean liability or to differences of variance of liability or to combinations of both.[7] If there is a difference between two populations in the pathology's heritability, then two or more (rather than only one) predisposing genotypes may be involved, and/or the lower heritability may be associated with environmental circumstances which increase the nongenetic differences of the liability. The relative contributions of heredity and environment for a trait may differ with different overall heredities (i.e. total genotypes) and with different environments. The reaction-norm concept is of fundamental importance to an understanding of gene action. Modification of a gene's environment, either by nongenetic environmental influences or through the effects of other genes, may affect the expression of any specific gene within the limits characteristic of that gene's phenotypic potential.

GENETIC RATIOS AND STATISTICAL SIGNIFICANCE

Inference or reasoning is a mode of thinking by which one starts from something known and proceeds to form a belief that there exists a fact hitherto unknown. The basic assumption is that one's methods of reasoning are reliable and lead to conclusions which correspond with facts. Many of the reasoned conclusions can then be checked and verified by perceptions of objective investigations. Every perception is received by a mind already made up of memories, interests, expectations, and bias. Therefore, the resultant new state of consciousness is derived in part from the observations and in part from what is in the mind already. Even at the times of observations, there are natural tendencies to confuse one's preconceptions with what is actually perceived.[8] The most useful hypotheses are those which are testable, and testability is associated with simplicity and directness. As hypotheses increase in their complexity and as they involve secondary hypotheses, their testabilities diminish. The contributions of additional hypotheses which cannot be subjected to conclusive tests, which cannot be proved or disproved, provide little more than exercises for their contributors. The many relevant variables cannot always be entirely controlled, and some of them may not even be recognized. One may measure the overall variation, which is due to all the uncontrolled variables, as well as the differences that can be related to the factors one is evaluating. Then, by comparing the two, one may judge whether the latter are statistically significant to provide confidence that they are not just the misleading expression of some uncontrolled, or even unrecognized, source of variation.[8] Scientific conclusions necessarily emerge as statements of probability. There is always the risk that particular results do not facilitate a sound or unbiased basis for a general statement, as reflected in the statement of probability which the analysis yields. The conclusions drawn from any study or combination of studies are not immutable or certain; but as the body of observational experience grows, the level of uncertainty is decreased.

A deviation from the expected ratio, in the observed ratio of unaffected and affected individuals, may be due to chance or to some relevant variable. The meaning of the *statistical significance* of a deviation can be calculated in terms of statistical probability that it may occur due only to chance. Statistical probability, the mathematical evaluation of chance, may be defined as the ratio of the number of occurrences of a specified combination of events to the total number of all possible combinations of the events. The term *statistical significance* relates to whether or not some specific reason or reasons, and not only chance, underlies the observed deviation. The probability level which is regarded as significant is somewhat arbitrary and depends upon the judgment of the investigator involved. If the probability that a deviation may occur between two populations, or between the observed and expected findings, is lower than 1 percent, then the findings would generally be regarded as statistically significant. A probability level above 1 percent but below 5 percent is sometimes considered significant, but more conservative investigators tend to consider such a figure as only of doubtful or questionable significance. The choice of which statistical test might most appropriately be utilized for any particular evaluation depends upon many factors, including the sizes of the samples involved and the parametric or nonparametric character of the data distributions.[9, 10] One of the simplest and most commonly used tests in genetics is the *chi-square test:* the value of chi-square equals the sum of the squares of the absolute differences between the observed and expected categories divided by the respective expected values being compared, and the probability value may be determined from a corresponding chi-square value listed in standard tables.[4] The happenings of nature involve probability laws rather than predetermined causality. A scientific investigator only determines his best posits from the available information.[11] The addition of more specifically relevant information facilitates more accurate judgments. A genetic factor may be manifested only by appropriate genotype-environment combinations. The determination of whether a gene is harmful or useful or indifferent is related to the bearer's environment. The genetic epidemiologist functions as a kind of ecologist seeking significant correlations between a disorder and one or another variable from the great array of environmental influences. His success in these efforts will be related to the uniqueness of the variable involved and the directness of its effect, to the frequency of the basic defect, and to the ease of detection of the disorder under consideration.

THE HARDY-WEINBERG PRINCIPLE

Hardy and Weinberg independently developed the principle that genotypic frequencies within a population in any generation depend solely upon the frequencies in the previous generation if none of the genotypes was affected by selective factors and if the population was random-mating.[12, 13] The fundamental idea of gene distribution in populations, based on the assumption of random mating and no differential selection and no significant contribution from new mutations, is a deduction from the Mendelian principles of segregation and recombination. In a large population, if A_1 and A_2 are alleles for one particular locus, and the frequency of the gene A_1 in the population is p_1, and the frequency of the gene A_2 is p_2, then the genotypic frequencies after one or more random matings will be: for genotype (A_1A_1), $(p_1)^2$; for genotype (A_1A_2), $(2p_1p_2)$; and for genotype (A_2A_2), $(p_2)^2$. The proportions may be determined by expansion of the binomial, $(p_1 + p_2)^2$. The genotypic frequencies may be determined for any number of alleles on a single locus by expansion of the appropriate binomial. Thus, for three alleles, with the genes A_1, A_2, and A_3 occurring in the population, respectively, at frequencies p_1, p_2, and p_3, $(p_1 + p_2 + p_3)^2$ equals $p_1^2 + p_2^2 + p_3^2 + 2p_1p_2 + 2p_1p_3 + 2p_2p_3$. The frequencies, then, would be: for (A_1A_1), $(p_1)^2$; for (A_2A_2), $(p_2)^2$; for (A_3A_3), $(p_3)^2$; for (A_1A_2), $(2p_1p_2)$; for (A_1A_3), $(2p_1p_3)$; and for (A_2A_3), $(2p_2p_3)$.

SELECTION, DRIFT, AND MUTATION

If individuals of differing genotypes should vary in their viabilities or fertilities, then there would be differences in the relative contributions of different genotypes to the succeeding generation. That is, *selection* would occur in different degrees for the different genotypes. The alleles in individuals whose genotypes are characterized by higher reproductive fitness will be represented in higher proportions than the alleles associated with the individuals of lower reproductive fitness. Selection can cause alterations in gene frequencies from one generation to the next.

The process of *genetic drift* may lead to the establishment of a trait which is neutral and nonadaptive or even unfavorable.[14] It involves the random fluctuation of gene frequencies due to chance and is particularly important for small isolated populations. The probability of genetic drift affecting the gene frequencies of a population depends largely upon the size of the population: with a greater number of parents, the likelihood of loss and fixation are lower because of the decreased probability that the same chance loss will occur consistently for the different parents. When an isolated population is small, the speed of drift may be high from one generation to the next; but when a population is large, the process of drift would tend to be a slow one.

Mutations are changes in the genetic material which are irreversible except for subsequent mutations. *Chromosomal mutations* involve genetic changes which are visible with present cytogenetic techniques and involve duplication, deletion, or translocation of a complete chromosome or discernible chromosomal fragment. *Gene mutations* or *point mutations* involve changes which are not cytologically visible and are identifiable only by their phenotypic manifestations. The estimated frequency of a gene mutation may be calculated. One method for such a calculation, primarily applicable to dominant mutations, is based simply on a census of the frequency of children with the particular dominant trait under study who are born to parents without the trait.[15] The underlying assumptions for such a direct calculation involve complete penetrance of the gene and the occurrence of no other genetic locus or mode of genetic transmission associated with the trait. An indirect method for calculating a particular mutation rate may be based upon the hypothesis that recurrent mutations from normal to abnormal balance the loss in fitness associated with the mutation. Thus, the number of new mutants would depend on the total number of normal alleles and the frequency with which they can mutate.[15] For a dominant mutant, the mutation rate equals $1/2(1 - f)x$ when f equals the fraction of mutant genes which are "lost" because of reduced reproductive fitness, and x is the frequency of the abnormality in the parent generation. The mutation rate for a recessive mutant may be approximated, using the same concept of equilibrium between new mutations and loss from the following formula: mutation rate equals $(1 - f)x$. An adaptation of the latter formula may be used for estimating the mutation rate for an X-linked recessive mutant. Since hemizygous males have one third of the X-linked abnormal alleles in the population, one third of the X-linked abnormal alleles are exposed to reduced reproducive fitness and some loss. Thus, in an equilibrium between mutation and elimination, the mutation rate equals $(1/3)(1 - f)x_m$ when x_m is the frequency of the abnormality among the males of the parent generation.[15]

LINKAGE

Genetic linkage is indicated by the greater association in inheritance of two or more nonallelic genes than would be expected with independent assortment. Genes are linked to each other when they occur on the same chromosome. In studying possible genetic linkage, it is first necessary to exclude the various other causes of positive or negative nonrandom associations of different traits. These causes include multiple effects of a gene, allelism, population heterogeneity, and inbreeding.

A high frequency of association between different traits is often simply due to multiple effects of the same gene. This is clearly so when the association is absolute, that is, when the two traits are always found together, for example, the excretion of phenylpyruvic acid and the absence of phenylalanine oxidase. The influence of allelism on associations of different traits is the converse of that of a gene with multiple effects. In a heterogeneous population, in which mating occurs at random with regard to the loci involved, and which is at equilibrium, associations of traits due to separate unlinked genes occur at random. On the other hand, nonrandom association of two genetic traits and the presence of a random association of either of these with a third, within a heterozygous population, may under certain conditions be indicative of linkage.[15] Inbreeding, more marriages between relatives than in panmixis, is another cause of a frequency of association of traits that exceeds random expectations in a population.

Direct analysis for linkage is possible if the genotypic phase of a double heterozygotic parent is known. The genotypes of the parents of sibships that provide information on linkage can sometimes be deduced with certainty from genetic information on the sibs' grandparents, investigations of the kinds of children in large sibships, and inspections of the parental phenotypes.

Indirect approaches to analyze for linkage make use of statistical phenomena that are consequences of linkage and can be found even in collections of pedigrees where the genotypes of parents are incompletely known. One can construct a table of mean values for the whole range of recombination values from complete linkage to independence, adjusted for any number of sibs in each sibship. Such a table can then be consulted for the detection and estimation of linkage.[15] The sib-pair method, conceived by Penrose and Burks, is another independent method for linkage analysis which utilizes data from a single generation.[16, 17] This method provides a means for distinguishing between independent recombination and linkage of different genes without knowledge of the parental genotypes. Accordingly, the entries of paired sibs in a four-fold table will show a random distribution when there is no linkage; but, when linkage exists, an excess of sib pairs will occur in those cells where the two sibs are alike in both traits or unlike in both. A third method for the detection and estimation of the strength of linkage, conceptually the simplest, was devised by Haldane and Smith.[18] This method consists of determining the amount of information available in a collection of data on two loci and comparing the probability of obtaining such data if the two loci are linked with the probability that they are not. The ratio of these two probabilities gives the odds for or against linkage. For example, when the probability of obtaining the observed distribution of two traits in a pedigree is much higher if the genes are linked than if they are not, then the odds are obviously in favor of linkage. Conversely, when the probability of obtaining the information given by a pedigree is much higher if the genes are not linked than if they are, then the odds are against linkage. The probability of obtaining the particular distribution of traits shown by a given pedigree or collection of pedigrees is expressed as a function of the recombination value, and different degrees of probability correspond to different degrees of linkage. The recombination value that gives the highest probability that the observed distribution would be obtained also provides the best estimate of the degree of linkage between the two loci.[15]

MULTIFACTORIAL INHERITANCE

Multifactorial, or polygenic, inheritance refers to the mode of genetic transmission involving traits which are associated with more than a single genetic locus, i.e. with more than a single pair of alleles. There is, as yet, no conclusive proof for multifactorial determination of alternative-trait development in man. Multifactorial inheritance could account for many abnormal traits that seem to have a genetic basis but have no regular sequence of generations or clear-cut Mendelian ratios. A multifactorial theory may be involved in an attempt to account for inheritance of graded characters within a basic Mendelian framework or in association with the concept of a genetic-threshold effect. Characteristically, such theories have been applied to problems which were obviously insoluble in terms of major genes. The principal fault associated with multifactorial theories is the fact that they are not specifically testable or refutable. A multifactorial device can be adapted to explain anything, with components which cannot be independently checked by observation.[19]

PAIR CONCORDANCE ANALYSIS

Traits or characters which can be dichotomized (e.g. into categories of affected versus not affected) may be subjected to analyses of parent-offspring and sibling-sibling pair concordance. A comparison of the concordance rates shown by different pair categories cannot, by itself, prove any particular hypothesized mode of genetic transmission. Nevertheless, such a comparison of the various types of parent-offspring pairs may be utilized to evaluate manifestations of sex influence, sex limitation, or sex linkage in transmission. In addition, the concordance proportions of the parent-offspring and sibling-sibling pairs may indicate whether or not a particular hypothesized mode of genetic transmission is consistent with the available data.

The parent-offspring pairs, each of which involves a parent affected with a particular trait, may be subdivided into four categories according to sex of the affected parent and sex of the offspring. The four categories of pairs are mother-daughter, mother-son, father-daughter, and father-son. If neither the transmission nor the manifestation of the trait is influenced by sex, then all four pair categories should show the same rate of pair concordance for the trait. Sex influence in manifestation of the genotype associated with the trait would not affect transmission of the genotype. Accordingly, in such a system, the mother-daughter and father-daughter concordance rates would tend to be equal to each other, and the mother-son and father-son concordance rates would tend to be equal to each other; and the pairs involving offspring of one sex would show concordance rates that differ from those involving offspring of the other sex. Sex influence in the trait's transmission, however, would be manifested as differences in the pair concordance rates of mother-daughter versus father-daughter pairs and mother-son versus father-son pairs.

The primary involvement of a single autosomal (dominant) gene associated with transmission of a parent's trait would have a probability of occurring, in that parent's offspring, equal to the product of the probability for transmitting the gene to the offspring and that gene's penetrance. The same value would be expected to occur for the rate of pair concordance based on pairs involving affected parents and their offspring. Sibships characterized by presence of the trait in at least one sibling would involve the same risk or probability of recurrence in each other (than the known affected) sibling. Accordingly, the sibling-sibling pair concordance rate, based only on first-affected members of sibships involving two or more siblings, would be the same as the parent-offspring pair concordance rate based only on affected parents and their offspring. Thus, the occurrence of similar pair concordance rates for parent-offspring pairs (based on affected parents and their offspring) and sibling-sibling pairs (based on first affected members of sibships involving two or more members) is consistent with the concept of transmission associated with a single autosomal gene. On the other hand, a trait associated with a genotype consisting of more genes than one would show rates of pair concordance that differ between the pairs of affected individuals and their offspring and the pairs of first-affected individuals and their siblings. A rare trait transmitted by a pair of autosomal recessive alleles would show a very low rate for pair concordance between affected parents and their offspring because the mates of the affected parents would characteristically not be carriers of the rare recessive allele involved. On the other hand, occurrence of the homozygous recessive genotype in an individual would characteristically (except in the relatively rare instances of occurrence of new mutations) indicate that both parents are carriers of the recessive allele involved. Accordingly, each offspring of that parental pair, each sibling of the affected individual, would represent a significant risk for inheriting the same trait. Thus, the sibling-pair concordance rate would be expected to exceed the parent-offspring concordance rate for a rare trait involving a pair of autosomal recessive alleles.

The occurrence of a marked difference

between concordance rates for affected parents and their offspring versus first affected members of sibships with at least one affected individual and their siblings, with the latter exceeding the former, would not be consistent with a single-gene mode of transmission. Similar rates of concordance for (1) the parent-offspring and (2) the sibling-sibling pair categories, particularly when the values approach 50 percent (and thereby do not require the ancillary hypothesis of a significantly reduced penetrance), support the concept of association between the trait and a single autosomal gene.

THE NATURE OF THE GENETIC MATERIAL

The genotype interacts continuously throughout development, maturity, and senescence with the nongenetic materials from within and without the cell. Pathology may develop in response to genotype, to environmental factors, or to a combination of both. The effects of the genetic material depend upon the structure and composition of the deoxyribonucleic acid (DNA) molecules which, together with proteins and histones, comprise the bulk of the chromosomal material within a cell's nucleus. Protein syntheses occur at the ribosomes in the cytoplasm, controlled by soluble ribose nucleic acid (RNA) particles formed by the nuclear DNA apparently acting as templates. Three classes of RNA molecules have been defined according to their activities in the transfer and translation of genetic information: messenger RNA, transfer RNA, and ribosomal RNA. A messenger RNA particle, formed on a segment of a DNA molecule in the nucleus, moves out into the cytoplasm, carrying in its molecular structure the information that determines the correct sequence of the assembly of amino acids in the synthesis of a specific protein. Ribosomal RNA molecules from the nucleus also move into the cytoplasm, and they combine with protein to form the ribosomes of the cytoplasm. Several ribosomes become attached to each molecule of messenger RNA to form the polysomes that are attached to the membranes of the cell's endoplasmic reticulum. There is a special transfer RNA for each of the twenty amino acids. Each one attaches to its own specific type of amino acid, selected from among those entering the cell from the blood, and transports this amino acid to a ribosome. There, the transfer RNA meets and attaches itself to the appropriate complementary site on the messenger RNA molecule and inserts its amino acid into the protein molecule developing on the ribosome. Each protein or polypeptide synthesis occurs at a ribosome site, after which the messenger RNA is broken down. Presumably, constant renewal of the messenger RNA is required for continued protein synthesis. Variations in the activity code of RNA, like those of DNA, are associated with variations in the sequence of purine-pyrimidine bases in the molecular chains. Mutations, which are irreversible changes (except for subsequent mutations) in the genetic material, may occur in any living cell.

HUMAN CYTOGENETICS

The normal human chromosome number, forty-six, was first demonstrated by Tjio and Levan in 1956.[20] Since then, numerous chromosomal anomalies have been recognized. At least 0.5 percent of all newborns have been described as being affected with morphological abnormalities associated with currently detectable chromosomal anomalies.[21, 22] Significant chromosomal peculiarities may occur in as much as 1.0 percent of the newborn population.[23]

A chromosomal disorder involving mosaicism, which is characterized by an individual's possessing both normal and abnormal cell lines (i.e. a cell line with normal and one or more cell lines with abnormal chromosome complements) may result from mitotic misdivision during embryogenesis.[24, 25] Such events generally appear to be isolated and without enhanced predisposition for repetition in future offspring produced by the same parents or by genetic relatives of the affected individuals. The probability of a chromosomal mosaic's transmitting his/her chromosomal anomaly to his/her offspring does not depend upon occurrence or extent of the phenotypic syndrome but upon whether or not the gonadal tissue has the chromosomal anomaly. If the gonadal tissue is affected, then the probability of transmitting the chromosomal anomaly and associated syndrome to his/her offspring depends upon the tendency to produce gametes (i.e. ova, spermatozoa) with the chromosomal anomaly. This would be essentially the same for the fertile mosaic whose gonadal tissue is affected as it would be for the individual whose tissues in general are affected with the chromosomal anomaly. The probability of transmitting an anomalous chromosome complement to a zygote directly, through an anomalous gamete, generally depends upon the probability of producing such a gamete. There are specific variations, however, regarding gamete viability and fertilizability and zygote viability. Thus, the probability of transmitting an anomalous chromosomal complement to a zygote via the gamete, *in some cases,* varies markedly from the probability of producing such a gamete and sometimes varies between the two sexes. Such variations regarding particular anomalies of specific chromosomes may be determined by examination of epidemiological and incidence data from relevant studies of the specific chromosome anomaly involved.

Numerical abnormalities of chromosome number (aneuploidies) most commonly result from *nondisjunction* (nonseparation) of a pair of chromatids during cell division. *Meiotic* nondisjunction produces gametes with abnormal chromosome complements (e.g. haploid number plus or minus the nondisjointed chromosome), and fertilization with such a gamete produces an abnormal zygote in which all of the cells have the same anomalous chromosomal complement. A *mitotic* error during embryogenesis, on the other hand, produces *mosaicism,* i.e. at least two distinct cell populations or lines in the affected individual. Mosaicism may also result from double fertilization. This may occur through fertilization with two independent spermatozoa, each one combining with one of two (i.e. already divided) nuclei within an undivided oocyte.

The viabilities of zygotes, fetuses, and children affected with particular chromosomal anomalies vary according to the severities of the anomalies. Chromosomal complements involving aneuploidies for the larger autosomes are generally inviable. At least 20 to 30 percent of spontaneous abortions or miscarriages have been shown to involve gross chromosomal anomalies,[26] and the incidence may be considerably higher.[27] Normal individuals with histories of repeated spontaneous abortions have been observed to contain significantly higher proportions of cells with chromosomal duplications and deletions than are generally observed.[28]

Aneuploidies in general, and *trisomies* in particular, do not occur as strictly random events. Their incidences have been associated with various pathological, physiological, and environmental variables.[29] The recurrence of chromosomal anomalies in some families may be related to the persistence of specific peculiarities affecting maternal or paternal gametogenesis. One possibility which may be rejected in each individual case only after karyotyping cells from cultured gonadal tissues of both parents is that one of the two parents may be a mosaic in whom the gonadal tissue is affected with the anomalous condition involved. Most karyotype analyses are based solely on leukocytes of peripheral blood, obtaining samples of which involves the least risk and the greatest ease compared to obtaining samples of other living tissues. The rare occurrence of a mosaic with generally normal phenotype and normal karyotypes based on cultured leukocytes, but with anomalous gonadal tissue, would not be diagnosed with standard karyotype analyses based only on leukocytes of peripheral blood. This rare possibility and the relevant recurrence risk is often overlooked or ignored.

Evidently, specific genes may sometimes affect mechanisms which regulate duplication of specific chromosomes. Chromosomal aberrations have been observed to occur at a high frequency in peripheral leukocytes in some individuals in the absence of exposure to known mutagenic agents.[30] Diverse chromosomal anomalies within the same family have been observed.[31] Parents of trisomic children have been observed to show a relatively high incidence of chromatid breaks, suggesting an impairment in their chromatin-breakage repair mechanisms.[32] Several studies have shown familial, and evidently genetic, differences in susceptibility or resistance to chromosomal changes or the disease processes associated with such changes.[29]

Sex Chromatin and the Sex Chromosomes

Sex-chromatin bodies occur in interphase nuclei of some female mammalian cells.[33] The maximum number in any particular nucleus is one less than the number of X chromosomes in that nucleus. After the single-X nature of the sex-chromatin body was demonstrated,[34] Lyon postulated that only one X chromosome continues to be active in each mammalian cell and that each sex-chromatin body is derived from one of the cell's inactivated X chromosomes.[35] Differences in morphology and in quantitative distribution of sex-chromatin bodies have been associated with antibiotic medication and other variables,[36,37] as well as with the different investigators' particular criteria for positive and negative designations.[38] Thus, interpolation of simple X-chromosome aneuploidies based solely upon qualitative sex-chromatin designations in buccal smears is not entirely reliable.[38]

Figure 2-2 is a photograph of the stained nucleus of a buccal cell showing two distinct sex-chromatin masses adjacent to the nuclear periphery. The occurrence of two intranuclear sex-chromatin masses in some of an individual's cells indicates the presence of three X chromosomes in those cells. The presence of only one sex-chromatin mass is consistent with a chromosome complement that includes two X chromosomes, the normal female complement. The occurrence of only a single X chromosome (e.g. in a chromosomally-normal male or in a female with the one-X or Turner syndrome) is indicated by the lack of any intranuclear sex-chromatin mass, as shown in Figure 2-3.

The X chromosome exists normally in the *monosomic* or *hemizygous* condition in the male (i.e. XY), whereas effects of deletion

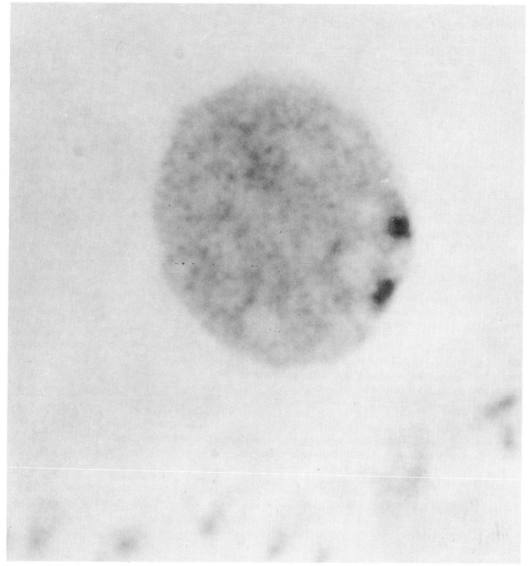

Figure 2-2. Photograph of stained nucleus of a buccal cell, showing two sex-chromatin masses adjacent to the nuclear periphery.

of one member of any autosomal pair are generally lethal or semilethal. The one-X condition is present in patients having Turner's syndrome with a count of forty-five chromosomes. (See Fig. 2-4.) A variant of this chromosomal syndrome occurs in the female who has one normal X chromosome and another one partially deleted. Many of the symptoms which characterize Turner's syndrome have been observed in affected individuals with X isochromosomes, i.e. anomalous structures involving deletion of the short arm and duplication of the long arm in some or deletion of the

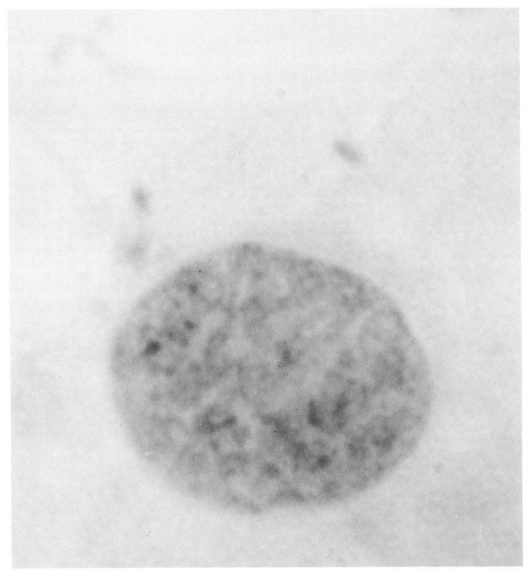

Figure 2-3. Nucleus of a buccal cell showing no clearly apparent sex-chromatin mass.

long arm and duplication of the short arm in others. Thus, factors on both the short and the long arm are needed in duplicate for normal female development. Mosaics have been observed whose tissues include cells with more than one genotype, e.g. both one-X and XX cells.

The XXX female has a total of forty-seven chromosomes. (See Fig. 2-5.) Mosaics have been observed whose cell lines include more than a single genotype, e.g. XX and XXX cells. Phenotypic females with three instead of two X chromosomes and two (instead of only one) sex-

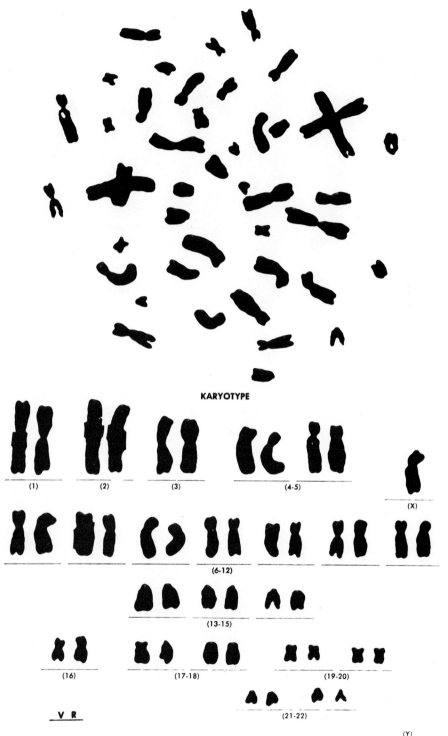

Figure 2-4. Photograph of the mataphase of a cultured peripheral-blood leukocyte (upper), and the karyotype of that cell (lower), from a patient affected with Turner's syndrome.

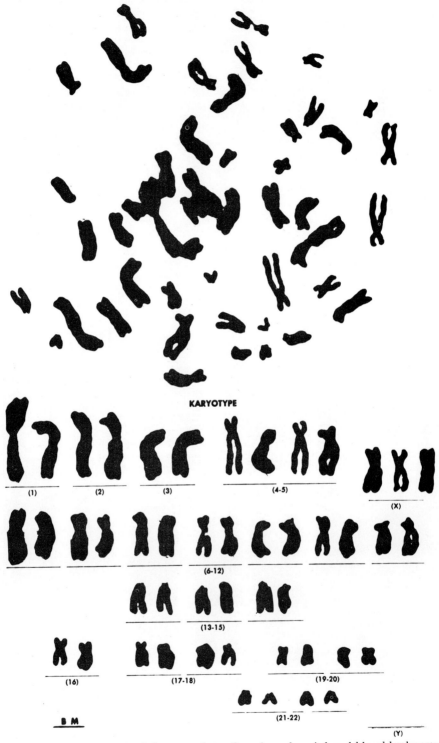

Figure 2-5. Photograph of the metaphae of a cultured peripheral-blood leukocyte (upper), and the karyotype of that cell (lower), from a patient with three X chromosomes.

chromatin bodies in some nuclei differentiate psychosexually as females and are not characterized by manifestations of gross morphological abnormalities. The reported clinical observations are variable, and most of the observed affected individuals are phenotypically normal females with pubescence, ovulation, and fertility. Incidence of the XXX syndrome is greater than that of the one-X syndrome, being approximately 1.2/1,000 among live-born females.[21] The XXX genotype is evidently the most common sex-chromosome abnormality observed in newborn females.

The occurrence of more than a single X chromosome in association with a Y chromosome is most commonly known as Klinefelter's syndrome, testicular dysgenesis, and chromatin-positive microorchidism. The XXY karyotype is the most usual variety associated with this syndrome. Combinations involving other multiples of X chromosomes with one or more Y chromosomes have also, but relatively rarely, been observed. Mosaics involving cells with characteristic Klinefelter karyotypes, as well as mosaics involving more than one kind of Klinefelter karyotype, have been described. Generally, no clinical abnormalities are observed in affected boys prior to puberty. Relatively long limbs have been described in many affected individuals, and enlarged breast development has been described in at least 25 to 30 percent of those diagnosed as affected with Klinefelter's syndrome; but there is no consistent pattern of stigmata, and many affected individuals are known to be normal in appearance. The outstanding clinical feature is the presence of small testes, which is generally detected only after puberty when the testes have failed to enlarge. MacLean, *et al.* found that the incidence of sex-chromatin positive individuals among unselected newborn males was about 2.0/1,000.[21] Nearly three fourths of the observed group had simple XXY karyotypes, and all but one of the remaining one fourth in the Scottish survey were XY/XXY mosaics.

The incidence of XYY individuals is not definitely known, but it is much lower than that of XXY individuals and is quite rare in the general population. Males who are characterized by possession of a supernumerary Y chromosome (e.g. XYY and XXYY males and mosaics who possess one or more cell lines with a supernumerary Y chromosome) or a duplication of part of a Y chromosome (i.e. in the "big Y" syndrome) tend to show an increased frequency of antisocial and aggressive behavior, very tall stature, and mild mental retardation.[25]

Autosomal Abnormalities

Figure 2-6 is a schematic representation of the various human chromosomes arranged according to their relative sizes. When cells are cultured for chromosome analysis, they are stimulated to start dividing; and they are arrested at metaphase when they can be stained and observed microscopically. Figure 2-7 is a schematic representation of the various human metaphase chromosomes arranged sequentially in standard groups.[39] Each metaphase chromosome depicted in Figure 2-7 is actually a pair of chromatids, duplicated daughter chromosomes still connected to each other at the centromere. The chromosomes, as schematized in Figure 2-6, are inferred from the visible metaphase structures of pairs of chromatids (as schematized in Fig. 2-7). Figures 2-4, 2-5, and 2-8 each include, in the upper half, an example of metaphase chromosomes as photographed from a stained cultured peripheral-blood leukocyte. The lower half of each of these fig-

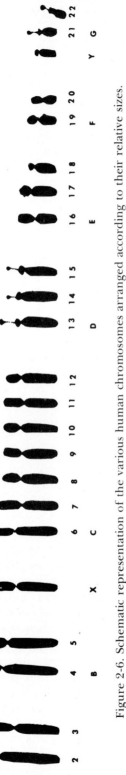

Figure 2-6. Schematic representation of the various human chromosomes arranged according to their relative sizes.

ures (2-4, 2-5, and 2-8) shows a karyotype derived from the metaphase plate in the upper half: the chromosomes shown in the enlarged photographic print were cut out and sorted according to relative size and shape. The metaphase plate shown in Figure 2-8 includes a normal complement of autosomes (twenty-two pairs), a normal female complement of sex chromosomes (two X chromosomes), and an abnormal pair of acentric chromosomal fragments.[38]

The first proven human autosomal anomaly associated with a well-known syndrome is the one associated with the G_1-trisomy syndrome (erroneously and anachronistically referred to as "mongolism" in English-speaking countries).[40] Numerous autosomal abnormalities have since been described, including various duplications, deletions, and translocations.[24, 29]

Chromosomal Abnormalities, Associations, and Clusterings

Inductions of chromosomal anomalies have been associated with various factors: genetically-transmitted pathological conditions, infectious agents, radiation, and numerous chemical agents. There have been reports of family clusters and of community clusters with disorders associated with cytogenetic abnormalities. The high correlation between increasing maternal age and increasing incidence of the autosomal aneuploidy involved in the G_1-trisomy syndrome (i.e. three, instead of the normal complement of two, number-21 chromosomes) has been well established. The association between increasing maternal age and increasing incidence of an anomaly is not confined to the G_1-trisomy syndrome but has also been observed for stillbirths, placenta previa, anencephaly, hydrocephaly, spina bifida,

Basic Human Genetics

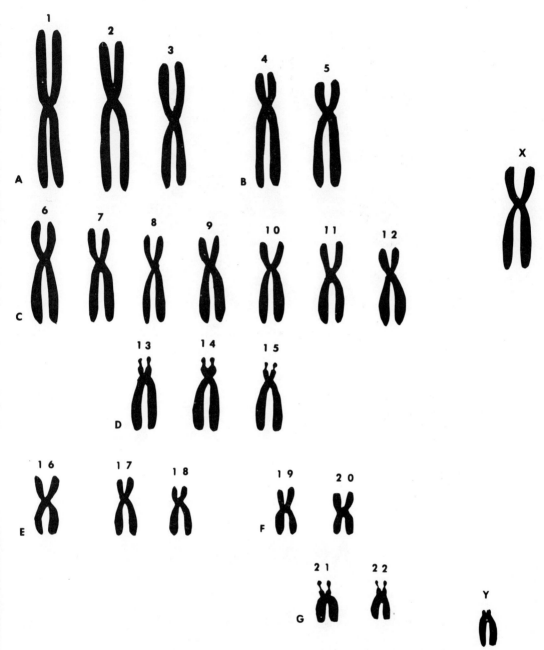

Figure 2-7. Schematic representation of the various human metaphase chromosomes (i.e. actually pairs of chromatids) arranged according to Denver system organization of groups.

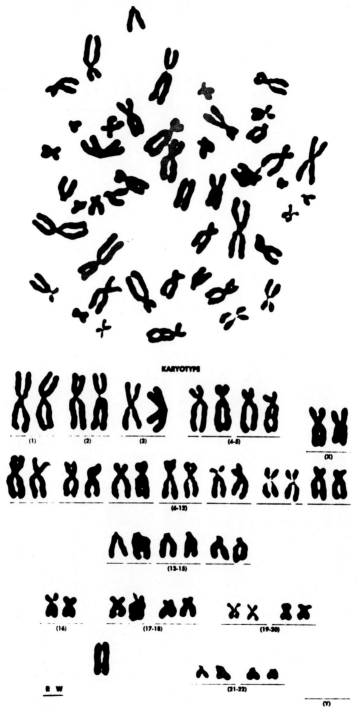

Figure 2-8. Photograph of the metaphase of a cultured peripheral-blood leukocyte (upper) and the karyotype of that cell (lower) showing a normal female chromosome complement plus an anomalous pair of acentric chromosomal fragments.

for pregnancies terminating in miscarriage, and for pregnancies producing normal infants who develop leukemia later in childhood.[24, 29] Epidemiological studies of the G_1-trisomy syndrome, which have shown annual fluctuations in incidence, multiannual periodicity, and regional clustering, may be interpreted as indicating the possible relevance of a virus or other infectious agent. Periodic fluctuations in incidence, which are characteristic of recurrent epidemics, have also been reported for other congenital anomalies such as anencephaly, spina bifida, and hydrocephaly.[24, 29] Observations of associations between the G_1-trisomy syndrome and leukemia incidence may suggest an etiological relationship between the two conditions. Similarly, associations have been observed between the congenital aneuploidies involved in Klinefelter's syndrome and leukemia and other malignancies.[41, 42] Increased incidence of cancer in general has been observed in families in which chronic lymphocytic leukemia and lymphosarcoma had occurred.[43] Patients affected with Bloom's syndrome and those with Fanconi's syndrome have shown abnormally high frequencies of chromosomal abnormalities in their somatic cells. They have also shown strikingly increased incidences of malignancies. These observations suggest that the latter two diseases (Bloom's syndrome and Fanconi's syndrome), each of which may involve simple Mendelian genetic transmission, induce abnormal frequencies of chromosomal rearrangements in the somatic cells of affected individuals. Abnormally high frequencies of chromosome breaks have been observed to occur in patients with serum hepatitis and in patients with aseptic meningitis.[44, 45, 46] The induction of chromosome breaks by viruses has been well-documented.[29] A specific virus, adenovirus 12, may be incorporated into the genetic material of a specific host cell and transform the normal cell into a malignant-type cell.[47] Chromosome anomalies in circulating leukocytes have been found in patients following their subjection to X-irradiation, gamma ray and proton exposure, and therapeutic doses of radioactive iodine.[29] Chromosome breakages have been induced by numerous chemical compounds.[48, 49, 50]

Various chromosomal abnormalities (i.e. chromosome and chromatid breaks and gaps, aneuploidies, translocations, duplications, deletions) represent effects of exogenous and/or endogenous mutagenic factors. Their associations with particular pathological entities need not necessarily involve cause-and-effect relationships: etiological factors which are relevant to increasing the incidence of a pathological category may also affect occurrence of discernible mutations. The demonstrated association of two abnormal effects, a particular clinical pathology with significantly-increased discernible chromosomal changes, may indicate a clue toward solving the problems involving etiology of the pathology. Laboratory methods are currently available for relatively simple tests of suspected mutagenic agents to screen factors for possible etiological involvement.[48, 49, 50]

As the genotype operates within its intracellular and extracellular environment, many different environmental factors may affect expression of the genotype as well as induction of changes in the genotype (i.e. mutations). Numerous environmental factors have been identified as mutagens. The continued elucidation of mechanisms by which these agents interact with the genotype will continue to expand our understanding of the dynamics of normal and pathological differentiation.[51]

CONCLUDING REMARKS

The modern view of heredity is not consistent with the old nineteenth-century concept of genetics as part of a simple dichotomy, nature versus nurture, genetics versus environment, constitutional or innate versus acquired characteristics. The contemporary concept of heredity is three-sided, involving: (1) the genetic material or genotype, (2) the various environmental factors, and (3) the critical periods. Interactions between factors of the genotype and factors of the environment acting at critical periods (and a critical period is often very limited) determine the phenomena that follow. This model applies to behavioral development as well as to morphological development. Numerous environments must be considered: the cellular environment; the intrauterine environment; the perinatal environment; and the postnatal environment, which includes those special environments of the body, the senses, and the social-cultural existence.

No given sample of behavior has exclusively genetic origins. Genotype cannot be expressed except in interaction with the environment, whether it be the environment created by neighboring cells in the embryo, the environment provided by the mother in the uterus, the perinatal environment, or the environment encountered in the home and community. Behavior genetics, therefore, involves ascertaining the environmental limits within which the genetic code can operate and unfold itself into a normal phenotype and the limits which environmental extremes may impose on the genetic code without destroying it but obligating it to unfold into a modified phenotype.

The development of any *trait* involves genetic and environmental determinants. The variation between different individuals for a particular trait may be almost entirely due to one or the other type of factors or to interactions of the two types of factors. The agglutinogens of red blood cells are examples of traits which are directly associated with simple genetic factors with no primary variance due to environmental influences. Customs and language are conventional examples of manifestations with little or no demonstrated simple genetic variance. Body size, resistance and predisposition to various specific infectious organisms, skin color, and hair color are several examples of traits which may be significantly affected by both genetic and environmental factors. In the realm of behavior, a trait cannot be classified as being genetic in the sense that many specific biochemical and morphological traits may be genetic. Except for gene-transmitted pathological syndromes, there are no known *specific* genes for specific aspects of behavior. Human behavior genetics deals primarily with inferences about genetic factors based upon population statistics involving associations between specific traits and genetic kinships.

REFERENCES

1. Kaplan, A.R.: Genetic counseling in mental retardation and mental disorder. In Lynch, H.T. (Ed.): *Dynamic Genetic Counseling for Clinicians.* Springfield, Thomas, 1969.
2. Morton, N.E.: Genetic tests under incomplete ascertainment. *Am J Hum Genet, 11*:1, 1959.
3. Morton, N.E.: Segregation and linkage. In Burdette, W.J. (Ed.): *Methodology in*

Human Genetics. San Francisco, Holden-Day, 1962.
4. Fisher, R.A., and Yates, F.: *Statistical Tables for Biological, Agricultural, and Medical Research,* 6th ed. New York, Hafner, 1963.
5. Hirsch, J.: Behavior-genetic analysis and its biosocial consequences. *Sem Psychiat,* 2:89, 1970.
6. Kaplan, A.R.: Concluding remarks: Genetics and schizophrenia. In Kaplan, A.R. (Ed.): *Genetic Factors in "Schizophrenia."* Springfield, Thomas, 1972.
7. Falconer, D.S.: The inheritance of liability to diseases with variable age of onset, with particular reference to diabetes mellitus. *Ann Hum Genet, 31*:1, 1967.
8. Mather, K.: *Human Diversity.* Edinburgh, Oliver and Boyd, 1964.
9. Mather, K.: *Statistical Analysis in Biology,* 4th ed. London, Methuen, 1965.
10. Burdette, W.J. (Ed.): *Methodology in Human Genetics.* San Francisco, Holden-Day, 1962.
11. Reichenbach, H.: Predictive knowledge. In Jarrett, J.L., and McMurrin, S.M. (Eds.): *Contemporary Philosophy.* New York, HR&W, 1954.
12. Hardy, G.H.: Mendelian proportions in a mixed population. *Science, 28*:49, 1908.
13. Weinberg, W.: Über den Nachweis der Vererbung beim Menschen. *Jahreshefte Verein f vaterl Naturk in Würtemberg, 64*:368, 1908.
14. Wright, S.: Classification of the factors of evolution. *Cold Spring Harbor Symp Quant Biol, 20*:16, 1955.
15. Stern, C.: *Principles of Human Genetics,* 2nd ed. San Francisco, Freeman, 1960.
16. Penrose, L.S.: The detection of autosomal linkage in data which consist of pairs of brothers and sisters of unspecified parentage. *Ann Eugen, 6*:133, 1935.
17. Burks, B.: Autosomal linkage in man. *Proc Natl Acad Sci, 24*:512, 1938.
18. Haldane, J.B.S., and Smith, C.A.B.: A new estimate of the linkage between genes for colour-blindness and haemophilia in man. *Ann Eugen, 14*:10, 1947.
19. 'Espinasse, P.G.: The polygene concept. *Nature (Lond), 149*:732, 1942.
20. Tjio, J.H., and Levan, A.: The chromosome number of man. *Hereditas, 42*:1, 1956.
21. MacLean, N., Harnden, D.G., Court-Brown, W.M., Bond, J., and Mantle, D.J.: Sex-chromosome abnormalities in newborn babies. *Lancet, 1*:286, 1964.
22. Robinson, A., and Puck, T.T.: Studies in chromosomal nondisjunction in man. II. *Am J Hum Genet, 19*:112, 1967.
23. Lejeune, J.: Chromosome studies in psychiatry. In Wortis, J. (Ed.): *Recent Advances in Biological Psychiatry.* New York, Plenum Pr Plenum Pub, 1967, vol. IX.
24. Kaplan, A.R.: The use of cytogenetical data in heredity counseling. *Am J Ment Defic, 73*:636, 1969.
25. Kaplan, A.R.: Chromosomal aneuploidy, genetic mosaicism, occasional acentric fragments, and schizophrenia: association of schizophrenia with rare cytogenetic anomalies. In Kaplan, A.R. (Ed.): *Genetic Factors in "Schizophrenia."* Springfield, Thomas, 1972.
26. Thompson, H.: Abnormalities of the autosomal chromosomes associated with human disease: Selected topics and catalogue. *Am J Med Sci, 250*:140, 1965.
27. Waxman, S.H., Arakaki, D.T., and Smith, J.B.: Cytogenetics of fetal abortions. *Pediatrics, 39*:425, 1967.
28. McKay, R.J., Witte, E.H., and Hodgkin, W.E.: Partially trisomic cells in chromosomal analyses. Paper read at Sixth Conference on Mammalian Cytology and Somatic Cell Genetics, Asilomar, California, 1967.
29. Kaplan, A.R., and Kelsall, M.A. (Eds.): *Leukocyte Chemistry and Morphology Correlated with Chromosome Anomalies. Ann NY Acad Sci, 155*: Art. 3, 1968.
30. Gooch, P.C., and Fischer, C.L.: High frequency of a specific chromosome abnormality in leukocytes of a normal female. *Cytogenetics, 8*:1, 1969.
31. Atkins, L., Bartsocas, C.S., and Porter, P.J.: Diverse chromosomal anomalies in a family. *J Med Genet, 5*:314, 1968.
32. Kahn, J., and Abe, K.: Consistent and variable chromosome anomalies in parents

of children with Down's syndrome. *J Med Genet, 6*:137, 1969.
33. Barr, M.L., and Bertram, E.G.: A morphological distinction between neurones of the male and female, and the behavior of the nuclear satellite during accelerated nucleoprotein synthesis. *Nature (Lond), 163*:676, 1949.
34. Ohno, S., Kaplan, W.D., and Kinosita, R.: Formation of the sex chromatin by a single X-chromosome in liver cells of *Rattus norvegicus. Exp Cell Res, 18*:415, 1959.
35. Lyon, M.F.: Gene action in the X-chromosome of the mouse *(Mus musculus L.). Nature (Lond), 190*:372, 1961.
36. Sohval, A.R., and Casselman, W.G.B.: Alteration in size of nuclear sex-chromatin mass (Barr body) induced by antibiotics. *Lancet, 2*:1386, 1961.
37. Kaplan, A.R.: Association of a quantitative variable, Barr body score, with length of confinement in state mental hospitals. *J Nerv Ment Dis, 143*:449, 1966.
38. Kaplan, A.R., and Cotton, J.E.: Chromosomal abnormalities in female schizophrenics. *J Nerv Ment Dis, 147*:402, 1968.
39. Denver Report: A proposed standard system of nomenclature of human mitotic chromosomes. *Am J Hum Genet, 12*:384, 1960.
40. Lejeune, J., Gautier, M., and Turpin, R.: Les chromosomes humains en culture de tissue. *C R Acad Sci (Paris), 248*:602, 1959.
41. Robinson, L. R.: Neoplastic liability in Klinefelter's syndrome. *Br J Psychiatry, 112*:713, 1966.
42. Miller, R.W.: Relation between cancer and congenital defects in man. *N Engl J Med, 275*:87, 1966.
43. Morganti, C., and Cresseri, A.: Neuvelles recherches génétiques sur les lencé mies. *Sangre, (Bare), 25*:421, 1954.
44. Aya, T., and Makino, S.: Notes on chromosome abnormalities in cultured leukocytes from serum hepatitis patients. *Proc Jap Acad, 42*:648, 1966.
45. Makino, S., Yamada, K., and Kajii, T.: Chromosome aberrations in leukocytes of patients with aseptic meningitis. *Chromosoma, 16*:372, 1965.
46. Makino, S., Aya, T., Ikeuchi, T., and Kasahara, S.: A further study of chromosomes in cultured leukocytes from aseptic meningitis patients. *Proc Jap Acad, 42*:270, 1966.
47. Fujinaga, K., and Green, M.: The mechanism of viral carcinogenesis by DNA mammalian viruses: Viral-specific RNA in polyribosomes of adenovirus tumor and transformed cells. *Proc Natl Acad Sci USA, 55*:1567, 1966.
48. Hollaender, A. (Ed.): *Chemical Mutagens.* New York, Plenum Pr Plenum Pub, 1971.
49. Epstein, S.S. (Ed.): *Drugs of Abuse: Their Genetic and Other Chronic Nonpsychiatric Hazards.* Cambridge, MIT Pr, 1971.
50. Vogel, F., and Röhrborn, G. (Eds.): *Chemical Mutagenesis in Mammals and Man.* New York, Springer-Verlag, 1971.
51. Jackson, L.G.: Genetics in neoplastic diseases. In Goodman, R.M. (Ed.): *Genetic Disorders of Man.* Boston, Little, 1970.

Chapter 3

THE FUTURE OF BEHAVIOR GENETICS

J. P. Scott

INTRODUCTION

THE GENETIC NATURE of man is a legitimate and basically important problem, for on it depends our understanding of individual variation and the extent of human capacities. At the same time, human behavior genetics is one of the most misused and least understood of any branch of scientific information. We can fairly assume that of all the attributes of man, behavior is most important; what men do is much more significant than their appearance. This implies that the genetic nature of human behavior should be a major concern of geneticists. Yet, in the early years of this century, following the rediscovery of Mendelian genetics and the development of the theory of the gene, the study of the genetic aspects of behavior was a stepchild of genetics simply because less complex and less time-consuming methods of studying variation were available. The color of the eye of a single fruit fly could be classified in seconds, whereas measuring its behavior might take several minutes. This resulted in the neglect of a basically important field of science which then became the domain of pseudoscientific attempts to apply the principles of genetics to behavior under the banner of the eugenics movement.

Apart from its practical significance, the interconnection of the two sciences of genetics and behavior provides some interesting theoretical problems. Genetics is, of course, the science of variation. Without variation, biological inheritance cannot be studied. Behavior, on the other hand, is the science of variability, for, without the capacity for varying behavior, adaptation to the changing conditions of the environment would be impossible. The distinction is that behavioral variability occurs within the individual, whereas genetic variation is primarily a matter of differences between individuals. Both sciences are concerned with a common topic, but applied to phenomena on different levels. Therefore, it is with the fascinating and extraordinarily complex problem of the genetics of variation in variability (of behavior) that the science of behavior genetics is chiefly concerned.

Rather than reviewing the general science of genetics from a classical and historical viewpoint, it is appropriate to the subject of behavior to begin with gene action.[1,2]

PHYSIOLOGICAL GENETICS: RELEVANCE TO BEHAVIOR

The basic units of heredity are genes, first discovered by inference from breeding experiments but now known to be very large organic molecules (of desoxyribonucleic acid, or DNA) that act chemically as enzymes. Since they reproduce themselves during the process of cell division, they have the nature of autocatalysts; that is, one of the chemical processes that a gene modifies is a process which reproduces the gene itself. Because of the specific chemical nature of genes it has long been held that for each gene there is one enzyme which is either the gene itself or a primary product of gene action.[3] Because the number of processes that can be modified by such an organic enzyme is quite limited, we can infer that in most cases a gene affects only one process directly.

An enzyme, or organic catalyst, can only act by speeding up or slowing down a chemical process. As with any chemical process, the raw material must be present before a gene can act. In cellular organisms these raw materials are present in the cytoplasm and eventually must be obtained from outside the cell through cellular processes. The nature of primary gene action thus intrinsically demonstrates the falseness of the dichotomy between heredity and environment and the meaninglessness of the "nature or nurture" controversy that is an outgrowth of this dichotomy. The genes act only in the presence of environmental materials. Likewise, the biochemical processes of living material cannot proceed without the genes. Neither heredity nor environment creates anything by itself. Rather, each modifies ongoing processes and thus produces changes and variation. The primary action of a gene, however, is restricted to processes going on at the molecular level, whereas environmental factors may directly affect processes on any level of organization.

As was demonstrated in the early years of the study of the genetics of variation in skin pigments and hair colors in various animals, many genes can modify the same process. These can either be alleles, that is, genes found in one locus on a chromosome and hence never present in any one individual in numbers of more than two; or they can be nonalleles, present in many different loci and hence present in larger numbers in any individual. The number of possible combinations rises rapidly according to the number of genes involved. With one allele only, one combination is possible; with two alleles, three combinations are possible; with three alleles, six combinations; with four alleles, ten combinations; etc. With four loci, each having two alleles, the number of genotypic combinations is 3^4 or 81.

Behavior itself, defined as the activity of an entire organism, must therefore be based on the combination of all the biochemical processes that constitute life. It follows that if variation exists in many gene loci, as it does in most natural populations, the corresponding behavioral variation will almost inevitably depend on multiple genetic factors. Also, as we begin to be able to analyze the biochemical processes that occur in living organisms, even the simplest sorts of activity, such as a muscle twitch, are dependent on long chains of chemical reactions. Each link in such a chain is potentially modifiable by a gene. Therefore, genes can coact with or counteract each other in a great variety of ways. For example, if a process is composed of a long chain of subprocesses, genes which knock out any one link could each destroy the whole process; and the effects of each would appear to duplicate each other even though they were actually quite different. On the other hand, genes which speed up various links in a chain would appear to act in an additive fashion. If we also consider cases in which the subprocesses are not connected in a chain-like fashion but are independent of each other except for the final effects, the numbers of ways in which genes can modify each other's actions are literally infinite.

Since behavior can theoretically reflect all of the processes that go on within the body, both in the present and in the past, it represents an extraordinarily difficult problem for genetic analysis. Where the genetic analysis of behavior has been most successful, it has been applied to limited segments of behavior which are presumably dependent upon a limited number of processes and where gene action is correspondingly limited. It also follows from the above considerations that the ultimate understanding of the effects of genes on behavior must rest on a knowledge of the biochemical processes that affect behavior and can be directly or indirectly modified by gene action.

Enough has been said to indicate that genes may affect any ongoing biochemical processes at any time in life. Gene action is therefore not limited to the early part of life. In fact, it may be first apparent quite late in life, as in Huntington's chorea, whose symptoms first appear in early middle age. However, many life processes are either confined to early existence or have their most important effects at that time; hence arises the importance of studying the effects of genetics from a developmental viewpoint.

DEVELOPMENTAL GENETICS AND BEHAVIOR

Development is a general process including a group of subprocesses, all of which have an organizing function. Again, while development proceeds most rapidly in early life, organizing processes that affect behavior may go on throughout life. Even in old age, degenerative changes in organizing processes may contribute to change. The chief reason why development is concentrated in the early part of existence is that this is when growth proceeds most rapidly.

Growth is a major developmental process. Sheer physical size has an effect on motor behavior and the kinds of adaptations that are possible for an individual. Some variation in the average size of cells is possible, but for the most part differences in size rest on the numbers of cells in an individual. From the viewpoint of physiological genetics, this is the kind of process in which additive effects of genetic and environmental factors are likely to be most apparent simply because the process of cell division is an additive one.

Different patterns of growth are found in different species and must rest primarily on a genetic basis. For example, the growth curve of young dogs proceeds in almost a straight line from the time when puppies are first able to take solid food (at about three weeks of age) until the time shortly after they develop their permanent teeth, around eighteen to twenty weeks.[4,5] At sixteen weeks they have achieved two thirds of their adult size and are then physically equipped to begin to make an independent living as hunters. In contrast, growth in human children proceeds at a very slow rate for many years and then is followed by very rapid growth in the usual preadolescent spurt. The patterns in the two species are quite different and must be determined largely by genetics. In addition, there is great individual variation within both the canine and human species.

As in most mammals where there are size differences between the sexes, human females on the average become sexually mature earlier than males, this being the process that limits total size. As we often forget, sex is inherited and is the most pervasive and important single factor affecting genetic variation, not only in size but also in behavior.

Within the sexes, one of the factors associated with size variation is the wide range of variation in the time of onset of puberty, a range in which there is a large genetic component, although most of the human studies have been made with comparisons between populations where there are many possibilities of cultural and nutritional differences as well.[6] One of the early studies, which deserves to be repeated with better environmental controls, showed a mother-daughter correlation of .40 in the age of menarche.[7] A character completely determined by heredity should show a correlation of .50.

Certainly, there are wide variations in age of maturity between dog breeds. The ancestral wolves do not become sexually mature until the end of the second year or later. Some dog breeds mature quite early, individual females becoming fertile as early as five months, whereas others only mature well into the second year. Basenjis have as annual breeding season based on declining day length; and, consequently, females regularly come into their first estrus at approximately nine months.[8] In both dogs and humans there is considerable variation with respect to the age of termination of fertility in both sexes, and presumably this also has a genetic basis.

As Jones and Bayley found with the University of California longitudinal study of human development, the time of onset of puberty had obvious and important effects on social development of the boys and girls involved.[9] At that cultural period, at least, early maturity in boys produced a favorable effect, while the opposite was the case in girls. This highly heritable characteristic deserves further investigation. More generally, the study of genetic variation in the timing of behavioral development is a very important one, particularly as it applies to the intelligence and motivational development of young children.

Likewise, there is an obvious genetic base for variation in longevity. Most dog breeds show signs of middle age beyond the fifth year. Great Danes rarely live more than two years longer, but individuals in other breeds may live fifteen or twenty years and still be relatively active. At first glance, longevity does not seem to be related to behavior, but it definitely determines the amount of behavior which an individual can exhibit during his lifetime; this may be very important in a highly social species such as man. A researcher who dies two or three years after he gets his Ph.D. has much less opportunity to affect the process of scientific information-getting than one who remains productive for thirty or forty years. Even in primitive man, longevity must have contributed to cultural continuity and preservation, although its primary selective advantage may have been more biological in nature, resulting in increased numbers of descendants.[10]

Another basic developmental process is that of *differential growth*. Bodily form is largely determined by unequal growth rates in various parts of the body, and considerable independent variation is possible. We found that different parts of the body such as the eyes, ears, and teeth, matured at different rates in the various dog breeds.[4, 11] However, all breeds eventually became quite well organized, even though the various processes did not appear to be well coordinated in early development.

With respect to behavior, various sorts of anatomical and physiological evidence indicate that the brain and central nervous system achieve their full size much earlier than the rest of the body. Again, there may be genetic variation in this process. Such variation would be difficult to study anatomically, but it could be measured physiologically and might correlate with differences in intellectual or cognitive abilities.

One consequence of differential growth rates is that the times at which either genetic or evironmental factors can exert maximum influence vary with particular organs and organ systems. This, in fact, is the basis of the critical period phenomenon with respect to growth processes. From a behavioral viewpoint, critical periods of growth are particularly important with respect to the development of sense organs, such as the eyes, where deficiencies or excellencies can have a profound effect on later behavior. Critical periods thus define the times during which genetic or environmental factors may act to produce such variation. There is, of course, an extensive literature on variation in human sensory capacities, and this should be one of the domains of the behavior geneticist.

More generally, the concept of critical periods will apply to any organizational process of any level of organization.[12] It states that any organizational process is modified most easily at the time when it is proceeding most rapidly. For a critical period to exist, the speed of the process must differ at different times in development; and the more the process is restricted to a particular time in development, the more precisely defined and important the critical period.

In summarizing the possibilities for behavioral genetic research with respect to the basic process of growth, we should point out that a major determinant of differences in body size is sex. In one middle-class white (and presumably equally well-fed) population measured in the early years of this century, boys averaged 12.5 cm taller and 8.7 kg heavier than girls at eighteen years of age.[13] The proportion of variance due to sex was at least equal to that generated by all other factors affecting size variation in either sex. Also, while size differences may contribute to differences in behavior, there are a whole host of other differences in behavior that are associated with sex. At the present time these differences are being studied seriously rather than being, metaphorically speaking, swept under the rug as in earlier years. A complicating factor is, of course, that many sex differences may consist either wholly or partially of cultural reactions to biologically determined sex rather than being directly determined biologically. One of the problems of this sort of study is that while it is possible to vary the major factors of cultural and socioeconomic status for analytical purposes, it is not similarly possible to vary sex, which comes in only two packages. There is, however, considerable variation in secondary sex characteristics, such as size and body conformation, which could be correlated with variation in behavior. It is to be hoped that research on sex differences does not become enmeshed in the false heredity-environment dichotomy that has given such sterile results in connection with the study of intelligence.

Another major organizing process of development is that of the *differentiation* of cells into tissues, organs, and organ systems. This does not involve growth, but changes in form and function. In fact, most highly specialized cells, such as those of the nervous system, are unable to grow through cell division. The basic biochemical processes underlying differentiation are still very incompletely known, but they must be enormously complex, especially if all of the differentiating processes in an organism are considered. The processes studied so far involve relatively superficial results, such as the formation of pigments or those which produce crippling defects through the modification or lack of differentiation of connective tissue.[14] These, of course, produce only indirect effects on behavior, and the genetics of differentiation of nervous tissue is still to be studied in the future.

From a behavioral viewpoint, one of the most interesting aspects of this subject is *functional differentiation*. The work of Hamburger has shown that early spontaneous behavior of the unhatched chick affects the development of muscles and joints,[15] and the work of Rosenzweig and Krech, *et al.* has demonstrated that brain size in rats is modified by placing them in enriched, rather than impoverished, environments.[16] These results imply that genetic differences in behavior may cause differences in structure, in contrast to the usual assumption that structure is a cause of behavior.

As the biochemical processes involved in development become better known, it will become increasingly possible to analyze the results of genetic variation in them. At the present time, however, a great many of the organizing processes can only be studied by measuring behavior and inferring the nature of underlying processes. A notable example is the process of learning which, at least in the higher vertebrates, is present soon after birth or hatching and in some cases even before the organism emerges into the outside world.

In one of our experiments with genetics and the development of behavior in dogs, we theorized that the ideal time to study the effect of genetics would be soon after birth, with the idea that developmental variation resulting from differential learning would be at a minimum. Studying the reactions of puppies to standard treatment while being weighed led to a far different conclusion. Far from showing less variation at the early ages, the puppies showed more variation and fewer limitations on their behavior. It was only after the puppies were exposed to the situation many times that their individual behavior became predictable and somewhat stereotyped according to the breed to which they belonged. This led to the conclusion that the limitations of behavior seen in particular breed populations are not the result of heredity per se, but of the process of habit formation interacting with genetic capacities. In other words, the organizing process which puts behavior into a strait jacket is that of learning and habit formation. This brings us to a consideration of the relationship between heredity and the process of learning, one which properly belongs with the study of behavior on an organismic level, as well as being a developmental process.

BEHAVIOR GENETICS, ORGANISMIC LEVEL

The process of learning is one of an individual's organizing his behavior with respect to variation in the environment. It is intimately connected with the process of behavioral adaptation, which may be defined as attempts to respond to change in ways which maintain life and well-being. While some adaptation may be stereotyped and mechanical, as in simple reflexes, this is useful only where the environment remains constant.

Most successful adaptation involves variability of behavior. When a successful adaptation is achieved, it is more economical for an organism to stereotype it by habit formation than to go through its whole repertory of variable behavior on each new occasion. This leads to the concept of two opposing processes, one of varying behavior even in identical situations, and the other of fixing or stereotyping the behavior through habit formation. Obviously, the relative behavioral fitness of an animal depends upon the degree to which it reaches the most appropriate balance between these two tendencies.

Variation and Habit Formation

We tested several breeds of dogs on a six-unit T-maze designed so that it would be difficult, but not impossible, for a three-month-old puppy to learn.[17] The usual behavior on the first few trials was a great deal of frantic searching involving an enormous amount of variation. All puppies eventually found their way out of the maze and, on subsequent trials, made fewer errors and wasted less time. After four or five trials, a few made their way through the maze slowly and deliberately, making perfect scores. However, none of them maintained perfection. Rather, all puppies soon reduced their behavior to a simple stereotyped habit, usually that of taking alternate right and left turns. This inevitably produced a certain number of turns into dead-ends, but the dogs dashed in and out of these and usually finished the run in fairly good time.

There was one outstanding breed difference. The beagles showed a much weaker tendency to stereotype their behavior and continued to vary it for many more trials. Consequently, although their initial scores were poorer than the other breeds, their final solutions which became fixed were better. We can conclude that heredity has in some manner shifted the balance between the processes of variation and habit formation in a way which is obviously adaptive for a hunting animal. On the other hand, the quick habit formation of some of the other breeds makes them easier to train to obey commands, and they have doubtlessly been selected in this direction. This suggests the possibility that a similar sort of genetic variation could exist in human populations.

Is There Genetic Variation in the Association Process?

This brings up the question of other possible ways in which hereditary variation could affect the learning process. Learning is a cumulative process as well as an integrative one. Since the results of learning are more or less permanent, each new experience of each individual adds something to his store of information, and each repeated experience strengthens the memory trace. Thus, even a relatively short-lived animal may accumulate thousands of items of stored information. A relatively minor basic difference in the capacity to learn could therefore eventually result in major differences in the amount of material learned. This cumulative aspect of the learning process is additive, but it should be pointed out that we have no evidence as to whether there are genetic differences in basic learning capacities which are correspondingly additive. As far as I am aware, no research has been done on genetic variation in the association process that is the basis of simple learning. It is quite possible that this process is so essential to survival that any variations are strongly selected against, with the result that only minor variation is present in any immediate result of experience. But even a small difference might result in a very large difference in the cumulative experience of a lifetime.

Variation in Integrative Learning

The process of learning is, at the same time, an adaptive, integrative, and creative one. As well as making simple associations, an animal may reorganize its behavior in new ways to meet new situations. It may also recognize similarities between situations that appear to be unlike. Neither of these is an additive process, and a basic question is whether individual genetic differences in creative abilities exist. Also, one of the consequences of creative capacities is that they may be used to organize behavior in ways which either facilitate or handicap the accumulative aspects of associative learning.

Unfortunately, the above question has become embedded in the IQ controversy, involving testing methods that obscure what ought to be a central problem. Intelligence tests such as the Binet-Simon and Stanford-Binet were usually developed in an empirical fashion by selecting problems that gave wide variation between age groups and between individuals within the same age group.[18] Such variation should theoretically have a large genetic component, provided that all children have been exposed to a standard environment. This latter condition is in part met by exposing children to a standardized school system. From a theoretical standpoint, one should be able to manipulate the relative size of the genetic component of variance by keeping environmental conditions more or less constant: that is, the greater the environmental variation the less important heredity is in determining individual differences, and vice versa.

Even under controlled environmental conditions such as we were able to maintain in an experimental animal colony at the Jackson Laboratory School for Dogs, the proportions of the genetic component (heritability) never approached 100 percent and were also highest in physiological responses rather than learning and other forms of complex behavior. How then, do the human IQ scores obtain such high indices of heritability?[19] The answer is that the tests were designed by selecting highly heritable items and omitting those with lower heritability. However, such a procedure has obvious drawbacks for genetic analysis. Each test consists of dozens of independent items that have been selected in such a way as to maximize a high level of heritability. Adding these together gives a score that reflects any capacities in which there may be genetic variation, irrespective of whether this is based on one basic capacity, a small number of capacities, or a very large number. The result is then a kind of genetic mix, or soup, which is difficult to analyze from the viewpoint of genetic mechanisms. It is as if, in studying physical inheritance, the experimenters had tried to represent eye color, hair length, and body size in one index of physique.

One obvious solution is to test whether or not the items on an intelligence test are correlated and thus attempt to determine how many basic capacities may be involved. For this purpose the technique of factor analysis was designed. When this technique is used, a very large number of factors can be obtained, some of which show high heritability coefficients while others do not.[20] Thus, these factors do not necessarily represent the basic processes that are modified by heredity but, rather, integrations of several processes which may or may not reflect common genetic causation. When these results are arranged on a scale, it is interesting that those factors which are most closely related to sensory and motor processes show the highest degree of heritability, while those factors which are purely cognitive show the lowest ones. These findings are reinforced by the study of differences in problem-solving abilities of animals, but these leads have largely been neglected in human studies.

One of the first experiments along these lines was that by Tryon on maze running ability in rats.[21] He used a mechanical maze designed to eliminate human handling as a possible variable modifying the results. After some seven generations of selection without inbreeding he produced two strains called "maze bright" and "maze dull," whose scores in the mechanical maze showed very little overlap. Later, Searle tested these same two strains on a variety of other tasks and found little correlation.[22] On a water maze, for example, the so-called maze dull animals performed better than the maze bright strain. He concluded that the primary difference between the two strains was that one was fearful of the somewhat noisy operation of the mechanical maze and the other was not.

One of the most ambitious attempts to demonstrate genetic effects on purely cognitive abilities was a selection experiment by W. R. Thompson carried out with rats on the Hebb-Williams maze.[23] In this maze, subjects are first trained on six practice problems under food motivation based on twenty-four-hour deprivation. Animals that did not become proficient in solving the first six problems were eliminated from the experiment. This would have the effect of selecting against those animals whose emotional and motivational characteristics interfered with maze running. Further selection was based on animals that did well or poorly as measured by errors. After six generations of selection, the mean scores were approximately 130 points apart with no overlapping between the two populations. Some other sorts of tests were given the rats, most of which showed no differences between strains except that the so-called dull rats tended to explore more in an enclosed maze than did the bright animals, although this difference was not seen on an elevated maze. Bright animals also seem somewhat more highly motivated by food, as one might expect. Of course, it would be extremely difficult to demonstrate a complete negative, i.e. that emotional and motivational characteristics were not involved in ability to solve complex problems. In Thompson's experiments, no effort was made to breed a third strain based on animals rejected during pretraining. Consequently, even if the differences between the dull and bright strains are accepted as cognitive, the experiment does not negate the hypothesis that genetic variation affecting emotional and motivational trials is important in populations of laboratory rats.

We obtained similar results to those of Tryon and Searle with our studies of performance in different dog breeds.[4] Unlike Tryon's original experiment, we measured all animals with a variety of testing and training procedures. Statistically significant and, in many cases, large differences were obtained on every test, although in every case there was some overlapping between breeds. However, there was no consistent correlation between performance on the different tasks, some breeds excelling in one and some in another. A factorial analysis, however, yielded one factor principally associated with the so-called intelligence or performance tests. This turned out to actually be a tendency for one particular breed to be afraid of strange apparatus. We concluded that, while the various dog breeds had been successfully selected for differential ability to solve problems and undergo training, the selection had operated principally upon basic variation in motivational and emotional responses. When it was possible to get dogs equally motivated and unafraid in a situation, the different breeds performed very much on the same level.

These results, together with the attempts to factor analyze human IQ tests, give a clear lead that human behavior geneticists might follow. If human populations are like those of other animals, there should be much more genetic variation in motivational and emotional responses than in cognitive ones, and the former should be the sorts of traits that would be productive to study. In any case, the study of particular and special traits should give more clear-cut and understandable results than a vague and global concept such as that of intelligence.

The importance of using a wide variety of tests and measurements in connection with any study of variation of performance cannot be overestimated. This should give a clearer picture of the ability or abilities being studied and, equally important, a definition of those capacities which are *not* involved.

On the other hand, human beings are different from other animals in their capacity for language. While there must be a strong selection for individuals who excel in verbal capacities, there must also be a great deal of variation yet remaining in this relatively new trait.

SOCIAL GENETICS AND BEHAVIOR GENETICS

Synergistic Relationship Between Genetic Variation and Social Selection

Man is, above all, a social animal, and his genetic structure and function must be highly influenced by social factors and selection pressures from the social environment. It has been pointed out elsewhere that the dog, through the process of domestication, has been adopted into various human societies and should reflect, in a somewhat modified but more sensitive fashion, the results of growth and changes in human cultures.[4] Being much shorter-lived than people, dogs are more rapidly responsive to changes in social selective pressures.

One of the characteristics of complex human societies is the multiplication and specialization of social roles, and we concluded that there is a synergistic relationship between genetic diversity and complexity of social organization. Genetic diversity makes for individuals having specialized abilities in many directions that are useful in a complex society. Reciprocally, a complex society can use and protect diversity. The result is that the dog and, to a lesser degree, man, comprise the two most variable animal species on earth.

Variation in Social Behavior

We found that 10,000 years or more of close association with man have not basically changed the organization of wolf-like behavior patterns of the dog. Rather, the basic patterns have been modified with respect to frequency and ease of elicitation. To take an example, wolves show only a moderate degree of agonistic behavior in a well-organized wolf pack. In domestic dogs, some of the terrier breeds are so aggressive that they are likely to kill each other even when raised in the same litter. At the other extreme, the hound breeds have been selected for their capacity to tolerate each other, and two male beagles can be placed together in the same pen with little danger of any fighting, other than vocal threats.

Again, we have a model for human behavior genetics. It is quite probable that such differences in agonistic behavior also exist among individuals in human populations and that they have important effects on the ease with which these individuals adjust to the conditions of civilized living and, consequently, upon their own mental health. Similar differences in other systems of social behavior would be expected in sexual behavior, care-giving behavior (especially in maternal care[24]), and others.

Maladaptive Behavior

Another social phenomenon which shows some genetic variation is the process of primary socialization or social attachment. While all dogs show this process in a well-developed form, as it is obviously necessary for survival that a young puppy become attached to either its own or its adopted species, there are differences in the rapidity and depth of these attachments. In particular, we found that our strain of basenjis formed attachments to people much less readily than did cocker spaniels and appeared to be emotionally tough with respect to separation. Emotional reactions given to separation were quite brief. Such a dog is difficult to manage when taken into human society as a pet and has a tendency to become a canine juvenile delinquent, wandering far from home, getting into frequent fights with other dogs, and even occasionally biting strange human beings. We need to look into the genetic basis of social attachment in human beings, not only as it might be related to the sociopathic personality, but also with respect to all disturbances of social attachment, such as those found in autistic children and possibly also in adult schizophrenia.

A perennial problem of human behavior genetics is that of the genetics of schizophrenia. While we have no precise nonhuman animal model of this condition, particularly because it has largely been described in terms of symptoms of verbal communication, there are two lines of work that suggest new lines of progress.

One is the conclusion that functional maladaptive behavior can be experimentally produced in nonhuman animals under the following conditions: (1) intense motivation or arousal (2) lack of opportunity to escape from the situation producing the motivation or arousal and (3) impossibility (for either personal or situational reasons) to react adaptively to the cause of the motivation or arousal.[25] Genetic systems could modify the outcome in at least three ways by determining: (1) the intensity of motivation or arousal produced by a given situation (2) the capacity of an individual to react adaptively when aroused, and (3) the nature of maladaptive symptoms expressed.

The second line of work is that arising from the study of social attachment and separation in both nonhuman and human animals.[26] Separation produces a wide spectrum of maladaptive symptoms, ranging from depression in monkeys to extreme timidity and fear-biting in dogs.[27,28] Separation from the rearing place of animals that have been previously reared in isolation produces autistic symptoms in both dogs and rhesus monkeys.[29,30] From this perspective it is obvious that human infantile autism is a disorder of the initial stages of the normal social attachment or bonding process, and Clancy, et al. have made fruitful therapeutic applications of this concept.[31]

Rosenthal has presented evidence linking infantile autism with schizophrenia.[32] Whether or not this view is correct, it is obvious that one aspect of adult schizophrenia is a disturbance of social relationships and social attachments; the verbal symptoms effectively prevent the development of any deep emotional relationships with others.

Taken together, these results suggest the search for at least two genetic systems affecting the expression of schizophrenia. The first is a system modifying susceptibility to stress. The most probable thing to look for would be variation in some form of emotional sensitivity, presumably that connected with attachment and separation (although it might be others). Such sensitivity would be an adaptive and useful capacity under ordinary conditions or in moderate degrees.

The second is a genetic system which would modify the symptoms expressed once an individual had been exposed to stress. Rosenthal analyzes data concerning the risk of schizophrenia in sibs of individuals that show neuroses and finds that sibs of persons with anxiety reactions, depersonalization, and oversensitivity show substantially higher rates.[32] This evidence supports the hypothesis of two kinds of genetic systems. Gottesman and Shields argue for polygenic inheritance, which would also be consistent with this view.[33]

At the very least, these facts suggest the genetic study of schizophrenia using more than one kind of test, especially tests that would measure emotional and motivational reactions. Anything as complex as behavior cannot be satisfactorily measured on only one dimension.

ECOLOGICAL GENETICS AND BEHAVIOR GENETICS

Ecological genetics may be defined as the genetic study of the forces of natural selection.[34] Hitherto, most of these studies have been done with anatomical rather than behavioral characteristics and with a variety of short-lived invertebrates that might be expected to show relatively rapid responses to selective pressures. However, there is every reason to suppose that behavioral patterns are responsive to selection, especially since these are the immediate ways in which most animals adapt to the environment, and that even the longest-lived vertebrates would be affected to some degree, although long life affords some protection against selection.

Interaction of Selective Pressures

Selective pressures may act on any level of organization and at any point in development, from conception to old age. With respect to selective pressures at different levels, these might be expected to act in either similar or opposite directions. In the former case, the effect should be to shift the genetic composition of the population rapidly; and, in the latter, some sort of state of equilibrium should result. Conflict between selection pressures should result in the preservation of variation.[35]

Social Selection

For any sort of highly social animal, social selection pressures should play a major role. These should be particularly important in man, for whom the development of a high degree of culture has blunted the effects of many selection pressures from other sources. However, man has developed the capacity for cultural change to an extent which is far beyond anything present in other social species, which means that social selection pressures vary rapidly over time; in fact, so rapidly that it is difficult to see how genetic factors can be greatly modified by them.[4]

By contrast, in other highly social animals the social environment is the most constant feature of any aspect of the environment. It is only under a constant environment that stereotyped behavior is more adaptive than variable behavior, and, indeed, most of the examples of stereotyped behavior patterns cited by instinct theorists come from social behavior. Man is, therefore, quite different from other social species. Even in his social behavior it is difficult to identify more than a few simple remnants of stereotyped behavior patterns, such as the social smile. Sexual behavior, which tends to be highly stylized in most nonhuman social animals, is more variable than in any other species.

Besides the theoretical problem of the genetic effects of changing social selection pressures in man, it should be interesting to study individual genetic variation in the capacity for emitting variable behavior itself.

SUMMARY

One of the basic problems for the future of behavior genetics, whether applied to man or the nonhuman animals, is where do we go from here—what will be the most important, fruitful, and relevant areas of work in the future?

Mendelian Genetics

This can be studied quite easily in human populations, provided genotypes can be reliably identified. The principal advantage of knowing that variation in a particular trait is determined by a simple one-factor or two-factor Mendelian mechanism is that it permits a relatively high degree of predictability for the offspring of parents whose genotypes are known. The possibility of simple Mendelian inheritance affecting behavioral variation should always be checked out. On the other hand, the principal theoretical aspect of Mendelian studies is as a test of the chromosomal theory of inheritance, about which there is no longer any question and for which studies on behavior can add no new information. Furthermore, there is now no doubt that genetics *can* affect behavior, and further studies of this sort tend to be redundant.

Physiological Behavior Genetics

What we lack are good experiments on *how* genetics affects behavior. How do genetic differences develop, and what are the physiological mechanisms that produce them? From this viewpoint, it is essential that important behavioral characteristics should be studied; and in this paper are listed some of the leads that are provided by animal models. To summarize these, work should be concentrated on basic processes which bear a reasonably close relationship to primary gene action. From the nature of gene action, the simpler the process, the more meaningful the results. If very complex processes are studied or if simple processes are randomly added together as they have been in studies of intelligence, little more can ever be determined than that heredity produces variation, with some estimates of its importance under special conditions. However, since heritability estimates only apply to special situations, their general value is questionable, except when two different kinds of behavior are investigated in the same situation.

Developmental Genetics

Studying the development of behavioral differences is an alternate method to that of physiological genetics for determining how genetics affects behavior. There is an obvious need for long-term longitudinal studies of related human individuals. Sex is a variable whose heritability is very nearly 100 percent; and the study of the development of sex differences in behavior should be a fertile field, particularly if it is done with due regard to the interaction of genetic and environmental influences. Of particular interest is the variable of the onset of puberty, but similar variation may exist in almost any other maturational process that has an effect on behavior.

Individual Genetics

The study of individual differences in any behavioral characteristic could be rewarding, but those areas in which a payoff is particularly likely are those most closely related to physiological processes. Among these are genetic variations in sensory and perceptual processes, and there is an almost entirely neglected field of genetic variation in motor capacities. As I have pointed out earlier, the area of intellectual or cognitive capacities will be fruitful in proportion to the degree that simple and separable capacities can be studied. An especially promising area in view of the new interest in psycholinguistics is the inheritance of language capacities.

Social Genetics

We need to study the basis of genetic variation in the important systems of social behavior, including agonistic behavior, sexual behavior, care-giving behavior, and the like. For example, the quality of parental care has very important social consequences, and we need to know how this is affected by genetic factors. Among other behavioral systems, that of allelomimetic behavior is closely related to the process of social attachment. Disturbances or aberrations of the attachment process have serious implications for mental health. Indeed, one of the obvious consequences of neurotic and psychotic behavior is that it has a disturbing effect on social relationships and disrupt normal social organization. We need to find out how to identify genetic variants as early in development as possible, so that appropriate preventive or facilitative actions can be taken.

Ecological Genetics

The study of the population genetics of behavior is still in its infancy, but we need to know the effects of various kinds of population organization and cultural change as they modify the genetic systems of these populations.

What Should We Study?

An obvious answer is those kinds of behavior that are potentially most important. From an evolutionary viewpoint, this means those kinds of behavior that strongly affect survival of groups or individuals. The differential onset of sexual maturity, with all its attendant effects of hormones on behavior, has already been mentioned. The emotional and motivational background of this and other important systems of social behavior should be studied, as these have potentially an enormous effect, not only upon social adaptation and adjustment, but also upon cognitive performance. With respect to cognitive activities themselves, we badly need studies on genetic variation in basic learning processes such as the tendencies toward rapid or slow habit formation, speed of association, and acquisition of language skills. Once genetic information is collected, it may give leads regarding the nature of underlying physiological processes and eventually the possibility of modifying them by other than genetic means.

Finally, the nature of gene action suggests that behavior genetics should become process-oriented rather than oriented toward causal determination. There is still a holdover of primitive thinking which urges us to look for final causes divided into dichotomous categories such as heredity and learning. We need to think, instead, in terms of ongoing processes which are modified by a large number of factors, including the genes, and to realize that no understanding of genetic determination is complete without a knowledge of the effects of environmental variation. Environmentalists and hereditarians can and should meet on the common ground of the study of variation.

REFERENCES

1. Wright, S.: Evolution and the genetics of populations. In Wright, S. (Ed.): *Genetic and Biometric Foundations.* Chicago, U of Chicago Pr, 1968, vol. I.
2. Caspari, E.: Gene action as applied to behavior. In Hirsch, J. (Ed.): *Behavior Genetic Analysis.* New York, McGraw, 1967.
3. Beadle, G.W., and Tatum, E.L.: Genetic control of biochemical reactions in *Neurospora. Proc Nat Acad Sci (USA), 27*:499, 1941.
4. Scott, J.P., and Fuller, J.L.: *Genetics and the Social Behavior of the Dog.* Chicago, U of Chicago Pr, 1965.
5. Corbin, J.E., Mohrman, R.K., and Wilcke, H.L.: Purebred dogs in nutrition research. *Proc Anim Care Panel, 12*:163, 1962.
6. Zacharias, L., and Wurtman, R.J.: Age at menarche. *N Engl J Med, 280*:868, 1969.
7. Popenoe, P.: Inheritance of age of onset of menstruation. *Eugen News, 13*:101, 1928.
8. Fuller, J.L.: Photoperiodic control of estrus in the basenji. *J Hered, 47*:179, 1956.
9. Jones, M.C., and Bayley, N.: Physical maturity among boys as related to behavior. *J Educ Psychol, 41*:129, 1950.
10. Neel, J.V.: Lessons from a "primitive" people. *Science, 170*:815, 1970.
11. Scott, J.P.: Critical periods in the development of social behavior in puppies. *Psychosom Med, 20*:42, 1958.
12. Scott, J.P.: Foreword and chapter on critical periods for the development of social behavior in dogs. In Kazda, S., and Denenberg, V.H. (Eds.): *The Postnatal Development of Phenotype.* Prague, Academia Press, 1970.
13. Bayer, L.M., and Bayley, N.: *Growth Diagnosis.* Chicago, U of Chicago Press, 1959.
14. McKusick, V.A.: *Heritable Disorders of Connective Tissue,* 2nd ed. St. Louis, Mosby, 1960.
15. Hamburger, V.: Embryonic motility in vertebrates. In Schmitt, F.O. (Ed.): *The Neurosciences: Second Study Program.* New York, Rockefeller, 1970.
16. Rosenzweig, M.R., Krech, D., Bennett, E.L., and Diamond, M.C.: Modifying brain chemistry and anatomy by enrichment or impoverishment of experience. In Newton, G., and Levine, S. (Eds.): *Early Experience and Behavior.* Springfield, Thomas, 1968.
17. Elliot, O., and Scott, J.P.: The analysis of breed differences in maze performance in dogs. *Anim Behav, 13*:5, 1965.
18. Anastasi, A.: *Psychological Testing,* 2nd ed. New York, Macmillan, 1961.
19. Erlenmeyer-Kimling, L., and Jarvik, L.F.: Genetics and intelligence: A review. *Science, 142*:1477, 1963.
20. Strandskov, H.H.: A twin study pertaining to the genetics of intelligence. *Proc 9th Int Congr Genet,* 1953, p. 811.
21. Tryon, R.C.: Studies in individual differences in maze ability. *Comp Psychol, 11*:145; *12*:1, 95, 303, 401; *28*:361; *30*:283, 535; *32*:407, 477; 1930-41.
22. Searle, L.V.: The organization of heredi-

tary maze-brightness and maze-dullness. *Genet Psychol Monogr, 39*:279, 1949.
23. Thompson, W.R.: The inheritance and development of intelligence. *Proc Assoc Res Nerv Ment Dis, 33*:209, 1954.
24. Levy, D.M.: *Maternal Overprotection.* New York, Columbia U Pr, 1943.
25. Scott, J.P.: Comment on psychosocial position. In Kruse, H.D. (Ed.): *Integrating The Approaches to Mental Disease.* New York, Hoeber, 1957.
26. Schaffer, H.R. (Ed.): *The Origins of Human Social Relations.* London, Acad Pr, 1971.
27. Spencer-Booth, Y., and Hinde, R.A.: Effects of six days separation from mothers on 18- to 32-week-old rhesus monkeys. *Anim Behav, 19*:174, 1971.
28. Senay, E.C., and Scott, J.P. (Ed.): *Separation and Depression; Clinical and Research Aspects. AAAS Symp,* Washington, D.C., 1973.
29. Fuller, J.L.: Experiential deprivation and later behavior. *Science, 158*:1645, 1967.
30. Harlow, H.F., and Harlow, M.K.: The affectional systems. In Schrier, A.M., Harlow, H.F., and Stollinitz, F. (Eds.): *Behavior of Nonhuman Primates.* New York, Acad Pr, 1965.
31. Clancy, H., Entsch, M., and Rendle-Short, J.: Infantile autism: The correction of feeding abnormalities. *Dev Med Child Neurol, 11*:569, 1969.
32. Rosenthal, D.: *Genetic Theory and Abnormal Behavior.* New York, McGraw, 1970.
33. Gottesman, I.I., and Shields, J.: Schizophrenia: Geneticism and environmentalism. *Hum Hered, 21*:517, 1971.
34. Ford, E.B.: *Ecological Genetics.* New York, Wiley, 1964.
35. Scott, J.P.: Evolution and domestication of the dog. *Evol Biol, 2*:244, 1968.

Chapter 4

ON THE GENETICS OF BEING HUMAN

Robin Fox

THAT HUMAN BEHAVIOR IS firmly rooted in genetic processes is no longer in dispute among those having any but the briefest acquaintance with the facts. In years to come it will appear as remarkable to our successors that this could ever have been doubted, as it appears to us remarkable that our ancestors doubted the obvious relationship of man to the great apes or the theory of natural selection itself. Today, only religious fundamentalists will challenge these particular truths, and perhaps we are rapidly moving into a phase where the genetic basis of behavior will be vehemently denied only by those scientific fundamentalists who (for such ideological reasons as the sociologists of knowledge will no doubt uncover) are determined to persist in the belief that "man has no instincts," that "all human behavior (or at best 99%) is learned," and that "there is no innate component of behavior," or even that whatever appears to be innate was in fact learned in the womb![1]

The work of the behavior geneticists and the ethologists (the latter dealing, in a sense, with "the behavior genetics of species") have provided frameworks for the investigation of the nature of behavioral variation within species and species-specific behavioral repertoires. It is important to note that while behavioral variation is not always wholly or partly genetic in character, it is the genetically based performance that is important for natural selection, since it is on the genotypic variations that selective pressures ultimately operate, even though this pressure is mediated through the phenotypic activity of the organism. Both behavior genetics, and ethological genetics, have to be placed in a framework of evolutionary genetics; that is, a phylogenetic framework. This involves seeing the present behavioral repertoire of a species and the variations in the repertoire within the species as outcomes of long runs of selection and adaptation and the investigation of the selection pressures over time that have contributed to the end result. Again, this involves cross-reference to another related discipline: population genetics. The shifts in gene frequencies that give rise to changes in the genetic heritage of a species, and eventually to new species, are the provenance of those concerned with gene flow and interaction within breeding populations. Ultimately, all these special disciplines refer back to molecular genetics on the one hand and the theory of natural selection on the other: twin guardians of the route of knowledge in the biosphere. (See Fig. 4-1.)

A complete picture of the behavior of any species is only possible with the combined efforts of all these disciplines, and

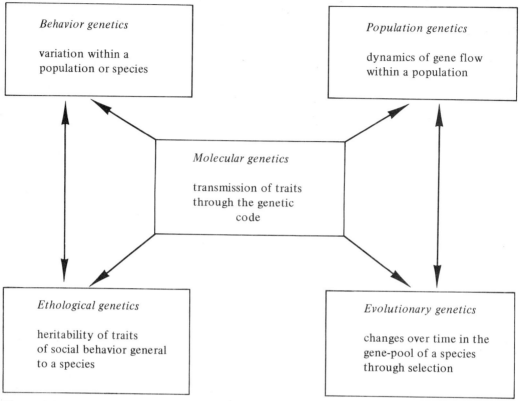

Figure 4-1. Intercorrelations among the genetic sciences with respect to the analysis of behavior.

indeed they overlap continuously. Thus, ethologically we can establish by observation of many members of a species in many varied populations, under a variety of conditions, and after close comparison with related species, that a certain behavioral trait is characteristic of the species under observation. This need not be an instinct; it could be a pattern of learning, a critical period of learning, a mode of communication depending on both innate and learned features, and so on. Having established that it is species-specific, we can then look at the variations it manifests. Like all other characteristics it is likely to fall along a more or less normal curve of distribution (whether the units be individuals or populations). Thus, some individuals will exhibit extremes of the trait while others will barely manifest it at all, and yet others will have it in moderation—the majority, in fact. At this point, behavior genetics can tell us something about the conditions for producing these variations by selective breeding. Population genetics can then examine the assortative mating techniques by which a shuffling and reshuffling of these variations can occur. The overall picture will be completed by a diachronic account from evolutionary genetics of the selection pressures which have operated over time to produce the trait in question. Ultimately, of course, it would be necessary to locate the actual instructions in the genetic code, but the difficulties in the way of doing this, even for physical structure, are enormous as yet. What matters is that we know that the code is capable of bearing extremely complex information about behavioral as well as structural features, not that these are unrelated.

We could take as a simple example some trait that was clearly universal in our own species, like language. Human language is a unique mode of communication and is the most usual criterion offered for the uniqueness of the human species. The idea that language (as opposed to languages) is a pure human invention, that is, is completely a learned phenomenon depending on the laws of conditioning for its development in each child afresh, has to be discarded.[2] It is pointless to reiterate the argument here, since it has been more than well ventilated: it is sufficient to say that the capacity for grammatical speech is clearly innate and matures as the child matures.[3] The close and detailed observations of linguists (who for our purposes are *ethologists*) have established this and the patterns of ontogenetic development from gurglings to well-formed sentences.[4, 5] If there is genetic variation in this capacity, behavior genetics should uncover it. There seems to be evidence that chromosomal abnormality produces severe speech impairment, for example, which not only gives us evidence of sources of variation, but also suggests that speech capacity may well be localized and not just the product of general intelligence or neural capacity.[6] What is remarkable about language, however, is that there seems to be very little variation in the capacity to acquire it; deviations about the mean are very small. This in itself is significant since it points to the evolutionary importance of the trait as an adaptive mechanism. We take this lack of variation for granted because it is so familiar to us, but it has to be explained in the same way that a trait with high variation has to be explained. Evolutionary genetics would be concerned with telling us how the mutations favorable to language acquisition came to be favored and how they would have gradually become dominant in the species, thus overlapping with the population genetic analysis.

We will return to language, but for the moment we can contrast it with another universal trait that shows wide variation, such as intelligence. Here the deviations about the mean are very large, the genetic basis of the trait is well understood, and significant clusterings occur both within and between populations. The evolutionary significance of the trait is obvious, although much remains to be done in determining why the specifically human mode of intellectual functioning was selected; the involvement of verbal skills is paramount, for example, with mechanical skills close behind. But it is here that population and behavior genetics come into their own, since a trait showing this degree of variation is profoundly influenced by breeding patterns.

Again it should be stressed that not all the variation found in the ability either to speak or to exercise the intellect is genetic by any means. In both cases, though in different degrees, there are different inherited capacities for both, but the manifestation of either is profoundly influenced by the stimulus the organism receives, the input from its environment. If there are critical periods, for example, in both concept and language learning, and if the organism is not stimulated properly at these times, then its inherited capacities, however excellent, may never be able to compensate for this loss.

Caution has been exercised in starting this essay on the genetics of being human by choosing two obvious examples of behavior—language and intelligence—that are genetically rooted, universal, and accessible to rigorous methods of investigation. When we stray into other areas such as territoriality, dominance, bonding, reciprocity, and superstition, we have to be more daring and speculative and hence open to both reasonable criticism and unreasoned attack (from those who wish to maintain that *nothing* in human behavior is innate). Let us first establish that, *in principle* at least, there is no reason why what is true of language and intelligence should not be true of gambling or gossip (to pick two seemingly outrageous examples). Such a principle does not seem outrageous, for example, to so distinguished a molecular biologist as Jacques Monod:

> When behavior implies elements acquired through experience, they are acquired according to a *program*, and that program is innate—that is to say, genetically determined. The program's structure initiates and guides early learning, which will follow a certain preestablished pattern defined in the species' genetic patrimony. Thus, in all likelihood, is the process to be understood whereby the child acquires language. And there is no reason not to suppose that the same holds true for the fundamental categories of cognition in man, and perhaps also for a good many other elements of human behavior, less basic but of great consequence in the shaping of the individual and society.[7]

He goes on to add that while the ethologist can isolate such elements by experiment, this cannot be done in human society for humane reasons. But often, in the names of efficiency and even humanity, the experiments are performed anyway; and where they cannot, careful comparison of differing instances of the same phenomenon in an attempt to get at the underlying elements that are shaped by the program can always be resorted to.

Let us take an example of an experiment. Man is a mammal, and the female of the species *Homo sapiens* is a particularly well equipped mammal. Throughout the mammalian order, for obvious reasons, close bonds between mother and offspring are forged by various means. Count has called this *the lactation complex* and has pointed out that much more than milk is involved.[8] The mother is the child's environment throughout the crucial period of early learning of which Monod speaks, and the program seems, in the higher mammals at least, to have a very definite instruction: "form a close and emotionally satisfying bond with the mother." On the basis of the successful completion of this phase, much of the future emotional capacity of the organism will depend.[9] This is known from a fairly simple experiment: deprive the organism of the mother during this period and it will develop symptoms which, at their worst, show total withdrawal and emotional breakdown with a complete failure in later life to make emotional attachments or even to reproduce in some cases. The experiments are done daily in orphanages. Controlled experiments with monkeys merely serve to confirm the worst horrors of such traumatic separation. Some research suggests that even small disruptions of the bond can have lasting consequences, although again, there is considerable variation about the mean in this instance. The deprivation experiment, the classic method of ethology, cannot, of course, always be applied in the human case as Monod observes, and for obvious reasons. No one knowing the consequences would willfully induce such symptoms. Unfortunate for humanity, but rich with information for science, however, are many instances of deprivation induced by ignorance. The example above illustrates that, like the capacity for language and the capacity for abstract reasoning, the capacity for love is dependent on the acquisition through experience of elements demanded by an innate program. A failure to match the output with an appropriate input results in a breakdown, usually only partially reversible, of the whole system of behavior. The analogy with physical growth is all too obvious: no matter how good its innate physical constitution, a child demands a certain input of protein, vitamins, etc., to match its output of energy. Should this not be forthcoming, it will be deformed, mentally retarded, and may even die. That careful attention is paid to physical input while virtually ignoring behavioral input is a consequence of a failure to understand that the program is as carefully mapped for behavioral as it is for physical needs. The minimum daily requirement of love may not have been established, but a total lack of it is as dangerous as a total lack of vitamin D. In this instance, of course, we are lucky to know what it is the child is being deprived *of*. Hundreds of similar deprivation experiments may be going on daily and on a large scale, but until the program involved and the input that corresponds to it can be identified, we are as much in the dark as we were in the days before the consequences of vitamin deficiency were known. Identifying the behavioral vitamins is the next major task that confronts us.

The best ethological approach to this is to look at the bonds that are associated with the human lifecycle. The obvious reason is that the lifecycle represents the most highly programmed aspect of human existence and its behavioral and physiological elements are so closely interwoven. The mother-child bond that has been briefly touched on is clearly the most basic, as well as the first in time of these bonds. But, as Lorenz demonstrated in the now classic paper on companions in bird life, the organism will respond to an appropriate person in a way appropriate to the stage of the cycle it has reached.[10] Thus, the human infant responds in a particular way to the mother when young (with a battery of innate equipment such as smiling and crying), to peers when adolescent, to members of the opposite sex in opposite ways during courtship and the preliminaries of mating, to offspring during the parental phase, to the young of each sex differentially when middle-aged, and so on. Again it must be stressed that there will be wide individual variations in response; but every social unit that is natural, that is, one that sees its members through all stages of their life cycles, must arrange some accommodation between the various blocks of persons that represent these different stages.

Using another ethological approach, the comparison with closely-related species, we can establish what the major blocks of these social systems must be. In all primate societies, but especially in the terrestrial primates (and even more especially in the savannah-dwelling primates who are so important for understanding human evolution), the major units seem to be these: the block of mothers-with-young (all primate females become pregnant as soon as they are able); the block of heirarchical males; and the block of peripheral males, largely adolescent and preadolescent males, but also some others, all having in common a failure to be in the central heirarchy. The blocks themselves are internally divided. The female-with-young block has older females past childbearing (a rarity in other primates than man, however), mothers, and young females about to be mothers. The male hierarchy has the senior males with established dominance positions and the younger cadets who are established as candidates for the heirarchy but as yet do not have a ranking place in it. The peripheral males are a more amorphous group but may have their own hierarchy. The social system, then, consists of the accommodations made between the blocks and within the blocks according to the interests of the members.[11, 12] That is, males want something from females and vice versa, while the older males wish to contain the younger; males want various things from each other, as do females, while the younger males want something from the older, and so on. Thus, sex, age, and status become the three variables one must observe.

Human primate societies must make accommodations among these sex-age-status units in the same way that nonhuman primate societies do, but the nature of the wants and the methods of accommodation are somewhat different because of the different evolutionary histories of man and the other primates. Let us take as an example, not the more obvious male-female or mother-child bonds, but those between older and younger males. (It is one contention, for example, that a good deal of the male-female relationship, including the male dominance of females, is a by-product of the male-male relationship. In other words, males dominate females in order to dominate other males). In many primate societies, the relationship between males is dependent on their competition for ovulating females. The more dominant the male, the more chance he has of copulating with the female during ovulation, and hence a greater chance of perpetuating his genes. Not all primate societies operate on this principle but, even in those that do not, relationships of dominance between the males are important. (Again, one must tediously reiterate that there is much variation here. The tendency to establish a dominance order among males is, however, seemingly universal, being much more pronounced in some species than in others.) This creates a considerable tension between the older established males and those younger males who are trying to establish a place in the ranking order. The younger males will begin to challenge the older ones when they come to maturity, after spending some years literally on the peripheries of the group. The older males will respond to the challenge with threats and, if necessary, with physical violence that can be lethal. The incautious and overpresumptuous young male may very well not survive this period. His more cunning and restrained colleague has a better chance. This process of timing one's responses and, in a sense, calculating the odds for and against a foray into the breeding arena, has been called *equilibration,* and is the route by which a young male gradually works his way into the central hierarchy.[13]

This is a crude and brief description of a complicated process. But the point trying to be established is that this is a kind of primate baseline from which the evolving hominids developed their own peculiar form of accommodation between the old and young males. The whole of hominid evolution to make the point cannot be recapituated here, but it should be sufficient to say that with the advent of systematic hunting, the great differentiating mechanism whose selection pressures profoundly influenced hominid development, the relations between the old and young males became even more tricky and the form of accommodation consequently had to be more sophisticated. Natural weapons in the form of fangs and sheer strength no longer counted for very much (as Freud observed) and manufactured weapons evened out the crudely physical competition. If the older males were to dominate the younger ones successfully, recruit them into the hunting group, and yet keep them from the females, then they had to produce a very elaborate threat gesture that would tap the equilibrational potential of the young. In other words, the young males, at the point in their life cycles when the production of testosterone increases so dramatically, are geared to a dual performance: they must on the one hand display aggressiveness and initiative and a desire for membership in the central hierarchy; on the other hand they must display caution and deferred gratification and ultimately identification with the senior males they both wish to displace and to be. The older males capitalize on these tendencies by various procedures of initiation: the human variant of the ancient primate process of equilibration.

Let us look carefully at this example and see what it is that is being claimed here. Following the ethological method, we have pinpointed a crucial social relationship, a bond, and looked at the problems associated with it. Thus, among many birds, the crucial bond is that between the mating pair. This involves a lowering of the potential for agression, a courtship ceremony, the acquisition of a territory, the making of a nest, etc. Depending on the species, there is a variety of means for accomplishing these ends, varying from simple appeasement movements to elaborate courtship and even the performance of duets. In the human case we have a crucial bond, that between older and younger males, and similar problems are associated with it: selection, lowering of aggression, identification, etc. Again, certain processes of a ritual kind are indulged in which seem to have the effect, if they succeed, of creating the dominant-cadet bond which, at least in the early hunting phase of our species' existence (and that is 99% of our history), was as crucial for survival as the male-female bond itself, if not more so.

With the last statement we have leaped forward a little to the phylogenetic reasons for the bond in question and its peculiarities. This is, of course, the question that lurks behind every ethological enquiry: as Lorenz puts it, the question is "How come?"—a teleonomic rather than a teleological demand. Again we must look sideways at our other disciplines. Capacities for bonding, for equilibration, for dominance, etc., must vary among individuals and even populations. The genetic basis for the variation can in principle be sought. Thus, not all birds have the same capacity for song, one supposes, and presumably the less effective ones will be less successful over time. Also, different populations of the same species can become reproductively isolated by developing differences in song and so eventually become subspecies and even in the end separate species entirely.

In the case of birds, however, the program has a much more fixed nature than it does with humans. The exigencies of human existence being much more varied, it makes sense that the program should allow considerable latitude here; greater, for instance than is the case with the mother-child bond. The point in the life cycle is reached where the postadolescent organism is moved to put out energy of a certain kind; often, seemingly, of a somewhat diffuse and negative kind. It is a high energy level and is sharply responsive to postadolescent females on the one hand and to older and dominant males on the other. The two responses are not unrelated, of course. The older male, having passed through this stage and achieved a dominance status, is equally moved to contain this challenging energy. The tension that exists in the relationship is then handled by ritual means of one kind or another, varying from extreme physical suffering and humiliation for the initiate to fairly mild exercises in self-control or rote learning. The outcome, if it all goes right, is young men whose sexual and aggressive energies are at the disposal of the society because they have identified with the older males in the initiatory process. Thus is the bond forged.

But it is not forged in the same way at the same time in every population of the species. The instructions in the program leave the question of means relatively open; one cannot predict exactly what the outcome will be in terms of the actual rituals employed. This outcome will depend on the particular conditions in each society, and how, in particular, the earlier bonding processes have been handled. In societies where young boys are gradually encouraged to identify with older males, the need for intensive initiation will be less than in those where such early identification is discouraged.[14] Many variations are possible. The program makes certain demands, and these are universal in the same way that the life cycle itself is universal. This is not in any sense a banal proposition: the demands could be different. One could in principle design a creature that lacked them; one in which senescence never occurred so that the old-young transformation presented no problems. Or one could design an animal on the insect model in which the transformation was automatically triggered off, producing a fully mature adult out of a chrysalis, and so on through the science-fiction possibilities. No. The program is what it is because of the selection pressures that have produced it. In its way it is quite specific, in its *demands*, as it has been put. Where it is not so specific (and with good reason, as we have seen) is in the range of means it allows to meet the demands.

It is the demands, then, that are in Monod's words, "innate, that is to say, genetically determined"; and that "will follow a certain preestablished pattern defined in the species' genetic patrimony." To those diehards who tiresomely ask, "Where are the genes for all this?" one can only reply "Where are the genes for bird song?"

The demands concern elements acquired through experience and these are acquired according to a program. The instructions are both to learn, on the part of the young, and to teach, on the part of the old. Initiation, as a process, is the *outcome* of these demands and the varying responses to them. Thus, in a sense, initiation is universal and inevitable, but not because there is an instinct for initiation. It is because we are programmed (or, as Lionel Tiger and Fox put it, going a stage further back in the computer metaphor, "wired") to put out certain kinds of energy and to respond to certain kinds of stimuli at a particular point in the life cycle.[15] Similarly, one could argue that some form of marriage, which at its broadest means an agreed-upon assignment of mates, is a necessary outcome of the tensions induced by the programs variously affecting the young males, the young females, and the older males, respectively. Kinship systems are elaborations of this tendency. The human imagination can seize upon these possibilities and weave many patterns, but the underlying program is the same everywhere and for everyone.

This is where the ethology of human behavior differs somewhat from its animal counterpart and is what makes analogies between the animal and the human situation tricky. There is no difference in principle, but a good ethologist should know all the details of all the variations in the behavior of all populations of the species he is studying. Unfortunately, a number of ethologists have not been as rigorous when they have come to pronounce on human behavior as they have been when looking at sticklebacks or greylag geese. The difference really is one of degree. There is a good deal of variation, for example, between societies of baboons of the same species living in different habitats. Being the same species, the program cannot differ, but the outcome can be quite different. The difference is largely the result of the same elements being combined in different ways, or some elements being emphasized to the exclusion of others, and so on. By looking at baboons in all possible habitats, we can still make up a list of the elements and work out the basic programs. This even goes across species boundaries. A lot has been made, for example, of the differences between baboon and chimpanzee social behavior. (People who deny that we can learn anything from primates about our own behavior nevertheless seem very happy to cite chimpanzee evidence to refute any resemblance between ourselves and the baboons. The logic is odd.) But all the elements of the baboon social system are present in the chimpanzee, and the more we learn of the chimp from research by the Japanese workers in Africa, for example, the more we see again that the differences are of emphasis and recombination. Chimpanzees who venture out onto the savannah, for example, begin to behave very much like savannah baboons.

Moving up to human societies, the same principle can be seen at work. Something as complex as the dominance-initiation process we have been discussing is made up of a number of elements which can occur in various combinations and recombinations, depending on the circumstances. It is indeed possible, to follow Wittgenstein's famous analogy of family resemblances, for no single element to be common to all the cases and yet for them all to be recognizable as belonging to the same family.[16] This is illustrated in Figure 4-2 where columns 1 through 6 are the cases and rows A through F are the elements. No two combinations are exactly alike, and no element is universal—that is, common to all the cases—and yet we would

	1	2	3	4	5	6
A	X	X	X	X	X	
B	X	X	X	X		X
C	X	X	X		X	X
D	X	X		X	X	X
E	X		X	X	X	X
F		X	X	X	X	X

Figure 4-2. Universals of the "family resemblance" model.

unhesitatingly recognize them as related and even in some way the same. The fact, then, that something in the list of elements is missing from some societies does not destroy the universality of the complex of which we are talking. But it is different from the case of songbirds where every songbird of a species certainly sings the same song, although even here it may be that some similar considerations apply.

It is seemingly different also from the case of language, since every human speaks. But the difference is more apparent than real. Every society, for example, initiates, in some way or other, according to the definition of initiation offered above. What differ are the means, as we have argued. In some way, every society speaks a language, but again the means differ. Thus, there is a limited set of distinctive features which can be used to make phonetic distinctions in human speech. But no two languages come out with the same list of phonemes. Figure 4-2 could represent the set of distinctive features (A through F) and the range of languages (1 through 6). Again, no one distinctive feature would be common to all languages, but the languages would have been selected from this common stock, and would thus resemble each other, and be certainly recognizable as the same phenomenon; that is, a human language. Any language falling outside this range would be unrecognizable as such.

Recognizing the elements of human social behavior that emanate from the program is a little like coming up with a list of distinctive features in language. But understanding the underlying programmatic instructions that put these into combinations is more like a search for the universal grammar of the rationalist linguists, recently revived by Chomsky and his associates.[17] And again it brings us to the point of universals, precisely what we are looking for when we look for items that are general in the behavior of the species. For what the universal grammar does is to specify *form*, not *content*, which is what we have been arguing regarding behavior. The universal grammar is, in its way, a program, and Chomsky believes it to be innate largely on the grounds that it would be impossible to conceive of any way in which a child could learn a particular language if it were not programmed to handle grammatical rules. But the universal deep structure carries no instructions about the specific content of the language. It carries instructions of the nature of the "A over A" rule ("no noun phrase can be extracted from within another noun phrase") or the principle of "structure-dependent operations" and so on.[17] These are part of the innate schematism applied by the mind to the data of experience. He makes the important point that we have made in connection with dominance-initiation above: they could be different; there is no obvious functional utility to them; one could have designed something simpler. They represent, in the sphere of language, a set of constraints. However much languages may differ in content and complexity, they cannot violate these basic principles without being something other than human languages. The organism is programmed to accept and process information that is compatible with them, and this is how, even with the fragmentary knowledge at its disposal, a child learns a language in the first place.

Comparable rules are not yet known in the behavioral sphere, but the principle involved is similar. The program dictates certain principles for the acquisition of behavior; it does not and cannot specify the actual forms the behavior takes. Thus, the linguistic universals do not specify actual elements that crop up in every language; but, rather, principles that all languages must observe whatever their elements.

If we consider the implications of the analogy further, we come up with the serious conclusion that, in the same way that an avoidance of these principles would render any attempts at language gibberish, so any failure by a society to observe the comparable principles in the behavioral sphere would result in a literally inhuman society. The range of variation is large, but it must have limits. To take a familiar example: failure to follow the program concerning dominance-initiation-identification, for example by the old abandoning all attempts to encourage the equilibrational functions of the young, may have disastrous results. While with language the problem does not arise since we do not legislate away or otherwise tamper at a deep level with the system, we constantly violate basic principles in the behavioral area: the traumatic separation of mother and child is one example that we have discussed. It thus becomes imperative to establish exactly what the innate programs are in order to work out the open range of variation. Many kinds of society are compatible with the basic programs in the same way that many kinds of language are compatible with the principles of the universal grammar. But they must have these principles in common if they are to count as languages at all.

Chomsky's answer to the problem of universals is that certain basic and innate schemata exist in the human mind, and these work on the material of linguistic experience in predetermined ways, the outcome being human languages. Similarly, we have posited the existence of innate demands which work on the material of experience, in this case social experience—the experience of other persons—and so produce human societies. In each case, the universal principles have to be located by a careful examination of cases, although with behavior in general, we have the advantage of related species to draw on. Chomsky does not think that studies of animal communication will help us to understand human language; the discontinuities are too great.

Before leaving the subject of universals, it is worth pointing out that, although these can be established without the identical items cropping up in all cases, this does not preclude the possibility that identical items can themselves be universal. But a universal item could, in fact, be relatively unrelated to any of the basic programs we have been talking about. The mere fact that something crops up in all societies, while constituting a *prima facie* case of a genetically based universal, still has to be shown to be so, not just assumed. One might observe, for example, that controlled defecation was characteristic of all human adults, but this is perhaps anything but an exigency of human living, since there appears to be no program for the learning of sphincter control comparable to the learning of language or the learning of love or the learning of various bonds. (The higher primates, for example, show no evidence of it.) One can teach an animal to do all kinds of things that are simply not part of its normal repertoire; things it is not programmed to learn. Singing dogs and dancing elephants testify to this; and, of course, it is what vitiates a good deal of the work of the behaviorist psychologists. An organism, suitably punished or suitably bribed, can be trained to do a lot of things, but these are extraneous to its nature; that is, to the program for acquiring elements through experience that is part of its genetic patrimony.

It is often the case, however, that a certain kind of universal pattern can appear which is not as random as that illustrated on Figure 4-2, and this should be noted. The model is that of the Guttman scale.

	1	2	3	4	5	6
A	X					
B	X	X				
C	X	X	X			
D	X	X	X	X		
E	X	X	X	X	X	
F	X	X	X	X	X	X

Figure 4-3. Universals on the "Guttman Scale" model.

Figure 4-3 illustrates this. Let us take columns 1 through 6 as types rather than individual cases. Type 1 has the whole range of elements associated with a particular complex, and whenever we come across a case with item A, we can predict that it will have all the others. Type 6 has only one item, and this item is common to all. It is the least common denominator, but not, therefore, the only universal in our sense, since what is universal is the program that determines the possibility of combinations of A through F existing at all. Thus, the same principle applies here as in Figure 4-1, even though the distribution of elements is very different indeed.

Let us recapitulate briefly. The various branches of inquiry into behavior that can be characterized as genetic are all needed if a complete picture of human behavior is to emerge. This picture will deal with the innate programs in terms of which behavior is acquired. The best way to isolate these programs is the ethological method of concentrating on the major social bonds as they are related to the life cycle and by observing on the one hand what is learned easily, and on the other hand what deprivations cause disruptions in behavior. A careful note of variations in input must be taken if we are to understand the outcomes, that is, human social systems. But, whatever the input, it has to operate in terms of the demands of the organisms which are programmed and, in consequence, only certain combinations of elements are possible, although the actual combinations may differ from case to case. There is no determinism here, then, at the level of content. The analogy with language illustrates that it is *forms of grammatical process* that are universal, not elements of surface grammar. One can invent a completely new language (Esperanto, for example), but it will have to fall within the range prescribed by the formal process or else not be a human language at all (a computer language, for example). In the other areas of behavior the same applies. Certain universal formal processes must be observed if social systems are to count as human. These processes are the genetic program for the acquisition of behavior that is peculiar to our species and are what make human behavior literally human. It is to the task of identifying and analyzing these processes that social scientists should now address themselves, with all the help from their colleagues in the genetic disciplines that they can get. Ultimately, this should mean a new framework for analyzing human behavior (biosociology), and perhaps even a saner perspective for appraising human values.

Bibliographical Note

The bibliography has been kept to a minimum. Extensive references to all the issues raised can be found in Tiger and Fox.[15]

The whole issue of innateness is ably dealt with by Lorenz.[1] The innate features of language and language learning are discussed by Lenneberg[3] and Chomsky.[2] Chromosomal abnormalities and language are discussed by Lenneberg,[6] and the structure of child language by Weir[4] and Bellugi and Brown.[5] The quotation from Monod[7] is from *Chance and Necessity*. The whole issue of the mother-child bond and deprivation can be found in Bowlby.[9] "Companions" in bird life are described in Lorenz.[10] The concept of "blocks" of individuals with like interests is developed in Fox,[12] and rests on the concept of "equilibration" put forward by Chance and Mead,[13] and elaborated by Fox.[11] Variations in initiation practices with respect to identification problems are examined in Whiting, Kluckhohn, and Anthony.[14] The idea of family resemblance which some philosophers think has solved the vexatious problem of universals is in Wittgenstein.[16] The latest statement of Chomsky's views on universal grammar from which the remarks quoted here are taken is Chomsky;[17] his discussion of the historical origins of the idea is in Chomsky.[17]

REFERENCES

1. Lorenz, K.: *Evolution and Modification of Behavior*. Chicago, U of Chicago Pr, 1965.
2. Chomsky, N.: *Language and Mind*. New York, HR&W, 1968.
3. Lenneberg, E.H.: *Biological Foundations of Language*. New York, Wiley, 1967.
4. Weir, R.H.: *Language in the Crib*. Hague, Mounton, 1962.
5. Bellugi, U., and Brown, R.: The acquisition of language. *Mongr Soc Res Child Dev,* 29:1, 1964.
6. Lenneberg, E.H. (Ed.): *New Directions in the Study of Language*. Cambridge, MITPr, 1964, pp. 63-88.
7. Monod, J.: *Chance and Necessity*. New York, Knopf, 1971, pp. 152-153.
8. Count, E.: The lactation complex: A phylogenetic consideration of the mammalian mother-child symbiosis, with special reference to man. *Homo,* 18:38, 1967.
9. Bowlby, J.: *Attachment and Loss: Volume I, Attachment*. London, Hogarth Press and Inst Psychoanalysis, 1969.
10. Lorenz, K.: Companions as factors in the bird's environment. In *Studies in Animal and Human Behavior*. Cambridge, Harvard U Press, 1970, vol. I.

11. Fox, R.: In the beginning: Aspects of hominid behavioral evolution. *Man: J Roy Anthrop Inst, 2*:415, 1967.
12. Fox, R.: Alliance and constraint: Sexual selection in the evolution of human kinship systems. In Campbell, B. (Ed.): *Sexual Selection and the Descent of Man.* Chicago, Aldine, 1972.
13. Chance, M.R.A., and Mead, A.P.: Social behavior and primate evolution. In *Evolution: Symposium of the Society for Experimental Biology 7.* New York, Jonathan Cape, 1953.
14. Whiting, J.W.M., Kluckhohn, R., and Anthony, A.S.: The function of male initiation ceremonies at puberty. In Maccoby, E.E., Newcomb, T.M., and Hartley, E.L. (Eds.): *Readings in Social Psychology.* New York, HR&W, 1958.
15. Tiger, L., and Fox, R.: *The Imperial Animal.* New York, HR&W, 1971.
16. Wittgenstein, L.: *Philosophical Investigations.* New York, Macmillan, 1953.
17. Chomsky, N.: *Cartesian Linguistics: A Chapter in the History of Rationalist Thought.* New York, Har-Row, 1966.

Chapter 5

MEASURING HUMAN BEHAVIOR

Richard Allon

It seems almost axiomatic that scientific advance cannot occur unless it is preceded by an advance in measurement technique. For proof of this axiom one need only look as far as the Table of Contents of this book. This would be a slim volume, indeed, were it written prior to the development of personality and intelligence measures, sphygmomanometers and skin electrodes, electroencephalographs, electrophoretic and multivariate data analysis techniques, or, for that matter, even the lowly yardstick. While this list is hardly exhaustive and represents a wide range of technological diversity and sophistication, the science of human behavior genetics would simply not exist in the absence of these and other human behavior measurement devices.

There is a corrolary to the above axiom: not only is scientific advance preceded necessarily by advances in measurement technique, but the magnitude of the advance will be in direct proportion to the sophistication of the technique which is developed. If one has a measurement device capable of rank ordering a given human behavior into two quantitative groups, e.g. high versus low, and another capable of quantifying that same behavior along a continuum, it is obvious that the latter measure will be superior to the former in that it will be more discriminating and, for this reason, when used in an experiment, will be more likely to yield meaningful and interpretable results.

A second look at the Table of Contents of this volume makes another thing clear: a competent discussion of the current state of development of measurement techniques and of measurement problems in such disparate fields as genetics, psychology, physiology, psychiatry, electroencephalography, pharmacology, biochemistry, and anthropology, all brought together in this volume under the common rubric of human behavior genetics, is beyond the scope of both this chapter and of the author's abilities. In view of this limitation, the major portion of this chapter will be devoted to a general discussion of the theory of human behavior measurement and of the principles for developing accurate devices for measuring human behavior rather than to a consideration of specific measurement problems. Some of the examples selected to illustrate certain points will be particularly relevant to human behavior genetics while others will be selected from more disparate areas of the behavioral and social sciences and will be presented more for their illustrative value than for immediate application. The use of such illustrations may put the nonpsychologist reader and particularly the reader with primary training in the biological rather than the social sciences at a disadvantage in seeing the immediate applications to his primary discipline. Such problems are not uncommon to newly developing interdisciplinary areas of study, and the author apologizes to the members of those disciplines neglected through illustrative application; such neglect is motivated by the author's ignorance rather than his chauvanism. It can only be hoped that readers of all disciplines will go on to find application of the principles put forward here to their specific fields for, while it may not be immediately obvious, the problems encountered by the endocrinologist seeking an accurate measure of adrenocorticosteroid level in subjects after insulin injection are more similar to than different from those of the primary school teacher seeking a valid and reliable measure of the amount of arithmetic learned by students after three months of classroom instruction. Obviously, the technological difficulties of measurement on the micro versus macro levels will require measures with different levels of sensitivity, and the devices used to obtain such measures will differ greatly in technological sophistication, but the need for accurate measures of human behavior are common to all the disciplines which contribute to human behavior genetics. Furthermore, while the specific needs and methodologies may be variable across disciplines, the principles for the construction and use of human behavior measures are constant. It is these principles that are the concern of the chapter.

One of the first steps in the development and growth of an interdisciplinary field of study should be to have the members of each participating discipline acquaint those of the various other allied disciplines with the terminologies, concepts, methodologies, and problems of their respective fields. It is to this end that this chapter is presented. Little that is new will be found for the reader who is well-skilled in measurement theory. The goal of this chapter is to acquaint the nonpsychologist user of measures with a general survey of measurement principles both for informational purposes and as an aid in the selection of proper measurement tools for research applications. The author will begin with a discussion of various ways in which the problem of human behavior measurement is conceptualized, followed by a consideration of the basic criteria which must be met in order to develop an accurate measure of human behavior, e.g. sampling, reliability, validity, norms, and standardization. Some specific areas of human behavior measurement commonly employed in human behavior genetics research, e.g. intelligence, personality, and psychopathology, will be discussed. Finally, some methodological problems affecting human behavior measurement will be considered. Statistical and theoretical development will be kept to a minimum and terms specific to statistics and to measurement theory are defined in the text or in footnotes and/or in the glossary.

CONCEPTION OF HUMAN BEHAVIOR MEASUREMENT: PURISM, "BALONEYISM," AND THE MIDDLE OF THE ROAD

It is possible to view the problem of human behavior measurement in many ways: for the purposes here, three viewpoints will be elucidated. The first might be termed the approach of the "purist" school, the membership of which is composed of strict adherents to classical measurement theory and to devotees of the exact fulfillment of the assumptions underlying the use of parametric statistical techniques.* To the proponents of the purist point of view the very thought of conducting a factor analysis on a thirty-item† rating scale with anything less than a sample of three hundred subjects, i.e. ten times as many subjects as items, would lead them to suffer massive attacks of acute indigestion; and they would never consider standardizing a set of data which was more than very slightly nonnormally distributed in its raw score form. It is not that the arguments for insisting on this approach are without some merit. A large part of the foundation of modern statistics and many of its current applications to human behavior measurement owes its development neither to a mathematician nor to a behavioral scientist but to a geneticist, Sir Ronald Fisher. Many commonly used statistical techniques were developed by him for the purpose of analyzing large samples of data in studies concerned largely with physical characteristics of organisms. Such characteristics tend to be normally distributed in natural populations. If one is examining a given aspect of human behavior, the natural distribution, (i.e. frequency of occurrence) of which is generally unknown, in fifty human beings, it seems only logical to endorse caution when the techniques used to analyze that behavior were originally developed to measure a given physical trait, the distribution of which is generally normal, in 500,000 kernels of corn. The question, of course, concerns just how much caution need be exercised in analyzing human behavior data with techniques originally developed for another purpose. (It is not the techniques per se that are in question here but rather the distributions of the populations of interest and the availability of samples which can be studied from those populations.) At the one extreme, the purists might hold that there are really very few statistical and measurement applications to human behavior research because the very nature of human behavior and the sampling procedures employed in collecting human behavior data are such that, to a major or minor degree, the assumptions underlying the applications of parametric statistical techniques cannot often be met. The author is reminded of one of his graduate school professors who, in preparing a human behavior research protocol, carefully laid out the conditions and proper experimental controls to insure that he would meet all the underlying assumptions of the statistics to be used in the data analysis. He used the patients in a 3,000 bed state psychiatric hospital as a pool from which to draw his sample. With rigorous application of his conditions and controls and the investment of many weeks of painstaking examination of hospital charts, he found a sample of six subjects, making it impossible to conduct the study. The old professor is not alone, however, in that a rigid application of the purist viewpoint will often lead one to the conclusion that it is impossible to study particular questions.

At the extreme opposite to the purist viewpoint is one which might logically be termed the impurist point of view. Due to the socio-religio-legal connotations of the

---

* There are a number of assumptions which must be made about the distribution of population data in order to analyze them using parametric statistical techniques. The reader is referred to Li for an excellent nonmathematical treatment of these considerations.[2] For purposes here let it simply be said that collecting large random samples of data will meet most of these assumptions.

† Item is used here and throughout the chapter as an expression for the basic unit of a measuring device. If thirty words appear in a vocabulary subtest of an intelligence measure, each word is considered to be an item. If galvanic skin response is recorded ten seconds after termination of an electric shock and every twenty seconds thereafter for five minutes, each GSR recording is considered an item.

latter, however, this approach might best be termed the *baloney* school, in the sense of "Any way you slice it, it's still. . . ." The adherents of this approach are aware, if somewhat vaguely, of the objections made by the purists, and they are perhaps aware of the practical complications of developing accurate measures of human behavior. The baloneyism reaction to such objections and difficulties is that, since it is so difficult to fulfill assumptions and meet adequate standards in human behavior measurement, the best approach is to ignore them. A proponent of this view may wish to study the genetics of affective disorder; obviously, he will need a device to measure degree of depression in some quantitative fashion. A list of behaviors felt to be typical of depressives is compiled—e.g. cries a lot, looks sad, has insomnia—and each behavior is set in a format so that a nurse-observer may then rate each item on a continuous scale from 0 (never) to 10 (frequently). The nurse-observer completes her quantification task by observing each of ten depressed patients on the ward and marking the scale to the best of her judgment. (Observing is defined here as watching the patient in between serving meals, bathing patients, handing out medications, attending ward rounds, completing progress notes, escorting patients to the laboratory for blood studies, etc.) Total scores are calculated for each patient rated, an arbitrary cut-off point is established, and scores are compared with a psychiatrist's independent judgment of the clinical condition of the patient. Given that there is some agreement between the ratings and the judgment, the conclusion often reached after all of the above is that the scale does indeed measure depression, and it is published in the literature as such. A variant of the baloney approach is to go to the literature in search of a scale for depression and, having found one and being particularly careful not to become bogged down with textual descriptions describing scale development, apply it to a research sample.

The descriptions of the above two approaches are obviously a biased effort on the author's part to point out some conceptual problems, although even a cursory glance at the literature on human behavior measurement brings one to the uncomfortable conclusion that there are many human behavior researchers who come dangerously close to matching one of these ideal types. To the purists, no human behavior can be measured; to the baloneyists, no human behavior cannot be measured. A third approach is suggested, a combination of these two extremes and which, for this reason, might be called the middle-of-the-road point of view. Those who hold this view understand the need for attention to the assumptions underlying statistical technique and measurement theory while concomitantly realizing that the human being is not a kernel of corn or a laboratory animal. It is not always possible to obtain a human sample of the desired size or quality or to apply rigid controls to it, if not for technical, then often for ethical reasons. Even if it is technically and ethically possible to obtain such a sample, to do so is quite costly, and the pragmatic concern of limited funds available from granting agencies often stymies such research at this point. There is the nonsensical and distasteful but ever-present, dictum of "publish or perish" hanging over the heads of most researchers which is not conducive to the large sample and long-term strategies demanded in developing measures of human behavior. The approach of one's holding the middle-of-the-road point of view is to be aware of the ideal conditions, to temper this idealism with the pragmatic realities of available samples and resources, and then to do one's best to meet the ideal under the given circumstances. Fortunately, as Nunnally points out, the majority of parametric statistical techniques are robust, i.e. as long as the data remain monotonic in statistical manipulation, the end result is not greatly affected if other conditions are not met.[1] It should be noted, however, that many with purist leanings would take violent exception to the above statement. Taking a middle approach does not mean that those who hold this point of view grant themselves license to do as they will as long as their intentions are pure. Another requirement of this approach is to realize the limitations placed on measures developed and results obtained under less than perfect conditions and to report the results of such work in accord with these limitations.

DEVELOPMENTAL REQUIREMENTS FOR HUMAN BEHAVIOR MEASURES

The basic requirements necessary for the development of an accurate measure of human behavior are relatively few in number. A human behavior measure must: (1) be developed on the basis of adequate sampling, (2) possess reliability, (3) be valid, (4) have adequate norms, and (5) be administered under standardized conditions.

SAMPLING IN HUMAN BEHAVIOR MEASUREMENT

The first requirement for the development of an accurate measure of human behavior is that the measure be based on a sample which is large and randomly selected.

The steps in the development of a human behavior measure are actually special cases of the procedures followed in conducting an experiment according to the classical scientific method. Observations are made and collected. (Certain observable behaviors are seen to be related to a larger behavioral construct. These behaviors are then ordered, if necessary, and presented in a measurement, i.e. test or scale, format.) Hypotheses are formulated. (For example, scores on the measure will indicate the presence or absence and the degree of the behavior or behavioral construct present in the subject being measured.) The hypotheses are tested. (The measure is developed according to the methods described in this chapter.) Finally, a conclusion is formulated on the basis of the results of the test of the hypothesis. (The measure is or is not an indicator of the presence of the behavior or behavioral construct.)

In standard research procedures and in the special case of measurement development, one is always limited to studying a sample and, on the basis of the findings, making inferences back to the population from which that sample was selected. The quality of the inference made is directly determined by how closely the sample mean and variance (called statistics) approximate the population mean and variance (called parameters). The reader may have heard or used the phrase, "sample statistics are unbiased estimates of population parameters." The word unbiased is an important one, for if the sample estimates of the population are biased in some fashion, the inferences made after statistical analysis of the sample data will be incorrect. Sample selection on a random basis increases the probability that the estimate of the parameter from the statistic will be made without bias.

The need for using large samples is intuitively obvious, but there are technical reasons for selecting large samples as well. Having collected a sample and having calculated its mean and standard deviation (the square root of the variance), it is possible to calculate the probability that the sample statistics are correct estimates of the population parameters. This is done by calculating the standard errors (S.E.) of the mean and standard deviation. Why and how this is done will not be presented here. The reader is referred to Li and to Guilford for a more complete discussion of this topic.[2,3] The important point is seen in the formulas for calculating the standard errors in which s = standard deviation of the sample and n = size of the sample:

$$\text{S. E. of the mean} = \frac{s}{\sqrt{n-1}}$$

$$\text{S. E. of the standard deviation} = \frac{s}{\sqrt{2n}}$$

Note that in both expressions, n, the sample size, is found in the denominator on the right. Because of this, the larger the sample size, the smaller the standard error. The smaller the standard error, the better the estimate of the parameter from the statistic.

There are other reasons why sampling procedures are of central importance to the development of a good behavioral measure. These will be considered in context as they specifically relate to other principles of measurement development discussed below.

There is one common source of sampling error which is so prevalent that it deserves special mention here. This is the use of "samples of convenience," or captive samples. It seems that there is no other category of humans which has been studied as much and measured as much as *studentis universitarum*. It is easy and tempting to collect data on a group so readily available for study. Cooperation is easily elicited by substituting a data collection period for a lecture, to which most students are happy to accede, or by making participation a course requirement, which students are generally unhappy about and which is potentially unethical, but which occurs anyway. All of this in spite of the fact that it is obvious that university students are totally atypical of the general population. For example, they score higher on IQ tests than a large majority of the general population and generally come from higher socioeconomic levels. Another manifestation of the same error occurs in medical applications to human behavior genetics, alluded to in the previous section, where samples for study are often selected on the basis of whatever group of patients are in the hospital at the time the study is being conducted. Such a sample may be representative of a given population, but representativeness is not randomness.

True random sampling becomes increasingly difficult with the advent of codes of ethics and legislation put forward to insure that the rights of human subjects are not violated in experimentation. For example, in many places it is now required that human subjects voluntarily consent to be included in research samples. One has to wonder whether or not a random sample selected from a group of volunteers can ever be equivalent to a random sample selected from a general population. The reader should note that the above comment is not meant to be a statement against human research ethics and legislation. There is no doubt, however, that new methodological difficulties will be introduced into human research with the introduction of such standards.

Finally, a word is in order regarding stratification in sampling. Many human behavior traits are known to be, or are suspected of being, differentially distributed in different subgroups of populations. For example, the distribution of the

IQ appears to be related to socioeconomic variables. In collecting a sample in order to give proper representation to such differential distribution of a characteristic among subgroups, the researcher might correctly decide to take certain steps in sample selection to insure their representation in his sample. For example, if a trait is known to occur twice as frequently in men as in women, the researcher might select his sample in the ratio of two males for every female. Such a procedure is known as stratifying a sample and, while it will guarantee representativeness, it will not insure randomness. If a choice must be made between representativeness and randomness, then randomness should come first. However, it is possible to collect what is known as a stratified-random sample. This may be done either by deciding on which strata one wishes to include in the sample and then selecting randomly from each stratum, or by selecting a large sample randomly and then stratifying the sample after it is collected.

The importance of large samples selected by random procedures at the onset of behavioral measurement research (or any type of research for that matter) cannot be overemphasized. The violation of this principle contributes more to problems with and errors in human behavior measurement than any other source.

Reliability

A second principle of measurement development is that of reliability. The reliability of a measure is expressed as a correlation coefficient which indicates quantitatively the degree to which a measure is replicable, i.e. the degree to which it measures the same thing each time it is administered. The emphasis in reliability is not on the trait, characteristic, or behavior which is measured, but on the "sameness" of the measure from trial to trial, irrespective of what is being measured. For example, if a subject scores 102 on an IQ test on Tuesday, 84 on the following Friday, and 114 a week later, it is obvious that the measure is not replicable; it has low reliability.

In order to develop the concept of reliability more fully, it is helpful to examine a single score that a person might obtain on an individual item from a given measure; this score will be designated as the observed score, or $score_o$. It is possible to theoretically partition this score into two components, the true score, or $score_t$, which is that part of $score_o$ due to the actual behavior which one is measuring, and the error score, or $score_e$, which is that part of $score_o$ due to errors in measurement perhaps caused by imperfections in the measuring device or by changes that have taken place in the subject between one measurement and another which artificially alter the behavior of interest (e.g. perhaps a subject develops a headache halfway through an IQ assessment) or by changes in the conditions under which the measurement takes place (e.g. a measure of vocabulary is printed in small type and ten minutes after the subject begins, the neon tube in the lamp above him begins to flicker). Thus, any individual item on a measure can be described as: $score_o = score_t + score_e$. If these item scores are summed over a measure for a given subject or subjects, it is then possible to calculate the mean and variance of the scores on the measure. Thus, using the same notation as above, the observed variance on a measure can be described as: $var_o = var_t + var_e$. It is possible to control some, even most, but generally not all, of the variance due to error in any given measure. The amount of error variance present in a measure is directly related to the reliability of that measure. Specifically, reliability =

$\frac{\text{var}_t}{\text{var}_o}$. The reader may question the utility of such a description of reliability, in that one cannot directly estimate that part of the observed score variance which is due to true score variance. However, if $\text{var}_o = \text{var}_t + \text{var}_e$, then $\text{var}_t = \text{var}_o - \text{var}_e$, so that reliability $= \frac{\text{var}_o - \text{var}_e}{\text{var}_o}$. It was stated above that reliability is expressed as a correlation coefficient, which can take a range of values between 1.0 and 0.0. If the reliability coefficient is near 1.0, it should be clear from the above that var_e must be very small. If the reliability coefficient is exactly equal to 1.0, then $\text{var}_e = 0$, and if the reliability coefficient is equal to 0.0, the measure must be composed entirely of error. Thus, without actually determining the numerical values of the components of var_o, it is still possible to estimate the relative effect of var_e. There are a number of different methods for obtaining reliability estimates.

The most general estimate of reliability is the correlation of scores from two different administrations of the same measure administered to the same sample at two different points in time. This method of estimating reliability is known as the test-retest method. The correlation coefficient will reflect that which is common between the two administrations. Whatever is left over, i.e. not common, must be due to error.

The test-retest method is not the preferred method of reliability estimation in that it tends to give unstable results. The reasons are that for most types of psychological measures, any given subject has the advantage of familiarity of the format and the content of items on the second administration and for this reason is likely to do better the second time he completes the measure than the first, i.e. he is likely to obtain a better score on the second administration. This will have the effect of decreasing the correlation between the two administrations and, thus, of artificially decreasing the reliability estimate of the measure. The amount that a score is affected by practice will vary from subject to subject and from sample to sample. One solution to this difficulty which has been proposed is to increase the length of time between the two administrations of the measure, thus reducing the practice effect. While this alternative will produce the desired effect of reducing error due to familiarity with the measure, it unfortunately introduces a new source of error. The longer the time between the two administrations of the measure, the greater the probability that changes in the subject will take place. Scores on the second administration of the measure may thus be different from those on the first, not because of the unreliability of the measure or because of practice effects, but because the behavioral variable which is being measured has changed within the subjects. To speak in terms of variances, var_e which would be affected by both the unreliability of the instrument and practice effects, has not changed; instead, var_t has changed. Because it is not possible to directly estimate the components of variance in reliability problems, the only noticeable effect is an overall decrease in the reliability coefficient which would be interpreted erroneously as the low reliability of the measuring device. Test-retest reliability works best with measures which are tapping behaviors not significantly modified by learning effects and which, at the same time, are relatively stable in a subject over time. That such behaviors must be few and far between seems obvious. There is no set convention on the maximal separation in time between the two administrations when test-retest reliability is employed, although for most psychological traits, one to two weeks seems to be an average.

An alternative to test-retest reliability estimation is referred to as parallel form reliability because it involves the development of two equivalent measures of the same behavior. A correlation coefficient is calculated between the scores of the same sample on the two measures taken consecutively. The advantages of parallel forms over retesting are obvious in terms of the reduction of error variance due to practice effects as well as the elimination of the possibility of changes in var_t due to changes in subjects over time. Its equally obvious limitations are found in the additional effort required to develop an exactly parallel form of the same measure. Particularly in the field of human behavior measurement where the manifestations of behaviors are often subtle and difficult to control from a practicable point of view, parallel form reliability estimation offers severe restrictions.

A third type of reliability estimation is a method referred to as split-half. This method, rather than requiring two administrations of the same form or consecutive administrations of two equivalent forms of a measure, compares scores obtained on one half of a measure with those obtained on the other half for each subject across a sample. This division into halves is best made by comparing odd-numbered with even-numbered items rather than items on the first half with items on the last half of the measure due to the possible error variance which may develop in time during the administration of a measure. For example, if fatigue sets in during the administration of a long measure of mental functioning, its effects are likely to be more similar on items at the end of the measure than they are between items at the beginning and at the end of the measure. If items on the first half of the measure are compared with those on the last half, the differential effect of fatigue, which is part of var_t, will artificially lower the reliability coefficient. It should be obvious that odd-even reliability is appropriate only for measures which tap a single behavioral dimension. If two or more behavioral dimensions are being assessed by the same measure, then the correlation between items will be spuriously low. For example, if a ten-item measure taps two behavioral dimensions with Items 1 through 5 assessing Behavior A and Items 6 through 10 assessing Behavior B, an odd-even reliability will compare Item 5 (Behavior A) with Item 6 (Behavior B). The correlation between Items 5 and 6 will be low and the overall reliability of the measure will be spuriously reduced. The same effect will occur if the items are not all equally difficult to answer, irrespective of whether the measure taps a single behavioral dimension or multiple behavioral dimensions. It should also be obvious that a measure with an odd number of items cannot be appropriately subjected to odd-even or split-half reliability estimation procedures.

The last method of reliability estimation to be discussed here is referred to as the internal consistency method. While the theory underlying the use of this method differs conceptually from those discussed above, its concern is still with the amount of variance in a measure which is due to error.* The theory underlying its use is that any given measure is actually composed of a finite sample of items drawn from an infinitely large pool of items. While it is obvious that in the development of a measure one does not first construct an infinite (or even an extremely large) set

───

*Actually, the underlying theory is the same for all methods of reliability estimation. It is only due to the developmental manner in which the various methods were introduced here that a difference may now seem apparent. See Nunnally for a more complete description of reliability theory.[1]

of items and select from these for the given measure, conceptually it can be seen that the particular items decided upon for inclusion in a measure, for example, a measure of perceptual acuity, are only a finite sample of a larger set of items. Depending on a number of variables (including the theoretical biases of the designer of the measure), another measure might consist of a totally different set of items and still be a reliable measure of perceptual acuity.

The method of internal consistency holds that the reliability of a measure is equal to the average correlation of every item with every other item in this infinite and theoretical pool of possible items. The reasoning behind this is as follows: as discussed above, it is possible to think of every item as having an error component associated with it; error due to imperfections in the measuring ability of the item itself. It is maintained that error is randomly distributed among items in this infinite pool. Therefore, in intercorrelating all the items in this pool, there are as many items which contribute to inflating the correlation as there are items which would tend to deflate it; the total effect of error on all items in the domain with each other must be zero. If the sum of the variances of all observed scores in the item pool is equal to var_t minus var_e, as defined above, and if, in this case, the sum of the error equals zero, then the average correlation of all items with all other items in the pool must be a correlation of all true scores in the item pool with each other. Again, this infinite pool of items is never available; therefore reliability coefficients are referred to as reliability *estimates*. Formulae such as coefficient alpha and the KR-20, employing sample data, provide good estimates of what the average correlation would be if all items in the theoretical domain of items were available for sampling.[4,5]

The internal consistency method of estimating reliability (of a single behavioral dimension measure) will provide an estimate of the upper limit of reliability, since it is not affected by error due to time intervals between administrations of the same measure or to that due to slightly different parallel forms. In addition, since internal consistency is based on an estimate of the average correlation of all items in the domain of a given behavior with each other, the error variance will be reduced more by this method than by the split-half method which employs only a finite set of items in the correlation formula. For these reasons, the internal consistency method is the one of choice in estimating the reliability of a measure. If the measure consists of more than one behavioral dimension and it is not possible to separate out those items common to each dimension, then either test-retest or parallel form methods must be employed. Of these two, that of parallel forms provides the more accurate estimate, but as pointed out above, in human behavior measurement it is often very difficult to develop an alternative form of a measure. If it is not possibe to develop a parallel form, then the only choice left is to use the test-retest method and report its limitations accordingly.

There are a number of factors which have a direct effect on the size of the reliability coefficient which are under the direct control of the designer and/or user of the measure. The first of these was alluded to above at the beginning of the discussion on reliability—conditions, both internal and external to the subject being measured, which may change over time, and thus affect the estimation of reliability. The action of some of these variables are known and can accordingly be controlled. For example, if one were assessing the reliability of a measure of cortiocosteroid secretion in humans and measured corticosteroid blood levels once a day, at different times during the day, the resultant reliability coefficient would be low, but not necessarily because the measure was unreliable. The diurnal curve of steroid secretion in man is well established.[6, 7] Thus, where it is known that such variability over time occurs, proper control of the time at which the measurement is taken must be built into the reliability study. In instances where the effect of such variables are not known, but expected, proper controls can be maintained to insure that the reliability of the measure is not spuriously low. Where the effect of such variables is both unknown and unexpected, the only solution is its discovery through continued experimentation with the measure.

The other two factors mentioned here which spuriously affect the reliability of measures are both related to the formula:

$$\text{reliability} = \frac{\text{var}_o - \text{var}_e}{\text{var}_o}$$

It can readily be seen that if the sample group on which the reliability of an instrument is being assessed has a restricted range of the behavior being measured, the reliability will be spuriously low. For example, if a new measure of intelligence is being assessed and the reliability study sample consists only of people with at least a college education, it is obvious that a large portion of the population with intelligence ranging from slightly above average on down will be excluded, as well as many highly intelligent people without formal education. In this case, the variability of observed scores will be small. Examining the above equation, if var_o decreases in both the numerator and the denominator, with var_e held constant, the reliability will decrease. Thus, in selecting a sample for a reliability study, the researcher must attempt to obtain a group which represents all extremes of the behavior in question. This condition can generally be satisfied by appropriately defining the population of interest and randomly sampling from that population.

The second factor affecting reliability has to do with the var_e term in the above equation. Remember that in an infinite population the sum of var_e is equal to zero. It is a statistical axiom that given a random sample, the larger that sample, the more it approximates the theoretical population distribution. Thus, the larger the sample, the more likely it is that the sum of var_e will approach zero. An important point, however, is that it is not the number of subjects in the reliability sample which is under consideration here, but the number of items in the measure. (Remember that the infinite distribution spoken of throughout is not a distribution of people, but rather a distribution of items.) Thus, a very important, and probably the easiest, way to increase the reliability of a measure is to increase the number of items in that measure.

A final consideration is: how much reliability is enough reliability? No set answer can be given. Kelley offers the following guidelines (based on a statistical consideration which will not be given here): to evaluate the level of group performance (i.e. to obtain meaningful information on the basis of a single administration of a measure to a group of subjects) the minimum reliability required is 0.50; to evaluate the level of group performance on two or more administrations (e.g. to compare the same group with itself on the same measure prior to and after the administration of an experimental variable or to compare two different groups) the minimum reliability required is 0.90. For individuals in the same two experimental paradigms given above (here one is concerned with the applied or clinical usefulness of a measure), the required minimal reliabilities are 0.94 and 0.98, respectively.[8] One now runs head-on into an argument with the purist school in that in terms of most measurement problems in human behavior research it seems rather the exception than the rule to find reliabilities much above 0.70 (although there are notable exceptions.) An eminently reasonable answer to this dilemma is to consider reliability in relative terms. If one has devloped a new measure of a specific behavioral trait, e.g. anxiety, with a reliability of 0.68 and the highest reliability among the previously established measures of anxiety is 0.55, then in spite of Kelley's criteria, one has a very reliable measure of anxiety which may be used and interpreted, with appropriate caution, until someone else develops a measure of the same behavior with a reliability significantly above 0.68.

A brief mention should be made here regarding the concept of interrater reliability. Interrater reliability is employed in instances where measure is completed by an external observer rather than by the subject himself. A large number of psychiatric behavioral rating scales, for example, are administered in this fashion. Interrater reliability coefficients are obtained by correlating the ratings of two or more raters, using the same measure, rating the same subject at the same time. Interrater reliability is not a direct estimate of the reliability of the measuring device itself in that the error variance, which is of primary interest here, is not due to imperfections in the measuring device or to changes in subjects while being measured, but rather to error variance associated with the persons doing the measuring, i.e. the raters. Interrater reliability provides an estimate of the raters' agreement as to what the true score is on any given measure rather than to an estimate of the actual true score, which is determined by the subject's behavior. This is not to suggest that estimates of interrater reliability are not important but to point out that only an estimate of interrater reliability is not sufficient to establish the actual reliability of a measure.

Validity

The validity of a measure is concerned with answering the question: Does the measuring device actually measure what it says it will measure? Is a measure of arithmetic ability actually assessing that ability or is it, as is often the case if the items are presented in textual form, also measuring verbal fluency? Is a galvanic skin response or a forearm plethysmograph actually measuring anxiety or are they measuring a generalized body arousal?

Perhaps the best place to start discussing validity is with a frequently used term, *face validity,* which is really not a measure of validity in the psychometric sense at all. Face validity is an attempt to have the measure look like, or appear to be measuring, what it is supposed to measure. It is often more important for establishing rapport and motivation in subjects than for any other reason. A face valid measure of arithmetic ability should contain items with numbers; a face valid measure of verbal ability should contain items with words; a measure of sexual preference should contain items about sexual behavior, and so on. Probably the most common use of face validity is found with human behavior measures used in business and industry to assess such things as employee suitability, management potential, and specific job skills. There are times when face validity may not be desired, e.g. in testing for many mental functions where face valid items might generate anxiety levels sufficient to interfere with the proper assessment of the variables in question. This is not to say that the subject completing such a measure is forever denied knowledge of the specific functions being tested. Provisions for subject protection in such circumstances are outlined in detail in the relevant professional codes of ethics.[9, 10]

Another procedure, often referred to as a type of validity, but really the first step in many instances to obtaining a valid measure, is what is referred to as *content validity*. In using this procedure, the domain of content to be sampled by the measure is made specific before the measure is designed. A common example, although by far not the only application of content validity, is in the classroom where the teacher is designing a final course examination. It may be decided that the course material can be broken down into three topic areas and that there are three specific goals of the course: recall of specific facts, data manipulation, and application of theory to new problems. The teacher would then devise a 3 x 3 table, with one axis labeled with the three course areas and the other with the three course goals. The teacher then decides how important each goal is to each topic area and labels the cells accordingly with appropriate percentages reflecting this relative importance. When the measure is constructed, the number and types of items are developed in accord with this predesigned plan. Using content validity will not insure that one has constructed a valid measuring device in a statistical sense, but it does provide a rational basis for then subjecting the instrument to other types of validity checks.

In terms of formally defined validity, there are two types: empirical, or statistical, and construct. The former type can be further divided into two types: predictive and concurrent.

The use of empirical, or statistical, validity requires the availability of an independent measure of the criterion behavior in question, i.e. independent of the measuring device which is being validiated. As an example, consider a problem from the field of business and industry. Suppose a company finds that it is expending large sums of money in retraining secretaries because a large percentage of those hired turn out to be unsatisfactory workers after six months on the job. A measurement specialist is brought in to study the problem and begins by designing a measure (probably using content validity) and assessing its reliability. Assuming the reliability is satisfactory, the specialist might then administer the measure to all applicants seeking secretarial positions with the company over a given time period, for example, one month. The measures are scored, and their results are filed away. Six months later, the specialist then asks supervisors to evaluate the performance of all secretaries who have completed the experimental measure. It is extremely important that the supervisors are not aware of the scores the secretaries obtained on the measure to avoid possible confounding. The scores on the measure are then correlated with the scores of the supervisors' ratings. If the scale is a valid one, the correlation between the two sets of scores should be high, i.e. secretaries who received high ratings should have obtained high scores on the measuring device and vice versa. Of course, it is possible that a negative relationship develops: secretaries who received low ratings might have obtained high scores on the measure and vice versa. This is not important (the scoring of the measure can simply be reversed) as long as the relationship between the measure and the criterion is high and consistent. Assuming that a high correlation is obtained, in the future the company need only administer the measuring device to all job applicants and hire only those who obtain high scores.

There is, however, a problem here. In order to validate a measurement device using the method of predictive validity, time must elapse between the initial scale administration and the occurrence of the behavior which the scale was designed to predict. A businessman would not likely wish to invest six months of secretaries' salaries, knowing that a large proportion of them will be fired at the end of that six months, in order to validate a scale. At least, he would not be willing to do this if an alternative were available. *Concurrent validity* provides this alternative.

The principles underlying concurrent validity are the same as those underlying *predictive validity*, but the method of obtaining the necessary validity coefficient is different. Instead of assessing all applicants for the secretarial positions, the measurement specialist would administer the scale to all secretaries presently employed by the company. At the same time, supervisors would be asked to independently rate secretarial ability, and the correlation between measure scores and supervisors' ratings can be computed immediately.

It would appear that the advantages of savings in time and other expenditures would make concurrent validity a superior choice over predictive validity, but this is not necessarily the case. Earlier, in discussing reliability, it was pointed out that restricting the range (i.e. the variability) of the reliability sample would artificially reduce the reliability coefficient. The same is true for the validity coefficient, and indeed for any product-moment correlation. In applying concurrent validity, the range is restricted for the following reason. To return to the example of the secretarial pool, while it is true that the company is having problems with secretaries who are not competent and therefore it is expected that some of these will be employed at any given point in time, it is equally true that the very worst of this secretarial group will have already been dismissed. In addition those who applied for jobs and were so unskilled as to not even have been hired will not be included. Thus, the distribution of the criterion group will be very skewed, with few or no representatives from the lower end of the distribution of secretarial skills. The range will be restricted, and the concurrent reliability coefficient will underestimate the actual reliability of the measure.

The decision as to whether to use predictive or concurrent validity will often depend largely on pragmatic variables. One must weigh the increased time and costs of collecting predictive validity against the required accuracy needed in the validity coefficient. This, in turn, is affected by the importance of the decision or decisions to be made on the basis of the measuring device.

In asking the question, "How much statistical validity is enough validity?" one finds again that the answer is relative. In part, it depends, as does reliability, on the validity of other measures already tapping the same dimension. If the validity of the new measure is low, but higher than any other available, then it is satisfactory until a more valid device is developed. Another way of looking at the degree of satisfactory validity is through the standard error of estimate, abbreviated σ_{est}. The standard error of estimate gives an indication of how much better the selection ratio, i.e. the prediction, is with the measure than without it. The formula for the standard error of estimate, presented without development, is:

$$\sigma_{est.} = \sigma_y \sqrt{1 - r_{xy}}$$

where, σ_y = standard deviation of the criterion scores
r_{xy} = validity coefficient

It can readily be seen that, if the test is perfectly valid, i.e., $r_{xy} = 1.0$, that the standard error of estimate equals 0. If the measure is totally invalid, i.e., $r_{xy} = 0$, then the

$\sigma_{est.} = \sigma_y$ and the information obtained from the new measure is no better than what one had to begin with, i.e. the criterion scores. Obviously, as $\sigma_{est.}$ approaches zero, the predictive ability of the measure increases.

The selection of the appropriate criterion in validity studies is the most important and, at the same time, the most difficult problem in assessing statistical validity. A most important step, one mentioned above and mentioned here again, is that the measures of the criterion and the measures from the instrument whose validity is being assessed must be kept separate and independent in order to avoid the confounding of one by the other.

A relatively common practice in selecting a validity criterion has been to use an already established measure as the criterion for a new one. For example, in developing a new measure of IQ, one might validate it by collecting independent IQ scores on the sample, using an already established IQ test. The specific reason why this should not be done will become clearer when measures of intelligence are discussed below, but the general reason is as follows: the validity of most presently available measures of human behavior is somewhat less than perfect. To validate a newly developed measure (bound to be less than perfect due to sampling, item, and unspecified random error) against another less than perfectly valid measure used as a criterion can only result, in the author's opinion, in less than perfectly clear interpretations of the true validity of the measuring device.

One final point should be made regarding statistical validity, and this is the relationship between it and reliability. In order to do so, one must begin with a formula known as the *correction for attenuation*. This formula provides an estimate of what the correlation between a measure and a criterion would be if both of them were made perfectly reliable. Again, the formula will be presented without development:

$$r_{\infty\omega} = \frac{r_{xy}}{\sqrt{r_{xx}} \sqrt{r_{yy}}}$$

where $r_{\infty u}$ = the correlation between a measure and a criterion, where both have perfect reliability

r_{xy} = the obtained correlation between a measure and a criterion

r_{xx} = the obtained reliability of measure x

r_{yy} = the obtained reliability of criterion y

If the measure were perfectly valid, $r_{\infty\omega}$ would equal 1.0. Assume that the obtained reliability of the criterion is perfect, i.e. r_{yy} = 1.0. Substituting these two terms in the above equation:

$$1.0 = \frac{r_{xy}}{\sqrt{r_{xx}} \sqrt{1.0}}$$

Solving for r_{xy}:

$$r_{xy} = \sqrt{r_{xx}}$$

Thus, given the above assumptions which imply nearly perfect measuring conditions, the maximum validity obtainable between a measure and a criterion can be no larger than the square root of the reliability of the measure.

It is important to stress that the above formula does not imply that given a measure with a reliability of 0.70, the validity of that measure would equal 0.837, i.e. $\sqrt{0.70}$. The formula states, rather, that 0.837 is the maximum possible validity the measure will have, given that the reliability of the criterion is perfect, i.e. 1.0, a phenomenon not often found. The formula is useful in that once the reliability of a new measure is calculated, it becomes possible to determine the theoretical maximum possible validity. Suppose that a measure of anxiety with a validity of .75 is available and a new measure of the same trait with a reliability of .49 is developed. Without the need for assessing the validity of the new instrument, it can at once be determined that the maximum possible validity obtainable under the best conditions of measurement is $\sqrt{.49}$, or .70. A decision can then be made as to whether or not it will be worthwhile to complete the development of the measure. One likely conclusion to be drawn is that the new measure must be made more reliable. As pointed out above, the basic way to increase the reliability of a measuring device is to increase the number of items. Thus, on the basis of the above information, the measurement designer might try to add more items to the measure.

A type of validity with particular relevance to human behavior measurement is that of *construct validity*. Science in general, and particularly the science of human behavior, is greatly involved with the specification of constructs and the study of the relationships among these constructs. Examples of the use of constructs in empirical research are found throughout this book in terms such as intelligence, anxiety, and depression. To some readers, it may be surprising to read that such terms are defined as constructs, because they are often thought of as behavioral entities. Yet, a moment's reflection will demonstrate that: (1) each of these terms is used to denote a set of interrelated behaviors rather than any one specific behavior, and (2) the set of behaviors specified by each is not fixed and tends to vary from measure to measure, study to study, and theory to theory. Construct validity is used as a methodology for the logical and empirical development and validation of constructs.

The first step in establishing construct validity is defining the construct by specifying its observable aspects. This generally involves drawing up a list of observable and measurable behaviors in as exhaustive a fashion as is possible. The next step is to select a sample of subjects and attempt to measure these behaviors in the sample. The measurement may involve a simple "present-absent" check list, a multiple-point rating scale, an audio- or videotape, a galvanic skin response recorder, indeed any type of measuring device. Of course, the reliability of the particular instrument which is used must be assessed prior to its being employed.

The next step is to specify the nature of the relationship among these various observable behaviors. Probably the most common approach applied here is that of factor analysis, particularly in psychological research, although a wide variety of multivariate statistical techniques which can be used to the same end are available. The reader is urged to consult Nunnally, Li, and Cattell for a more complete discussion of such techniques.[1,2,11]

Given that the construct has been defined in terms of observables and the relationship among these observables has been specified, the final step in construct validity is to determine whether these observables, now constituting a measure, act as though they were measuring the construct. As an example, suppose that one were interested in the construct of depression. One might select a sample of psychiatrists and ask them what they look for, in terms of behavioral indices, when they make a diagnosis of depression. In addition, one might go to the relevant literature in the field and draw up a list of depressive behaviors recorded there. The next step is to scale these behaviors in some fashion and apply them to a random sample of psychiatrically diagnosed depressives. The ratings obtained might then be factor-analyzed. Having established theoretical factors of depression, one might then administer them to another group of depressed subjects on admission to hospital and again at discharge. If the factor scores are significantly lower on discharge than on admission, one has construct validated the scale. It is more exact to say that the construct validation of the scale has begun, for there are other criteria which the scale should be validated against as well; for example, functioning outside the hospital and subjective reports of psychological state. In addition, different samples from different hospitals in different geographic regions might also be studied. Jones has put forward the interesting suggestion that, through the analysis of twin data, the heritability of a behavior may be selected as one of the possible criteria to be used in the construct validation of behavioral measures.[12] If the measure developed continues to yield results in expected directions, then it continues to accumulate construct validation. In the final analysis, a construct will be used as long as it continues to demonstrate heuristic and empirical usefulness. When it ceases to do so, it will be replaced by another construct.

Construct validity is the most nebulous of the several types of validity discussed here. At the same time, it is most important to the continued development of the field of human behavior measurement. More complete treatments of construct validity can be found in other publications.[1, 13, 14, 15]

NORMS

The third requirement of a good measure of human behavior is adequate normative development. As an example of its importance, consider the development of a hypothetical College Success Scale (CSS). Items are written; a measure is constructed; its reliability is assessed; and it is found to be satisfactory. The author of the scale decides to use predictive validity and administers the scale to a large sample of suburban high school students. Four years later, he discovers that any student who scored 60 or higher on the scale graduated from a university with at least a B average. The development of the scale is complete, and it is published. A teacher in the central area of a major city decides to administer the scale to students in his class and is disappointed to discover that the highest score in the class is only 40 points. The obvious and erroneous conclusion to be drawn is that no student from a central city school should go on to a university. The teacher is not satisfied, however, and keeps the scores on file; four years later, he contacts the schools to which his former students have gone. To his delight he finds that any student who scored 30 or higher on the CSS received at least a B average in the university.

The variable performance of different groups on the same measure is not necessarily, and not at all in the example above, an indication of inaccurate measurement. There are myriads of reasons why such results might be obtained. In the example above, the students from the two schools are likely to differ in attitude toward education, educational experience, socioeconomic status, cultural and ethnic norms, and, perhaps most of all, in general attitude toward test-taking.

It is because of factors like these that the collection of normative data on different groups is a very important step in the proper development of a measure. Those normative variables which might differentially affect test performance can often be hypothesized by the test developer, but in many instances they may be relatively subtle and discoverable only by administering the measure to samples differing normatively from one another. Some of the more common variables on which one might encounter normative differences are sex, age, socioeconomic status (generally consisting of occupation and education), urban versus rural residence, physical and physiological differences, psychological differences (intelligence and personality type, for example), and diagnosis, be it medical or psychiatric.

In using a measuring device, it seems that one can often estimate its general value (without looking at its reliability and validity) by noting the number and types of normative groups to which it has been applied. There is no mathematical relationship between norms and other characteristics of test development, and the relationship is not invariable; but if the author has taken the time and effort to establish good norms for his measure, it can be generally assumed that his work has been careful and diligent in other respects as well.

In spite of good norms, it is often the case that the particular group being studied might not be represented among the normative groups. The researcher then has two choices: one is to find the normative group most similar to his experimental group and use those norms for interpretation, but with caution. If the study is to be reported in the literature, an explicit statement regarding differences between the normative and experimental groups should be made. The second choice is to revalidate the measure on the experimental group. Obviously, because of the extra work involved, this is not often done.

STANDARDIZATION

The final requirement of a good measure of human behavior is standardization of administration and scoring. In discussing test-retest reliability, it was pointed out that there are numerous variables which may change during the time elapsed between the first and second administrations of a measure made on the same subject or group of subjects. Some of these changes may reflect actual differences which have taken place in the subject(s) between measurements, while others may be due to artifacts of the measuring device itself or to the measurement process, e.g. practice effects. Little can be done regarding subject variation over time. If the variable being measured is likely to change in subjects over time, however, then subject variability is a problem in reliability estimation. One would not want to use a test-retest method in this instance; but after the reliability is established through other methods, subject variability can be controlled to a large extent through standardization.

To cite an earlier example, if one is attempting to measure change in symptomatic depression over the course of hospitalization, then one expects a difference in measurement scores between prehospitalization and posthospitalization measures for any given subject. If it is established that the measure is reliable and valid, then the researcher wants to be sure that any change in score over time is due to actual changes in his subjects and not due to artifacts. If some subjects are measured for depression on the morning of admission while others are assessed in the evening, and the same is true for measures collected on discharge (i.e. the time of assessment is random for any subject), then an artifact has been introduced, in that the level of depression may fluctuate over a twenty-four-hour period in any one subject. Thus, time of administration becomes an important variable in the standarized administration of a measure.

Other standardization variables of import are as follows: (1) Instructions to the subject (or, if the subject is relatively passive in the measurement situation; for example, if his behavior is being observed and rated by another, instructions to the rater): instructions should be prepared beforehand, written out, and, if the instructions are presented verbally, memorized so that each subject (rater) has the same understanding of the measurement task. The reason for memorizing verbal instructions is simply one of establishing rapport with subjects. Reading instructions to a subject might not inspire great confidence in the researcher's competence or interest to a naive subject. (2) Scoring: this variable becomes more important as the variable being measured becomes more nebulous. For example, in measuring a person's temperature, scoring consists of nothing more than a careful reading of the scale on the thermometer; in measuring a personality construct on the basis of a response to a Rorschach card, the scoring becomes much more vague, e.g. is a superficial response due to anxiety on the part of a subject or to a defense against revealing oneself or because it is the best response which the subject can make, possibly indicating subnormal intelligence; in rating the presence of affiliative or sociable behavior, given a person who spontaneously initiated a conversation twice in a given observation period, say over eight hours, is that person to be rated as 0 (unsociable) or 1 (less sociable than average) or 2 (average)? The difficulties in such measurement are obvious and it is for this reason that proper instructions as to scoring procedures be decided on, either empirically through prior test, or operationally before the measure is to be scored. (3) Conditions of administration: if two people are being measured on an IQ test and one assessment is done in a well-lighted, well-ventilated space with adequate room for work while the second is done in a dimly lit, stuffy basement laboratory on the corner of a workbench, it is likely that the final scores on the two measures will reflect the physical conditions of the administration, artificially depressing the true score of the second subject.

Standardization of measurement conditions, as the reader might recognize, are no more than the application of proper experimental controls to the measurement situation. The potential problem lies in the fact that it is very easy to look on the measurement process as something separate from the experiment of which the measurement is a part. Certainly, in terms of the present development and sophistication of human behavior measurement techniques, the measurement process must be considered an experiment in itself.

IQ MEASUREMENT

It is beyond the scope of this chapter to provide a general review of human behavior measurement devices presently available. The best source for obtaining information on published tests and measures of human behavior is the *Mental Measurements Yearbook,* edited by Oscar K. Buros. The *Yearbook* is revised every several years and contains basic information regarding the development, reliability, validity, norms, and standardization of each measure. In addition, it contains information such as administration, time, cost, and publisher plus a critical review or reviews of the tests which are written by invited clinicians and researchers. The most recent edition of the *Yearbook* is the sixth, published in 1965.[16] For tests not published by a test publishing company, the best source of information will, of course, be found in the journal literature.

As mentioned above, no general review of human behavior measures will be conducted here. It may be informative to consider, in a general way, some strong points and difficulties associated with some of the areas of human behavior measurement which are commonly used in studies of human behavior genetics, specifically in the areas of intelligence, personality, and psychopathology.

Intelligence is a very easy word to use but a very difficult one to define. Indeed, a well-known tautology has arisen regarding measurement in this area; namely, that intelligence is what is measured by intelligence tests.

There are two important historical developments in the measurement of intelligence which heavily influenced intelligence measurement as it is known today. The first is the work of Binet and Simon, who set out to develop a measure which would identify children with learning problems in the Paris school system.[17] Binet and Simon developed a series of tests graded according to difficulty and based on age norms so that, for example, there were a series of four-year-old tests and five-year-old tests. Terman later standardized the test on a sample of United States children and used these age tests in a ratio of the highest correctly answered age-graded test divided by the chronological age, i.e. mental age of the subject/chronological age.[18] He multiplied the ratio by 100 and termed this the Intelligence Quotient (IQ).

A second development sprung from the factor analytic work of Spearman and Thurstone.[19,20] Spearman felt that intelligence is comprised of two factors: a general factor of g, common to all intellectual functions, and several specific factors unique to the particular aspect of intellectual functioning tapped at any one time. He later posited the presence of group factors which are comprised of several interrelated specific factors which may be all common to a specific intellectual function. Spearman's theory of intellectual organization appears today to find most of its support among European, and particularly British, psychologists. Thurstone, on the other hand, felt that there was no common factor in the organization of intellectual functioning, but rather that intelligence is comprised of a series of independent broad group factors, each of which is differentially weighted (i.e. combined in different ways with other group factors) in the organization of any specific intellectual function. It is this latter theory which receives most support from North American psychologists.

Given the above developments, one finds today that almost all measures of intellectual functioning are based on some factor analytic rationale and that differences among intelligence measures depend on whether the test author subscribes to Spearman's or to Thurstone's or to another theorist's concept of the organization of intelligence and also on the specific factor or factors which any given measure taps. These differences have structured development of the phrase mentioned above, i.e. intelligence is what intelligence tests measure.

Regarding the current use concept of IQ, it is important to note that over the years the IQ seems to have developed some sort of special mystic quality. The fact is that an IQ is nothing more or less than a test score. To emphasize this point, consider some problems in the development of IQ measures. For example, the IQ is no longer calculated as a ratio of mental to chronological age. The reason for this is that it became increasingly difficult to develop age-appropriate tests after the adolescent years. If one considers the ontological development of intelligence, it should be clear that the basic types of skills assessed on most IQ tests are either present at birth or well-developed through learning by the mid-adolescent period. After this time, of course, learning continues, as does the development of specific skills, but it is unusual to see development of basic skills such as abstract reasoning or verbal facility developing in persons in later life periods. Because of the difficulty in developing age-specific tests for the adult years, a thirty-two-year-old person might complete every item correctly on an IQ test and receive a mental-age score of sixteen years. Applying these ages to the IQ ratio would leave this individual with an IQ of fifty points.

If the ratio of mental age to chronological age is used to calculate IQ, another consideration must be made. In order to have ratio IQ's comparable at different ages, the standard deviation of mental age must increase in proportion to the mean of mental age. As an example, consider two subjects who are both one standard deviation above the mean of IQ test scores for their respective age groups. If the standard deviation of mental age is constant and equal to two years, then a ten-year-old subject would have a mental age of twelve (10 years [mean mental age] + 2 years [one standard deviation of mental age] = 12 years) and an IQ score of 120 (12/10 x 100 = 120). A fifteen-year-old subject at the same position as the twelve-year-old, relative to his age group, would have an IQ of only 113 (fifteen years [mean mental age] + 2 years [one standard deviation of mental age] = 17 years; 17/15 x 100 = 113). If the standard deviation of mental age increases in proportion to the mean, this problem does not exist. If the standard deviation of mental age remained at two years for a ten-year mean and increased to three years at a fifteen-year mean, then the respective IQ scores in the example above would be (10 + 2)/10 x 100 = 120 and (15 + 3)/15 x 100 = 120. One of the few presently-used IQ tests which meets this requirement of proportionally increasing mental age standard deviations is the 1937 revision of the Stanford-Binet;[21] most other tests do not, and thus cannot be scored by a ratio IQ method.

A solution to both of the above problems has been the standardization of IQ scores. Raw scores on IQ tests for each age group have been mathematically treated so that the mean IQ equals 100 and the standard deviation is equal to approximately 16, making IQ's comparable for different age groups on the same test and, in certain cases, to IQ's on other tests. The exact method for making this correction is discussed by Anastasi, Cronbach, and Guilford.[13, 14, 3] A further complication, however, is that while by convention all IQ tests are based on distributions with means of 100, not all of these distributions have standard deviations of sixteen points. Scores from such tests are not comparable with scores from tests which do fulfill this agreed-upon standard. Again, a short example should suffice. Suppose two IQ tests have mean scores of 100, but test A has a standard deviation of fifteen points and test B has a standard deviation of twenty points. Suppose, further, that an individual has a level of intelligence two standard deviations above the mean. Taking test A, his IQ will be 130, but with test B his IQ is 140, a sizeable discrepancy.

Concerning the above, it should be clear that the IQ is not magical or mystical. The IQ is a score on an IQ test, and like scores on other types of tests, is subject to the same difficulties and vagaries in interpretation. It should also be obvious that the fact that two or more tests are said to measure IQ does not mean that they are therefore automatically comparable from one study to another.

Some general comments regarding other characteristics of IQ tests might be in order here. General is emphasized because of the exceptions which are found to the various statements made here. IQ tests can be divided into two classes: individual tests and group tests. Individual tests such as the Stanford-Binet and the Wechsler Adult Intelligence Scale are the author's preference for research purposes in that generally the reliability quotients are higher than, and the normative and sampling developments superior to, those of group IQ tests.[22, 23] In addition, if one subscribes to Thurstone's theory of intellectual structure, the individual IQ tests tend to be composed of multiple factors as opposed to the general unifactor construction of group tests. Group tests offer advantages which individual tests lack. The administration of individual tests is time-consuming (up to several hours per administration) and only one person can be assessed at a time. In addition, the proper administration of individual IQ tests requires a fair amount of training and experience with the instrument. Group tests, on the other hand, can be administered quickly (in 10 to 20 minutes) to many subjects at once, often by a competent technician.

The choice of whether to use an individual or group test depends very much on the degree of accuracy needed for the study in question. If one is looking for general trends in the data and is able to tolerate a fair amount of error, group tests seem to be the choice. If the study in question demands accurate estimates of IQ, then individual measures are in order.

A final word on the measurement of IQ. The reader may note that in most places the author spoke of IQ rather than intelligence testing. The reason for this is that, obviously, the definition of what is intelligence has not been completed in any final sense. Note also that, depending on which particular factor or factors an IQ test taps, the operational definition of intelligence may differ. It is nonetheless surprising that, in spite of these qualifications, many studies using many different IQ measures have found certain consistencies. For example, heritability estimates of intelligence tend to be rather high and consistent from study to study. Also, IQ tends to correlate highly and positively with social class, occupation, and education. In spite of such similarities, however, it is not clear that IQ tests measure something akin to pure intelligence. It is more likely that they all tap one or more dimensions of the same construct of intelligence. In view of this, it seems best, for the present, to refer to the measurement of and the heritability of IQ rather than of intelligence.

PERSONALITY MEASURES

What was said of intelligence is equally true of the term *personality;* it is a term that is easy to use and difficult to define. Again, one is dealing with a construct which will ultimately have to be validated. It is still to be determined, for example, whether personality consists of a number of independent traits with each person possessing a given number of these traits in different degree or whether the basic unit of personality is something larger and more complex, composed of a differential relationship between and among something akin to personality factors. Of course, there is the more difficult question of whether personality is, in its largest part, inherited or learned, ingrained in childhood and permanent or modifiable throughout life. It seems clear that the answers to such questions will not be found until better measures of personality than those presently available are developed.

Measures of personality can be divided into two general types: objective and projective. Objective personality measures are those with set formats (e.g. a series of behavioral statements to which the subject responds by indicating whether they are true or false about him) and fixed scoring and interpretation procedures. Projective techniques are those in which a set of standard stimuli are presented to a subject whose task it is to free associate to those stimuli. Obviously, the scoring and interpretation of these latter responses is very subtle and often idiosyncratic to the person administering the measure.

The area of personality measurement is not nearly as well developed as that of IQ assessment; discussing which measure of a particular aspect of personality is best to use is not often a problem. Instead, the major problem is often one of finding a valid and reliable measure of a specific aspect of personality which one wishes to study. (As an example of the axiom that measurement development precedes scientific advance, the reader is invited to compare the number of studies on the heritability of IQ with the number of studies on the heritability of personality conducted to date.)

In considering the reliability and validity of objective personality measures, the techniques are relatively straightforward, as elucidated in the above section, and the various tests can stand or fall on their merit. When one begins, however, to discuss projective personality techniques, the problem becomes much more hazy. Formal attempts at assessing the reliability and validity of such measures have been frought with difficulties, and the results have not always been satisfactory. Two arguments have been proposed to account for this difficulty. One put forward largely by those interested in clinical measurement holds that personality is a complex unity with each aspect of personality interdependent with the other; therefore, any attempt to reduce this complex whole to its parts is a difficult, if not impossible, task. The result is that one has to be content with less formal estimates of the usefulness of these measures, relying on such concepts as clinical validity, i.e. when used on an individual clinical basis, one determines whether or not the results of the tests are generally in keeping with other clinical insights into the personality structure and functioning of the individual being assessed. Whether or not a result is "in keeping with" other findings often depends on the particular theory of personality to which the user of the measure subscribes. The opposite argument comes from more formal measurement theorists who might hold that if the variables in question cannot be accurately quantified and assessed for their reliability and validity, then they are simply not very good measures. The author's opinion tends to run with the latter group in that he prefers to speak of projective techniques as clinical tools or aids rather than as tests in the formal sense of the word. In the hands of a well-trained and sensitive clinician, projective techniques can often be useful in providing insights into a client's functioning. In the hands of an unskilled and/or insensitive individual who might use the clinical techniques as if they were more formally valid and reliable measures, they can be dangerous in that decisions affecting a significant part of an individual's future are often made on the basis of findings from projective techniques. Obviously, the author does not feel that such devices are appropriate for research work. Despite this opinion, however, some researchers, working with great care and effort, have obtained what are apparently meaningful research results using projective techniques.[24, 25]

DEFINING PSYCHOPATHOLOGY

The third area of human behavior assessment to be considered here is more a problem of classification of human behavior than of its measurement, although much of what was said above regarding the concepts, rather than the specific techniques, of reliability and validity still are relevant. This area is the classification of mental disorder. Psychiatric genetics is a rapidly growing field; yet, the presently used diagnostic classification in psychiatry is beset with problems in terms of its reliability and validity. For example, studies have shown that there is little agreement among clinicians in terms of diagnosis and this disagreement tends to become more pronounced with increasing clinical experience.[26, 27] Other work has suggested that there is little correspondence between observations of symptomatic behaviors and diagnoses assigned to psychiatric patients.[28] It would seem that cross-study comparisons must be done only with caution because it is quite likely that diagnostic terms and procedures vary not only from clinician to clinician, but from hospital to hospital, region to region, and most particularly, across international boundaries. (Regarding the latter, a recent paper has suggested that with proper analysis such diagnostic differences can be reduced to manageable levels so that cross-study comparisons can take place;[29] but it seems obvious that if special analytic procedures must be introduced in order to compare such data, a significant defect in the presently used diagnostic system must exist.) For example, there is increasing evidence that the presently used diagnostic term, schizophrenia, consists not of eight properly defined diagnostic subtypes, but rather of what appear to be two different types of disorders.[30, 31, 32] While the problem of the validity of diagnosis in psychiatry will be a long time in solving, proper experimental controls can at least insure higher levels of reliability. Using only first-admission patients seems to be one important control. Mednick and Goffman, among others, have spoken of the debilitating effects of institutionalization resulting from multiple hospitalizations which can lead to diagnostic confusion.[33, 34] Having subjects for a study diagnosed by a single clinician at the time the study is being conducted, rather than the more usual reliance on the diagnosis found in the patient's chart, can also reduce error in the definition of the independent variable. If the single diagnostician is biased, at least all bias in the study will be in the same direction. The ultimate solution to the diagnostic problem would seem to involve a problem of construct validation in measuring a wide variety of isolatable and well-defined observable behaviors in psychiatric patient samples, using multivariate techniques to investigate the groupings of these variables and then relating them to relevant criteria; for example, whether the individual remains out of hospital on discharge or returns after a period of time. Such work is obviously detailed, time-consuming, and expensive to maintain, but it seems imperative in order to facilitate solution of the problem.

FACTOR ANALYSIS

One area of final consideration is the wide application of multivariate techniques, particularly factor analysis, in human behavior genetics research. Thompson reviews some of this work and introduces some interesting potential applications of multivariate techniques to human behavior genetics studies.[35] The application of factor analysis, however, is not as straightforward as it may seem. Many researchers see factor analysis as a final step in measurement development, when, in actuality, it is only an intermediate one. The fact that a measure, or group of measures, has been factor-analyzed, and that the factors which are derived have been given names, does not insure that meaningful information will be obtained if the factor scores are then applied to another sample on which the same measures have been collected. A factor analysis provides nothing more than a series of linear combinations of variables. It is an intermediate step in construct validation. A factor has no final interpretable meaning, outside of a statistical one, until it is validated by studying its performance in relation to other constructs.

To give a concrete example, the usual procedure is to factor analyze a measure or group of measures, examine the item content of the factors obtained on this basis, apply names to the factors, and publish the results. Another investigator comes along, applies the measure to an experimental and a control sample, conducts the experiment, calculates the subjects' scores on the different factors, and draws conclusions accordingly. If significant results are not obtained, the experimenter then concludes, erroneously, that the behavioral dimensions supposedly tapped by the factors were not affected by the experimental treatment. This is not necessarily the case. The behaviors which the factors claim to measure may actually be related to the experimental treatment. The problem may be that the factors were not actually measuring those behaviors. While the experimenter thought he was conducting an original piece of research, what he was actually doing was completing a final step in a construct validation study. In short, factor analysis, of and by itself, cannot convey any magical interpretation to behavioral measures. It is an intermediate technique used in construct validation. Therefore, it is to the behavioral researcher's advantage to select factor analyzed measures for research with some caution, using only those factors whose validity have been demonstrated through use in construct validation studies.

A second reason for being cautious in the use of factor analysis techniques is that the method is not without its statistical imperfections. Some of these difficulties, such as parameter estimation, have only recently been solved, while others, such as indeterminancy, are still problematic.[36, 37]

The above discussion is not meant to imply that the use of factor analysis should be avoided or that results obtained by its use are inaccurate. Factor analytic techniques, while still not perfectly exact even when used properly, are invaluable in providing parsimony to much human behavioral data and are indispensible to construct validation. The important point is that they should be used properly. Surveys of factor analytic procedures can be found in other papers.[2, 37]

SUMMARY

This chapter provides an overview of the basic principles of human behavior measurement. There are three general ways, in the author's conception, in which the problem of human behavior measurement may be approached. One is through the formal and rigid application of the assumptions underlying parametric statistical techniques. A second is not to be concerned with statistical assumptions except in a very general sense. The first approach leads to the conclusion that little human behavior can be measured accurately; the second, that all human behavior can be measured, given sufficient imagination. A third approach is to modify statistical assumptions on the basis of pragmatic considerations and to use the measures developed on this basis with appropriate reservation and selectivity.

There are five basic principles of human behavior measurement. The first is the use of large, randomly selected samples. The second is reliability. The reliability of a measure may be estimated by the test-retest, parallel forms, split-half, and internal consistency techniques. Internal consistency estimation is the method of choice in assessing the reliability of a measure. The third principle of human behavior measurement is validity. Face, content, predictive validity, concurrent validity, and construct validity are defined and considered. The maximum statistical validity of a measure is limited by the square root of the reliability of that measure. The fourth principle is the adequate normative development of a measure; and the fifth principle involves standardization of techniques of administration and scoring.

The current level of development and the limitations of measures of IQ, personality, and psychopathology are discussed, and the potential uses and abuses of factor analytic applications to human behavior data are considered.

REFERENCES

1. Nunnally, J.C.: *Psychometric Theory.* New York, McGraw, 1967.
2. Li, J.C.R.: *Statistical Inference.* Ann Arbor, Edwards, 1964, vol. I and II.
3. Guilford, J.P.: *Fundamental Statistics in Psychology and Education,* 4th ed. New York, McGraw, 1965.
4. Cronbach, L.J.: Coefficient alpha and the internal structure of tests. *Psychometrika, 16:*297, 1951.
5. Richardson, M.W., and Kuder, G.F.: The calculation of test reliability coefficients based upon the method of rational equivalence. *J Educ Psychol, 30:*681, 1939.
6. Bliss, E.L., Sandberg, A.A., Nelson, D.H., and Eik-Nes, K.: The normal levels of 17-hydroxycorticoids (17-HC) in the peripheral blood of man. *J Clin Invest, 32:*818, 1953.
7. Perkoff, G.T., Eik-Nes, K., Nugent, C.A., Fred, H.L., Nimer, R.A., Samuels, L.T., and Tyler, F.H.: Studies of the diurnal variation of plasma 17-hydroxycorticosteroids in man. *J Clin Endocrinol Metab, 20:*1259, 1960.
8. Kelley, T.L.: *Interpretation of Educational Measurements.* Yonkers, World Book, 1927.
9. American Psychological Association: Ethical standards of psychologists. *Am Psychol, 18:*56, 1963.
10. American Psychological Association: *Standards for Educational and Psychological Tests and Manuals.* Washington, D.C., American Psychol Assoc, 1966.
11. Cattell, R.B. (Ed.): *Handbook of Multivariate Experimental Psychology.* Chicago, Rand, 1966.
12. Jones, M.B.: Heritability as a criterion in the construction of psychological tests. *Psychol Bull, 75:*92, 1971.
13. Anastasi, A.: *Psychological Testing,* 3rd ed. New York, Macmillan, 1968.
14. Cronbach, L.J.: *Essentials of Psychological Testing,* 3rd ed. New York, Har-Row, 1970.
15. Helmstadter, G.C.: *Principles of Psychological Measurement.* New York, Appleton, 1964.
16. Buros, O.K. (Ed.): *The Sixth Mental Measurements Yearbook.* Highland Park, Gryphon, 1965.
17. Binet, A., and Simon, T.: Méthodes

nouvelles pour le diagnostic du niveau intellectuel des anormaux. *Ann Psychol, 11*:191, 1905.
18. Terman, L.M.: *The Measurement of Intelligence*. Boston, H-M, 1916.
19. Spearman, C.: "General intelligence" objectively determined and measured. *Am J Psychol, 15*:201, 1904.
20. Thurstone, L. L.: *The Vectors of Mind: Multiple-factor Analysis for the Isolation of Primary Traits*. Chicago, U of Chicago Pr, 1935.
21. McNemar, Q.: *The Revision of the Stanford-Binet Scale: An Analysis of the Standardization Data*. Boston, H-M, 1942.
22. Terman, L.M., and Merrill, M.A.: *Stanford-Binet Intelligence Scale: Manual for the Third Revision, Form L-M*. Boston, H-M, 1960.
23. Wechsler, D.: *Manual for the Wechsler Adult Intelligence Scale*. New York, The Psychological Corp, 1955.
24. Kantor, R., Wallner, J., and Winder, C.: Process and reactive schizophrenia. *J Consult Psychol, 17:* 157, 1953.
25. Becker, W.C.: A genetic approach to the interpretation and evaluation of process-reactive distinction in schizophrenia. *J Abnorm Soc Psychol, 53:*229, 1956.
26. Arnhoff, F.N.: Some factors influencing the unreliability of clinical judgments. *J Clin Psychol, 10*:272, 1954.
27. Ash, P.: The reliability of psychiatric diagnosis. *J Abnorm Soc Psychol, 44*:272, 1949.
28. Phillips, L., Broverman, I., and Zigler, E.: Sphere dominance, role orientation, and diagnosis. *J Abnorm Soc Psychol, 73*:306, 1968.
29. Shields, J., and Gottesman, I. Cross national diagnosis and the heritability of schizophrenia. Paper read at the Fifth World Congress of Psychiatry, Mexico City, 1971.
30. Goldstein, M.J., Judd, L.L., Rodnick, E.H., and LaPolla, A.: Psychophysiological and behavioral effects of phenothiazine administration in acute schizophrenics as a function of premorbid status. *J Psychiatr Res, 6*:271, 1969.
31. Magaro, P., and Vojtisek, J.E.: Embedded figures performance of schizophrenics as a function of chronicity, premorbid adjustment, diagnosis, and medication. *J Abnorm Soc Psychol, 77*:184, 1971.
32. Rappaport, M., Silverman, J., Hopkins, H.K., and Hall, K.: Phenothiazine effects on auditory signal detection in paranoid and nonparanoid schizophrenics. *Science, 174*:723, 1971.
33. Mednick, S.A., and McNeil, T.F.: Current methodology in research on the etiology of schizophrenia. *Psychol Bull, 70*:681, 1968.
34. Goffman, E.: *Asylums*. Garden City, Doubleday, 1961.
35. Thompson, W.R.: Some problems in the genetic study of personality and intelligence. In Hirsch, J. (Ed.): *Behavior Genetic Analysis*. New York, McGraw, 1967.
36. Jöreskog, K.G.: Some contributions to maximum likelihood factor analysis. *Psychometrika, 32*:443, 1967.
37. Harman, H.H.: *Modern Factor Analysis*, 2nd ed. Chicago, U of Chicago Pr, 1967.

Chapter 6

TWIN STUDIES

STEVEN G. VANDENBERG

THE TWINS
by Henry Sambrooke Leigh

In form and feature, face and limb,
  I grew so like my brother,
That folks got taking me for him,
  And each for one another.
It puzzled all our kith and kin,
  It reached a fearful pitch;
For one of us was born a twin,
  Yet not a soul knew which.

One day, to make the matter worse,
  Before our names were fixed,
As we were being washed by nurse,
  We got completely mixed;
And thus, you see, by fate's decree,
  Or rather nurse's whim,
My brother John got christened me,
  And I got christened him.

This fatal likeness even dogged
  My footsteps when at school,
And I was always getting flogged,
  For John turned out a fool.
I put this question, fruitlessly,
  To every one I knew,
"What would you do, if you were me,
  To prove that you were you?"

Our close resemblance turned the tide
  Of my domestic life,
For somehow, my intended bride
  Became my brother's wife.
In fact, year after year the same
  Absurd mistakes went on,

And when I died, the neighbors came
  And buried brother John.

Long past the point at which new concepts demand new approaches, new techniques and new experiments, many of us, through inertia or ignorance, because of hypercaution, or of reluctance to give up the familiar for the sake of the unknown, continue to operate within the framework of the old levels. Thus, though the modal point has been passed in time, we are apt to devote our energy and attention to matters made obsolete by it. —I. Michael Lerner[1]

INTRODUCTION

Most animals give births to a large number of offspring at one time, whether they are insects, fishes, or frogs. Even in mammals large litters are still the rule. Primates, however, usually give birth to one young at a time, but on occasion they have twins. Guttmacher has reviewed twinning in Unipara, including man, and Schultz summarized data on multiple births in primates.[2,3]

Many myths are associated with this phenomenon. In some cultures it is regarded animal-like and therefore something to be ashamed of. In other cultures it is considered a special blessing. A number of novels, plays, and, most recently, films, have exploited the possible confusions and the special psychological situation of twins. For a detailed account see Gedda.[4]

The use of twins for research into the role of heredity in shaping human psychological traits was first proposed in 1875 by Sir Francis Galton[5] in a paper entitled, "The History of Twins as a Criterion of the Relative Powers of Nature and Nurture." This title suggests that nature (genetic factors) and nurture (environmental influences) are opposing factors, almost like the protagonists in a classical drama, when in reality they are, of course, cooperating and equally essential prerequisites for normal development. Much of the research done in the century that has passed has been influenced by this false dichotomy. To some extent that is due to the fact that the "classical" twin method cannot elucidate genetic mechanisms.

What the classical twin method does is to analyze the observed variance of a trait into several components. The observed variance may be compared with the tip of an iceberg which emerges above water, while the submerged mass of the ice resembles the tremendous invariance that is present in any human being, no matter how deformed or retarded: there is a heart, lungs, liver, nervous system, and so on. The twin method is a useful first device which allows one to see whether the observed differences between individuals in some trait, behavioral or other, are in part due to hereditary factors. We will later see that the method may also be modified to break such traits up into components which may be more amenable to Mendelian analyses. The twin method becomes almost useless when specific genetic factors are found that can be demonstrated in pedigrees as mutant genes or assigned to abnormalities in the chromosomes (duplicated or missing parts or whole chromosomes).

As a first screening device the twin method is very economical compared to pedigree studies. Because of the high visibility of twins, they are rather easily located and because of the natural tendency of parents, teachers, and friends to compare twins, it is rather easy for them to gain an understanding of the value of studying twins and thus secure cooperation.

In the 100 years since Galton, the twin method has been applied in many countries for the study of ever increasing numbers of human diseases and normal traits, so that today it is practically impossible to keep track of the numerous publications about twin studies, whether methodological or substantive. In this chapter we will generally not review the findings from such studies because these will be dealt with in separate chapters on hereditary factors in abilities, in personality, and so on. For a brief survey of these see Mittler.[6] Rather, we will look at some methodological issues in the classical twin studies, try to specify the assumptions underlying twin studies, and attempt to review to what extent these assumptions have been subjected to empirical validation. Finally, we will consider some other uses of twins. We will start with the zygosity diagnosis.

TWO KINDS OF TWINS

There are, of course, two kinds of twins: one-egg, identical, or monozygotic (MZ); and two-egg, fraternal, or dizygotic (DZ). Because MZ twins originate from one fertilized egg, they will have the same set of genes except for occasional errors in the original mitotic division.

Nielsen has summarized the literature on eleven MZ twin pairs with chromosomal differences including two pairs in which phenotypically one was male, the other female, five pairs in which one had Down's syndrome and the other was normal, and two pairs in which one twin had Turner's syndrome and the other was a normal female.[7] Table 6-I is taken from his paper.

Because twins in most MZ pairs have the same genotype, they will resemble each other to a degree not found in ordinary sets of brothers or sisters. Nevertheless, a particular pair may on occasion differ markedly in their phenotype because of prenatal or postnatal environmental factors. A few examples may illustrate this point. In the twin study conducted at the University of Michigan there was a set of twins in which one of the boys had had tuberculosis in one leg and, as a result, this leg was shorter and less well developed than the other. In contrast to his brother he did not participate in athletics but exerted himself more in his academic studies, obtaining good grades to the point of being considered a "grind." He was less popular with girls than his brother and somewhat shy. When first meeting the two boys, I guessed them to be DZ but the blood group tests and detailed study of their ears, iris patterns, and fingerprints suggested that they were MZ.

A pair of girls also studied in Michigan again seemed DZ at first sight: the two girls differed in hair color and weight, while only one girl wore glasses. It turned out, however, that one girl had bleached her hair, controlled her weight, and refused to wear glasses, though she was as nearsighted as her sister. The other girl was considerably overweight due to continual snacking. The first girl had been "going steady" for some time while the other one

TABLE 6-I
MONOZYGOTIC TWINS WITH DIFFERENT PHENOTYPE OR CHROMOSOME CONSTITUTION

Authors	Sex	Phenotype	Chromosome Condition
Turpin, et al.	M	Normal	46, XY
	F	Turner's syndrome features	45, XO
Fanconi, et al.	F	Down's syndrome	–
	F	Normal	–
Wolff, et al.	M	Down's syndrome	47, trisomy 21
	M	Normal	46, XY
Turpin	M	Down's syndrome	47, trisomy 21
	M	Normal	46, XY
Bruins, et al.	M	Normal	46, XY
	M	Down's syndrome	47, trisomy 21
Mikkelsen, et al.	F	Normal	45/46, XO/XX
	F	Turner's syndrome	45/46, XO/XX
Benirschke and Sullivan	F	Normal	45/46, XO/XX
	F	Multiple congenital abnormalities, webbed neck	–
Dekaban	F	Down's syndrome	47, trisomy 21
	F	Normal	46, XX
Bigozzi, et al.	F	Gonadal hypoplasia	46, XY
	F	Normal	45, XX
Edwards, et al.	M	Short stature	45, XO
	F	Turner's syndrome	45/46, XO/XY
	F	Normal	45/46, XO/XX
	F	Turner's syndrome features	45/46, XO/XX
Shine and Corney	F	Normal	46, XX
	F	Multiple abnormalities	Chromatin-negative

From J. Nielsen, "Inheritance in Monozygotic Twins," *Lancet*, 2:717 (1967).

had no boy friend and pretended not to be interested. The personalities of the two girls seemed to be quite different but teachers said that, up to about a year earlier when the first girl started to bleach her hair, they had seemed very similar. Here, also, the blood group tests, folds of the outer ear, iris patterns, and fingerprints suggested that the girls were MZ.

In Louisville, Falkner described one MZ twin pair in which one twin had been pumping blood into the other one through a shunt in the venous anastomosis in the placental mass.[8] As a result the twins differed markedly in weight at birth although, in time, this difference disappeared gradually.

On the other hand, because some members of DZ twin pairs share many more than the expected average of 50 percent of their genes, they can sometimes resemble each other as much as do most MZ twins. This can still be further enhanced by a strong family resemblance such as is occasionally also seen between two children of different ages. For this reason it is important to use blood groups and other genetic markers to establish zygosity, especially when the samples of MZ and DZ twins are small or if considerable precision in the assignment of each pair to the correct classification is desired because the measurements of the traits studied were expensive or because the population from which the twins came was rather unusual or not previously studied.

PRECISE ZYGOSITY DETERMINATION

In fact, with the presently available markers, almost any desired degree of accuracy in the diagnosis of zygosity can be achieved if one is willing to pay the price. This can be seen clearly from Table 6-II, which is taken from a recent text on genetic markers in human blood by E. R. Giblett.[9] It shows the probability for two *unrelated* individuals to have the same alleles

TABLE 6-II
SEVENTEEN BLOOD GENETIC SYSTEMS LISTED IN ORDER OF THEIR USEFULNESS (i.e. MNSs IS THE MOST USEFUL) FOR DISTINGUISHING BETWEEN TWO RANDOM SAMPLES OF BLOOD FROM WESTERN EUROPEANS. PARENTHESES DENOTE THE ANTIGENS TESTED IN A GIVEN SYSTEM

Genetic System	Probability That Two Randomly Selected Persons Have the Same Phenotypes	Combined Probability
MNs	0.16	0.16
Rh (CC^{w}cDEe)	0.20	0.032
ABO (A$_1$A$_2$B)	0.33	0.011
Acid phosphatase	0.34	3.7×10^{-3}
Kidd (Jk^{a}Jk^b)	0.38	1.4×10^{-3}
Duffy (Fy^{a}Fy^b)	0.38	5.0×10^{-4}
Haptoglobin	0.39	1.95×10^{-4}
GM (1,5)	0.40	7.8×10^{-5}
Gc	0.45	3.5×10^{-5}
PGM	0.47	1.6×10^{-5}
Lewis (Le^{a}Le^b)	0.57	9.3×10^{-6}
P (P$_1$P$_2$)	0.67	6.2×10^{-6}
Adenylate kinase	0.82	5.1×10^{-6}
Pseudocholinesterase, E$_2$	0.82	4.2×10^{-6}
Kell (Kk)	0.84	3.5×10^{-6}
Lutheran (Lu^{a}Lu^b)	0.86	3.0×10^{-6}
6PGD	0.92	2.8×10^{-6}

From E.R. Giblett, *Genetic Markers in Human Blood* (London, Blackwell, 1969).

for a series of loci. Of course, such probabilities would be much greater for a pair of fraternal twins because their "choice" of alleles is limited by the genotypes of their parents. For instance if both parents were homozygotic for blood groups B, the twins could also only be homozygotic B, so that the probability of concordance for that locus would be 1.00. If one parent was AB and the other AA, the probability of chance concordance of the twins as AB would be 0.25. Calculations of this kind require that information about the blood groups of each parent is available. This is often not worth the additional time and expense, although such information, once obtained, may also be used for linkage studies if the parents can be persuaded to be measured for the traits under study.

If only data are available for the twins, the best one can do is to make deductions about the possible genotypes of the parents, given the information about the twins and to assign each of these genotypes a probability based on the population frequencies for each of these genotypes. Then one can work back towards the twins and obtain the probability of concordance for the particular phenotype observed at that locus. (In some cases two or more different genotypes in the parents may be indistinguishable such as AO and AA or A_1O, A_2O, A_1A_2. The probabilities for concordance for each of these will have to be added, and the probability that the twins are different be taken into account.) For further details about such procedures, see Smith and Penrose,[10] Sutton et al.,[11] and Bulmer.[12] Wilson has calculated the probabilities of concordance in DZ twins, not knowing parental phenotypes, for a number of blood groups, refining the procedure to exclude unlike-sexed pairs and ignoring hypothetical twin pairs from the deduced parental types in which neither twin has the observed phenotype.[13] It would be relatively easy to write a computer program to calculate such probabilities. It could be readily updated as new genetic markers and improved frequency estimates become known. As a final step in the procedure, the probabilities for the various sets of alleles are then multiplied by each other. The product obtained equals the probability that this set of twins is concordant at all the loci considered, even though they are fraternal. For most concordant twins these values will be quite low and they can safely be considered MZ. If the value for a certain pair is greater than some previously accepted limit, one can exclude these twins, conduct further tests, or, if one wants to be very conservative, one could include these with the DZ twins. Of course, the finding of any discordant genetic marker suffices to classify the twin pair as DZ.

ZYGOSITY DETERMINATION BY QUESTIONNAIRE OR BY PHYSICAL SIMILARITY

If no greater precision is needed and a more economical procedure seems indicated, it is possible merely to ask twins whether they were frequently mistaken for their twins by parents, teachers, friends, or casual acquaintances.

Cederlöf, et al. reported agreement between the results of five blood group systems and a brief mailed questionnaire asking about being mixed up or being as similar as two peas in a pod, for seventy-two out of seventy-three MZ and ninety-nine out of 108 DZ cases.[14] For another nineteen pairs no assignment could be made from the questionnaire. The median probability of chance concordance for pairs really dizygotic was estimated to be about 4 percent.

Nicols and Bilbro applied a similar

technique to the twins in the National Merit Scholarship Corporation sample.[15] They used graded criteria based on physical similarities plus reports about mistaken identity and found this method to be effective in 95 percent of the cases with a greater than 90 percent accuracy. Jablon, *et al.* described zygosity diagnosis from the data on twins in Veterans Administration records.[16] When fingerprint patterns plus height and weight were combined, the diagnosis was correct, compared to blood group information, in 87 percent of the cases. Use of a questionnaire alone gave 90 percent correct classification.

Even better results were obtained by Hauge, *et al.*[17] They found that 170 out of 173 twins pairs classified as MZ on the basis of questions concerning physical similarity and difficulty of telling the twins apart were serologically concordant, while 143 of 147 classified as DZ were discordant for one or more blood groups—better than 97 percent agreement.

Nichols and Bilbro found a large quantity of misinformation about zygosity even in the highly intelligent group they studied: of eighty-two MZ twins 24 percent believed they were fraternal, and of forty-one DZ twins 19 percent believed they were identical with an additional 10 percent uncertain. This shows that one *cannot* ask twins whether they believe themselves to be identical or fraternal because these terms are frequently misused as when a boy-girl pair announces that they are identical because they have a marked family resemblance, or when a pair of girls said that because they were female they could not be fraternal. In addition, parents frequently have been given incorrect information about the zygosity at the time of the birth of the twins. Such mistakes are easily made because, contrary to common opinion, there is no perfect correlation between monozygosity and a single placenta. Actually, some MZ twins have been reported with two chorions (the chorion is the outer membrane in the placenta which contains the amnion, the inner membrane around the embryo), while monochorionic DZ twins have also been seen. Whether there is one or two chorions probably depends on the size of the embryos at the time of their implantation on the wall of the uterus and on their distance at the time. Because MZ embryos resulted from one fertilized egg that split into two, they will, of course, often stick together and therefore have one chorion. Besides the rare monochorionic DZ twins there are many instances of dichorionic DZ twins with the two placentas fused, i.e. some of the tissue is grown together so that it would take detailed examination to see that there are two chorions, so that these DZ twins are frequently thought at birth to be MZ twins.

Although judgments about physical similarity can be influenced by the twins' psychological traits, it may still be necessary to go back to the old method of relying on judgments about the twins' general physical resemblance when neither the questionnaire method nor the blood groups method seems practical. It is advisable to objectify this procedure as much as possible by singling out certain features for intensive study after photographing them. Such details as the patterning of the external ear, the patterns of the iris, and the fingerprints and handprints lend themselves particularly to the following procedure. After these items are photographed, taking care that pictures of both twins are on the same color film to control for differences in the film or in its processing, one asks several judges to sort the photos of, for instance, the four eyes into two piles: two belonging to one twin and two belonging to another. If the majority of judges can do so correctly, the twins may be considered fraternal; if they cannot, the twins are probably identical. The same procedure can be used for photos of the ears and for fingerprint ridge counts. For a detailed description of dermatoglyphic methods, see Cummins and Midlo;[18] for the morphology and genetics of the iris patterns, see Ritter.[19]

FREQUENCY OF TWINNING

Uninformed individuals often believe that twins are rather rare. This is not the case. The frequency of twin births is approximately one in eighty-three deliveries in the United States or about twelve per thousand. It is possible to estimate the frequency of MZ and of DZ twin births from birth records by Weinberg's method which begins by noting the number of boy-girl pairs. When this number is doubled we have a close estimate of the total number of DZ twins on the assumption that there is no differential prenatal loss for pairs of either sex or for unlike-sexed pairs compared to like-sexed pairs. It turns out that the MZ twinning rate is close to 3.5 per thousand for all ethnic groups, while the DZ twinning rate varies considerably even within Europe, as can be seen from Table 6-III taken from Bulmer.[12] For United States Whites the DZ twinning rate is about 8.5 per thousand. Among United States and African Negroes the twinning rates are considerably higher: about sixteen per thousand in the United States and about twenty per thousand in Africa with considerable regional variation. On the other hand the DZ twinning rate is probably no more than 2.5 per thousand in most Asiatic peoples. The cause(s) of twinning is not understood, but some factors influencing it are known.

Nutrition seems to affect the DZ twinning rate, as can be seen from Figure 6-1,

TABLE 6-III
THE TWINNING RATE PER THOUSAND IN EUROPE, STANDARDIZED FOR MATERNAL AGE

Country	Period	Dizygotic	Monozygotic
Spain	1951-53	5.9	3.2
Portugal	1955-56	6.5	3.6
France	1946-51	7.1	3.7
Belgium	1950	7.3	3.6
Austria	1952-56	7.5	3.4
Luxembourg	1901-53	7.9	3.5
Switzerland	1943-48	8.1	3.6
Holland	1946-55	8.1	3.7
West Germany	1950-55	8.2	3.3
Norway	1946-54	8.3	3.8
Sweden	1946-55	8.6	3.2
Italy	1949-55	8.6	3.7
England and Wales	1946-55	8.9	3.6
East Germany	1950-55	9.1	3.3

From M.G. Bulmer, *The Biology of Twinning in Man* (Oxford, Oxford University Press, 1970).

also taken from Bulmer,[12] which shows that there was a marked drop in the DZ rate during the famine winter of 1944 to 1945 in Western Holland when the German occupying forces cut off most supplies of food from the farming areas in the Central, Eastern, and Northern parts of Holland after the Allied invasion took place.

Two other factors that appear to influence the DZ twinning rate are the mother's age and the number of previous births (or parity), as shown in Figure 6-2, taken from a study of twinning in California by Shipley, *et al.*[20] By contrast, there is no effect on the MZ twinning rate, as is clearly shown in Figure 6-2.

These effects are quite similar for Negroes and Whites, as can be seen from Figure 6-3 taken from a study by Myrianthopoulos.[21]

There may also be genetic factors. In a review of the research up to that time, MacArthur concludes in 1953 that there has been general agreement about an inherited tendency towards DZ twinning but disagreement about a similar factor in MZ twinning.[22]

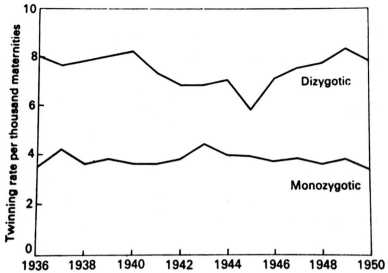

Figure 6-1. The twinning rate in Holland during the Second World War, standardized for maternal age. (From M.G. Bulmer, *The Biology of Twinning in Man* [Oxford, Oxford University Press, 1970].)

The statistic, one twinning in eighty-three deliveries in the United States, has the practical consequence that one should find roughly one pair of twins for every one hundred pupils in a school or wherever one has a reasonably complete geographical sample, of which one-third would be expected to be MZ. Of course, there is the usual sampling fluctuation around that average. Thus, there were at one time seven pairs of twins in a small rural school in Kentucky with just over one hundred pupils, while there were none in a high school in Detroit with well over one thousand students, but these were rare instances, for both in Eastern Michigan and in Louisville the number of twin pairs in a school was usually close to the expected frequency at the time of the Michigan and Louisville Twin Studies.

DEPRESSED NUMBER OF MALE DZ PAIRS

It has been observed by a number of investigators that the number of male DZ pairs in the study is lower than the number of female DZ pairs. The reason for this is not yet clear, although two explanations, which possibly are both correct, have been advanced. It may be that male DZ pairs really are less frequent perhaps due to an exaggeration in twins of the greater mortality, at birth and later, of male babies, but it is not clear why this should be restricted to DZ pairs and not affect MZ pairs. In fact, the opposite has been reported by Barr and Stevenson and by Osborne and DeGeorge.[23,24] Barr and Stevenson reported the following frequencies per thousand: for stillborn in males, 123 for

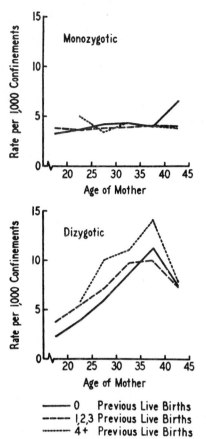

- 0 Previous Live Births
- - - 1,2,3 Previous Live Births
- ⋯⋯ 4+ Previous Live Births

Figure 6-2. MZ and DZ twinning rate for 1000 births for whites in California by age of mother and number of previous live births, 1940 to 1959. (From P.W. Shipley, J.A. Wray, H.H. Hechter, M.G. Arellano, and N.O. Borhani, "Frequency of Twinning in California, Its Relationship to Maternal Age, Parity and Race," *American Journal of Epidemiology*, 85:147 [1967].)

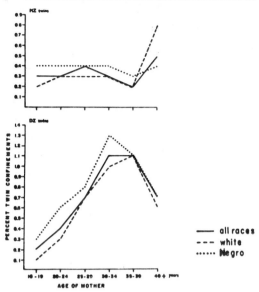

Figure 6-3. Frequency of twin births in relation to maternal age by race and zygosity. (From N.C. Myrianthopoulos, "A Survey of Twins in the Population of a Prospective Collaborative Study," *Acta Geneticae Medicae et Gemellologiae* [Roma], 19:15 [1970].)

single born, 44 for one DZ twin, 92 for one MZ twin; and, in females, respectively, 21, 42, and 72. For stillbirth plus infant death the figures were 58, 159, and 260 in males, and 47, 140, and 177 in females. These data were for England and Wales in 1949 to 1950. Bock, *et at.* found a deficiency of male twin pairs discordant for Rh+, Rh−) in a sample of 308 like-sexed DZ pairs from Louisville, Kentucky, which may in part explain the usual deficit of male DZ pairs.[25] The problem is complicated because in twin studies one generally does not know about a twin pair of which one member died prior to the study.

The other explanation sometimes suggested is that differences in ability are frequently rather large in DZ pairs and that the less able twin is more likely to have dropped out of school, to be unwilling to cooperate, or to be in another school. Such occurrence would be more frequent for boys than for girls as is often noted in the single birth population. In addition, boys usually are less compliant and less likely to compensate lower ability by harder work than are girls. This raises the issue of sampling problems. Most twin studies have given relatively little thought to questions regarding the representativeness of their sample or what the characteristics are of the population to which one can generalize from the twin sample. A few studies probably included almost all twins falling into a certain category such as Swedish recruits (Husen[26]), or high school students in Louisville (Vandenberg, et al.[27]). Nevertheless, even in those studies there must have been pairs missing because one or both co-twins were too retarded to be in the sample. The occurrence of mental retardation in both twins would probably be somewhat more frequent in MZ than in DZ pairs, but the occurrence in *one* only would be considerably more common in DZ pairs. The last tendency would lead to underestimation of DZ within-pair variance and thus of hereditary variance.

In the introduction it has already been suggested that twin studies and heritability estimates are no final goals but only convenient ways to obtain a first idea about the importance of genetic factors for a certain variable before engaging in pedigree analysis. For that reason the precision of heritability estimates is not too important; therefore, sampling is also less important than is usually the case. However, the eventual goal of behavior genetics is the allocation of variance in psychological traits to specific loci, and for this purpose the more dramatic cases of big twin differences are generally more informative.

What can be learned from studying twins? This depends somewhat on the particular method chosen. We will discuss the following four major methods, although further subdivisions are possible.
1. Comparison of MZ and DZ twin samples.
2. Study of MZ twins raised apart.
3. Co-twin control studies.
4. Comparison of like-sexed and unlike-sexed DZ twins.

CLASSICAL TWIN STUDY AND ITS HERITABILITY INDEX

By any count the most common use made of the natural experiment provided by the occurrence of human twins has been the comparison of the percentage of concordant MZ and DZ twins (for all-or-none traits) or the comparison of MZ and DZ pair differences or intraclass correlations (for continuous variables). This is usually followed by the calculation of a heritability coefficient. In recent years there have been various developments which fall outside the scope of this discussion. R. B. Cattell has proposed a comprehensive method of assessing the importance of genetic and environmental components of variance, as well as correlations within and between families of genetic and environmental influences, by the Multiple Abstract Variance Analysis (MAVA).[28, 29] This method has been reviewed by Loehlin.[30] Jinks and Fulker have also proposed an integrated approach to biometrical genetic analysis.[31] Both methods go beyond the twin method for their data.

The concept of heritability was introduced into the genetics of animal breeding to guide the selection of sires and dams for economically important traits such as egg laying, milk production. Heritability was the name given to the proportion of the total variance due to genetic differences. Lush proposed calling this "heritability in the *broad* sense" to distinguish it from that fraction of the total variance due to additive genetic effects which is "heritability in the *narrow* sense."[32]

Additive genetic effects are those common to a parent and a child and do not include the effects of a particular combination of genes, such as those due to dominance (effect at one locus) or due to epistasis (effects of alleles at various loci on one another). Monozygotic twins share these nonadditive effects and, therefore, the "broad" heritability estimated from twin studies is higher than heritability estimated from parent-child correlations. In fact, if data are available on twins as well as their parents, possible discrepancies between the heritabilities calculated by the two methods can be used to see how important such nonadditive effects are for the traits in question.

In 1929 Holzinger introduced the concept of heritability into twin studies and proposed two formulae for use with continuous traits which we will discuss.[33] They were quite faulty and are mentioned only to allow understanding of the older literature in which they were used uncritically.

HERITABILITY FOR CATEGORICAL DATA

First, we will discuss the formula for heritability (in the broad sense) for all-or-none traits such as diabetes or schizophrenia. The usual formula for concordance in categorical traits is:

$$h^2 = \frac{C_{MZ} - C_{DZ}}{1 - C_{DZ}} \quad (1)$$

where C is the percentage of concordant twins. This can also be used for more than two categories.

Several improvements and elaborations have been suggested. Falconer described a threshold model of disease and devised a method for calculating correlations between relatives for a disease from information about its incidence in the general population and in the relatives of probands or "index cases," the affected persons who come initially to the geneticist's attention.[34] The model assumes that genetic and environmental factors determine an individual's liability, with manifestation of the disease only if the liability exceeds a certain (unknown) threshold value. If the presence of common environmental factors among relatives can be ruled out, the method will provide an estimate of the heritability of liability to that disease.

Smith extended Falconer's method to concordance rates in MZ twins.[35] He showed that in rare diseases the concordance rates for MZ twins can only be high if the heritability is very high. Figure 6-4 is taken from his paper. The author further emphasizes the importance of controlling the effects of environmental similarities. Allen, *et al.* have proposed a refinement which takes into account the age of the twins; this is necessary because in most diseases there is a certain spread in the age of onset.[36] They also suggest use of a *proband concordance rate* which shows the proportion of concordant cotwins of previously identified twins (probands) showing a certain disease. This rate will generally be

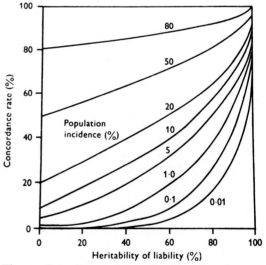

Figure 6-4. Concordance in N twins for a threshold variable as a function of heritability of liability and population incidence. (From C. Smith, "Heritability of Liability and Concordance in Monozygous Twins," *Annals of Human Genetics*, 34:85 [1970].)

lower than the usual *pairwise* concordance rate because concordant pairs have a double chance of being ascertained compared to discordant pairs.

HERITABILITY FOR CONTINUOUS TRAITS

The original two formulae proposed by Holzinger for heritability (in the broad sense) were for continuous variables. At that time there was much use made in psychology of first-order correlations and multiple and partial correlations, rather than of analysis of variance. The first formula was:

$$h^2 = \frac{r_{MZ} - r_{DZ}}{1 - r_{DZ}} \quad (2)$$

where r is the intraclass correlation which can be calculated by the usual product moment correlation

$$r = \frac{N\Sigma XY - \Sigma X\Sigma Y}{\sqrt{N\Sigma X^2 - (\Sigma X)^2} \sqrt{N\Sigma Y^2 - (\Sigma Y)^2}} \quad (3)$$

provided that each pair of twins is entered twice: once with twin A as X and twin B as Y, and once with twin A as Y and twin B as X, or by:

$$r = \frac{\text{between-pair variance} - \text{within-pair variance}}{\text{between-pair variance} + \text{within-pair variance}} \quad (4)$$

The two values obtained will differ by a factor of (N-1)/N. The variance between pairs is:

$$\frac{\Sigma(X+Y)^2 - 1/N(\Sigma X + \Sigma Y)^2}{2(N-1)} \quad (5)$$

and the variance within pairs is:

$$1/N[\Sigma X^2 + \Sigma Y^2 - 1/2\Sigma(X+Y)^2] \quad (6)$$

where N is the number of pairs and X is the value for twin A, Y the value for twin B.

Formulae (5) and (6) are merely adaptations of the usual partitioning of variance as can be seen from Table 6-IV.

In actual calculations formula (6) can be simplified to:

$$\sigma_w^2 = \Sigma(x-y)^2/2N \quad (6a)$$

The other formula Holzinger proposed was:

$$h^2 = \frac{\text{DZ within-pair variance} - \text{MZ within-pair variance}}{\text{DZ within-pair variance}} \quad (7)$$

where the within-pair variance is the same given by formula (6). While neither formula is very meaningful, the latter is preferred because genetics deals with differences, and formula (2) is only equivalent to formula (7) if the sum of the within-pair and between-pair variance, which is the total variance, is equal for both types of twins as shown by Clark.[37]

Formula (7) has the advantage that it can be related to a test of the significance of the hereditary variance by:

TABLE 6-IV
PARTITIONING OF VARIANCE FOR TWIN STUDIES

A. General Formulas

Variation	Sum of Squares	d.f.
Between k groups	$1/n \Sigma(\text{group total})^2 - 1/N (\Sigma x)^2$	$k - 1$
Within k groups	$\Sigma\Sigma x^2 - 1/n \Sigma(\text{group total})^2$	$N - k$
Total	$\Sigma\Sigma x^2 - 1/N (\Sigma x)^2$	$N - 1$

B. Adapted for Use With p Groups of n=2 (Twin Pairs)

Variation	Sum of Squares	d.f.
Between p pairs	$1/2 \Sigma(\text{pair sum})^2 - 1/2p (\Sigma x)^2$	$p - 1$
Within p pairs	$\Sigma x^2 - 1/2 \Sigma(\text{pair sum})^2$	p
Total	$\Sigma x^2 - 1/2p (\Sigma x)^2$	$2p - 1$

$$F = \frac{\text{DZ within-pair variance}}{\text{MZ within-pair variance}} \quad (8)$$

with degrees of freedom N_{DZ} and N_{MZ}.

F and h^2, provided it is calculated by formula (7), are related by:

$$h^2 = 1 - \frac{1}{F} \quad (9)$$

and

$$F = \frac{1}{1 - h^2} \quad (10)$$

Other advantages of formula (7) are that it can readily be generalized to include consideration of repeated testings, differences between right- and left-handed performance, performance on parallel tests or of measurement errors, as shown by Lundstrom.[38] An interesting attempt to illustrate in a diagram some of the relations between several observed and hypothetical variance components was made by Lundstrom. This diagram is shown in Figure 6-5 in which ninety-degree angles are indicated by the usual arc connecting the two sides of the right angle. The length of the lines denote the square root of the variance and the Pythagorean formula $c^2 = a^2 + b^2$ is used, where a and b are the two sides of the right angle and c is the hypotenuse.

Finally, formula (8) and therefore formula (7) can be generalized to the multivariate case. Suffice it to say that the within-pair variance is replaced by the within-pair covariance matrix, and the division is replaced by its equivalent in matrix algebra which consists of finding a matrix which, when multiplied by the divisor, yields the dividend.

At this point we have to consider the rationale upon which formula (7) is based. The MZ within-pair variance, or the within-pair differences from which it is derived is, by the definition of monozygosity, devoid of genetic differences and thus

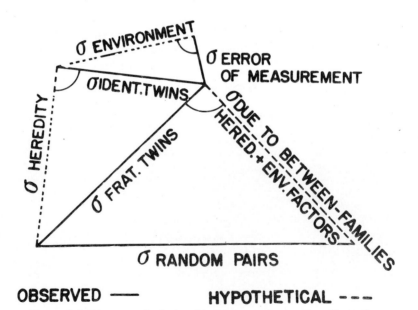

Figure 6-5. Diagram of relationship between observed and hypothetical variance components. (From A. Lundstrom, *Toothsize and Occulusion* in Twins [New York, Karger, 1948].)

solely due to within-family *environmental* effects on the MZ twins. In other terms

$$\sigma^2 wMZ = \sigma^2 env.w \qquad (11)$$

In contrast, the DZ within-pair variance is due to within family hereditary differences as well as within-family environmental effects, *plus heredity-environment interaction and correlation:*

$$\sigma^2 wDZ = \sigma^2 env.w + \sigma^2 her.w + \sigma^2 her.env. \qquad (12)$$

where σ^2her. env. stands for both correlation and interaction. If we assume that the variances of within-family environmental effects are the same for MZ and DZ twins, we can subtract (11) from (12) to get

$$\sigma^2 her.w + \sigma^2 her.env.w = \sigma^2 wDZ - \sigma^2 wMZ \qquad (13)$$

Note that this does not mean equality of individual MZ or DZ family environmental differences but only comparability of distributions of these differences. We can thus see that Holzinger's coefficient (formula 7) does include the heredity-environmental term as well as the effects of dominance and epistasis mentioned earlier. It is usually also assumed that the effects of interaction and correlation are negligible, so that

$$\sigma^2 her.w = \sigma^2 wDZ - \sigma^2 wMZ \qquad (14)$$

We should stop referring to Holzinger's formulae as heritability estimates, since they are too imprecise to deserve this name, nor should they be used any longer.

RECENT DEFINITIONS OF HERITABILITY ESTIMATES FOR CONTINUOUS TRAITS

Several authors have proposed formulae for estimating heritability from twin data which come closer to the goal. First of all, it is fairly common to adjust the identical and fraternal correlations upward to correct for the lack of perfect reliability in the measurement of the trait. This procedure makes most sense when one wants to compare the heritabilities of variables which differ markedly in the accuracy with which they can be measured.

Falconer developed a new formula for twin studies based on the argument summarized in Table 6-V.[39] It shows that the quantity $r_{mz} - r_{dz}$ only considers half the genetic variance, and therefore he proposed doubling this value to obtain

$$h^2 = 2(r_{mz} - r_{dz}) \qquad (15)$$

where no division is needed because the r's are already proportions.

Nichols constructed a very useful diagram, shown in Figure 6-6, to represent the sources of variance in twin data.[40]

Next he also suggested that because fraternal twins have on the average half their genes in common, the difference between the intraclass correlations for identical and fraternal twins, which under the assumptions listed before equals the proportion of the total variance due to hereditary differences within fraternal twins, should be doubled to obtain a better estimate of the hereditary portion of the total variance.

Finally, he suggested that r_{mz} rather than $1-r_{dz}$ be used for the denominator for the following reason: in Holzinger's formula the quantity $1-r_{dz}$ equals the within-pair hereditary variance (DH in the diagram) plus the within-pair environmental variance (DE). (We are assuming that the correlations were corrected for unreliability so that the error variance term drops out.) Nichols argued that the within-family environmental differences are minor compared to the between-family environmental differences, especially for ability measures. On the other hand r_{mz} equals the variance due to common environment (or between-family environmental variance) plus the hereditary variance. He feels that this provides a closer estimate

TABLE 6-V
SOURCES OF CORRELATION BETWEEN AND WITHIN TWIN PAIR

Type of Correlation	Between Pairs	Within Pairs
MZ twins	V genetic + V common environment	V different environment
DZ twins	1/2 V genetic + V common environment	1/2 V genetic + V different environment
Difference	1/2 V genetic	1/2 V genetic

From D.S. Falconer, *Introduction to Quantitative Genetics* (Edinburgh, Oliver and Boyd, 1960).

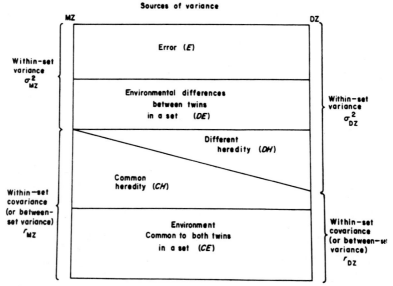

Figure 6-6. Schematic representation of sources of variance in twin data. (From R.C. Nichols, "The National Merit Twin Study, in S.G. Vandenberg, ed., *Methods and Goals in Human Behavior Statistics* [New York, Academic Press, 1965].)

of the quantity needed, i.e. the total variance.

His final formula is, therefore:

$$HR = \frac{2(r_{mz} - r_{dz})}{r_{mz}} \quad (16)$$

It is called HR to distinguish it from Holzinger's index, which has sometimes been called H (Holzinger himself proposed t²). Nichols' formula frequently gives values greater than 1.00.

Jensen suggested an improvement in Falconer's formula by noting that assortative mating will raise the correlation between siblings and DZ twins from the value of 0.50, expected for two children from parents who assorted at random, to a value of 0.66 for two children from a hypothetical self-mated mother.[41] If ρ_{oo} is the correlation between siblings due to genetic factors and ρ_{pp} is the correlation between siblings due to genetic factors then:

$$\rho_{oo} = \frac{1 + \rho_{pp}}{2 + \rho_{pp}} \quad (17)$$

and

$$h^2 = \frac{r_{mz} - r_{dz}}{1 - \rho_{oo}} \quad (18)$$

If ρ_{oo}, the genetic correlation between siblings is 0.50, this formula reduces to formula (15).

This can be more clearly seen if we introduce a coefficient called C which is the reciprocal of $1-\rho_{oo}$:

$$h^2 = C(r_{mz} - r_{dz}) \qquad (19)$$

The value of C for a given trait will at present have to be guessed at from data in other studies because as yet no twin studies have also measured the parents, but it must lie between 2.00 and 2.94. If one wants to include the correction for unreliability into the formula, it would become:

$$h^2 = C(r'_{mz} - r'_{dz}) \qquad (20)$$

where r' is the intraclass correlation corrected by the formula

$$r = r' (1-\sigma^2\text{error}) \qquad (21)$$

J. C. Loehlin has proposed a further elaboration based on a procedure aimed at calculating (from other data) an estimate k_1 of the increased similarity of the environment (or of the reduction in the within-pair environmental variance) of MZ twins compared to DZ twins. It should be noted that this estimate is based on assumptions, even though they may be reasonable ones. The assumptions concern the comparability of data from different studies and obtained with different instruments; as well as the appropriateness of these instruments for obtaining reasonably hard data on within-family environmental influences on twins. We will consider this problem a little farther on. He also proposed k_2 as an index of the increased similarity of environments of fraternal twins compared to ordinary siblings and k_3 a correction equal to $r_{sg}/1-r_{sg}$ for the increase in sib correlations due to assortive mating since this makes the within-family genetic variance smaller than the between-family genetic variance. He then suggested use of the following formula adapted from Fisher[42] to calculate r_{sg}:

$$r_{sg} = \tfrac{1}{4} (1 + C_2 + 2C_1C^2{}_2 m) \qquad (22)$$

where C_2 is the proportion of the genetic variance that is additive, C_1 is the proportion of the total variance that is genetic, and m is the observed correlation between spouses. Finally, he calculates the within and between families hereditary and environmental variances by the following formulae:

Environmental variance within family,
$$V_{we} = k_1 (1-r_{mz}) + (1-k_2)r_{dz}$$

Genetic variance within families,
$$V_{wg} = 1-k_2 r_{dz} - V_{we}$$

Genetic variance between families,
$$V_{bg} = k_3 V_{wg}$$

Environmental variance between families,
$$V_{be} = 1-V_{we} - V_{wg} - V_{bg}$$

Since these are all proportions it follows directly that:

$$h^2 = V_{wg} + V_{bg} \qquad (23)$$

J. C. Loehlin also made the useful suggestion that sometimes one may wish to correct the intraclass correlations for errors of twin diagnosis, if the study did not use blood groups. His assumption that these errors will be equally likely in both directions, i.e. MZ misclassified as DZ and DZ as MZ, seems rather a strong one on which more information is needed. Calling DZs "MZ" will lower r_{mz}, calling MZs "DZ" will raise r_{dz}, so both types of errors will decrease the difference $r_{mz} - r_{dz}$.

The time may have come for an evaluation of the usefulness of these improved heritability estimates by applying these to a large sample of twins on whom data are available on a variety of tests. Such an effort would allow a reinterpretation of some older studies on a firmer basis. Much of this effort would be unnecessary if we were dealing with variables in which psychological processes can be ruled out, such as imitation or attempts to develop a unique personal identity.

In future work it would be advantageous to collect data not just on twins, but also on other siblings and on their parents so that specific questions can be asked about nonadditive genetic variance, assortative mating, etc., for the same sample. Kempthorn and Osborne have listed eleven forces acting on members of a MZ twin pair and raise doubts about estimating genetic variance from twin data only.[43]

Elston and Gottesman have described how data on twins, their sibs, and their parents can be used to obtain heritability estimates that are calculated through the use of a modified least squares estimation procedure.[44]

They have chosen the following sources of variation for inclusion in a set of equations with five unknowns:
1. Between families

Within families:

2. Within MZ twin pairs
3. Within DZ twin pairs
4. Between sibs and twins versus sibs
5. Parent versus offspring

Formulae are given for the sum of squares for these five sources of variation, as well as the appropriate degrees of freedom.

Estimates of the variances for these five are then obtained by $\hat{\sigma} = M^{-1}s$, where s is a column vector of the five sums of squares, $\hat{\sigma}$ is a column vector of unbiased estimates of the five parameters, and where M^{-1} is the inverse of M, a matrix of the expected values of the sums of squares.

Returning for a moment to Falconer's heritability index and its refinements by Jensen and Loehlin, two questions remain to be discussed. Nichols' diagram makes it clear that the need persists to consider the total variance for the two types of twins, regardless of the final formula chosen for the estimation of heritability.

If one wants to generalize from one's twin study, one must be able to regard both the MZ and DZ sample as representative of the population. This implies that the total variance for the MZ twins should be roughly equal to the total variance for the DZ twins and both should be estimates of the population variance, the best estimate of which is the combined variance. It would be best to check this separately for the males and the females because their variances may differ.

If for whatever reason the total variance for DZ twins was significantly lower than for the MZ, we would tend to underestimate $V_{her. w.}$, yet it does not seem warranted to merely correct upward the total variance (and the within-pair and between-pair variance) for the DZ twins.

In my opinion no conclusion about heritability (or genetic variance) is warranted if the DZ total variance is lower because this indicates that something is peculiar about the sample. Unless there is convincing evidence that no bias exists, as for instance when a twin registry has been used, no thought should be given to ways of "correcting" the estimate of the within-family variance.

It would also be desirable to have a method which would permit statements about the significance of a difference between two heritability estimates.

It is perhaps possible, although awkward, to calculate confidence limits for a given heritability estimate by using the standard error of z, Fisher's transformation of r. One should first test that $r_{mz} = r_{dz}$ by $(z_{mz} - z_{dz})/\sqrt{\dfrac{1}{n_{mz}-3} + \dfrac{1}{n_{dz}-3}}$. This value should exceed a preselected significance level.

The variance of z, $s^2_z = \dfrac{1}{n-3}$ is distributed almost normally so that confidence limits at a desired level of significance can be set around the observed level of z. It is then possible to see whether there is overlap for the values of r_{mz} and r_{dz} that correspond to $z \pm l_\alpha$, where l_α is the value required to exceed a desired significance level.

After it has been established that there is no overlap of the two ranges of values, confidence limits can be set at still more significant levels: .001, .0005, etc., until the two ranges do touch. That value may perhaps be regarded as the significance level for $r_{mz} - r_{dz}$.

Alternatively, one could set a at a predetermined value k and use the distance from the lower limit of r_{mz} to the upper limit of r_{dz} as the quantity to be compared between variables. As a third possibility, one could perhaps regard the distance between the lower limit of r_{dz} and the upper limit of r_{mz} at some predetermined significance level as confidence limits on $r_{mz} - r_{dz}$.

Such values for different variables studied can then be compared. A given confidence limit may be calculated by $\lambda a = ca/\sqrt{n-3}$ where a is the significance level and ca is the distance from the mean of the confidence limit.

For example, when $a = .025$, ca is 1.96 or when $a = .005$, ca is 2.5758 so that for N = 19, la will be .490 if a is .025 or .644 if a is .005 and for N = 52, l .025 = .280 and λ .005 = .368. The detailed tables of Normal Probability Functions of the National Bureau of Standards will have to be consulted to obtain the values of a and ca needed for this approach.

Confidence limits for selected values of r and N at significance levels of .05 and .01 are given in Tables 6-VI and 6-VII.

TABLE 6-VI

CONFIDENCE LIMITS FOR CORRELATION ($\alpha = .05$)

r	.90	.85	.80	.75	.70	.65	.60	.55	.50	.45	.40	.35	.30	.25	.20	.15
N 15	72 97	60 95	49 93	39 91	29 89	21 87	13 85	05 83	00 81							
16	73 97	61 95	50 93	41 91	31 89	23 87	15 84	08 82	01 80							
17	74 96	63 95	52 93	42 91	33 88	25 86	17 84	09 82	03 79							
18	75 96	64 94	53 92	44 90	35 88	26 86	19 83	11 81	04 78							
19	75 96	64 94	54 92	45 90	36 88	28 85	20 83	13 80	06 78							
20	76 96	65 94	55 92	46 90	37 87	29 85	21 82	14 80	07 77	01 74						
22	77 96	67 94	57 91	48 89	40 87	32 84	24 82	17 79	10 76	04 73						
24	78 96	68 93	59 91	50 89	41 86	33 84	26 81	19 78	12 75	06 72						
26	79 96	69 93	60 91	51 88	43 86	35 83	28 80	21 77	14 74	08 71	02 68					
28	79 95	70 93	61 90	52 88	44 85	37 82	29 80	22 77	16 74	09 71	03 67					
30	80 95	71 93	62 90	53 87	46 85	38 82	31 79	24 76	17 73	11 70	05 67					
35	81 95	72 92	64 90	56 87	48 84	40 81	33 78	27 75	20 71	14 68	08 65	02 61				
40	82 95	73 92	65 89	57 86	50 83	42 80	36 77	29 74	22 70	16 67	10 63	04 60				
45	83 94	74 92	66 89	59 86	51 82	44 79	37 76	31 73	24 69	18 66	12 62	06 58	01 55			
50	83 94	75 91	67 88	60 85	52 82	45 79	39 75	32 72	26 68	20 65	14 61	08 57	02 53			
55	83 94	76 91	68 88	61 85	53 81	47 78	40 75	33 71	27 68	21 64	15 60	09 56	04 52			
60	84 93	76 91	69 88	61 84	54 81	47 78	41 74	34 71	28 67	22 63	16 59	11 56	05 52			
70	84 94	77 90	70 87	63 84	56 80	49 77	43 73	36 70	30 66	24 62	18 58	13 54	07 50	02 46		
80	85 94	78 90	70 87	64 83	57 80	50 76	44 72	38 69	32 65	26 61	20 57	14 53	09 49	03 45		
90	85 93	78 90	71 86	64 83	58 79	51 76	45 72	39 68	33 64	27 60	21 56	15 52	10 48	05 43		
100	86 93	79 90	72 86	65 83	58 79	52 75	46 71	40 67	34 63	28 59	22 55	17 51	11 47	06 43	00 38	
125	86 93	79 89	73 86	66 82	60 78	54 74	47 70	41 66	36 62	29 58	24 54	19 50	13 45	08 41	03 36	
150	86 93	80 89	73 85	67 81	61 77	55 73	49 69	43 65	37 61	32 57	26 53	20 48	15 44	09 39	04 35	
175	87 93	80 89	74 85	68 81	62 77	56 73	50 69	44 65	38 60	32 56	27 52	21 47	16 43	11 38	05 34	00 29
200	87 92	81 88	74 85	68 81	62 77	56 72	50 68	45 64	39 60	33 55	28 51	22 47	17 42	12 38	06 33	01 28
250	87 92	81 88	75 84	69 80	63 76	57 72	51 67	46 63	40 59	35 54	29 50	24 45	18 41	13 36	08 33	03 27
300	88 92	82 88	76 84	70 80	64 75	58 71	52 67	47 62	41 58	36 54	30 49	25 45	19 40	14 35	09 31	04 26
N r	.90	.85	.80	.75	.70	.65	.60	.55	.50	.45	.40	.35	.30	.25	.20	.15

TABLE 6-VII

CONFIDENCE LIMITS FOR CORRELATION ($\alpha = .01$)

r	.90	.85	.80	.75	.70	.65	.60	.55	.50	.45	.40	.35	.30	.25	.20	.15
N 15	62 98	47 96	34 95	23 94	12 92	03 91										
16	64 98	49 96	37 95	25 93	15 92	06 90										
17	66 97	51 96	39 95	28 93	18 92	09 90	01 88									
18	67 97	53 96	41 94	30 93	20 91	11 89	03 88									
19	68 97	55 96	43 94	32 92	22 91	13 89	05 87									
20	69 97	56 96	44 94	34 92	24 90	15 89	07 87	00 85								
22	71 97	58 95	47 93	37 92	27 90	18 88	10 86	03 84								
24	72 97	60 95	49 93	39 91	30 89	21 87	13 85	06 83								
26	73 96	62 95	51 93	41 91	32 89	23 87	16 84	08 82	01 80							
28	74 96	63 94	53 92	43 90	34 88	25 86	18 84	10 81	03 79							
30	75 96	64 94	54 92	44 90	36 88	27 85	20 83	12 80	05 78							
35	77 96	66 94	57 91	48 89	39 87	31 84	23 82	16 79	09 76	03 74						
40	78 96	68 93	59 91	50 89	42 86	34 83	26 81	19 78	13 75	06 72	00 69					
45	79 95	70 93	61 90	52 88	44 85	36 83	29 80	22 77	15 74	09 71	03 68					
50	80 95	71 93	62 90	54 87	46 85	38 82	31 79	24 76	17 73	11 70	05 66					
55	81 95	72 92	63 90	55 87	47 84	40 81	32 78	26 75	19 72	13 69	07 65	01 62				
60	81 95	72 92	64 89	56 87	48 84	41 81	34 78	27 74	21 71	14 68	08 64	02 61				
70	82 95	74 92	66 89	58 86	50 83	43 80	36 77	30 73	23 70	17 66	11 63	05 59				
80	83 94	75 91	67 88	59 85	52 82	45 79	38 76	31 72	25 69	19 65	13 62	07 58	02 54			
90	83 94	75 91	68 88	60 85	53 82	46 78	40 75	33 71	27 68	21 64	15 60	09 57	03 53			
100	84 94	76 91	68 88	61 84	54 81	47 78	41 74	34 71	28 67	22 63	16 59	10 56	05 52	00 48		
125	85 94	77 90	70 87	63 84	56 80	50 77	43 73	37 69	31 65	25 62	19 58	13 54	08 50	02 45		
150	85 93	78 90	71 87	64 83	58 79	51 76	45 72	39 68	33 64	27 60	21 56	15 52	10 48	04 44		
175	86 93	79 90	72 86	65 82	59 79	52 75	46 71	40 67	34 63	28 59	22 55	17 51	11 47	06 42	01 38	
200	86 93	79 89	72 86	66 82	59 78	53 74	47 71	41 67	35 63	29 58	24 54	18 50	13 46	07 41	02 37	
250	86 93	80 89	73 85	67 81	61 77	55 74	49 70	43 65	37 61	31 57	25 53	20 49	15 44	09 40	04 35	
300	87 93	80 89	74 85	68 81	62 77	56 73	50 69	44 65	38 60	32 56	27 52	21 47	16 43	11 38	05 34	00 29
N r	.90	.85	.80	.75	.70	.65	.60	.55	.50	.45	.40	.35	.30	.25	.20	.15

THE GENERALITY OF HERITABILITY ESTIMATES

It has already been mentioned that heritability estimates should not be seen as the final goal of behavior genetics because they are only correct in a very limited way: they apply only to the particular population to which one can legitimately generalize from the sample that was studied and only for the particular measure used under the usually rather particular conditions of *that* study. Frequently, these conditions have not been specified in as much detail as is needed for replication or meaningful interpretation. Time pressure, size of group, number of proctors, etc. are all known to influence performance on tests. For these and other reasons it is highly desirable to obtain data on other sibs as well as the twins parents so that other estimates of heritability can be obtained on the same sample, such as those based on correlations between parents and offspring or full sib correlations.

That a heritability estimate will vary is dramatically illustrated in Figure 6-7 which shows the changes with age in heritability for height for boys and girls reported by Furusho.[45]

A related problem which has become a major topic in current discussions concerns the applicability of conclusions based on studies of European and American Whites to other populations. This problem is probably more serious for ability measures than for personality traits and more serious for personality traits than for mea-

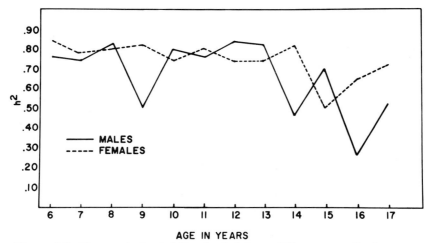

Figure 6-7. Changes in heritability for height at different ages for Japanese boys and girls. (From T. Furusho, "On the Manifestation of Genotypes Responsible for Stature," *Human Biology*, 40:437 [1968].)

sures of motor skills and for performance on sensory tasks, and it possibly affects data on mental illness relatively little (if the culturally conditioned content of specific symptoms in mental illness can be disregarded).

The question of the generality across racial groups of the findings from twin studies (and generally of quantitative genetic studies) has been sharply brought into focus by a paper by Jensen in which he suggests that the observed difference between Negro and White intelligence scores are in part genetic in origin, rather than being entirely attributable to disadvantaged environmental conditions for many Negroes.[46]

He based this conclusion mainly on the fact, which he documented extensively, that the variability in the population is to a very large extent genetic in nature. Jensen estimated that the genetic proportion of the total observed variance was 80 percent, which is somewhat higher than others have suggested. (This is heritability in the broad sense as indicated before.) However, the studies cited by Jensen were generally of White twins and probably mostly of middle-class White twins. The question can be raised whether heritability estimates for intelligence would be equally high for Negroes. (Even if it were, this would still not prove that the difference between the means was genetic in origin as was shown by DeFries.[47]) We are beginning to get some answers to this question.

Vandenberg reported a small study of Negro twins in which he found much lower values for F, suggesting a smaller hereditary variance component in a battery of tests for Negro than for White twins as shown in Table 6-VIII.[48] However, Osborne reached the opposite conclusion, using formulae (2) and (16) because r_{mz} and r_{dz} were lower for the Negroes but r_{dz} more so than r_{mz}.[49,50]

Recently Scarr has provided further evidence for the lack of generality of heritability estimates in her analysis of test scores in the Philadelphia Public School on record for lower and middle class Negro and White twins enrolled in 1968. The

TABLE 6-VIII
SIGNIFICANCE OF F RATIO BETWEEN DZ AND MZ WITHIN PAIR VARIANCES FOR NEGRO AND WHITE STUDENTS

F is less than 1.00	F is greater than 1.00							Total Significant
	n.s.	p<.10	p<.05	p<.02	p<.01	p<.005	p<.001	
5	9	2	1	3	0	0	0	6
2	5	2	2	0	1	1	7	13

analysis was complicated by the fact that MZ and DZ twin correlations had to be estimated from correlations of opposite-sexed and same-sexed twins because zygosity diagnoses through blood typing or other means were not available. This is now being corrected, but no data are available from that analysis at this time.

She reported the heritabilities for lower class (disadvantaged) Negroes and Whites and for middle class Negroes and Whites shown in Table 6-IX. The dashes indicate that the same-sexed and opposite-sexed correlations were not sufficiently different (in several cases the latter were higher), so that the heritabilities could not be estimated but were probably close to zero. The heritabilities for the Whites are unusually low, underlining the need for a reanalysis.

A different analysis based only on the opposite-sexed pairs was used to estimate the percentages of between-families genetic, between-families environmental, and within-families genetic and within-families environmental variances for the verbal and nonverbal test scores for the four groups. The results of this analysis are displayed graphically in Figures 6-8 and 6-9.

ASSUMPTIONS OF TWIN STUDIES

How realistic are the assumptions underlying the classical twin study method? There has not been as much research directly focused on this topic as one might expect, partly because it is rather difficult to design a study which asks a really crucial question. Much of the literature on this topic grew out of "after the fact" reservations about the assumptions; while suggestive, it does not provide the hard data needed for a final answer.

Besides the sampling problems one can distinguish several other, separate, though related problems in the classical twin study. They can be stated as follows:

1. Do parents, on the average, influence MZ twins in the direction of greater similarity, as compared to DZ twins? If this is indeed true, it would mean that Loehlin's coefficient k_1 is not zero. The same question can be asked with respect to the influence of persons other than the parents, but systematic study of this factor would be even more difficult and probably an analysis of parental influences would allow conclusions about a proportionally reduced effect of other persons.

TABLE 6-IX
HERITABILITIES FOR VERBAL AND NONVERBAL TEST SCORES OF LOWER CLASS AND MIDDLE CLASS NEGROES AND WHITES

	Verbal Test Scores	
	Negro	White
Lower Class	.309	—
Middle Class	.651	.436
	Nonverbal Test Scores	
	Negro	White
Lower Class	—	—
Middle Class	.580	.038

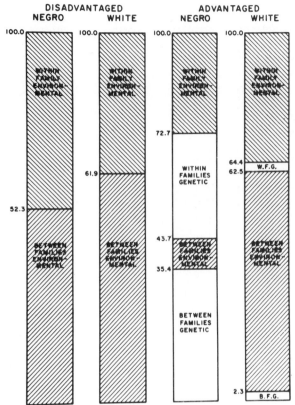

Figure 6-8. Percentages of genetic and environmental variance within families and between families for nonverbal ability. (From S. Scarr-Salapatek, "Race, Social Class and IQ. Population Differences in Heritability of IQ Scores Were Found for Racial and Social Class Groups," *Science,* 174:1285 [1971].[51])

2. Do twins influence one another in the direction of a kind of division of labor such that one twin frequently is more dominant or venturesome and the other more submissive or imitative? More importantly (a) are these tendencies more pronounced in MZ twins, which would lead to an *under*estimation of hereditary effects, or (b) are they more pronounced in DZ twins, in which case an *over*estimation of hereditary effects would be the result.

Let us first look at the assumption that the within-pair environmental variance is expected to have the same value for MZ as for DZ pairs.

At first sight this seems an unlikely assumption to make. One always hears of some pair of identical twins who seem to be subject to many influences favoring greater similarity, and then remembers in contrast some pairs of fraternal twins who seem to share very few common experiences and who are very different.

On closer examination, matters are not

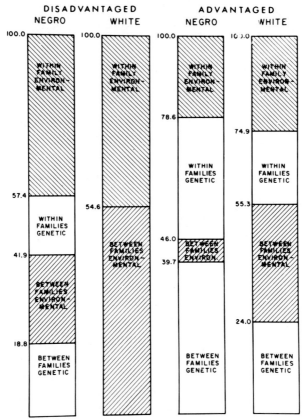

Figure 6-9. Percentages of genetic and environmental variances within families and between families for verbal ability. (From S. Scarr-Salapatek, "Race, Social Class and IQ. Population Differences in Heritability of IQ Scores Were Found for Racial and Social Class Groups," *Science,* 174:1285 [1971].)

so clear. Many of the environmental factors that seem to make for similarity or differences between twins cannot be considered to be causal or antecedent but are rather the result of, and at the very least subsequent to, the development of similarities or differences because to a large extent each twin chooses his or her own environment. This environment may or may not include the other twin to varying degrees, depending on the presence of other sibs. Whether this kind of confounding of heredity and environment should more properly be grouped with hereditary or with environmental variance is debatable and would tend to betray one's bias. To demonstrate or perhaps even measure such effects it would be necessary to study the twin-relationship for a number of MZ and DZ twins over an extended period in much more detail than has been accomplished so far.

Let us now look at a few of the studies directly dealing with these assumptions.

Most of them have had to rely on questionnaires with either the twins or associates as respondents in what Jones and Wilson aptly called the study of "reputation differences" because they were generally based on opinions.[52]

Schäfers performed an interesting study that deserves to be better known.[53] He determined the popularity ("Beliebtheit") of twins in elementary school classes and compared seventeen MZ and fourteen DZ pairs. His sociometric index based on rank orders of nominations by classmates could run from 0 to 1.00 with the latter the high end. The mean for the MZ was .72, for the DZ .57. However, one DZ twin was more often in first or second place with the other twin somewhat lower, while the two MZ children tended to be in third, fourth, or fifth place. Incidentally, there was a significantly greater similarity in this trait for the MZ than for the DZ twins (F = 2.387, p = .05) It would, however, seem that being chosen is not so much the result of arbitrary environmental forces than of an active interplay between the twin's genetic makeup and his world and of the other children's reaction to each twin's phenotype.

Is it possible to penetrate beyond these rather superficial, hypothesized "outside" influences on twins?

The investigator who has made the most subtle effort to analyze the special psychological nature of the relation between twins and of the within family influences that affect twins is René Zazzo of the University of Paris. He interviewed and tested twins studied by the medical geneticist, Dr. Turpin, from about 1941 to 1955. After a while he lost interest in the comparison of concordance rates or in other possible determinations of the degree of hereditary control of psychological traits. Instead he concentrated more and more on the special situation in which twins are placed. To study this phenomenon he relied heavily on a questionnaire which he devised after a number of interviews with twins. A translation of it is found in an appendix at the end of this chapter. Zazzo reported his findings on this questionnaire in Volume 2 of a book entitled, *Les Jumeaux, Le Couple et La Personne (Twins, The Couple and The Individual)*.[54]

Before summarizing some of Zazzo's findings, it may be worthwhile to remember that there are undoubtedly some important differences between France and the United States which will affect the situation in which twins grow up. For instance, Zazzo found that approximately 75 percent of his twins slept in one bed at least until their teens. This is a rather rare event in the United States; even if twins share a room, they have double-decker bunk beds or separate beds on different sides of the room. Although no direct comparison is possible, information about the number of children in France make it also seem likely that in more cases the French twins had no other siblings close to them in age, whereas many United States twins do.

ZAZZO'S STUDY

Zazzo reports strong reactions by twins to separations and gives anecdotal data about even stronger reaction by the single twin to the marriage of the other twin. Even though such cases may be somewhat unusual in France, they may well be even rarer in the United States where social mixing is so much encouraged in high school. At least in my studies I have never encountered such strong attachments between twins. Some twins solve this problem by marrying twins. Recently, there was a report by Taylor on fifty such marriages.[55] Children from such unions are especially interesting in the case of MZ twins marrying MZ twins because they are extremely closely related, but even when the two pairs are DZ the children still are double first cousins.

A contrary trend can be deduced from the fact that in Zazzo's studies a lower percentage of twins was reported to be dressed alike at higher ages than found in the United States. Zazzo reports 19.3 percent up to five years of age, 33.9 percent from six to ten, 40.1 percent from eleven to sixteen, and 6.7 percent for twins seventeen years and older. The figures from another study were 4.0 percent, 18.7 percent, 65.0 percent, and 12.0 percent.

Zazzo also reports a high proportion of cases in which one twin was reported to be dominant over the other twin. Table 6-X summarizes his findings. It should be kept in mind that this was based on reports by parents, teachers, and the twin themselves. No comparable data have been obtained by me, but the responses to several questions in which an attempt was made to obtain information about dominance gave no clear indication of a general polarization of this type within pairs.

The other phenomenon reported by Zazzo was the existence of a special language or the presence in twins of unusual retardation of language development which could have resulted in or be interpreted as a special language. Table 6-XI shows these results. No data of comparably high frequencies have ever been reported in the United States, although the existence of a less marked but somewhat delayed verbal development has been reported by many investigators.

Zazzo does not report a statistical breakdown for the responses to many of the most interesting items in his questionnaire. One wonders whether his objections to the

TABLE 6-X
PERCENTAGES OF PAIRS WITH ONE TWIN DOMINANT OVER THE OTHER, BY AGE GROUPS.

Age of twins	N	Percentages of Pairs in Which...		Dominance alternates	There was no dominance	There was no answer
		One twin is dominant				
		boy	girl			
1-5 years	34	23.5	44.7	2.9	5.9	23.5
6-7	21	19.6	33.3	4.8	0	42.9
8-10	36	19.4	63.9	5.6	2.8	8.3
11-15	60	35.0	43.9	3.3	6.7	11.7
16-20	30	16.7	48.7	6.6	13.3	16.7
20+	36	27.8	30.6	5.5	11.1	25.0

From R. Zazzo, Les Jumeaux, Le Couple et La Personne (Paris, Presses Universités France, 1960).

TABLE 6-XI
PERCENTAGE OF INDIVIDUALS WITH SPECIAL LANGUAGE OR WITH LANGUAGE RETARDATION

Type of Twin	N	Special Language	Language Retardation	Both Symptoms	All Cases
MZ f	141	19.8	9.9	11.3	41.1
m	144	26.3	14.5	16.6	57.6
DZ f	82	12.2	12.2	5.5	27.7
m	94	14.8	13.8	7.2	41.8
f-m	164	14.6	9.7	9.1	33.5

From R. Zazzo, Les Jumeaux, Le Couple et La Personne (Paris, Presses Universités France, 1960).

classical twin studies are not unduly influenced by a few typical cases.

KOCH'S STUDY

Another person who studied the twin situation extensively is Helen Koch of the University of Chicago. Some years ago she decided to make the questioning of the assumption of similar environmental influences for MZ and DZ twins the focus of a study.[56] She decided not to analyze within-pair differences as they relate to hereditary factors, but to study the differences between twins and singletons, between identical and fraternal twins, and between like-sexed and unlike-sexed fraternal twins. She located twin pairs who were the only children in a family and matched each pair with four single born children of the same age and sex from two-child families of the same socioeconomic status. The siblings of these single born children were up to six years older or younger. (About equal proportions were under two, two to four, and four to six years older or younger.) Because each pair of twins was separately matched to four single born children, there was a different comparison group for the identical twins and for the like-sexed fraternal twins, separately for boys and for girls and yet another one for the boy-girl fraternal twins. Because each twin group had its own comparison group, the presentation of results was rather complicated, especially because the results frequently were not similar for all five groups. It is not stated whether some of the same individuals were used in the various control groups.

In addition the socioeconomic status of the parents was considered. Even though only a three-level distinction was used, it further complicated the design and reduced the numbers in a given category.

A total of ninety pairs of twins aged between fifty-nine and eighty-six months were located with the sex and zygosity distribution shown in Table 6-XII.

The zygosity of the twins was based on physical similarity to avoid the difficulties of obtaining blood samples.

While there were ninety pairs of twins, it was not possible to obtain complete information on all pairs. The number of cases on which data were available varied somewhat for different variables.

The main areas studied were: (1) size—birth weight, present height and weight; (2) mental abilities—four of the five scores of the Primary Mental Abilities test for age five to seven; (3) language behavior—ratings by mother, teacher, and Dr. Koch, as well as several scores derived from taped stories told by children in the Children's Apperception Test (CAT); (4) left-handedness and stuttering; (5) personality and attitudes—studied in interviews with mother, teacher, and each of the children; (6) effect of closeness and dominance in twins and twin separation in

TABLE 6-XII
NUMBER OF PAIRS OF VARIOUS GROUPS OF TWINS BY SOCIAL CLASS, SEX, AND ZYGOSITY

	Class	Boys	Girls	Total
Pairs of MZ twins	I	8	5	13
	II	4	3	7
	III	6	9	15
		(18)	(17)	(35)
Like-sexed DZ	I	6	6	12
	II	9	5	14
	III	3	7	10
		(18)	(18)	(36)
Total like-sexed	I	14	11	25
	II	13	8	21
	III	9	16	25
		(36)	(35)	(71)
Unlike-sexed DZ	I		7	7
	II		5	5
	III		5	5
			(19)	(19)

From H. Koch, *Twins and Twin Relations* (Chicago, University of Chicago Press, 1966).

school. A brief summary of the results will permit some discussion of the implications of the study.

SIZE: There was no difference between MZ and DZ twins in birth weight and only a significant difference in present weight between male DZ and their control single borns. The heavier of a pair of twins at birth was still heavier at the time of study in 94 percent of the girls and 67 percent of the boys. The intraclass correlation for birth weight was .49 for male and .84 for female fraternal twins, and .82 for male and .65 for female identical twins. It was .73 for the unlike-sexed fraternal twins. No explanation is offered for the unusually high value for the female DZ.

MENTAL ABILITIES: The main findings were that twins scored lower on three of the PMA subtests than the control single born children, but *higher* on the Perceptual subtest.

Dr. Koch believes that the test conditions favored the twins especially on this part of the test (both boys and girls from unlike-sexed pairs did better on the spatial test than their control single born). All groups were above average. The mean IQ for all twins was 105.9, and 109.2 was the average of the mean IQ's of the various control samples of single born. The intellectual handicap of the twins was therefore only a relative one and a rather minor one. The most significant difference between twins and single born was on the Verbal subtest, with the MZ scoring lowest, the like-sexed DZ next lowest, the unlike-sexed DZ somewhat higher, and the single born the highest. No scores were obtained for the twins on the motor subtest. Incidental findings were that the children from the lowest social class did poorer on all PMA subtests except on the Spatial test and that the girls did better on the Perceptual test than the boys.

LANGUAGE BEHAVIOR: Data were collected on single born for only some of the variables in this part of the study. The correlations between the ratings were fairly high, .55 between mother and teacher, .73 between mother and Dr. Koch, and .49 between teacher and Dr. Koch. Correlations between mother, teacher, and Dr. Koch and the number of syntax errors in the stories were, respectively, .37, .35, and .38. Only in girls did the DZ rate higher than the MZ. Twin closeness (described below) did not correlate with the various language measures, nor did involvement with adults or number of friends. As Dr. Koch concludes, these results do not consistently lend support to any theory advanced to explain the retarded verbal development in twins, such as less contact with adequate language models.

LEFT-HANDEDNESS AND STUTTERING: Depending on whether actual left-handed drawing or only some left-handed tendencies in other tasks were used as a criterion, the percentages were 2.8 and 0.9 or 11.4 and 7.5 for MZ and DZ pairs respectively. Hypotheses concerning the inheritance of left-handedness advanced by Newman and others are reviewed, but no conclusion based on the present data could be reached. Dr. Koch favors Dennis' idea that the baby's early experience with placement of the bottle on the left or the right side helps to establish position habits, but wavered back and forth between this position and one of biological determination and ended by calling for more research.[57] She found no relation between stuttering and either left-handedness or prematurity.

PERSONALITY AND ATTITUDES: Significant differences were infrequent, and when measures of closeness to, or of dominance of, one twin over his partner were related to personality items, significant relations were rare. The closer co-twins did not, for instance, have fewer playmates or poorer social skills.

It may be asked whether it is meaningful to compare the responses of twins with those of singletons to questions about play with the sib because in the one case there is no age difference, but in the other the age difference was up to six years for a third of the cases, up to four years for another third, and up to two years for the remainder. Playing with a brother or sister who is four to six years younger (practically a baby) would be quite a different situation than playing with someone the same age, and play with an older sib is not the same thing as play with someone of the same age.

The measure of closeness was derived from twelve items in which responses reflecting a favorable attitude to the sib were counted. They dealt with whether the child preferred to play alone or with the sib, whether he would rather play with others than with his sib, whether he would prefer to go visiting alone or have the sib go with him, whether he preferred to have a bedroom of his own or share one with the sib, whether he played much with his twin or little, whether he would be happier with than without the sib, whether he wanted the sib in his class in school, whether he desired to dress like the sib, whether he would prefer a twin to an older sib, whether he would prefer a twin to a younger sib, whether he quarreled seldomly or frequently with the sib, whether he liked being a twin. It is reported that the corrected split-half reliability for this measure was .93.

Identicals were closer than like-sexed fraternals, who were closer than boy-girl pairs; but there was no difference between MZ and DZ in number of playmates or in judged involvement with adults; some related personality differences were found, but these were not consistent between male and female pairs. About 500 correlations were computed between measures of closeness and personality variables and only forty-two (or slightly more than 5%) were significant at the 5 percent level or better.

Dr. Koch tried to see if it made any difference whether or not twins had been placed in different rooms in school. Few consistent trends could be noted. It seemed to be more the case that troublemakers or poorer students got often separated compared with well-behaved or achieving students.

DOMINANCE: The question was asked whether one twin was usually dominant, and if so, was this relatively stable and was it general or rather limited to certain roles, with the two twins perhaps each being dominant in their own sphere?

Dominance was judged by mother, teacher, and the experimenters. Mothers of DZ twins seldom hesitated but mothers of MZ were less certain. In somewhat over half the cases dominance seemed to be fairly stable; it was more likely in girls than in boys. No information on dominance was obtained for around one fifth of the cases seen before this variable was added to the study.

The dominant and nondominant MZ twins did not differ significantly on any personality measure beyond the few values one can expect by chance. For the DZ twins the dominant one tended to be rated more aggressive, quarrelsome, uncooperative with peers, moody, given to exhibitionism and alibis, and less apprehensive and less obedient.

There was no relation in either group with intelligence, popularity, leadership, involvement with adults or children, emotionality, speech performance, originality, etc.

It would have been interesting to compare for all the variables studied the sex differences for the boy-girl fraternal pairs with those for the remainder of the sample. Does having a sister of his own age make a boy more of a sissy and is the girl more tomboyish? There is no systematic treatment of this, but, here and there, a few sentences do throw light on this question. Dr. Koch mentions that at school entrance the girl tends to be the dominant one in the boy-girl pairs, with the boy's trying to keep up with the sister. In the PMA test boys from boy-girl pairs did better than MZ and like-sexed DZ boys on the Verbal, Spatial, and Perceptual subtests, while girls from boy-girl pairs did better than MZ and like-sexed DZ girls on the Spatial subtest only.

The overall conclusion that may be drawn from the study is that twins are not importantly different from single-born children, nor identical twins from fraternal, so that the twin method has not been invalidated either by this study. It seems possible that the fashionable tendency to question the twin method gives too much weight to dramatic instances of a twin pair with a large difference in dominance or a very shy pair. Parker has made the same point with regard to the incorrect idea that twins might be more subject to mental illness.[58]

OTHER STUDIES OF ASSUMPTIONS

Other data that suggests the twin situation has some effect comes from a comparison of MZ twins separated during the first year or later (Vandenberg and Johnson[59]). Twins who were separated sooner had a longer period of different environment at an age when environmental stimulation is thought to be most important. They might, therefore, be expected to show greater pair differences. Yet, this was not the case. On the contrary, it was found that twins separated later showed greater differences. One likely explanation is that some polarization has occurred.

Similarly, Shields found greater concordance for extraversion in MZ pairs raised apart than in pairs raised together.[60] Then there are the occasional negative correlations between twins' partners, usually DZ.

Finally, there are the findings of Wilde that members of DZ pairs living apart for five years or more were more concordant on measures of neuroticism, extraversion, test-taking attitude, and masculinity-femininity than DZ twins still living together.[61] No such differences was found for MZ pairs.

To account for these differences, Wilde suggested that members of DZ twin pairs respond in part to expectations on the part of parents, teachers, and friends that they will behave differently but that MZ tend to respond to expectations that they will behave similarly.[62] He performed a small study with the following hypotheses:

1. MZ twins more than DZ twins believe that they are more similar than they actually are.
2. MZ twins imitate each other more than DZ twins.

To test the first hypothesis, Wilde administered the Eysenck Personality Inventory to twelve MZ pairs and nine DZ pairs once the regular way and once with instructions to try to answer the questions as the co-twin would respond. MZ twins were indeed expecting a greater similarity of their co-twin's responses to their own than did the DZ twins.

To test the second hypothesis, Wilde used a procedure derived by Milgram in which the subject is asked to judge which of two lights has stayed on longer.[63] An additional pair of lights supposedly showed the response of another subject, while actually a preprogrammed sequence is used which controls whether the right or left one of these lights comes on.

Because one twin will thus observe twenty out of sixty instances in which his co-twin appears to make a judgment which differs from his own, he will be tempted to yield; that is, he may consider that his co-twin is probably correct and respond in a similar way, even though this is an incorrect answer.

The number of such responses is a measure of conformity.

In Wilde's experiment MZ twins yielded more than did DZ twins, but the increase (in concordance) was not significant.

Let us finally look at a few United States studies which are frequently cited in the more recent literature. The first is a study done in Baltimore, reported in 1965.

Smith obtained reports from 184 pairs of high school age twins about their activities and habits.[64] He compared the proportions concordant for various behaviors in MZ and DZ twins separate for the males and the females and for the combined sample. He obtained the results shown in Table 6-XIII.

It can be seen that for males there were only two significant differences, five for the females and five for the combined sample, each time out of a possible fourteen. Smith concluded that "the assumption of a common environment for MZ and DZ twins is of doubtful validity...." But it can surely be argued that many of the behaviors he studied are not really clear-cut indicators of environmental pressures towards similarity, but instead that many of these concordances can be seen as expressions of already existing similarities within-twin pairs, probably due to identical genetic makeup.

Scarr performed one of the most crucial analyses on this topic.[65] She exploited the fact mentioned above, that parents frequently have incorrect information about the zygosity of their twins, and used this situation to compare the opinions about their twins of mothers of MZ and DZ who had classified their twins correctly and incorrectly as far as zygosity is concerned.

Her results are shown in Table 6-XIV.

TABLE 6-XIII
PROPORTIONS CONCORDANCE IN SELECTED HABITS AND ACTIVITIES FOR MZ AND DZ TWIN PAIRS, BY SEX

	Males		Females		Both	
	MZ	DZ	MZ	DZ	MZ	DA
a. Number of Pairs	40	34	50	40	90	74
1. Eating between meals	72.5	66.7	77.1*	57.5	75.0*	61.6
2. Snack before bedtime	50.0	66.6	60.4	62.5	55.7	64.4
3. Time usually go to bed	59.0	72.7	72.9	57.5	66.7	64.4
4. Time usually get up	60.0	53.1	72.9	65.0	67.0	59.7
5. Dressing alike	59.0	57.1	68.7*	33.3	64.4*	40.3
6. Study together	23.1	9.1	54.2*	20.5	40.2*	15.3
7. Active in school groups	28.2	32.3	58.7	57.9	44.7	46.4
8. Active in other clubs	37.5	41.4	43.5	30.0	40.7	34.8
9. Number of sports played	23.1	25.0	20.8	25.0	21.8	25.0
10. Active in other outdoor sports	60.5	59.4	57.4	45.0	58.8	51.4
11. Attend sports games	79.5*	59.4*	66.7	60.0	72.4*	59.7*
12. Attend movies together	55.0	39.4	63.8	55.0	59.8	47.9
13. Play musical instrument	15.0	14.7	34.0*	17.5	25.6	16.2
14. Same close friends	55.0*	30.3*	59.6*	35.0	57.5*	32.9

a. Proportions are based on varied number of pairs depending on response item.
*This proportion is significantly greater at $p \leq .05$.
From R.T. Smith, "A Comparison of Socio-environmental Factors in Monozygotic and Dizygotic Twins, Testing an Assumption," in S.G. Vandenberg (ed.), *Methods and Goals in Behavior Genetics* (New York, Academic Press, 1965), pp. 45-61.

TABLE 6-XIV
PERCENTAGE OF CORRECTLY AND INCORRECTLY CLASSIFIED PAIRS RATED AS SIMILAR FOR SEVERAL CHARACTERISTICS

Mothers' Ratings of Co-Twins	Correctly Classified		Misclassified	
	MZ (N=19)	DZ (N=22)	MZ (N=4)	DZ (N=7)
Similar now	79	9	75	43
Similar social maturity expected	95	67	75	43
Dressed alike	74	45	75	57
Similar early behavior problems	79	59	100	57
Similar early development	79	54	50	71
Adjective Checklist Scales	Mean Differences			
n Affiliation	4.1	8.5	5.7	6.6
n Change	5.6	15.8	7.3	9.7
Anxiety	4.7	12.0	4.0	5.9
Vineland Social Maturity	0.4	1.1	1.3	1.8

From S. Scarr, "Environmental Bias in Twin Studies," *Eugenics Quarterly*, 15:34 (1968).

It can be seen that in general the percentages and reported mean differences are comparable for the correctly and incorrectly classified twins.

The author concluded that "environmental determinants of similarities and differences between MZ and DZ co-twins are not as potent as the critics charge. Differences in the parental treatment that twins receive are much more a function of the degree of the twins' genetic relatedness than of parental beliefs about 'identicalness' and 'fraternalness' (p. 40)." She also expressed the hope that the same method will be used with larger samples.

One other study suggested that greater DZ than MZ differences are mainly due to hereditary factors and not the results of parental influences.

Freedman and Keller obtained mental, motor, and social development scores with the Bayley test and filmed twenty pairs of same-sexed twins every month before their first birthday.[66] Eleven of the pairs were found to be DZ and nine MZ by extensive blood group determinations. Even though parental influence and/or within-pair imitation can be ruled out at this early age there were significant differences for all three scales.

NAMES OF TWINS

Nevertheless, some of the parental influences may start early to affect MZ and DZ twin differences for pairs that appear more or less similar at birth. Plank studied the names given to twins.[67] Many times twins are given names starting with the same initial, while other names of twins are even more similar, such as John and Joan, or Paula and Paul. What he found is summarized in Table 6-XV. He did not report the number of MZ and DZ pairs but, assuming that they were in roughly equal numbers, the difference is significant at $p < .0125$ ($\chi^2 = 7.418$, df 2). This prompted me to make a similar analysis for the twins studied in Louisville. Table 6-XVI shows a tabulation.

TABLE 6-XV
PERCENTAGES OF MZ AND DZ TWINS WITH SIMILAR NAMES

	MZ	DZ
Same initial	67%	58%
Other similarity	21%	15%
Dissimilar names	12%	27%

From R. Plank, "Names of Twins," *Names*, 12:1 (1964).

TABLE 6-XVI
FREQUENCY OF SIMILAR FIRST NAMES FOR UNLIKE-SEXED TWINS, MZ AND DZ TWINS

TWINS' FIRST NAMES	DZ ♂♀	DZ ♂	DZ ♀	MZ ♂	MZ ♀	Totals
Rhyme	29	40	44	23	27	163
Are a joke	1	1	0	1	0	3
Sound similar	19	10	12	7	4	52
End differently	35	1	11	1	10	58
Second name rhymes	15	6	11	5	14	51
Have same initial	62	35	25	19	30	171
Are different	86	34	34	25	27	206
Totals	247	127	137	81	112	704

Some examples of the categories used are:
rhymes: Royce and Joyce, Donald and Ronald, Karen and Sharon
jokes: Jack and Jill, Tom and Jerry
similar sounds: John and Joan, Marvin and Melvin, Jane and Jean
different ending only: Paul and Paula, Orlander and Orlester, Sherry and Cheryl
second name rhymes: John Dale and Mary Gail, Jim Kerry and Tom Derry, Beth Linda and Lois Brenda
same initial: Phillip and Paula, John and James, Mary and Martha

A comparison of the unlike-sexed pairs, the MZ pairs, and the like-sexed DZ pairs, by chi-square, gave a value of 47.706 for twelve degrees of freedom which is significant beyond the .001 level. When the boy-girl pairs, the male pairs and the female pairs are compared, chi-square is 57.92, df 12, p < .001. However, this is mainly due to fewer rhymes and more names with only different endings such as Paul and Paula or Carl and Carla for boy-girl pairs.

When categories one through four and five plus six are combined, we obtain the results shown in Table 6-XVII, and a different conclusion would be drawn. This time the main discrepancy is that there is more similarity in the names of like-sexed DZ pairs than for unlike-sexed DZ or MZ pairs. The same conclusion applies to a comparison of same-sexed and different-sexed pairs. It is clear that whether or not twins are given similar

TABLE 6-XVII
FREQUENCY OF SIMILAR OR DISSIMILAR FIRST NAMES IN TWIN PAIRS

	♂♀	DZ	MZ	♂♀	♂	♀	Different Sex	Same Sex
Very similar	84	119	73	84	84	108	84	192
Some similarity	77	77	68	77	65	80	77	145
No similarity	86	68	52	86	59	61	86	120
		chi-square 10.08, df 4, p <.025			chi-square 7.70, df 4.10, p <.025			chi-square 6.65, df 2, p <.025

names is not influenced by the zygosity of the twins. This makes good sense because most parents have chosen the names for possible boys or girls before the twins are born.

Table 6-XVIII shows a comparison only of the two kinds of same-sexed twins. It shows that there is no significant difference left. Apparently, it is the birth of unlike-sexed twins that occasionally, but much more often than chance, leads to the choice of names such as Patrick and Patricia or Carl and Carla, which gives rise to most of the significance in these tables. This leads one to speculate whether this occurs when only one child is expected and its name has already been chosen and the same name is now used for both twins.

DYADIC RELATIONSHIP OF TWINS

Let us next look at the second question.

Do twins develop a division of labor or is one twin more dominant and the other more submissive?

Much has been written about the special relationship between twins and even more anecdotal material comes to anyone studying twins. We already touched on this when discussing Zazzo's study in which he reported anecdotes about strong emotional attachment between twins. There can be no doubt that close relationships do sometimes occur, but the question is how lasting these influences are or whether these are more frequent in MZ or DZ. A comparison of heritabilities for twins of both types judged to have a close relationship and those that do not would be very informative.

Von Bracken, who contributed to the literature on the twin relationship, provided a good summary of this kind of research in the second part of Volume I of the German *Handbook of Genetics*.[68] (He used the expressions "minister of the interior" and "minister of the exterior" to indicate that one twin may manipulate the other, so that this second one appears to take the initiative more often and therefore may incorrectly be seen as more dominant.) He distinguishes between: (1) shared activities, (2) efforts toward similarity or difference, and (3) role division.

We have already discussed shared activities as part of the question whether the two types of twins experienced similar environmental influences. Let us now look at role division and specifically the dominant, submissive polarization. We have already discussed Zazzo's and Koch's studies which give some support to this idea.

In a large retrospective study of adult twins, Tienari[69] found, for MZ twins, relationships between being reported heavier at birth and: stronger in childhood (p <.001); psychologically more dominating in childhood (almost significant); scoring lower on Aggression in a personality inventory (p <.002); and scoring more normal on the Word Connection test, a modified word association test in which a choice of normal and abnormal responses are provided (p <.10). The heavier twin also tended to have done a little better in school, to have been more of a leader in youth, and to have scored higher on the

TABLE 6-XVIII

Twins First Names	DZ	MZ
Rhyme	84	50
Are joke	1	1
Sound similar	22	11
End differently	12	11
Second name rhymes	17	19
Have same initial	60	49
Are different	68	52
chi-square is 5.022, df 6		

Names Are:	DZ	MZ
Very similar	119	73
Somewhat similar	77	68
Different	68	52
chi-square is 2.75, df 2		

Autonomy scale of a personality inventory. Some of these correlations are shown in Table 6-XIX.

INFLUENCES OF BIRTH WEIGHT DIFFERENCES

In DZ, the fact that differences in birth weight may be correlated with later personality differences is not proof that the personality differences are solely due to effects of the weight difference, and still less that they are not partly hereditary. But, if such a relationship could be demonstrated for MZ twins, where they must be nonhereditary, then they must, at least in part, also be nonhereditary in DZ twins. One way to obtain information on this would be to see if the correlations of birth weight differences with personality variables are different for MZ and DZ twins. If they were to display, with time passing, an increasingly discrepant pattern for the two types, this would be especially informative.

Perhaps the personality differences are due to differences in intelligence.

The idea that there may be a relationship between physical size or height and intelligence is by no means a new one. There are small but consistent differences in height and birth weight between the upper class, the middle class, and the lower class, even in countries where malnutrition is rare, such as Belgium. Figure 6-10 shows the results of a study by Cliquet of 6170 subjects selected from nineteen- and twenty-year-old men examined for military service.[70] A group of young men who had obtained postponement of military service to finish a higher education were also included.

It has also long been known that severe prematurity tends to lead to lower intelligence. Of course, severely premature children also have very low birth weights, but this is perhaps a kind of spurious relationship.

Drillien found an increased frequency of later maladjustment for babies with birth weights under four pounds eight ounces compared with babies heavier at birth.[71] This was further accentuated by stressful conditions. Though Drillien's study was concerned with single birth, premature babies, its findings may have relevance for full-term babies and for twins.

Is there in fact a relationship between the birth weight of full-term children and their later intelligence? Allen and Kallmann found that in MZ twins discordant for retardation, the retarded twin usually had the lower birth weight.[72] In fact, they concluded that it was a causal factor in the defect. The same was not true in discordant DZ pairs. Churchill did a similar study in which only a small number of twins were seriously retarded, and he excluded six pairs with one or both twins who were neurologically abnormal.[73] He found mean IQ's of 88.8 and 87.1 for the heavier and lighter co-twin of thirteen pairs of monochorionic twins, 85.2 and 80.9 for those of nine pairs of MZ twins, 79.6 and

TABLE 6-XIX
CORRELATIONS WITH BEING HEAVIER AT BIRTH

Physical strength	.48
Psychological dominance	.23
Being the spokesman	.08
School achievement	.11
Leadership in youth	.11
Amount of schooling	.18
Socioeconomic status	.24
Disturbance in interview	−.07
Disturbance as measured by tests	−.08
Introversion as measured by tests	.02
IQ	−.05
Reserve in interview	−.07
Aggression score of Personality Inventory	−.26
Ambition score of Personality Inventory	−.11

From P. Tienari, "On Intrapair Differences in Male Twins. With Special Reference to Dominance-Submissiveness," *Acta Psychiatrica Scandinavica Supplement*, 188 (1966).

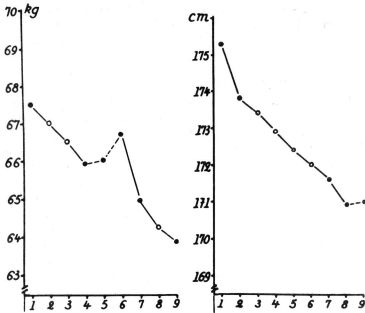

Figure 6-10. Mean weight in kilograms and mean height in centimeters for Belgian recruits from nine socioeconomic classes. (From R.L. Cliquet, "Bijdrage Tot de Kennis van Het Verband Tussen de Sociale Status en Een Aantal Anthropobiologische Kenmerken [Contribution to the Knowledge of the Relation Between Social Status and a Number of Anthropobiological Traits]" in *Verhandelingen van de Koninklijke Vlaamse Academie voor Wetenschappen, Letteren en Schone Kunsten. Klasse der Wetenschappen no. 72* [Brussels, Akademie voor Wetenschappen, 1963].)

75.9 for those of fourteen pairs of like-sexed DZ twins. There also were fourteen boy and girl pairs with mean IQ's of 79.3 and 76.8. Willerman and Churchill reported on another twenty-seven MZ twin pairs in which, again, the twins with the lower birth weights had the lower IQ's.[74] Scarr added another twenty-five MZ pairs with means of 105 and 96.[75] Nevertheless, the correlation is very small, as displayed in Figure 6-11. In this diagram, the birth weight and IQ of the lighter twin is connected by a line to the birth weight and IQ of the heavier one. It can be seen that these lines do not always go up as they would if the relationship were always positive.

It seems quite clear that when one twin is seriously underweight, it is usually retarded or predisposed toward other problems. Kalmar found that in twelve out of twenty-five pairs with large birth weight differences, the smaller twin had neurological problems and/or retardation.[76] Similarly, Pollin found that in 86 percent of the MZ pairs of twins who were discordant for schizophrenia (studied at the National Institutes of Health), the affected twin had the lower birth weight.[77] Mednick, *et al.*

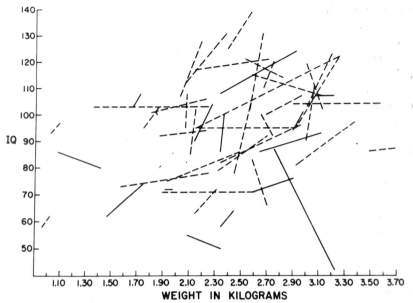

Figure 6-11. The relationship between differences in birth weight and IQ differences in MZ twins.

have also found that those children of schizophrenic mothers (but raised in a different home) who suffered a severe psychiatric breakdown generally had low birth weights.[78]

However, Gottesman and Shields compared the birth weight of MZ pairs discordant for schizophrenia and found that the schizophrenic twin was the heavier one in half the cases.

When the difference in birth weights is minor the effect may be postnatal, as McKeown has argued, because the IQ scores of twins raised alone (because their co-twin died) are higher than those of twins raised together.[79] Record, McKeown, and Edwards came to the same conclusion.[80]

Matheny and Brown reported similar findings for other variables: in twin pairs with small birth weight differences there were few significant differences but for those with large differences in birth weight ten of eighteen behavioral variables also showed differences; these included problems with sleeping, feeding, toilet training.[81]

It seems safe to conclude that differences in dominance are probably not due to birth weight differences unless the difference is a large one, in which case it causes more general differences.

TIENARI'S STUDIES

Let us now return to the direct study of dominance. We mentioned earlier the retrospective Finnish study of adult MZ twins. Although Tienari in that study did not explicitly raise the issue whether the dominance-submission polarization limits the twin method, he made the twin relationship the major object of his study. He used a selected sample of the adult male twins investigated for another study sponsored by the Finnish Foundation for Alcohol Studies in which 902 pairs were interviewed (201 MZ, 612 DZ, and 89 undiagnosed).

In the earlier study, most Finnish-speaking MZ were tested and an equal number of DZ. Of these Tienari examined in more detail eighty MZ pairs who were all living in a specified Southern part of Finland, plus forty-six pairs with mental problems, ulcers, alcoholism or complete abstention from alcohol, plus some pairs separated from birth. The final number of DZ pairs examined was forty-five.

Tienari asked whether dominance is related to being larger and stronger or is it independent of this? He also asked whether the concepts of *Aussenvertretung* (spokesman) and *Innenvertretung* (strength within dyad) represent two different and meaningful dimensions.

Tienari performed a factor analysis of twin differences on twenty-six variables that might reflect the dominance-submissiveness relationship, using only the information on the MZ pairs. Further correlations with eight ability tests, thirteen scores of the personality inventory, and the Word Connection test were excluded because they were generally not significant.

Among the variables included were birth order, birth weight, physical strength, psychological dominance, being the spokesman, leadership, and within-pair comparisons on items such as age of marriage, school achievement, military rank. The factor analysis resulted in five uncorrelated factors which Tienari interpreted as follows: Factor I represents leadership in childhood and youth; Factor II, liveliness and being the spokesman; Factor III, achievement of socioeconomic independence; Factor IV, neurotic withdrawal; and Factor V physical superiority. The complete list of variables and their loadings on each of the five factors is given in Table 6-XX.

It is clear from this analysis that there was no general one-dimensional polarization within the twin pair in these retrospective data. Nor was it merely a matter of two dimensions, with one twin being the more influential within the pair or *Innenminister* while the other was the spokesman or *Aussenminister*. while these two aspects are rather clearly separated in the data as Factors I and II, an additional three factors emerged. Factor V was based on birth weight and physical strength; Factor IV was based on adult mental health; and Factor III represented the earlier establishment of economic and residential independence.

Because this was a retrospective study one might have expected a strong general halo effect. When five independent factors appeared in spite of such an expectation, Tienari's conclusion that there exists a comparatively constant, significant dominance-submissiveness dimension seems too strong. The last sentence in the monograph which follows this conclusion should therefore be underlined. It reads: "Nevertheless, certain later intrapair differences appear to have connections that cannot be accounted for by such a dominance-submissiveness dimension."

In conclusion I believe that the dyadic relationship is probably more important for social psychology than for behavior genetics. It is obvious that, in some pairs, be they MZ or DZ, there occurs a marked polarization, but apparently it is not a consistent nor a lasting phenomenon in many pairs. It seems likely that the degree to which polarization occurs will depend on the particular pair, the particular trait, and the age at which the twins are being studied. Such phenomena belong, in general, in the category genotype-environment interaction, which will not be taken up in this chapter because there is no good evidence so far for this in human

TABLE 6-XX
THE FIVE PRINCIPAL FACTORS OF MZ INTRAPAIR DIFFERENCES ROTATED ACCORDING TO THE VARIMAX CRITERION

1. Birth order	37	13	−19	−02	35
2. Birth weight	14	−04	04	01	53
3. Physical strength	21	−02	02	06	66
4. Psychological dominance	68	39	21	−13	16
5. Being the spokesman	31	76	04	−08	11
6. School achievement	51	17	18	00	07
7. Liveliness	11	81	−10	−01	04
8. Fieriness or temper	−08	47	06	−05	−10
9. Leader as child	76	21	32	−04	09
10. Outside employment	16	02	51	10	−07
11. Left home earlier	03	06	74	03	37
12. Married earlier	09	−03	62	−02	47
13. More education	54	−07	07	−10	09
14. Occupational status	52	−26	02	10	23
15. Higher military status	35	−11	36	−24	17
16. Use of alcohol	09	15	−07	−20	−10
17. Socioeconomic status	48	−20	37	−03	22
18. Disturbance judged in interview	−40	−29	−47	12	06
19. Disturbance from tests	−62	−18	−14	17	−02
20. Neuroticism rated from tests	−11	01	03	62	03
21. Lack of energy rated from tests	−12	05	−27	57	07
22. Introversion rated from tests	−08	−02	07	61	−16
23. Constriction rated from tests	−06	−15	−37	38	08
24. Infantilism rated from tests	−21	05	−44	−08	28
25. Sociopathy rated from tests	−26	21	−05	−52	−05
26. Total IQ	20	−12	39	−26	−12
Proportions of common variance accounted for	.296	.182	.224	.149	.139

P. Tienari, "On Intrapair Differences in Male Twins. With Special Reference to Dominance-submissiveness," *Acta Psychiatrica Scandinavica Supplement,* 188 (1966).

behavioral traits. This is not to say that it is not there, only that it has not been investigated. However, tests for genotype-environment interaction have been proposed by Jinks and Fulker.[31] The reader is referred there for a discussion. Suffice it here to say that the twin sums ($X_{iA} + X_{iB}$) are correlated with the twin differences ($X_{iA} - X_{iB}$).

SOME UNPUBLISHED FINDINGS ABOUT ATTITUDES OF TWINS TOWARDS BEING A TWIN

Because so much has been written about the special relationship between twins, I decided some years ago to administer to twins a special questionnaire designed to inquire about their attitude towards being a twin and towards one another as well as to obtain some information from the mother. No report about this has been published before. At the time the study was done the results were rather disappointing to me, because out of some two hundred questions only a small number gave significant results. When preparing this chapter, it seemed appropriate to summarize my findings. The twins were of high school age.

The following questions were asked of the mother:

1. Are the twins emotionally attached to each other? (a) not at all; (b) some; (c) very much. The MZ were more often reported to be very much attached to one another; chi-square was 7.481, $p < .05$ and the females more often than the males, chi-square was 3.55, $.10 \, p < .20$.
2. What is the sleeping arrangement

for the twins? (a) each has own room; (b) they share one bed; (c) they share a room but have different beds; (d) they share a room with other sib. No differences were found between MZ and DZ, chi-square was 2.82, df 3, n.s. Alternatives (a) and (c) were most common.

3. Have the twins a way of communicating which nobody but the twins understand? (a) yes; (b) no; (c) not known. Chi-square was .557 for the MZ-DZ comparison. Not significant.

4. Do the twins have the same friends? (a) yes; (b) no. Chi-square for the MZ-DZ comparison was 12.00, df 1, $p < .01$. Chi-square for the male-female comparison was .616. Not significant.

5. When you have to scold one of the twins for something, what does the other usually do? (a) nothing; (b) stays out of it, but feels badly; (c) takes parents' side; (d) takes twin's side. The MZ twin took the twin's side more often, chi-square was 8.941, df 3, $p < .025$.

6. Do the twins argue a lot with one another, or do they get along well? (a) they rarely argue; (b) they frequently argue; (c) they usually argue. There was no difference between MZ and DZ ($\chi^2 = 1.437$) or between the sexes ($\chi^2 = .906$).

7. Do the twins try to be different from one another? (a) no; (b) at times; (c) frequently; (d) definitely; (e) they are different without having to try. MZ differed from DZ, chi-square was 27.77. However, when the last alternative was removed, chi-square dropped to a nonsignificant value of 2.21. Upon inspection, there was no consistent trend for the remaining four alternatives.

8. Up to what age were the twins dressed alike? For seven DZ pairs no age was given because after infancy the twins were not dressed alike. When the remaining thirty DZ pairs were compared with the forty-five MZ pairs the following means were found. Male MZ 14.12, female MZ 13.04, male DZ 11.8, female DZ 10.45. The value of t for the MZ-DZ comparison was 4.387, $p < .001$, for the male-female comparison 2.318, $p < .02$.

9. Has one of the twins ever told you that they should not be dressed the same way any longer? (a) yes, (b) no. The "yes" answer was given more often for the MZ. Chi-square was 8.81, $p < .01$.

10. Do the twins prefer to dress alike? (a) yes; (b) no; (c) sometimes. MZ were reported to dress alike more often. Chi-square was 6.08, $p < .05$.

A total index of "twinness" was calculated for each pair, based on the mother's report. Its value ran from two for one pair to thirty for another pair. The mean value for this index was:

male MZ 20.92, female MZ 20.10, male DZ 16.57, female DZ 15.96.

The t-test for the difference between MZ and DZ was 2.698, $p < .01$, between male and female .784, which is not significant. This index correlated moderately with two indices based on reports by the twins themselves, as will be noted later.

These questions were asked of the twins themselves:

1. When you go to a show, do you generally go with your twin? Four graduated answers. MZ went more often together, chi-square was 13.63, df 3, $p < .01$.

2. Who decides what to do in the evening or on weekends? Five answers. No difference was found between

MZ and DZ, but boys more often said that their twin chose or that they discussed it together, girls more often said that they themselves chose.

3. Do you double date with your twin? MZ twins said yes more often; however, many twins reported that they did not date, so chi-square was 2.01, not significant.
4. The twins were also asked how they reacted when their co-twin was being scolded or punished. This time no significant difference was found, although MZ more often took their twin's side. Chi-square was 4.77, df 3.
5. Would you rather be a twin or single? DZ twins chose the first alternative more often, MZ the second, but chi-square was only 1.72, not significant.
6. Would you like to have twins yourself when you have children? No significant difference was found.
7. Whom do you talk to when you are worried? Among the several alternatives listed were the other twin, each parent, a friend, and some combinations. MZ twins chose their co-twin more often, chi-square was 9.29, df 5, so this was not significant, and girls mentioned their co-twin more regardless of zygosity, chi-square was 18.37, $p < .01$.
8. Do you keep any secrets from your twin? (a) yes; (b) no; (c) on some occasions. DZ answered yes more often, chi-square was 2.34, not significant, and girls also answered yes more often, chi-square was 3.51, $.10 < p < .20$.
9. Can you generally tell how your twin feels? There was a nonsignificant difference in favor of MZ twins and girls. Chi-square was 1.34 and 1.97.
10. Does your twin understand you pretty well? Chi-square values—in favor of MZ and girls—were 6.20 and 5.32, df 2, $p < .05$ and $p < .10$.
11. What is the best thing about being a twin? This was an open-ended question, and the reponses were coded into five categories: (a) positive reaction to twin as a person (companionship, someone to go places with, someone to depend on, help in trouble, fight, unspecified fun); (b) positive reaction to social advantages (more friends, get people's attention, help in school, get to go more places); (c) positive reaction to material advantages (more clothes, presents, can share expenses, help in chores); (d) negative (nothing good about it, cannot think of any advantages; (o) don't know or no data. The distributions are shown in Table 6-XXI. Chi-square was 11.82, df 4, $p < .025$, or omitting the last category $\chi^2 = 8.171$, df 3, $p < .025$.
12. What is the worst thing about being a twin? Four categories were used in coding the answers: (a) object to twin as a person (fights, lack of pri-

TABLE 6-XXI
FREQUENCY OF VARIOUS RESPONSES OF INDIVIDUAL TWINS BY ZYGOSITY AND SEX

What is the Best Thing About Being a Twin?	MZ ♂	MZ ♀	DZ ♂	DZ ♀
Positive reaction to twin as a person	29	24	23	32
Positive reaction to social advantage	9	11	0	5
Material advantages	5	3	2	3
Nothing good about it	4	0	0	4
No response	6	3	3	9

vacy, too much dependence, competition); (b) negative reaction to attitudes of others (mistaken identity, people's questions, same opinions expected, object to looking alike); (c) negative reaction to material disadvantages (must share clothes, presents); (o) don't know or no data. The distributions are shown in Table 6-XXII. Chi-square was 15.98, df 4, p < .005, or omitting the last category $\chi^2 = 11.809$, df 3, p < .0025. It is interesting to note that in *both* questions the DZ gives the first kind of answers more often.

Out of forty questions about eating habits and preferences, bed time, getting up in the morning, chores at home and school, ownership of tools for hobbies, interpersonal behaviors and so on, only a few were significant when analyzed for agreement or twin concordance.

One of these questions was, which one of you is: taller?; a better athlete?; more social?; better looking?; the better speaker?; more popular?; more ambitious?; more nervous?; better liked by father?; better like by mother?; more stubborn?; better liked by teachers?; better liked by other students?; in better health?; more helpful around the house?; helping the other with schoolwork more often?; dating more frequently? There was no significant difference between MZ and DZ in agreement on single items; but, for the total question, a higher agreement was scored for MZ than for DZ. Chi-square was 15.07, df 2, p < .001.

The other question was, which one of you does the following: orders from waiter in restaurant or drugstore?; talks to clerk when buying in stores?; arranges dates when meeting new people?; makes plans for Saturday nights and weekends?; does the introductions when meeting new people?; takes care of money?; finds some ways to earn money?; finds new friends?; gives in when the two of you disagree what to do?; gives in when you argue whose opinion is right? Again, there was a higher agreement, for the total question only, by MZ. Chi-square was 17.21, df 2, p < .001.

The twins were also asked whether their mother and their father frequently spoke of them as "the twins." Only for the father did the MZ, more often than the DZ, report that this happened, chi-square was 4.73, df 2, p < .025.

Two further indices of "twinness" were developed, one based on the similarity of answers to questions about hobbies, interests, and other relatively factual items, the other based on items requiring a more subjective answer. The mean values for the two indices were:

male MZ, 22.21 and 34.48;
female MZ, 22.00 and 36.75;
male DZ, 19.14 and 32.07;
female DZ, 19.04 and 33.17.

The differences between the MZ and DZ or between the sexes were not significant.

TABLE 6-XXII
FREQUENCY OF VARIOUS RESPONSES OF INDIVIDUAL TWINS BY ZYGOSITY AND SEX

What is the Worst Thing About Being a Twin?	MZ♂	MZ♀	DZ♂	DZ♀
Object to twin as a person	12	5	13	12
Negative reaction to attitudes of others	23	17	3	12
Material disadvantages	4	8	2	13
Nothing bad about it	9	9	6	9
No response	3	3	4	7

As mentioned earlier, the three indices of how close the twins were to one another were correlated. For the MZ twins the values were r_{12} .27, r_{13} .35, r_{23} .54, (N = 45); the DZ r_{12} .28, r_{13} .17, r_{23} .49, (N = 37); and for the combined group r_{12} .36, r_{13} .29, r_{23} .53.

Winestine speculated that twins who experience themselves less clearly as separate from their co-twin would do poorer on Witkin's Rod and Frame test.[82] For thirty pairs aged nine to thirteen she found a correlation of .50 which is significant at the 1 percent level.

I was unable to replicate these findings. However, my index of "twinness" was based on different items.

The results of my questionnaire study of the relationship between the twins go in the same direction as Zazzo's study, but they are less damaging to the twin study method and less consistent. Really close relations between twins are rare, but when they occur they are almost equally likely between like-sexed DZ and MZ twins.

In future work it would be possible to use an index of the closeness of the twin relationship to obtain a rationally constructed value for Loehlin's k_1 (above).

As an alternative it might be interesting to compare twin concordance values on ability and personality traits for MZ (and DZ) pairs that had a rather close twin relationship and those that did not. Of course, this would require rather large samples. My personal expectation is that there would not be much of a difference.

This completes our discussion of the classical twin method. Let us now see what other research opportunities are provided by twins.

STUDY OF MZ TWINS REARED APART

There is another method available for the demonstration of hereditary factors which is far easier to comprehend and much more dramatic in its simplicity. It is the comparison of identical twins reared apart.

Apart from descriptions of isolated pairs, there have only been four studies of MZ twins who were separated early in life and raised in different homes: by Newman, Freeman and Holzinger;[83] by Shields;[60] by Juel-Nielsen;[84] and by Burt.[85] The results for these four studies are shown in Table 6-XXIII. They can provide us with another independent estimate of the heritability of intelligence, provided that these pairs can be considered to be representative samples and that we can assume that there was no systematic effort to match the foster homes with the expected intelligence of the child or with the socioeconomic condition of the biological mother. This means that we assume no heredity-environment correlation due to the placement.

We will discuss the method primarily in terms of the findings with regard to general intelligence, not so much to contribute

TABLE 6-XXIII
RESULTS FROM FOUR STUDIES OF MZ TWINS RAISED APART

Study	N (Pairs)	Mean IQ	SD	$\bar{d}$*	r
Burt	53	97.7	14.8	5.96	.88
Shields	38	93.0	13.4	6.72	.78
Newman et al.	19	95.7	13.0	8.21	.67
Juel-Nielsen	12	106.8	9.0	6.46	.68
Combined	122	96.8	14.2	6.60	.82

*$\bar{d}$ is the mean absolute difference between twins.

to the discussion of the inheritance of intelligence, but to make discussion of the methodological issues more concrete.

Jensen has recently reanalyzed these four studies and this section relies heavily on his paper.[86]

If no biased placement has occurred, the intraclass correlation between MZ twins reared apart will give a direct estimate of heritability because the correlation would be solely due to the common genetic makeup of the twins. The correlation between the partners in the 122 MZ pairs reared apart is illustrated in Figure 6-12. The value of the correlation is .82.

Unfortunately, there probably is some correspondence between the socioeconomic status (SES) of the foster family and that of the twins' parents or at least that of the mother (for illegitimate children) so that this value may be an overestimate. However, at least for the forty-six pairs studied by Burt, the correlation between the SES of the foster parents of the

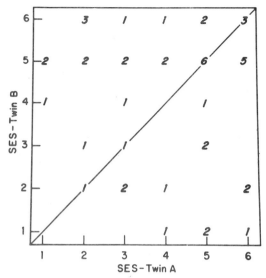

Figure 6-13. Scatterplot of the socioeconomic status (SES) of MZ twins raised apart, showing absence of correlation. (From A.R. Jensen, "IQ's of Identical Twins Reared Apart," *Behavior Genetics*, 1:133 [1970].)

co-twins was .03, as shown in Figure 6-13 taken from Jensen's paper. This means that at least there can be no *high* correlation between the SES of *all* the biological and *all* the foster parents.

Jensen has dealt with another question regarding these twins. Should these twins be regarded as highly abnormal, perhaps mostly derived of the lower half of the IQ distribution, because they probably come mainly from the lower SES groups? Figure 6-14 shows the distribution of the IQ scores for all 244 individuals. The mean is 96.8, somewhat below that in the population, but the standard deviation is 14.2 which is only slightly below that of the general population. Jensen found that the distribution does not depart significantly from normality. Except for the slightly lower mean which has been reported before for verbal or general intelligence tests, the sample is quite representative of the population as a whole.

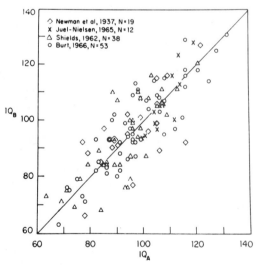

Figure 6-12. Correlation between IQ's of twins from 122 MZ pairs raised apart. (From A.R. Jensen, "IQ's of Identical Twins Reared Apart," *Behavioral Genetics*, 1:133 [1970].)

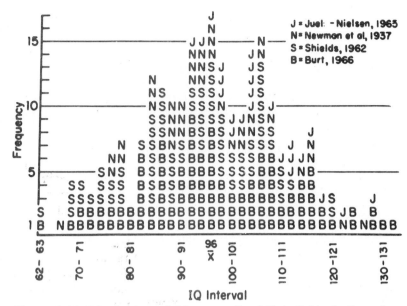

Figure 6-14. Distribution of IQ scores for 240 individuals from four studies of MZ twins raised apart. (From A.R. Jensen, "IQ's of Identical Twins Reared Apart," *Behavior Genetics,* 1:133 [1970].)

For other variables such as personality, the assumption of no correlation between the new environment and the original one is even easier to make. Only if the child were placed with close relatives would one expect any correlation for personality traits between biological and foster parents.

The conceptual simplicity of these studies and their resemblance to a "real" experiment makes it all the harder to accept the fact that at present most social agencies handling adoptions or foster placements are opposed to research by outsiders, since they want to protect the privacy of their records. Systematic collection of data, perhaps in connection with a follow-up of the success of the placement, could furnish invaluable information. It would also be worthwhile to include DZ twins, especially if some information on the biological parents could be obtained.

The next method to be discussed does not require an extensive search for subjects, but it does require an unusual degree of cooperation from the parents of a number of twins. It is the method of co-twin control, in which an experiment is conducted with two perfectly matched sets of subjects: one twin is placed in the experimental group and his co-twin in the control group.*

CO-TWIN CONTROL STUDIES

This method has generally been used for the purpose of studying the relative importance of maturation and learning in a child's development. Of course this is

---

*As the father of a set of twins said to the godmother when she asked why only one twin was being baptized: I am keeping one for a control.

really the same as the nature-nuture or heredity-environment issue.

Probably the first such study was by Gesell and Thompson.[87] One identical twin was given practice in climbing steps every day for six weeks when forty-six to fifty-two weeks old. Her co-twin received no training until she was fifty-three weeks old when she had two weeks practice. At fifty-five weeks of age the second twin did much better even though the first one had practiced three times as much, showing that maturation was more important. The first twin also practiced cube handling and, even though the second received no practice at all, the two performances at age fifty-five weeks were practically identical. Later, Strayer gave language training to the first twin for five weeks when she was eighty-four weeks old, while the second twin received four weeks of training when she was ninety-two weeks old.[88] Although the second twin learned faster, she did not catch up to the first one. Nevertheless, the experiment showed that the later training was more effective due to the additional maturation which had occurred. Next, Hilgard gave earlier practice with memory and motor skill tests to the first twin and later practice to the second twin.[89] Again, the later practice led to quicker improvement but when tested three and six months later both twins did equally well. The effects of training on choice of toys to play with was studied when the same twins were three and a half years old. The effects again proved to be temporary only. When the twins were seventeen years old, Gesell and Thompson tried to determine whether the same pair of twins were less similar than they had been as babies.[90] They concluded that their techniques had not been sensitive enough to answer this question, although it appeared as if small differences had become more noticeable with age.

While these studies were interesting in demonstrating the overriding importance of maturation in those performances which would tend to emerge anyway by themselves, it would be interesting to continue similar separate training much longer, to see whether permanent differences can be produced.

The same limitation applies somewhat to the study of Jimmy and Johnny by McGraw in which Johnny was given practice in crawling, sitting up, walking, hanging from a rod, jumping, roller skating, tricycle riding.[91] At the age of six Johnny still was better coordinated and confident.[92] One wonders to what extent Johnny might have developed more confidence and motor skill anyway. It has been suggested that they were DZ.

The next study used more pairs and made a systematic selection of the twin to be trained. Mirenova used four pairs of MZ and six DZ pairs of four-year-old twins and trained the inferior one in jumping, throwing a ball at a target, and rolling a ball towards a target.[93] After eighteen weeks of practice they did much better than their co-twins. She concluded that maturation played less of a role in complex motor skills and special training is more important.

Luria and Mirenova gave one member of each of three twin pairs, aged seven years, training in constructing some buildings from blocks of wood, in copying drawings, in the Embedded Figures Test, in transposition of movements, and in verbal reasoning problems.[94] The control twins generally did much poorer. When retested a year and a half later, the experimental twins, i.e., the ones who received the training, still did much better than the control twins. This experiment suggests that it may be possible to produce quite long-lasting differences.

For the next few decades there were no

more co-twin control studies. It is first difficult to understand why this kind of research is so rare. Attempting to organize it supplies the answer readily. For this reason it seems worthwhile to indicate two ways in which such research could be fostered: (a) a summer camp for twins and (b) a school for twins. The former was organized on the German island of Nordeiney by Gottschaldt but interrupted by the takeover of the camp by the Hitler Jugend, the National Socialist Youth Organization.[95]

Gottschaldt's camp was apparently organized in such a way that the camp counselors and other adults on the staff helped to collect observations about the personality of each twin under realistic living conditions with the real stresses of competition, teasing, fatigue, fear of the dark, and so on that occur during a stay in a camp with a number of new companions. Perhaps another effort should be made to organize such a camp with the help of adult twins or of some fraternal service organization.

A school for twins has, to my knowledge, never been tried beyond an initial experiment by Naeslund. He separated ten pairs of identical twins during school hours for the first three school semesters in 1953 to 1955.[96] The twins were placed at random in two different classes, in one of which they were taught reading by the synthetic (phonics) or the analytic (whole word) method. The parents were cautioned against discussing the reading methods at home, and the twins promised at the beginning to treat this period of the day as a game of which the secret had to be kept.

The same teacher taught reading by the two different methods in the two classrooms, during which time another teacher conducted classes in different subject matter in order to eliminate the possibility of differences due to the teacher's personality.

At the end of the year and a half the twins were tested for speed and accuracy of reading by a modified tachistoscope, a modified memory drum, and for reading comprehension and interest by oral reading of meaningful text recorded on tape with the performance graded by three teachers.

The zygosity of the twins was determined by blood groups, fingerprints, iris patterns, etc. Their IQ's were tested with the Terman-Merrill test. The results are

TABLE 6-XXIV
INTELLIGENCE QUOTIENTS ON THE TERMAN-MERILL FOR THE TWENTY CHILDREN STUDIED BY NAESLUND

Pair	IQ Synth.	Anal.	Diff. S-A
	Better Half		
5	129	129	0
2	113	118	−5
10	100	93	7
1	99	91	8
7	98	91	7
	Lower Half		
8	93	95	−2
4	92	88	4
3	87	87	0
6	85	83	2
9	83	81	2

From J. Naeslund, *Metodiken Vid den Forsta Lasundervisningen* (Uppsala, Almquist & Wiksell, 1956).

shown in Table 6-XXIV. Unfortunately, the more intelligent twin was more often in the synthetic group.

No advantage was found for the analytic method either in reading speed and accuracy or in reading comprehension, but a nonsignificant difference was found for interest in reading.

When the results were divided into those for the more intelligent half and those for the less intelligent half, it was found that the synthetic or phonics method was superior to the analytic or whole word method in teaching children of lower ability, while for the more intelligent there was no difference between the methods.

This method seems tailor-made for the evaluation of the effectiveness of various Head Start type programs, of "new math," in fact of any kind of innovation of curriculum proposed.

There remains one more research use of twins to consider.

Study of Sex Differences in Male-female Twin Pairs

One potential use of twins which has been almost completely overlooked so far is the comparison of sex differences in unliked-sexed twin pairs with those found in the general population, to see whether having an opposite sex sib of the same age would minimize sex differences.

Lichtenberger, in a touching monograph dedicated to his deceased mother, analyzed the social interactions of a pair of male-female twins in their first year.[97] In spite of their remarkable similarity, clear sex differences appeared from the start.

In Louisville the scores were compared of boys and girls from unlike and like-sexed twin pairs on a preliminary version of the Comrey Personality Questionnaire.[98] The results are shown in Figure 6-15.

It is clear that there were no significant

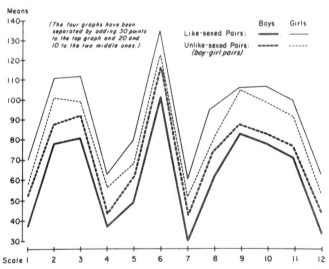

Figure 6-15. Mean values for boys and girls from like-sexed and unlike-sexed twin pairs on twelve personality scales from the Comrey Personality Questionnaire.

differences between the four groups. In fact, the four profiles would have been almost indistinguishable if they had not been artificially separated in the figure by adding constants. This does not only suggest that Comrey has been rather successful in eliminating sex differences but also that, in general, personality questionnaires may not be sensitive enough to pick up what are perhaps rather subtle differences. It would be particularly interesting to perform this type of analysis with spatial visualization tests on which there generally are large sex differences to see if girls from such pairs do better than girls from like-sexed pairs.

Before closing the chapter, it may be worthwhile to take a look into the immediate future.

Future Directions For Research

After having looked back, one is tempted to try to guess about future work and to make some recommendations.

A number of suggestions have already been made in the preceding text, either explicitly or implied in the criticism. Let us summarize these.

If at all possible, zygosity should be diagnosed by blood groups. This will also permit a search for linkage between these genetic markers and the continuous traits. Haseman and Elston have devised a procedure for this, including a computer program.[99] The method outlined by Loehlin for dividing the total variance into between and within-family hereditary and environmental variance should be used. Further work needs to be done testing the assumptions of twin studies and to provide a firmer basis for estimating values for k_1, k_2, and k_3, the indices of the degree to which the within-family environmental influences on MZ and DZ twins differ from one another or from those of single children. In future studies it would be advantageous to collect data on parents and other sibs as well because this permits cross-checking of heritability estimates and calculations of indices of assortative mating and a comparison of father-son and father-daughter correlations, à la Stafford.[100]

Environmental Assessment

There has been much written about environmental factors which influence intellectual functioning; see McCandless or Vernon.[101, 102] We can specify some of the gross factors such as the socioeconomic status of the parents, the number of children in the home and their spacing, and we have some idea about more subtle ones such as the amount of time the mother spent with the child and the extent to which she stimulated it and answered questions. In fact, Wolf was able to develop a set of scales for rating three variables concerned with psychological factors favoring intellectual development,[103] while Dave developed similar scales aiming at educational achievement.[104] They used 60 fifth-grade students selected to represent the full range of socioeconomic status found in the school system. The school system (which is not named) included urban, suburban, and rural areas, but it seems likely that it was (lower) middle class Midwestern, because both Wolf and Dave were students of Bloom at the University of Chicago.

The scales in both studies were based on the same sixty-three-item interview schedule, but there were some differences in the variables being rated in the two studies. The three variables in Wolf's study were: (1) press for achievement motivation, (2) press for language development, and (3) provision for general learning. The reliability for the total scale was .899 and the three parts had the following correlations r_{12} .69, r_{13} .73, and r_{23} .66. The total scale correlated .69 with the Henmon-Nelson group intelligence test, whereas an index of social class for the same children did not correlate with intelligence (r −.024).

Dave rated six variables and obtained correlations between his total index of Educational Environment and various achievement scores ranging from .55 to .77. The correlation with the Total Achievement Score was .80.

Ferguson and Maccoby studied relations between PMA scores and personality characteristics assessed by self-reports and peer ratings in fifth graders.[105] They found high verbal ability to be associated with continued dependency on adults and lowered social interaction with peers; but numerical ability was associated with assertiveness, interpersonal competence, and less dependency on adults. High spatial ability, when coupled with low verbal or number ability, was related to sex-inappropriate behavior: passivity in boys and aggressivity in girls.

There has been somewhat less discussion on the relation between environmental factors and personality, especially of a systematic kind. However, Schaefer has developed a parental attitude research instrument (PARI) which has been shown to be related to some child variables other than intelligence.[106, 107, 108] Heilbrun has also related maternal child rearing to personality as well as cognition in the child.[109] Baumrind and Black related parental attitudes to childrens' behaviors rated by Q-sort and reported different intercorrelations for mothers of girls, fathers of girls, mothers of boys, and fathers of boys, which suggests that data ought to be collected from both parents.[110]

Several years ago I made some brief comments about the possibility of a more general relationship between parental child rearing practices and the personality of children.[111] I was reviewing the evidence on hereditary factors in personality traits as measured by questionnaires administered to twins. Before doing so, I attempted to organize the traits measured by these questionnaires in some systematic manner. I was impressed by the recurrence of three major themes which varied between questionnaires, partially because they were named differently, partly because particular questionnaires appeared to be measuring different combinations of these three qualities. They were suggested by the Dutch psychologist, Heymans, in publications based on empirical studies which appeared between 1906 and 1918. He called his three personality dimensions: activity, emotionality, and primary or secondary function.

By activity, Heymans meant vital energy as shown in the achievement by an individual of work, whether mental or physical; by emotionality, he meant a normal expression of feelings rather than a lack of balance. The distinction between primary and secondary function has to do with the relative predominance of new experiences or of the past. When the primary function is stronger, a person is more influenced by the impressions of the moment; when the secondary function is stronger, the influence of past experiences is greater. Lack of concentration and perseveration might be seen as the two extremes of this continuum, but Heymans did not intend a value judgment in favor of one end or the other of this continuum. The average person would lie in between the two extremes. In this earlier paper this system was compared with the conceptions of Jung, Eysenck, and Vernon.

The three dimensions of parental behavior proposed by Becker,[112] Heilbrun,[109] and Schaefer[106] are firm discipline versus lax, inconsistent discipline; rejection versus acceptance; and encouragement of autonomy versus intrusiveness and overcontrolling.

The relationships which might be expected are as follows: parents that show love and acceptance of their children would tend to have children who would show easier expression of feelings while parental rejection and hostility would lead to the blocking of emotional expression or depression. Parental encouraging of autonomous behavior in the child would lead to development of activity and initiative, while intrusive overcontrolling would lead to dependence, lack of initiative, and the formation of safe, minimal, routine habits. Firm, consistent discipline by the parents would lead to the development of more long-range planning and a dependable character in which the secondary function predominates while lax, inconsistent discipline may lead to opportunistic and unpredictable, undependable behavior in the child.

This is remarkably similar to the findings of Hewitt and Jenkins in a study cited by Eysenck.[113, 114] They studied case histories of five hundred problem children and related the occurrence of three types of abnormalities with parental attitudes. They reported the correlations shown in Table 6-XXV. One wonders whether the same relationships hold in an attenuated form for normal children.

It would be especially valuable to check such concrete predictions from environmental variables to children's characteristics in a study in which hereditary components were being estimated at the same time. Meredith has developed a model for the analysis of such data.[115]

This leads us to a brief consideration of the other direction in which the twin methodology can be profitably expanded; namely, through multivariate methods.

We have already suggested earlier that by a multivariate analysis the amount of information that can be obtained from twin studies in which a number of traits are measured can be considerably extended beyond a comparison of the separate heritabilities of each of the single measures.

Provided that the samples are large enough, one can see whether different hereditary components of ability of personality or of physique were being tapped by the various measures, so that by carefully choosing the measures, one can begin to explore the nature of the underlying genetic factors.

Perhaps the simplest way to start such an exploration is to remove any general factor that may be present and to calculate re-

TABLE 6-XXV
CORRELATION BETWEEN BEHAVIOR OF PROBLEM CHILDREN AND PARENTAL ATTITUDES

	Unsocialized Aggression	Socialized Delinquency	Over Inhibited Behavior	Number of Cases
Parental rejection	.48±.07	.02	−.20	101
Parental negligence	.12	.63±.07	−.17	78
Parental repression	.10	−.12	.52±.06	106
Physical deficiency	−.23	−.31	.46±.06	95

From L.E. Hewitt and R.L. Jenkins, *Fundamental Patterns of Maladjustment. The Dynamics of Their Origin; a Statistical Analysis Based Upon 500 Case Records of Children Examined at the Michigan Child Guidance Institute* (Springfield, Stephen Green Press, 1946).

sidual scores for which then another set of heritability indices is calculated, as done by Saunders and by Nichols.[116, 40]

Next, the variance-covariance matrices can be partitioned into within and between-pairs environmental and genetic components in the same way as was discussed above for the variance of a single variable. Examples of this can be found in Loehlin and Vandenberg,[117] Bock and Vandenberg,[118] and in Partanen, et al.[119] The results from these studies suggest that the variance in those mental abilities that are most often thought of as rather independent, such as verbal ability, number ability, spatial ability, etc., are at least partially controlled by separate genetic factors.

Of course a more definitive demonstration of this would require at least a two-generation segregation analysis.

SOME PRACTICAL INFORMATION

1. There exists a National Organization of Mothers of Twins Clubs and their address in 1970 was: Box 109, Main Post Office, Toledo, Ohio 43601. This national organization has the addresses of local Mothers of Twins Clubs, which meet in most large and many smaller cities. Such clubs can be very helpful to any person who wants to locate younger twins. He may be asked to give a few lectures in exchange. Plank discussed some problems raised by mothers of twins.[120]
2. Some of the local clubs publish pamphlets providing useful hints to new mothers of twins. These pamphlets may be useful to twin studies either for distribution as a partial reward for cooperation or as a source of material for talks to cooperating parents. Some local clubs hold picnics at which prizes are given for the most similar twins. Investigators of twins are sometimes asked to serve as judges in such contests.
3. Two of the pamphlets described above are: M.J.B. Gaul, P.K. Harer, and M.P. Meyer: *For Two, Please!;* and J.G. Baker, J.L. Fanelli, M.P. Meyer, and J.V.J. Moten: *Twins, A Guide to Their Education.* Both can be ordered from: Main Line Mothers of Twins Club, c/o Mrs. Harper Brown, 441 Bair Road, Berwyn, Pennsylvania 19312.
4. The Minneapolis War Memorial Blood Bank has facilities for doing blood group tests on fairly numerous samples sent to them by air mail. They will provide price estimates on request as well as details about the collection of blood samples. Their address is: The Minneapolis War Memorial Blood Bank, 2304 Park Avenue, Minneapolis, Minnesota 55404 (Attention: Dr. Herbert F. Polesky, Director).
5. The World Health Organization has published a list of twin registries and twin studies.[121, 122]
6. There are also some clubs of twins. These are generally somewhat older with a predominance of MZ twins. The address of the national organization is: International Twins Association, P. O. Box 121, Edmund, Oklahoma 73034. The association holds an annual convention on Labor Day weekend in a different city each year. It may be possible to collect some data at this convention.

REFERENCES

1. Lerner, I.M.: *Population Genetics and Animal Improvement.* Cambridge, Cambridge U Pr, 1950.
2. Guttmacher, A.F.: The incidence of multiple births in man and some of the other unipara. *Obstet Gynecol, 2*:22, 1953.
3. Schultz, A.H.: The occurrence and frequency of pathological and teratological conditions and of twinning among non-human primates. In Hofer, H., Schultz, A.H., and Starck, D. (Eds.): *Primatologia, Handbuch der Primatenkunde.* Basel, Karger, 1956, pp. 965-1014, Vol. I.
4. Gedda, L.: *Twins in History and in Science.* Springfield, Thomas, 1961.
5. Galton, F.: The history of twins as a criterion of the relative powers of nature and nurture. *Fraser's Mag, 12*:566, 1975.
6. Mittler, P.: *The Study of Twins.* Baltimore, Penguin, 1971.
7. Nielsen, J.: Inheritance in monozygotic twins. *Lancet, 2*:717, 1967.
8. Falkner, F.T., Banik, N.D.D., and Westland, R.: Intra-uterine blood transfer between uniovular twins. *Biol Neonate, 4*:52, 1962.
9. Giblett, E.R.: *Genetic Markers in Human Blood.* London, Blackwell, 1969.
10. Smith, S.M., and Penrose, L.S.: Monozygotic and dizygotic twin diagnosis. *Ann Hum Genet, 19*:273, 1955.
11. Sutton, H.E., Clark, P.J., and Schull, W.J.: The use of multi-allele genetic characters in the diagnosis of twin zygosity. *Am J Hum Genet, 7*:180, 1955.
12. Bulmer, M.G.: *The Biology of Twinning in Man.* Oxford, Oxford U Pr, 1970.
13. Wilson, R.S.: Blood typing and twin zygosity. *Hum Hered, 20*:30, 1970.
14. Cederlöf, R., Friberg, L., Jonson, E., and Kaaij, L.: Studies on similarity diagnosis in twins, with the aid of mailed questionnaires. *Acta Genet, 11*:338, 1961.
15. Nichols, R.C., and Bilbro, W.C.: The diagnosis of twin zygosity. *Acta Genet, 16*:265, 1966.
16. Jablon, S., Neel, J.V., Gershowitz, H., and Atkinson, G.F.: The NAS-NRC twin panel: Methods of construction of the panel, zygosity diagnosis and proposed uses. *Am J Hum Genet, 19*:133, 1967.
17. Hauge, M., Harvald, B., Fischer, M., Gottlieb-Jensen, K., Juel-Nielsen, N., Racbild, I., Shapiro, R., and Videbech, T.: The Danish twin register. *Acta Genet Med Gemellol (Roma), 17*:315, 1968.
18. Cummins, H., and Midlo, C.: *Fingerprints, Palms and Soles.* New York, Blakiston, 1943 (Reprinted by Dover Publications, 1961).
19. Ritter, H.: Zur Morphologie und Genetik normaler mesodermaler Irisstrukturen. *Z Morph Anthropol, 49*:148, 1958.
20. Shipley, P.W., Wray, J.A., Hechter, H.H., Arellano, M.G., and Borhani, N.O.: Frequency of twinning in California, its relationship to maternal age, parity and race. *Am J Epidemiol, 85*:147, 1967.
21. Myrianthopoulos, N.C.: A survey of twins in the population of a prospective collaborative study. *Acta Genet Med Gemellol (Roma), 19*:15, 1970.
22. McArthur, N.: *Genetics of Twinning, A Critical Summary of the Literature.* Social Science Monographs, 1. Canberra, Australian National University, 1953.
23. Barr, A., and Stevenson, A.C.: Stillbirths and infant mortality in twins. *Ann Hum Genet, 15*:131, 1961.
24. Osborne, R.H., and DeGeorge, F.V.: *Genetic Basis of Morphological Variation.* Cambridge, Harvard U Pr, 1959.
25. Bock, R.D., Waller, M.I., and Vandenberg, S.G.: Effects of blood group discordance in dizygotic twins. *Nature (Lond),* in press.
26. Husen, T.: *Tvillingstudier.* Stockholm, Almqvist & Wiksell, 1953 (An English

translation, *Psychological Twin Research,* appeared in 1959 from the same publisher).
27. Vandenberg, S.G., Stafford, R.E., and Brown, A.M.: The Louisville Twin Study. In Vandenberg, S.G. (Ed.): *Progress in Human Behavior Genetics.* Baltimore, Johns Hopkins, 1968.
28. Cattell, R.B.: The multiple abstract variance analysis. Equations and solutions for nature-nurture research on continuous variables. *Psychol Rev, 67*:353, 1960.
29. Cattell, R.B.: Methodological and conceptual advances in evaluating hereditary and environmental influences and their interaction. In Vandenberg, S.G. (Ed.): *Methods and Goals in Human Behavior Genetics.* New York, Acad Pr, 1965.
30. Loehlin, J.C.: Some methodological problems in Cattell's Multiple Abstract Variance Analysis. *Psychol Rev, 72*:156, 1965.
31. Jinks, J.L., and Fulker, D.W.: Comparison of the biometrical genetical, MAVA, and classical approaches to the analysis of human behavior. *Psychol Bull, 73*:311, 1970.
32. Lush, J.L.: *Animal Breeding Plans.* Ames, Iowa St U Pr, 1945.
33. Holzinger, K.J.: The relative effect of nature and nurture influences on twin differences. *J Educ Psych, 20*:245, 1929.
34. Falconer, D.S.: The inheritance of liability to certain diseases, estimated from the incidence among relatives. *Ann Hum Genet, 29*:51, 1965.
35. Smith, C.: Heritability of liability and concordance in monozygous twins. *Ann Hum Genet, 34*:85, 1970.
36. Allen, G., Harvald, B., and Shields, J.: Measures of twin concordance. *Acta Genet, 17*:475, 1967.
37. Clark, P.J.: The heritability of certain anthropometric characters as ascertained from measurement of twins. *Am J Hum Genet, 8*:49, 1956.
38. Lundstrom, A.: *Toothsize and Occlusion in Twins.* New York, Karger, 1948.
39. Falconer, D.S.: *Introduction to Quantitative Genetics.* Edinburgh, Oliver & Boyd, 1960.
40. Nichols, R.C.: The National Merit Twin Study. In Vandenberg, S.G. (Ed.): *Methods and Goals in Human Behavior Genetics.* New York, Acad Pr, 1965.
41. Jensen, A.R.: Estimation of the limits of heritability of traits by comparison of monozygotic and dizygotic twins. *Proc Nat Acad Sci [USA], 58*:149, 1967.
42. Fisher, R.A.: The correlation between relatives on the supposition of Mendelian inheritance. *Trans R Soc Trop Med Hyg, 52*:399, 1918.
43. Kempthorne, O., and Osborne, R.H.: The interpretation of twin data. *Am J Hum Genet, 13*:320, 1961.
44. Elston, R.C., and Gottesman, I.I.: The analysis of quantitative inheritance simultaneously from twin and family data. *Am J Hum Genet, 20*:512, 1968.
45. Furusho, T.: On the manifestation of genotypes responsible for stature. *Hum Biol, 40*:437, 1968.
46. Jensen, A.R.: How much can we boost IQ and scholastic achievement? *Harv Educ Rev, 39*:1, 1969.
47. DeFries, J.: Quantitative aspects of genetics and environment in the determination of behavior. In Ehrman, L., and Omenn, G. (Eds.): *Genetic Endowment and Environment in the Determination of Behavior.* New York, Acad Pr, 1972.
48. Vandenberg, S.G.: A comparison of heritability estimates of U. S. Negro and White high school students. *Acta Genet Med Gemellol (Roma) 19*:280, 1970.
49. Osborne, R.T., and Miele, F.: Racial differences in environmental influences on numerical ability as determined by heritability estimates. *Percept Mot Skills, 28*:535, 1969.
50. Osborne, R.T., and Gregor, A.J.: Racial differences in heritability estimates for tests of spatial ability. *Percept Mot Skills, 27*:735, 1968.
51. Scarr-Salapatek, S.: Race, social class and

IQ. Population differences in heritability of IQ scores were found for racial and social class groups. *Science, 174*:1285, 1971.
52. Jones, H.E., and Wilson, P.T.: Reputation differences in like sex twins. *J Exp Educ, 1*:86, 1932.
53. Schäfers, F.: Uber die Beliebtheit von Zwillingen. *Arch Ges Psychol, 105*:482, 1940.
54. Zazzo, R.: *Les Jumeaux, Le Couple et La Personne (Twins, the Pair and the Individual).* Paris, Presses Univ France, 1960.
55. Taylor, C.C.: Marriages of twins to twins. *Acta Genet Med Gemellol (Roma), 20*:96, 1971.
56. Koch, H.: *Twins and Twin Relations.* Chicago, U of Chicago Press, 1966.
57. Dennis, W.: Laterality of function of early infancy under controlled developmental conditions. *Child Dev, 6*:242, 1935.
58. Parker, N.: Twin relationships and concordance for neurosis. *Proc 4th World Congr Psychiatry, Madrid, 2*:1112, 1966.
59. Vandenberg, S.G., and Johnson, R.C.: Further evidence on the relation between age of separation and similarity in IQ among pairs of separated identical twins. In Vandenberg, S.G. (Ed.): *Progress in Human Behavior Genetics.* Baltimore, Johns Hopkins, 1968.
60. Shields, J.: *Monozygotic Twins Brought Up Apart and Brought Up Together.* London, Oxford U Pr, 1962.
61. Wilde, G.J.S.: Inheritance of personality traits. *Acta Psychol, 22*:37, 1964.
62. Wilde, G.J.S.: An experimental study of mutual behavior imitation and person perception in MZ and DZ twin pairs; implications for an experimental-psychometric analysis of heritability coefficients. *Acta Genet Med Gemellol (Roma), 19*:273, 1970.
63. Milgram, S.: Nationality and conformity. *Sci Am, 205*:45, 1961.
64. Smith, R.T.: A comparison of socio-environmental factors in monozygotic and dizygotic twins, testing an assumption. In Vandenberg, S.G. (Ed.): *Methods and Goals in Behavior Genetics.* New York, Acad Pr, 1965, pp. 45-61.
65. Scarr, S.: Environmental bias in twin studies. *Eugen Quart, 15*:34, 1968.
66. Freedman, D.G., and Keller, B.: Inheritance of behavior in infants. *Science, 140*:196, 1963.
67. Plank, R.: Names of twins. *Names, 12*:1, 1964.
68. von Bracken, H.: Humangenetische Psychologie. In Becker, P.E. (Ed.): *Humangenetik.* Stuggart, Thieme, 1969, pp. 409-561.
69. Tienari, P.: On intrapair differences in male twins. With special reference to dominance-submissiveness. *Acta Psychiatr-Scand [Suppl], 188,* 1966.
70. Cliquet, R.L.: Bijdrage tot de kennis van het verband tussen de sociale status en een aantal anthropobiologische Kenmerken. (Contribution to the knowledge of the relation between social status and a number of anthropobiological traits.) In *Verhandelingen van de Koninklijke Vlaamse Academie voor Wetenschappen, Letteren en Schone Kunsten. Klasse der Wetenschappen no. 72.* Brussels, Akademie voor Wetenschappen, 1963.
71. Drillien, C.M.: *The Growth and Development of the Prematurely Born Infant.* Baltimore, Williams, 1964.
72. Allen, G., and Kallmann, F.J.: Etiology of mental sub-normality in twins. In Kallmann, F.J. (Ed.): *Expanding Goals in Psychiatry.* Baltimore, Grune 1962, pp. 174-191.
73. Churchill, J.A.: The relationship between intelligence and birth weight in twins. *Neurology, 15*:341, 1965.
74. Willerman, L., and Churchill, J.A.: Intelligence and birth weight in identical twins. *Child Dev, 38*:623, 1967.
75. Scarr, S.: Effects of birth weight on later intelligence. *Soc Biol, 16*:249, 1969.
76. Kalmar, Z.: Angaben zur Entwicklung von mit grosser Gewichtsdifferenz

geborenen Zwillingen (Data on the development of twins with large birth weight differences). *Acta Paediatr-Hung, 6*:411, 1966.
77. Pollin, W., and Stabenau, J.R.: Biological, psychological and historical differences in a series of monozygotic twins discordant for schizophrenia. In Rosenthal, D., and Kety, S.S. (Eds.): *The Transmission of Schizophrenia.* New York, Pergamon, 1968.
78. Mednick, S.A., Mura, E., Schulsinger, F., and Mednick, B.: Perinatal conditions and infant development in children with schizophrenic parents. *Soc Biol, 18*:103, 1971.
79. McKeown, T.: Prenatal and early postnatal influences on measured intelligence. *Br Med J, 3*:63, 1970.
80. Record, R.G., McKeown, T., and Edwards, J.H.: An investigation of the difference in measured intelligence between twins and single births. *Ann Hum Genet, 34*:11, 1970.
81. Matheny, A.P., and Brown, A.M.: The behavior of twins: effects of birth weight and birth sequence. *Child Dev, 42*:251, 1971.
82. Winestine, M.C.: Twinship and psychological differentiations. *J Am Acad Child Psychiatry, 8*:436, 1969.
83. Newman, H.H., Freeman, F.N., and Holzinger, K.J.: *Twins: A Study of Heredity and Environment.* Chicago, U of Chicago Pr, 1937.
84. Juel-Nielsen, N.: Individual and environment: A psychiatric-psychological investigation of monozygous twins reared apart. *Acta Psychiatr Scand [Suppl], 183,* 1965.
85. Burt, C.: The genetic determination of differences in intelligence: A study of monozygotic twins reared together and apart. *Br J Psychol, 57*:137, 1966.
86. Jensen, A.R.: IQ's of identical twins reared apart. *Behav Genet, 1*:133, 1970.
87. Gesell, A., and Thompson, H.: Learning and growth in identical twins: An experimental study by the method of co-twin control. *Genet Psychol Monogr, 6*:1, 1929.
88. Strayer, L.C.: Language and growth: The relative efficacy of early and deferred vocabulary training, studied by the method of co-twin control. *Genet Psychol Monogr, 8*:209, 1930.
89. Hilgard, J.R.: The effect of early and delayed practice on memory and motor performances studied by the method of co-twin control. *Genet Psychol Monogr, 14*:493, 1933.
90. Gesell, A., and Thompson, H.: Twins T and C from infancy to adolescence: A biogenetic study of individual differences by the method of co-twin control. *Genet Psychol Monogr, 24*:3, 1941.
91. McGraw, M.B.: *Growth: A Study of Johnny and Jimmy.* New York, Appleton, 1935.
92. McGraw, M.B.: Later development of children specially trained during infancy: Johnny and Jimmy at school age. *Child Dev, 10*:1, 1939.
93. Mirenova, A.N.: Psychomotor education and the general development of preschool children: Experiments with twin controls. *J Genet Psychol, 46*:433, 1935.
94. Luria, A.R., and Mirenova, A.N.: Experimental development of constructive activity (differential training of identical twins). *Publications M. Gorki Inst Med Genet, 4*:487, 1936.
95. Gottschaldt, K.: Das Problem der Phänogenetik der Persönlichkeit. (The problem of the genetics of personality phenotypes.) In Lersch, P., and Thomae, H. (Eds.): *Persönlichkeitsforschung und Persönlichkeitstheorie. Handbuch der Psychologie IV.* Gottingen, Hogrefe, 1960.
96. Naeslund, J.: *Metodiken Vid den Första Läsundervisningen (Methods for the Elementary Teaching of Reading).* Uppsala, Almqvist & Wiksell, 1956.
97. Lichtenberger, W.: *Mitmenschliches Verhalten eines Zwillingspaares in seinen ersten Lebensjahren (Social Behavior of a Twin Pair in Its First Year of Life).* Munchen,

Reinhardt, 1965.
98. Comrey, A.L.: Comparison of personality and attitude variables. *Educ Psychol Measurement, 26*:853, 1966.
99. Haseman, J.K., and Elston, R.C.: The investigation of linkage between a quantitative trait and a marker locus. *Behav Genet, 2*:3, 1972.
100. Stafford, R.E.: Sex differences in spatial visualization as evidence of sex-linked inheritance. *Percept Mot Skills, 13*:428, 1961.
101. McCandless, B.R.: Relation of environment factors to intellectual functioning. In Stevens, H.A., and Heber, R. (Eds.): *Mental Retardation.* Chicago, U of Chicago Pr, 1964.
102. Vernon, P.E.: *Intelligence and Cultural Environment.* London, Methuen, 1969.
103. Wolf, R.M.: *The Identification and Measurement of Environmental Process Variables Related to Intelligence.* Unpublished doctoral dissertation, U of Chicago, 1964.
104. Dave, R.H.: *The Identification and Measurement of Environmental Process Variables That are Related to Educational Achievement.* Unpublished doctoral dissertation, U of Chicago, 1963.
105. Ferguson, L.R., and Maccoby, E.E.: Interpersonal correlates of differential abilities. *Child Dev, 27*:549, 1956.
106. Schaefer, E.S.: Development of a parental attitude research instrument. *Child Dev, 29*:339, 1958.
107. Schaefer, E.S., and Bayley, N.: Maternal behavior, child behavior and their intercorrelations from infancy through adolescence. *Monogr Soc Res Child Dev, 28*:1963.
108. Bayley, N., and Schaefer, E.S.: Correlations of maternal and child behaviors with the development of mental abilities. *Monogr Soc Res Child Dev, 29*:1964.
109. Heilbrun, A.B., and Orr, H.K.: Maternal child-rearing control history and subsequent cognitive and personality functioning of the offspring. *Psychol Rep, 17*:259, 1965.
110. Baumrind, D., and Black, A.E.: Socialization practices associated with dimensions of competence in preschool boys and girls. *Child Dev, 38*:291, 1967.
111. Vandenberg, S.G.: Hereditary factors in normal personality traits (as measured by inventories). In Wortis, J. (Ed.): *Recent Advances in Biological Psychiatry.* New York, Plenum Pr 1967, vol. IX.
112. Becker, W.C., and Krug, R.S.: A circumplex model for social behavior in children. *Child Dev, 35*:371, 1964.
113. Hewitt, L.E., and Jenkins, R.L.: *Fundamental Patterns of Maladjustment. The Dynamics of Their Origin; A Statistical Analysis Based Upon 500 Case Records of Children Examined at the Michigan Child Guidance Institute.* Springfield, Green, 1946.
114. Eysenck, H.J.: *The Structure of Human Personality,* 3rd ed. London, Methuen, 1970.
115. Meredith, W.: A model for analyzing heritability in the presence of correlated genetic and environmental effects. *Behav Gen, 3*:271, 1973.
116. Saunders, D.R.: Evidence for the relative primitivity of certain traits. *Am Psychol, 18*:411, 1963.
117. Loehlin, J.C., and Vandenberg, S.G.: Genetic and environmental components in the covariation of cognitive abilities: An additive model. In Vandenberg, S.G. (Ed.): *Progress in Human Behavior Genetics.* Baltimore, Johns Hopkins, 1968.
118. Bock, R.D., and Vandenberg, S.G.: Components of heritable variations in mental test scores. In Vandenberg, S.G. (Ed.): *Progress in Human Behavior Genetics.* Baltimore, Johns Hopkins, 1968.
119. Partanen, J., Bruun, K., and Markkanen, T.: *Inheritance of Drinking Behavior, A Study on Intelligence, Personality and Use of Alcohol of Adult Twins.* Helsinki, Finnish Foundation for Alcohol Studies, 1966.

120. Plank, E.N.: Reactions of mothers of twins in a child study group. *Am J Orthopsychiatry, 28*:196, 1958.
121. World Health Organization. The use of twins in epidemiological studies, report of the WHO meeting of investigators on methodology of twin studies. *Acta Genet Med Gemellol (Roma), 15*:109, 1966.
122. World Health Organization. *WHO International Survey of Twin Registers and Studies*. Geneva, WHO, 1966.

APPENDIX

Zazzo's Questionnaires

PART ONE, General Questionnaire. Note that some questions call for an answer for each twin.

Nature of the respondent (parent, teacher or twin) _____

Identification:
1. Full names of the twins: _____
2. Address: _____
3. Birthdate: _____
4. If one of the twins died, indicate the date: _____

Family:
5. Father's birthdate: _____
6. Mother's birthdate: _____
7. Father's occupation: _____
8. Mother's occupation: _____
9. Birthdates of other brothers: _____
   Birthdates of other sisters: _____
10. Are there other twins in the family (specify if possible the degree of relationship):
    On father's side: _____
    On mother's side: _____
11. Which twin resembles the father most physically? _____
12. Which of the two resembles the father most psychologically?
    (in personality or abilities) _____
13. Which twin resembles the mother most physically? _____
14. Which twin resembles the mother most psychologically? _____

Biological Data:
15. a. Was the delivery normal? _____

b. Was the pregnancy of normal duration?
   If not, how long was the duration?
c. Was there a single placenta?

	TWIN A	TWIN B

16. a. What was the birth weight of the twins?
       (identify A and B)
    b. What were their lengths at birth?
17. What are their present weights?
18. What are their present heights?
19. Which twin was born first?
20. Which twin is physically the stronger?
21. Are the twins right- or left-handed?
22. First serious illnesses for each twin, with the approximate dates
23. How similar are the twins in appearance (Are the twins mistaken for one another and up to what age did this occur)
24. Do the twins have the same color of hair?
25. The same eye color?
26. Other physical features that are similar or different?
    a. Same skin color and tone?
    b. Same dentition?

Community of Living:
27. Were the twins weaned at the same time (breast or bottle?)
28. Did they have, as infants, a language that only they understood? (give examples if possible)
29. At what age did they speak (first phrases) in an understandable way?
30. At what age did they walk?
31. Were they ever separated from each other?
    At what ages?
    For how long?
32. Did they suffer much from the separation?
33. Do they sleep in the same bedroom?
    Are they alone in the same bedroom?
34. Do they sleep in the same bed, or did they in the past?

	TWIN A	TWIN B
35. Did they go to the same school?	_____	_____
Up to what age?	_____	_____
36. Were they in the same classroom?	_____	_____
Up to what age?	_____	_____
37. Do they belong to the same groups, the same clubs or societies?	_____	_____
38. Do they or did they have a tendency to stay by themselves, the two of them; to form their own little group of two?	_____	_____
39. Did they always have the same friends or companions?	_____	_____
40. Are they each other's most intimate, confidential friend?	_____	_____
Are they emotionally attached to one another (a little, a normal amount, or a lot)?	_____	_____
41. Which twin takes the initiative more often for the pair or directs the pair (the dominant twin)?	_____	_____
42. Were they always dressed alike? If not, up to what age?	_____	_____
43. Did they wish no longer to be dressed alike? (at what age?)	_____	_____
44. Do they try to be different? In what way?	_____	_____

Personality:

	TWIN A	TWIN B
45. Do they have the same personality?	_____	_____
Which one serves as the model for the other?	_____	_____
Which one bosses the other?	_____	_____
Are one or both twins:		
timid, shy	_____	_____
closed, self-sufficient	_____	_____
relaxed, at ease	_____	_____
brash, self-confident	_____	_____
46. Education level (for children, indicate the actual grade, for adults, indicate the level reached by mentioning the diploma or degree).	_____	_____
47. Which of the twins does or did best in school?		
Report grades if possible.	_____	_____
48. Which school subject matter does each twin do best in?	_____	_____
49. Which school subject matter or course does each twin do poorest in?	_____	_____

	TWIN A	TWIN B
50. Preference for school subject?	_____	_____
51. Which occupation does each twin consider?	_____	_____
52. Are they married or single (give date of marriage)?	_____	_____
53. Did the marriage lead to a difficult separation for the twins?	_____	_____
54. Is the spouse also a twin?	_____	_____
55. Present occupation of each twin.	_____	_____
56. a. Number of children (are there any twins born to the twins?)	_____	_____
b. Would they like to have twins?	_____	_____

Ask each twin, if possible, these questions:
57. What is your earliest memory?
58. How old were you then?

PART TWO: The "Secret Hideout" questionnaire:
  For use with adolescent and adult twins.
  1. Family constellation, living conditions, sharing of room and bed.
A. *Reactions to separation:*
  2. Cause, date and duration of separations.
  3. How did you react to the separation?
  4. How did your twin react?
  5. Did you have the feeling that something was missing, that you were no longer complete?
  6. Did you have an even stronger, more painful feeling of being torn?
  7. Would a separation today be more or less painful now than then?
  7a. Have you ever thought about your twin's dying?
B. *Confusion, closeness and pair relationship:*
  8. Have you ever had the impression that your reflection in the mirror was your twin?
  9. Feeling of ownership with respect to the twin, in the past and now? (toys, books, collections, clothes, etc.)
  10. Secret language of the pair.
  11. Possible confusion of first names and personal pronouns.
  12. Do you occasionally pass yourself off as your twin, for fun or other reasons?
  13. Rivalries and competition between the twins (for affection of parents, school grades, friendships, dates, etc.)
  14. Did you ever have the feeling that your twin was better or inferior than you? In what respect? Was that the opinion of your parents?
  15. Which one of you takes the initiative or makes the decisions?
  16. Which one of you dominates the other, which one leads the pair? What do you think is the reason?
  17. Who in your personal circle of relatives and friends is the one person you are closest to?
  18. Are you annoyed or upset when one mistakes you for your twin?
  19. Was there ever a time that you two ganged up together against the world?

20. In your work, which person is closest to you?
21. In your hobbies and leisure time, which person is closest to you?
22. Do you have personal friends who are not friends of your twin?
23. Do you have different tastes or similar tastes about books, films, fashions, likes, and dislikes about people?
24. Do you have different or similar opinions?
25. Do you mix in different crowds, clubs, or cliques?

C. *The pair and the demands of adult life:*
26. Romantic or sexual awakening, first love, virginity.
27. Do you plan to marry one of these days? or if the twin is married: What do you think of marriage?
28. Does your twin plan to marry?
29. If you had the choice between living with your twin or living apart what would you prefer?
30. How do you see your job future?

D. *Resemblance to parents and identifications:*
31. Who do you resemble more: your father or your mother?
32. Who does your twin resemble more: father or mother?
33. Is your resemblance to that parent mostly physical or in personality?
34. Is the resemblance of your twin to that parent physical or mental?
35. Which person exercises the most authority at home (in your parents' home)?
35a. Do you have a definite signature? Since when?
35b. Do you have a model, a person you would like to copy or follow?

E. *The secret hideout:*
36. Do you have a secret hideout?
   (first explain the idea)
36a. Do you now or have you kept a diary?
37. Do you have secrets that you keep from your twin?
37a. Are there things that you do not talk about or would rather not talk about with your twin?
38. Do you have the impression that your twin has secrets that are kept from you?
39. Is your secret hideout closed to your twin as much as other people?
40. Do you have close friends you confide in other than your twin?
41. Is your twin your closest friend in whom you confide?
41a. Do you think that after marriage the same intimacy can exist with your twin?

F. *Physical intimacy:*
42. Do you feel a reluctance or aversion to use the same things as your twin (glass, fork, toothbrush, clothes, etc.)?
43. Does nudity embarrass you before your twin?
44. Does nudity embarrass you before other people?
45. Does nudity embarrass you even when alone?
46. Involuntary physical contacts with your twins. Erotic situations.

G. *Final questions:*
47. Would you like to have twins yourself? Why?
47a. Are there disadvantages to being a twin? What are they?
48. Would you be embarrassed if your twin learned of your answers?

Chapter 7

BEHAVIORAL INDIVIDUALITY IN CHILDHOOD

Alexander Thomas

IT IS A TRUISM to say that children show individual differences in behavior from the first weeks of life onward. This fact of individuality raises several questions requiring systematic investigation. How can behavioral individuality be categorized? What are the origins of such individual differences? What is the significance of any category of behavioral individuality in childhood for normal and deviant psychological development?

A number of schemes have been suggested for the categorization of behavioral individuality. A useful one divides behavior into ability, which is concerned with the *what* and *how well* of behaving; motivation, which seeks to account for *why* a person does what he is doing; and temperament or behavioral style, which refers to the *how* of behavior. The analysis of behavior into the *why, what* and the *how,* has been suggested especially by Guilford[1] and Cattell,[2] the latter of whom identifies "the three modalities of behavior traits" as: (1) "dynamic traits or interests... (including) basic drives plus acquired interests such as attitudes, sentiments," etc.; (2) "abilities, shown by how well the person makes his way to the accepted goals"; and (3) temperament, "those traits which are unaffected by incentive or complexity... like highstrungness, speed, energy and emotional reactivity, which common observation suggests are largely constitutional."

Until recently, systematic investigations of behavioral individuality had concentrated on the *what* and *how well,* as seen in the measurement of intellectual level and cognitive and perceptual characteristics; and the *why,* as elaborated especially in the psychodynamic formulations of psychoanalysis. By contrast, studies of the *how* were scattered and fragmentary,* perhaps because the concept of temperament appeared related to discredited earlier constitutionalist views that had ascribed heredity and constitution as exclusive causes for complex personality structures and elaborate psychopathological syndromes.

In the early 1950's my co-workers, Stella Chess and Herbert Birch and I became convinced that individual differences in behavior in the infant or older child could not be adequately categorized unless an analysis of temperament was included. Two children may each eat skillfully or throw a ball with accuracy and have the same motives in so doing. Yet, they may differ with respect to the intensity with which they act, the rate at which they move, the mood which they express, the readiness with which they shift to a new

---

*See Thomas, Chess, and Birch[3] for a review of these studies.

activity, and the ease with which they will approach a new toy, situation, or playmate.

Furthermore, it appeared to us that a consideration of temperamental individuality might help to explain certain facts which could not be encompassed by prevailing theories. These facts were: (1) the lack of simple relationship between environmental circumstances and their consequences; (2) individual differences in susceptibility to stresses and pressures; and (3) differential responses to similar patterns of parental care.

In order to obtain detailed information on specific patterns of temperament and their influence on psychological development, we began a longitudinal study which could identify temperamental patterns and follow the dynamics of the interaction of temperament and environment over a significant segment of the developmental course. This study was begun in 1956 and comprises a sample of 136 children who have been followed from the first few months of life onward. Sample characteristics and methods of data collection and analysis were previously detailed by Thomas et al.[3,4]

CATEGORIES OF TEMPERAMENT

Through inductive analysis of the behavioral data, nine categories of temperament were defined. These could be scored for each child in the sample from the age two to three months onward. Descriptive behavioral data obtained by other workers in the United States and abroad could also be rated for the same categories.[5,6] In addition, the same temperamental characteristics could be identified in samples of mentally retarded children and of children with congenital rubella.[7,8]

The nine categories of temperament and their definitions are as follows:

1. *Activity level.* The motor component present in a given child's functioning and the diurnal proportion of active and inactive periods. Data on motility during bathing, eating, playing, dressing, and handling, as well as information concerning the sleep-wake cycle, reaching, crawling, and walking, are used in scoring this category.
2. *Rhythmicity (regularity).* The predictability and/or the unpredictability in time of any function. It can be analyzed in relation to the sleep-wake cycle, hunger, feeding pattern, and elimination schedule.
3. *Approach/withdrawal.* The nature of the response to a new stimulus, be it a new food, new toy, or new person.
4. *Adaptability.* The child's responses to new or altered situations. One is not concerned with the nature of the initial responses, but with the ease with which they are modified in desired directions.
5. *Threshold of responsiveness.* The intensity level of stimulation necessary to evoke a discernible response, irrespective of the specific form that the response may take or the sensory modality affected. The behaviors utilized are those concerning reaction to sensory stimuli, environmental objects, and social contacts.
6. *Intensity of reaction.* The energy level of response, irrespective of its quality or direction.
7. *Quality of mood.* The amount of pleasant, joyful, and friendly behavior, as contrasted with unpleasant, crying, and unfriendly behavior.
8. *Distractibility.* The effectiveness of extraneous environmental stimuli in interfering with or in altering the direction of the ongoing behavior.
9. *Attention span and persistence.* The length of time any particular activity is pursued by the child (attention span) and the continuation of an activity in the face of obstacles to maintenance of the activity direction (persistence).

Because we found a variety of intercategory correlations, we have also found it helpful to characterize the children on the basis of their cluster of temperamental attributes. One common type of temperamental constellation found in the children we have studied comprises regularity, positive approach responses to new stimuli, high adaptability to changes, and a preponderantly positive mood of mild to moderate intensity. We have called youngsters with this cluster *easy* children. At the opposite end of the temperamental spectrum is the child with irregularity in biological functions, predominantly negative (withdrawal) response to new stimuli, nonadaptability or slow adaptability to change, frequent negative mood and predominantly intense reactions. Youngsters with these characteristics have been called *difficult* children.

Another important temperamental constellation comprises the combination of negative responses of mild intensity to new stimuli with slow adaptability after repeated contact. This characteristic sequence of response has suggested *slow to warm up* as an apt if inelegant appellation for such children. We have presented a detailed discussion of these clusters and their specific patterns of interaction with environmental influences elsewhere.[3]

After our initial attempts to use preponderance ratings or percent rank models as the basis for characterizing the children proved inadequate, we adopted a weighted score model. This method takes into account the full distribution of all individual scores on the three-point scale we used and decreases the likelihood of tied scores.

We found that the scores for six of the nine categories were distributed symmetrically across a fairly wide range. In two of the remaining categories without symmetrical distribution, approach/withdrawal and persistence, there is distribution of the scores over twelve or more scale intervals so that substantial differentiation of the group is achieved. Only in the category of distractibility is there an unsatisfactory differentiation of the children, with a marked skewing of the distribution in the distractible end of the scale. Thus, analysis of the frequency distribution of the scores indicates that for the first five years of life, with the possible exception of distractibility, characterization of the children by the weighted score method does permit their differentiation in terms of individuality in temperament.

TEMPERAMENT IN THE NEONATE

Most studies of the neonate have sought to investigate isolated components of behavior or psychophysiological functions.[9] The definition of behavioral categories which could allow for the comparison of the characteristics of the neonate with those of older children has been approached by a few investigators who have also reviewed some of the methodological difficulties involved in such studies.[10, 11, 12]

Our own study of neonatal behavior was taken up under the direction of Dr. Avram Katcher, chief pediatrician of the Hunterdon Medical Center, Flemington, New Jersey. The aims of the study included: (1) the collection of detailed factual behavioral data over substantial periods of time in a group of fifty neonates; (2) the identification of those features of temperament which could be compared to features of temperament in the older infant; (3) the determination of the relationship of temperamental characteristics on the second and the fourth days of life; (4) the determination of the relationship of temperamental characteristics in the neonatal period and at three and six months of life; and (5) the evaluation of the nursery nurse as a source of valid information on the neonate's behavior.

The study population consisted of neonates with birth weights over 2500 grams, gestation between thirty-eight to forty-two weeks, vaginal delivery, and an absence of any significant abnormality on the newborn physical and neurological examination. Any infant with a history of major disturbance in the mother's pregnancy, labor, or delivery was excluded.

For each neonate, 4 two-hour periods of behavior were recorded. These consisted of the 9 to 11 AM and 1 to 3 PM periods on the second and fourth days of life. The observer had been trained in making and recording descriptive, detailed, narrative accounts of behavior. She was instructed to include enough of the environmental circumstances so that the baby's behavior could be related to the context in which it occurred. No special aspects of newborn behavior were isolated for selective attention.

The nursery nurse was asked to pay special attention to the behavior of the specific baby being studied. On the second and fourth days she was interviewed by an independent interviewer and a detailed descriptive account of the infant's behavior on each of these days was obtained from her.

When the infant was three months of age and six months of age, a trained interviewer entered the home and interviewed the mother for the facts of the child's behavior, using the same protocol that has been used in our other studies of temperament.

Temperamental categories for the neonates were derived from an inductive analysis of the behavior protocols. For each child the following categories could be identified: (1) Activity level; (2) Adaptability; (3) Sensory threshold; (4) Response to stimulation other than sensory; (5) Intensity of response; (6) Mood; (7) Distractibility; and (8) Persistence. The data did not permit the establishment of satisfactory criteria for rhythmicity and approach-withdrawal. Scoring criteria were developed, three scorers were trained, and a high level of interscorer reliability was determined. As can be seen, these categories, mainly correspond to, but are not identical with, the nine categories developed for the later infancy period.

For the purpose of this study, sequential scoring was done. The record of infant behavior was broken down into separate events, each of which was individually scored. An event was defined as a procedure of baby care or handling or an interrupted activity. Examples would be carrying an infant to the scales for weighing, placing him on the scales and weighing, and carrying him back to his crib. Each would be a single event for which the original record would be searched for items descriptive of infant behavior. All information contained in the description of the event and the infant's response would be scored within the categories listed above. In the first cases, an average of 130 scorable items was obtained from each one of a nurse's reports and 350 items from the direct observations.

The following are detailed descriptions of the scoring categories.

Activity Level

This dimension describes the extent to which a motor component exists in the child's functioning; data on motility during time in crib, bathing, feeding, diapering-dressing, weighing and all other handling. For this category, these characteristics are formulated as opposing extremes, with a middle value which includes all behaviors pertinent to this category, but not at either extreme. These scores reflect high, moderate, or low activity levels.

Adaptability

This category describes the extent to which a child facilitates or impedes a procedure or circumstance of handling. The behavioral settings or events to be scored are feeding, combing of hair, application of petroleum jelly to the hair, eye washing, bathing, drying, diapering-dressing, weighing, wrapping in a blanket, burping, and any other special procedures that occur. In each case, adaptability would be shown where a child accepts the procedure without crying or resisting or, in case of a feeding, where the child takes the feeding well or with vigor. The absence of a mood component would indicate adaptability. On the other hand, failure to take a feeding well, or the presence of the negative mood component, would indicate nonadaptability. If there is no behavior suggesting adaptability at first, and then such behavior appears soon without intervening change of stimulation, this would also be adaptive.

In the case of feeding description, adaptability is scored on each occasion that the mother offers breast or bottle until the child sucks well and is scored as adaptable. Thus, if an infant is offered the bottle four times but does not suck well, he would be given four scores of nonadaptable. If, on the fifth trial, he sucks well, he would be scored then as adaptable, making a total score of 4 nonadaptable and 1 adaptable. Once an infant has been given one score of adaptable, he is not scored further on that particular procedure. If the infant discussed above had not had a fifth trial, his score for that feeding would have been 4 nonadaptable.

In other procedures, it is the initial reaction to the procedure which is scored, except when a previous ongoing activity continues briefly after the onset of the procedure this is counted. For example, if a crying infant is placed in the scale for weighing and after a moment quiets he would be scored as adaptable.

Sensory Threshold

This category refers to the intensity level of sensory stimulation that is necessary to evoke a discernible response. These characteristics are formulated as opposing extremes only. When there is a description of a single or predominantly single, discrete, brief sensory stimulation and a statement describing subsequent behavioral change or lack of change, a score may be made as Low Threshold for change and High Threshold for no change. All references to stimulation of the lips by a nipple during feeding are considered in the next category: Response to Stimulation Other than Sensory.

Response to Stimulation Other than Sensory

This category deals with behavioral change or lack of it directly subsequent to more complex types of stimulation than occur in the category of Sensory Threshold. An example would be stimulation produced by handling of the infant. This change can be in the direction of activity to nonactivity or nonactivity to activity.

Intensity of Response

In this category, interest is in the intensity of the mood and energy content of the response, irrespective of its direction. There should be a score in this category for each threshold and stimulation response, except where the description of the response is too vague to score. Scoreable items for this category are provided by descriptions of behavior occuring in relation to sensory stimuli, internal states, feeding, diapering-dressing, bathing, weighing, and any other procedures.

Mood

Since we could not define a description of positive mood for the neonate, only negative mood statements can be scored. (It is inaccurate to score "smiling expression" as an indication of a positive mood for children in this age group.) The component is expressed as the number of statements of crying.

Distractibility

This category refers to the effectiveness of extraneous environmental stimuli in interfering with or in altering the direction of the ongoing behavior. For this age group, the only clearly scoreable area would be that of comforting.

Persistence

This category refers to the continuation of a direction of functioning and to the difficulty with which such an established direction of functioning can be altered. These infants are scored as persistent if, and only if, there is a description of continued sucking after a nipple is removed from the baby's mouth. They are scored as nonpersistent if, and only if, there is a description of noncontinuing sucking after a nipple is removed from the baby's mouth.

A number of analyses of these temperamental data in the fifty neonates are currently in progress in an attempt to answer the following questions: What is the correlation between the observer's and nurse's reports? What is the correlation between the ratings of temperament on the second and those on the fourth days of life? What are the correlations between the ratings of temperament in the neonatal period and those at three and at six months of age?

ORIGINS OF TEMPERAMENT

Through the years, two major theories have been advanced to explain individual differences in the behavioral development of children. One, the constitutionalist approach, was most influential from antiquity to the late nineteenth century. This theory explains psychological development on the basis of characteristics already present at birth and usually assumed to be inherited.

Contrasting with this view is the environmentalist approach, wherein individual differences are seen as resulting exclusively from varying environmental and experiential influences. This approach was perhaps first formulated systematically by John Locke in the seventeenth century in his *"tabula rasa"* concept. In recent years, this extreme environmentalist view has been advocated especially by the behaviorist school of psychology. A similar approach has been taken by most American psychiatrists, who have emphasized postnatal effects, and particularly intrafamilial influences, as the major or even exclusive source of individual differences in behavioral development. Classical psychoanalysis, in spite of its concepts of inborn instinctive drives and Freud's occasional references to constitutional differences in libidinal energy levels, focused its attention on early intrafamilial relationships. This led to the basic psychoanalytic concepts of the Oedipus complex, regression to oral and anal levels of development, penis envy,* sibling rivalry, etc. This environmentalist emphasis of psychoanalysis, together with the similar focus of be-

---

*Though Freud asserted that "anatomy is destiny" in referring to the psychology of women, his postulate of penis envy, the basic determinant of feminine characteristics, was presumed to arise from environmental influences, namely the girl's comparison of herself to her brother and father.

haviorism, solidified the dominance of this approach in American psychology and psychiatry in the first half of the present century. The concern of some works, beginning in the 1930's, for the importance of social and cultural factors in influencing individuality in development, served to broaden the scope of this environmentalist view, but did not change it.

However, in recent years there has been increasing dissatisfaction with this exclusive attention to environment. Experience has shown that many children growing up within families whose child-care patterns are considered etiological for diverse conditions such as schizophrenia, homosexuality, and delinquency do not end up with the expected pathology. For example, Kudushin, in his follow-up study of children who became available for adoptions at age five years or older as a result of court action terminating parental rights because of neglect and/or abuse of the children, found that approximately 80 percent of the adoptions were assessed as successful despite the children's early harsh experiences.[13] He suggests that biological individuation might explain some of the resiliency shown by these youngsters in the face of environmental adversity.

As a result of findings such as these and others, there is now a gradual return to a study of genetic and constitutional factors which may influence psychological development. This is not, it should be noted immediately, a resurgence of the old nature-nurture controversy. Rather, the tenets of modern biology, in which physical growth and development are considered to be a product of the *interaction* between constitutional factors and environment, are being incorporated into the behavioral sciences. This concept is summarized as "genotype in interaction with environment yields phenotype."[14]

In general, it would appear that there are basically four kinds of influences to consider in determining the source of individualized behavior patterns, although none of these alone is sufficient by itself. In the broadest terms, these are: (1) hereditary or genetic factors, (2) prenatal or intrauterine factors, (3) paranatal factors, and (4) postnatal or environmental factors. Investigators have been proceeding in many directions in an attempt to study the roles of these factors as they combine with one another in influencing both normative development and the etiology of specific psychiatric syndromes. Examples are the identification of family clusters of schizophrenia (hereditary studies), of increased incidence of retardation and autism in children with congenital rubella (prenatal studies), and of specific dysfunctions following specific birth trauma (paranatal studies). In all of these investigations it has been apparent that the final behavioral outcome cannot be predicted on the basis of the genetic or physiological factors alone. These have developmental meaning only insofar as they interact with elements in the child's environment, in particular, as they both elicit parental handling techniques and define and limit the child's response to this handling.

The above considerations have guided our judgments as to the origin of temperament. Our reports have stated that the concept of temperament, *ipso facto*, contains no inferences as to genetic, endocrine, somatologic, or environmental etiologies. We have further emphasized that, whatever its origin, temperament is not a fixed immutable characteristic but can show differences in its expression and modification of its quality in different life situations or age-stage levels of development.

While our data are not sufficient to permit any definite conclusions to be drawn as to the etiology of temperament, some of the findings do bear on the possibility of constitutional-hereditary origin. The neonatal studies summarized above indicate that a number of characteristics of behavioral individuality which appear similar to temperamental attributes in the older infant can be identified and rated in the newborn infant. These neonatal characteristics are clearly not the result of postnatal environmental influences. If these attributes can be shown to correlate highly with temperament in the three-to six-month-old infant, this would be highly suggestive evidence for a constitutional origin. On the other hand, if a lack of consistency between temperament in the neonate and at six months were to be found, this would not rule out such a constitutional origin. It might rather indicate that temperament is not immutable and fixed but that its expression is rather, at all age periods, the result of a constantly evolving process of interaction with environmental influences.

TWIN STUDIES

Our longitudinal study sample included three pairs of monozygotic twins, five pairs of dizygotic twins, and a number of families in which two or more sibs were enrolled in the study. The data from these children were examined for any evidence of a possible genetic basis for temperament, keeping in mind that the methodological problems in twin studies and the small size of our sample make any inferences at best very tentative. The findings of this twin study, which included comparisons with twenty-six pairs of sibs, are reported in detail by Rutter, Korn, and Birch and will be summarized briefly here.[15]

Our twin study postulated that if a temperamental characteristic had a significant genetic basis: (1) monozygotic pairs would be more alike in infancy than dizygotic pairs, (2) dizygotic pairs would be no more alike than sibs, and (3) no large differences in temperamental characteristics should be found within monozygotic pairs. These postulates were tested, utilizing a series of mean rank difference comparisons for all nine temperamental characteristics during years 1, 2, and 3, except persistence and distractibility. The number of scores in these latter two categories were too small in some cases to permit the same systematic analysis as with the other seven categories. The results of these analyses are summarized in Table 7-I, which suggests that for no temperamental category do all methods agree in indicating a preponderantly genetic basis. The strongest evidence for a genetic component was present for activity level, approach/withdrawal, and adaptability. Intensity, threshold, and mood appeared to have a genetic basis in

TABLE 7-I
CATEGORIES IN WHICH EACH GENETIC CONSEQUENCE IS FULFILLED

Consequence 1 (monozygotic pairs more alike than dizygotic)	*Consequence 2* (dizygotic pairs no more alike than sibs)	*Consequence 3* (no large difference within monozygotic pairs)
Approach/withdrawal	Approach/withdrawal	Intensity
Activity level	Activity level*	Threshold
Adaptability	Adaptability*	
	Mood*	

*Asterisk indicates consequence fulfilled only in two out of three of the age periods.

only one of the three analyses, and by no method of analysis did rhythmicity appear to be genetically determined.

The behaviors from which our scores were derived were those of the children in their natural settings and were molar in character. Therefore, findings that rhythmicity, for example, appeared to be largely determined by nongenetic factors do not prove that genetic factors related to rhythmicity do not exist. Another possibility is that in the molar behavior of this group of children, such possible genetic influences were obscured by the greater effect of environmental variables.

An opportunity to examine the possible influence of nongenetic factors in the twin pairs was afforded by the existence of special circumstances in the life histories of four of the eight twin pairs. Two of these pairs (A and B) were monozygotic and were treated as such by their parents. The mother of pair B, however, although treating the twins very similarly, was concerned with whether or not it would be better to emphasize their differences by differential treatment. One pair (C) was monozygotic but the twins were separated at birth and brought up in different families fully isolated from one another. Although each family knew of the existence of a twin, neither had any knowledge of the placement. The last pair (D) was indubitably dizygotic, but for the first three years the twins had been regarded as identical by their parents. They had been dressed alike and the treatment of each by the mother was so similar that if one became wet both were changed.

If parental handling influenced the patterns of behavior studied, then one would except that the monozygotic twins reared apart would be closely alike initially but would diverge increasingly as they grew older. The fraternal pair reared as identical would be expected to start less alike but should converge. The monozygotic pairs "A" and "B" would be expected to start alike and to remain alike.

Table 7-II summarizes the results of the analysis of the above inferences. As may be seen from the table, in the first year the rank differences between members of each of the three pairs of monozygotic twins are all less than the mean rank difference found between the dizygotic twins reared as identical. By the end of the second year the dizygotic twins reared as identical have markedly reduced their mean rank difference, whereas the mean rank difference on the monozygotic individuals separated at birth and reared apart has increased to a level equal to that found in the dizygotic pair at the end of the first year. The monozygotic twins reared in one household started out with low mean rank differences and continued to have only small differences throughout the first three years. Unfortunately, this set of comparisons must remain semi-quantitative be-

TABLE 7-II
CONTRAST ANALYSIS OF TWIN PAIRS' MEAN RANK DIFFERENCES WITHIN PAIRS

		Twin Pair		
Year	A	B	C	D
1	9.4	11.6	10.6	16.1
2	8.9	11.2	16.4	5.5
3	1.9	8.0	–	9.1

Key: A = Monozygotic twins reared in same household.
B = Monozygotic twins reared in same household.
C = Monozygotic twins reared apart.
D = Dizygotic twins reared as identical.

cause the small number of cases precludes more detailed statistical analysis.

A qualitative analysis of the behavioral characteristics of the one pair of monozygotic twins separated at birth and reared by different adoptive parents has been done for their early years of life. Some striking similarities have been found, as well as differences which appear related to marked differences in child-care practices and attitudes of the adoptive parents.

Each of the identical twins adopted at birth into separate households was highly irregular in her functioning from birth onward in the area of sleep, with variable sleepiness at bed- and naptime and frequent night awakening. Both twins also began to scream loudly with their night awakening during the second year of life. Denials of demands in the second year also produced intense reactions in both children. In other respects, too, their temperamental characteristics and behavioral development were very similar. Both were moderately active and adaptable, distractible, and persistent.

With regard to parental handling, the twins' environments differed markedly. The mother of twin A tended to be inconsistent and to discipline her by shouting. The father also tended to be inconsistent in his behavior. At times he teased the child by pretending to be angry and at other times gave her special privileges in contradiction of the mother's rules. In contrast, the parents of twin B were calm and warm individuals whose basic attitudes were consonant. Both were consistent in distinguishing legitimate from improper demands of the child.

The consequences of this differential handling were exemplified by statements from the two sets of parents. The parents of twin A said that, at age twenty-eight months, "She rules the household because she screams so we have to give her everything she wants," while the parents of twin B said that, at age twenty-four months, "When we tell her "no" she puts up a fight, but if we are firm and definite that is the end of her fight." This difference was also apparent in the parents' responses to the twins' night awakening and screaming. The parents of twin A commented that they could not bear to hear her scream and gave her what she wanted. The parents of twin B, on the other hand, ignored the screaming. Twin A's screaming at night continued throughout her second year of life while twin B's eventually disappeared.

Twin A came to clinical attention because of the continuing sleep problem which reached its height at age two years when she awakened several times nightly and screamed until given a bottle of milk. At the same period she had tantrums when not given her desires. Psychiatric examination of twin A at twenty-eight months revealed no disturbance in her functioning aside from marked clinging to her mother and a slow and gradual involvement in the play session. Her free play was functional but lacked organization. The clinical diagnosis was mild reactive behavior disorder, and it was felt that her behavior could be altered by parental handling that combined clear and consistent demands with refusal to appease her tantrums.

The parents at first carried through on the advice in regard to the sleep problem, and this was temporarily allayed. But this modification in parental functioning was not sustained and, in follow-up a year later, it was found that the sleep problem had recurred. Furthermore, the parents continued to handle her as a demanding child who must be humored in all areas.

In summary, the data from the neonatal and twin studies suggest that temperament may at least in part have a constitutional basis and that some of the categories may be influenced genetically. In any case, environmental factors are decisive in shaping the expression of temperament and its influence on psychological development.

SIGNIFICANCE OF TEMPERAMENT

A number of children in the longitudinal study developed manifestations of behavioral deviations at various ages. In each case psychiatric evaluation was done by the staff child psychiatrist. Whenever necessary, neurological examination or special testing, such as perceptual tests, were done. In forty-two of the children, the behavioral disturbance was considered sufficient to warrant the clinical judgment that a behavior disorder existed.

In each of these forty-two children with behavior problems the psychiatric assessment was followed by a detailed culling of all the anterospective data from early infancy onward for pertinent information on temperament, environmental influences, and the sequences of symptom appearance and development. It has been possible, in each case, to trace the ontogenesis of the behavioral disturbances in terms of the interaction of temperament and environment. Temperament alone did not produce behavioral disturbance. Instances of children of closely similar temperamental structure to the children with behavior problems were found in the normally functioning group. Rather, it appeared that both behavioral disturbance as well as behavioral normality were the result of the interaction between the child with a given patterning of temperament and significant features of his developmental environment. Among these environmental features, intrafamilial as well as extrafamilial circumstances, such as school and peer group, were influential. In several cases, additional special factors such as brain damage or physical abnormality were also operative in interaction with temperament and environment to produce symptoms of disturbed development.

The above analyses have previously been reported in detail by Thomas, et al.[3] The findings challenge the validity of the currently prevalent assumption that a child's problem is a direct reaction of a one-to-one kind to unhealthy maternal influences. The slogan, "To meet Johnny's mother is to understand his problem," expresses an all-too-frequent approach in which a study of the mother is substituted for a study of the complex factors which may have produced a child's disturbed development, of which parental influences are only one. Elsewhere, we have described this unidirectional preoccupation of psychologists and psychiatrists with the pathogenic role of the mother as the "*Mal de Mère*" syndrome.[16] The harm done by this preoccupation has been enormous. Innumerable mothers have been unjustly burdened with deep feelings of guilt and inadequacy as a result of being incorrectly held exclusively or even primarily responsible for their children's problems. Diagnostic procedures have tended to be restricted to a study of the mother's assumed noxious attitudes and practices with investigations in other directions conducted in a most cursory fashion, or not at all. Treatment plans have focused on methods of changing maternal attitudes and ameliorating the effects of presumed pathogenic maternal attitudes on the child and have ignored other significant etiological factors.

Our data on the origin and development of behavior problems in children emphasize the necessity to study the child: his temperamental characteristics, neurological status, intellectual capacities, and physical handicaps. The parents should also be studied, rather than given global labels such as rejecting, overprotective, anxious. Parental attitudes and practices are usually selective and not global, with differentiated characteristics in different areas of the child's life and with marked variability from child to child. Parent-child interaction should be analyzed, not only for parental influences on the child but, just as much, for the influence of the child's individual characteristics on the parent. The influence of other intrafamilial and extrafamilial environmental factors should be estimated in relation to the interactive pattern of each specific child with his individual characteristics rather than in terms of sweeping generalizations.

Our longitudinal data have also permitted the examination of the relationship of temperament to school functioning and academic achievement. Here again, as with parental child-care practices, we have found that there is no one single approach that is optimal for all children. Different temperamental characteristics can influence, sometimes substantially, the response to specific learning techniques and classroom schedules.[17,18]

One of our findings was that several children with the slow-to-warm-up pattern were misjudged by teachers to be of lower intellectual level than they actually were. This finding suggested that a tendency might exist for some teachers to confuse the temperamental issue of the quality of participation in new situations with the issue of intellectual level. This hypothesis was explored systematically through a cooperative study of another cohort of children in a suburban school system with Dr. Edward Gordon.[19] Ninety-three kindergarten children were rated by their teachers as to their pattern of participation in new situations and activities. The teachers also made a judgment of each child's intelligence. Correlations between the two ratings were determined, as well as between each rating and the child's IQ score obtained the following year in first grade. The findings have been reported in detail elsewhere and can be summarized briefly here.[19] The child who plunged into new situations and activities positively and unhesitantly was likely to be judged as being more intelligent by his teacher than the slow-to-warm-up child, even when the two kinds of children were of the same measured intelligence. There was also a tendency for the teachers to overestimate the intellectual level of the child who involved himself quickly in new activities and to underestimate the level of the child who became involved slowly and gradually.

Inasmuch as a teacher's estimate of a child's intelligence can influence her expectancies and her demands and the child's responses to these expectancies,[20] a bias introduced by the misinterpretation of temperamental characteristics can have significant consequences for a child's academic achievement and school functioning.

REFERENCES

1. Guilford, J.P.: *Personality*. New York, McGraw, 1959.
2. Cattell, R.B.: *Personality: A Systematic and Factual Study*. New York, McGraw, 1950.
3. Thomas, A., Chess, S., and Birch, H.: *Temperament and Behavior Disorders in Children*. New York, NYU Pr, 1968.
4. Thomas, A., Birch, H., Chess, S., Hertzig, M.E., and Korn, S.: *Behavioral Individu-*

ality in Early Childhood. New York, NYU PR, 1963.
5. Carey, W.B.: A simplified method for measuring infant temperament. *J Pediatr, 77*:188, 1970.
6. Marcus, J., Thomas, A., and Chess, S.: Behavioral individuality in Kibbutz children. *Isr. Ann Psychiatry, 7*:43, 1969.
7. Chess, S., and Korn, S.: Temperament and behavior disorders in mentally retarded children. *Arch Gen Psychiatry, 23*:122, 1970.
8. Chess, S., Korn, S.J., and Fernandez, P.B.: *Psychiatric Disorders of Children with Congenital Rubella.* New York, Brunner/Mazel, 1971.
9. Brown, J.L.: States in newborn infants. *Merrill Palmer Behav Dev, 10*:313, 1964.
10. Ottinger, D.R., and Simmons, J.E.: Behavior of human neonates and prenatal maternal anxiety. *Psychol Rep, 14*:391, 1964.
11. Weller, G.M., and Bell, R.Q.: Basal skin conductance and neonatal state. *Child Dev, 36*:647, 1965.
12. Yarrow, L.J., and Goodwin, M.S.: Some conceptual issues in the study of mother infant interaction. *Am J. Orthopsychiatry, 35*:473, 1965.
13. Kadushin, A.: Reversibility of trauma: a follow-up study of children adopted when older. *Soc Work, 12*:22, 1968.
14. Freedman, D.G.: The impact of behavior genetics and etiology. In Rie, H.E. (Ed.): *Perspectives in Child Psychopathology.* Chicago, Aldine, 1971.
15. Rutter, M., Korn, S., and Birch, H.: Genetic and environmental factors in the development of primary reaction patterns. *Br J Clin Soc Psychol, 2*:161, 1963.
16. Chess, S.: Editorial: *Mal de Mère. Am J Orthopsychiatry, 34*:613, 1964.
17. Thomas, A.: Significance of temperamental individuality for school functioning. In Hellmuth, J. (Ed.): *Learning Disorders.* Seattle, Special Child Publications, 1968.
18. Chess, S.: Temperament and learning ability of school children. *Am J Public Health, 58*:2231, 1968.
19. Gordon, E.M., and Thomas, A.: Children's behavioral style and the teacher's appraisal of their intelligence. *J Sch Psychol, 5*:292, 1967.
20. Rosenthal, R., and Jacobson, L.: Teachers' expectancies: determinants of pupils' IQ gains. *Psychol Rep, 19*:115, 1966.

Chapter 8

GENETIC FACTORS IN INTELLIGENCE

THOMAS J. BOUCHARD, JR.

THE GENETICS OF INTELLIGENCE has been so thoroughly discussed in recent publications[1,2,3,4] that the orientation of this chapter will not be to cover the literature comprehensively or exhaustively. Rather, the focus will be restricted to outlining a widely accepted genetic model and discussing selected studies which demonstrate the data for which the model accounts. Complexities will be introduced into the model as the data demand them. Within this framework an attempt will be made to clarify most of the important concepts relevant to the nature-nurture issue and develop an understanding of the important methodological problems that characterize research in this area.

The range of the chapter will be further restricted by omission of any extensive discussion of twin studies and methodology or mental retardation discussed elsewhere in this book. The issue of heritability will also only be discussed in a limited way.

THE NATURE OF HUMAN INTELLIGENCE

Many of the fundamental distinctions necessary for a clear understanding of the concept of intelligence were made over two thousand years ago. Only relatively recently, however, have these distinctions been operationalized and the concept rendered measurable in a way that will allow the construction of theories that adequately deal with the phenomena.

Plato contrasted "nature" and "nurture," and distinguished roughly between the domains of cognition, affection, and conation. Aristotle introduced the idea of an ability or faculty when he distinguished between an actual or concrete activity and the hypothetical capacity underlying it. Cicero introduced the term *intelligentia* when translating a related Greek word.[5] *Intelligentia* derives from the Latin *intelligere* which means "to choose from among . . ., hence to understand, to know, to perceive."[6] Nevertheless, the word intelligence as used by psychologists is recent. As Wechsler points out, it did not rate a separate entry in 1901 in Baldwin's encyclopedic *Dictionary of Philosophy and Psychology*.[7] It was listed as an alternative to intellect. This neglect reflected psychology's resistance to the concept of individual differences, a resistance faced by Galton and still faced by psychologists and geneticists today.

The position taken in this chapter is that intelligence can be usefully construed as a single, relatively coherent trait best summarized as "g." The evidence for such a

---

Preparation for this chapter was supported in part by P.H.S. Grant No. 1RO1 HD05600-01.

claim comes from a variety of sources and will not be discussed here.[5,8,9,10] Contemporary theorists, however, attribute intelligence to as few as one and as many as 120 factors. Cattell and Horn argue for two main intellectual factors,[11,12] Vandenberg for six,[4] and Guilford for 120 plus.[13]

If we concern ourselves with the data generated by "successful" tests of intelligence, we bypass a great deal of unproductive controversy and quickly realize there is major agreement among researchers as to many of the facts about intelligence. The development of useful tests of intelligence was a great breakthrough achieved primarily by avoiding problems of definition and pursuing the problem of measurement.[14,15] Binet's successful resolution of the problem in 1905 to 1908, the Metrical Scale of Intelligence, was the culmination of a longtime French interest in the practical problems of education, psychopathology, and mental deficiency. The Binet scale and its successors predict educability in its broadest sense, effectiveness in dealing with the total culture.[16] Binet recognized that the word "intelligence" was vague and comprehensive and that all mental functioning could be forced under its aegis. He also realized that even the normal functioning of all the organs of sense were not absolutely necessary for the manifestation of intelligence, an insight dramatically validated by Furth's work with deaf children.[17,18] The touchstone of intelligence for Binet was judgment.

It seems to us that in intelligence there is a fundamental faculty, the alteration or the lack of which is of the utmost importance for practical life. This faculty is judgment, otherwise called good sense, practical sense, initiative, the faculty of adapting one's self to circumstances. To judge well, to comprehend well, to reason well, these are the essential activities of intelligence.[19]

These are also the skills one needs to succeed in the world of work and life in general, but they are even more crucially necessary to succeed in school. Fortunately or unfortunately, schools are the mechanisms which postindustrial revolution societies use to shape their talent pool in order to keep their social systems functioning. Whether our conception of intelligence would differ radically from what it is today if we had a different social structure, educational system, etc., is a moot point. Modern industrial society, with its enormous number of vocational roles, requires tools for differentiating men and women and training and fitting them to their work. Intelligence tests do this better than any other kind of instrument.[20] Simpler societies with their smaller number of roles and less complex demands surely require less differentiation and less emphasis on training and selection. The intelligence test is a *sine qua non* of modern industrial societies. Thus in such societies, intelligence is an important social parameter, and its nature and nurture are of great social concern. As we will show in this chapter, intelligence is also a biological parameter. In our opinion, therefore, it can only be thoroughly understood when both its biological and social determinants are taken jointly into account. A treatment of only genetic factors or only social and environmental factors would be inadequate and misleading. We will attempt to deal with both but, due to space limitations and the nuture of this book, the emphasis is on the genetic control of intelligence. We will also discuss in some detail the influence of genetic factors on an important aspect of the environment, namely social class.

A GENETIC MODEL FOR THE INHERITANCE OF INTELLIGENCE

There is considerable consensus among geneticists that normal intelligence is under polygenic or multifactorial control.[21, 22, 23, 24, 25, 26, 27, 28, 29, 30, 31] The polygenic model has been devised to account for variation in continuous (quantitative), as opposed to discontinuous (qualitative), characteristics. The model assumes that there are a number of genes involved whose effects are small, similar, and cumulative.

Consider a simple system with three pairs of genes (Aa), (B,b), and (C,c). Each capital letter is assigned a value of 1 and each lower case letter is assigned a value of 0. The array of genetic classes is given by the coefficients of the binomial $(p+q)^{2n}$ where n is the number of pairs of genes acting additively, p = .5 and q = .5 (each gene in a pair has an equal probability of being 0 or 1). In our example the phenotypic array is:

$$p^6: 6p^5q^1: 15p^4q^2: 20p^3q^3: 15p^2q^4: 6p^1q^5: q^6$$

Thus, there are seven classes or phenotypes and twenty-seven genotypes:

				AaBBCc			
		AABBcc	aaBbCC	aabbCC			
		AAbbCC	aaBbCc	aaBBcc			
Genotype			aaBBCC	AAbbCc	AAbbcc		
		AaBBCC	aAbBCC	AabbCC	AaBbcc	Aabbcc	
		AABbCC	aABBCc	AABbcc	Aabbcc	aaBbcc	
	AABBCC	AABBCc	AABbCc	AaBBcc	aaBbCc	aabbCc	aabbcc

| Phenotypic value | 6 | 5 | 4 | 3 | 2 | 1 | 0 |
| Frequency | 1 | 3 | 6 | 7 | 6 | 3 | 1 |

Ten pairs of genes very closely approximate the normal distribution of IQ.

The model has some important consequences. All classes except the extremes, which represent homozygous genotypes, contain sufficient genetic diversity within themselves to regenerate the entire array of genotypes. Thus on genetic grounds alone, we would expect considerable phenotypic diversity within a population of individuals from a single phenotypic class breeding *inter se*. Even within a family the range can be quite wide when a small number of genes is involved.[32]

Another phenomenon generated by the model is regression. Under the assumption of complete hereditary control, random mating, and no environmental effects, if we measure a number of subjects, we would expect the average score of their first degree relatives to be half-way between themselves and the mean for the population. The regression is 0.50. Thus, a group of men with a mean IQ of 130 who marry a group of women *chosen at random* (expected $\bar{X}$ IQ = 100) should yield a group of offspring with a mean IQ of 115. The further away the relationships, the greater the regression, as shown in Figure 8-1.

The reason for regression is quite simple. In random mating population, first-degree relatives have half their genes in common, and these common genes make them similar. The other half of their genes are different and, on the average, resemble those of the general population; these genes make relatives different from each other and more similar to the general population. More distant relatives share fewer genes, show less similarity, and regress further.[29, 33]

Assortive mating reduces regression and increases the correlation between relatives. At the extreme, there is no regression; if a group of men with a mean IQ of 130

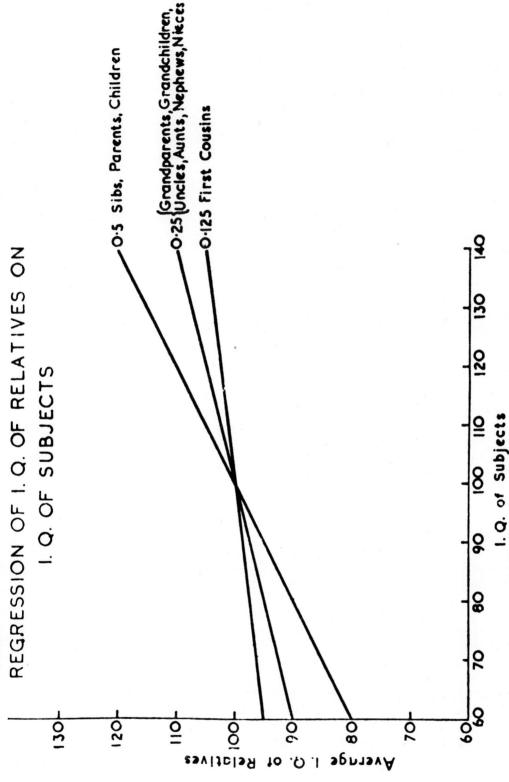

Figure 8-1. Regression of IQ of relatives on IQ of subjects. (From J.A.F. Roberts, "The Genetics of Mental Deficiency," *Eugenics Review*, 44:71 [1952].)

marry a group of women with a mean IQ of 130, their offspring will have a mean IQ of 130.

Let us now turn to the question, "How well does the basic model work, does it help us to understand the phenomenon of intelligence, and what additional modification might it need?"

The Distribution of Intelligence

There are two extremely common misconceptions about the distribution of IQ. It is sometimes assumed that the distribution is normal or Gaussian and, on analogy from physical traits such as height, it is assumed that this is evidence for the inheritance of the characteristic. On the other hand, it is also sometimes assumed that IQ scores are normally distributed because of the methods used to construct the tests. Consequently, the normal distribution of IQ is seen as a statistical artifact.[34, 35, 36] Neither of these arguments is correct. The existence of a normal curve for a trait does not by itself establish anything about inheritance. Although a polygenic system generates a normal curve, evidence of genetic control accrues when it can be shown that the trait follows the consequences of the genetic model. If these consequences follow, it does not matter how the scale was constructed. Nevertheless, IQ scales have been constructed using scaling procedures that do not assume normality.[34, 35, 36]

The distribution of IQ scores of most large samples is not normal, but only approximately normal. The difference between the actual distribution of IQ scores and the Gaussian curve is of both theoretical and practical significance when large populations are involved. Most unselected samples (e.g. the total population, not just school children) yield distributions of IQ scores that are excessively peaked and whose tails are higher and extend further out than a normal curve fit to the same mean and standard deviation. The lack of fit is extensive at the lower end of the distribution. Part of this latter lack of fit is due to the existence of a number of severe mental retardates whose condition can be attributed to major gene effects that override the polygenic system.[25, 38, 39] Additional cases are due to nongenetic pathological factors (e.g. injury at birth, brain injury). Figure 8-2, taken from Burt shows an observed distribution (N = 4,523) with the nongenetic low-IQ cases culled out.[35] A normal curve and a Pearson type IV curve have been fitted to the data.

The normal curve clearly underestimates the percentage of cases at both extremes. Notice that at the lower tail there is no discontinuity between those cases of mental retardation due to major gene effects and the familial retardates who represent the lower tail of the distribution generated by a polygenic system. Strong support for the distinction between major gene effects (responsible for severe mental retardation) and polygenes (responsible for mild or familial mental retardation) comes from a study by Roberts.[40] He plotted the IQ distribution of siblings of imbeciles (IQ less than 50) and siblings of feeble minded subjects (IQ = 50-75). Figure 8-3 shows the IQ distributions of these siblings. In spite of the extremely low IQs of their brothers and sisters, the imbeciles' siblings generate a distribution which is similar to that found in the normal population. This would be predicted from a polygenic theory. They represent a sample of cases from the entire population and reflect the distribution of that population. Their sibs must be retarded for other than polygenic reasons. The siblings of the feebleminded cases, on the other hand, generate a distribution with a mean twenty points below the imbecile siblings and ap-

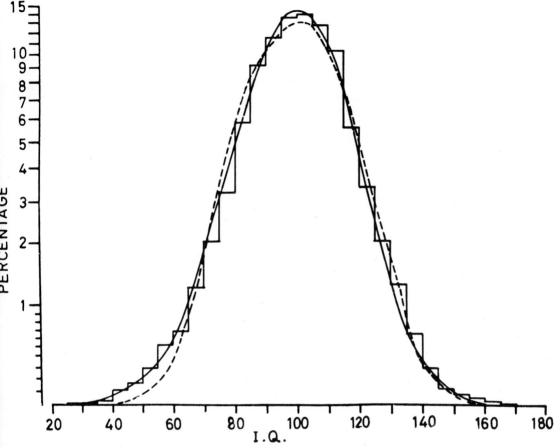

Figure 8-2. Observed and theoretical distributions: Type IV continuous line, normal curve dotted line. (From C. Burt, "Is Intelligence Normally Distributed?" *British Journal of Statistical Psychology,* 16:175 [1973].)

proximately halfway back to the population mean (regression coefficient = .51). Such a distribution is predicted from a polygenic model when the cases represent a sample from the lower tail of the normal distribution. We can therefore conclude that the feebleminded subjects are a part of the normal distribution and are retarded for polygenic reasons.

Although it looks small, the excess of cases at the high end of the distribution in Figure 8-2 is substantial. The reason for fitting the type IV curve (a special class of curves developed by Pearson) to the data is to allow estimation of the proportion of individuals having very high IQs. Table 8-I shows how misleading the normal curve would be if it were used for these purposes.[37] It shows that for large populations a sizeable number of people are involved. There has not been an adequate accounting for this excess of cases over that predicted by the normal curve. Burt has called these individuals "super-normal" and elsewhere suggested that this may be due to major gene effects similar to those that cause severe mental retardation.[37,41] Jensen has pointed out that they may also be

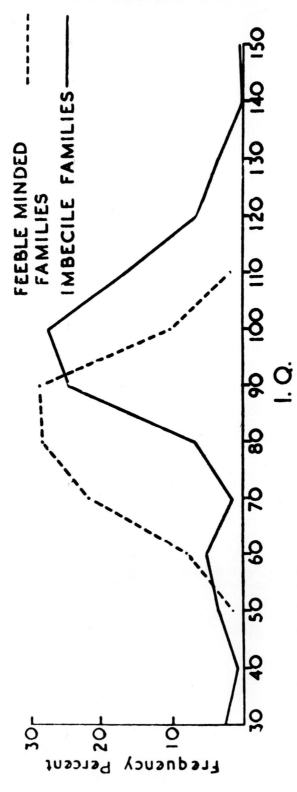

Figure 8-3. Frequency distribution of the IQs of sibs of feebleminded and imbeciles of the IQ range 30 to 68. (From J.A.F. Roberts, "The Genetics of Mental Deficiency," *Eugenics Review*, 44:71 [1971].)

TABLE 8-I
THE ESTIMATED PROPORTIONS OF THE POPULATION REACHING OR EXCEEDING
THE BORDERLINE SPECIFIED

Borderline	Number per Million	
	Normal Distribution	Type IV Distribution
160	31.7	342.3
175	3.3	76.8
190	0.1	19.4
200	<0.001	6.2

From C. Burt, "Is Intelligence Normally Distributed?" *British Journal of Statistical Psychology*, 16:179 (1963).

TABLE 8-II
CORRELATIONS FOR INTELLECTUAL ABILITY, OBTAINED AND THEORETICAL VALUES

Correlations Between	Number of Studies	Obtained Median r*	Theoretical Value†	Theoretical Value‡
Unrelated Persons				
Children reared apart	4	−.01	.00	.00
Foster parent and child	3	+.20	.00	.00
Children reared together	5	+.24	.00	.00
Collaterals				
Second Cousins	1	+.16	+.14	+.063
First Cousins	3	+.26	+.18	+.125
Uncle (or aunt) and nephew (or niece)	1	+.34	+.31	+.25
Siblings, reared apart	3	+.47	+.52	+.50
Siblings, reared together	36	+.55	+.52	+.50
Dizygotic twins, different sex	9	+.49	+.50	+.50
Dizygotic twins, same sex	11	+.56	+.54	+.50
Monozygotic twins, reared apart	4	+.75	+1.00	+1.00
Monozygotic twins, reared together	14	+.87	+1.00	+1.00
Direct line				
Grandparent and grandchild	3	+.27	+.31	+.25
Parent (as adult) and child	13	+.50	+.49	+.50
Parent (as child) and child	1	+.56	+.49	+.50

*Correlations not corrected for attenuation (unreliability).
†Assuming assortative mating and partial dominance.
‡Assuming random mating and only additive genes, *i.e.*, the simplest possible polygenic model.
From A.R. Jensen, "How Much Can We Boost IQ and Scholastic Achievement?" Harvard Educational Review, 39:1 (1969).

the result of genetic-environmental interaction and/or differential assortative mating.[27] The negative skewness in the curve may also be due to directional dominance. These concepts are discussed later in this chapter.

CORRELATIONS BETWEEN RELATIVES

Almost all of the kinship correlations in the scientific literature up to 1963 were summarized by Erlenmeyer-Kimling and Jarvik.[24] Jensen has since updated their work and his table is presented as Table 8-II.[27]

The authors of both articles conclude that the data are compatible with a polygenic hypothesis. The theoretical values derived from the simple polygenic model described earlier are given in the last column. The theoretical values expected from a model which assumed assortative mating and partial dominance are also given. The more complex model fits the data somewhat better. As we noted earlier, the adjustment for assortive mating would also help account for the slightly elevated tails of the empirically obtained IQ distribution because positive assortive mating increases variance. Since positive assortive mating for IQ occurs,[42] and it can be roughly estimated, there seems to be no reason why it should not be included in the model. Burt and Howard used an assortive mating coefficient of .386.[21] Higgins reports a figure of .33 for 1,061 couples,[43] Conrad and Jones report a figure of .52,[44] and Jensen suggests that the average is probably close to .60.[27] As will be shown later, there is evidence to suggest the existence of some dominant gene action for IQ. Thus, Burt's theoretical correlations are probably a reasonable approximation to the underlying genetic complexity.

Unfortunately, the findings in Table 8-II are not definitive. In most cases, both the direction and to some extent relative magnitude of the correlations, would be expected on the basis of environmental similarity. Clear-cut environmental effects are also demonstrated by the correlations between foster parent and child and between unrelated children raised together. These will be discussed below. The less than perfect correlations (even when corrected for attenuation) between monozygotic twins also index environmental effects. Nevertheless, the data are so strikingly congruent with the point predictions made by the genetic model that it is impossible to dismiss.

Foster Children Studies

Impressive indirect evidence for genetic control of IQ comes from studies which show that the IQ of orphans or foster children correlates more highly with that of their biological parents who had nothing to do with raising them than with that of their adoptive parents.

The use of adopted children designs to demonstrate the effects of heredity and environment is quite old but has seen very little usage. Rosenthal, in a series of brilliant studies, has recently revived interest in the method in the area of psychopathology.[45, 46] Surprisingly, the best study of this type, dealing with IQ, was conducted in 1928 by Barbara Burks. It is a model study and remains to be improved upon.

Burks examined two carefully matched groups, a set of foster children (N = 214) and their parents and a control group of parents and their biological children (N = 105).[47] The socioeconomic status of the parents was distributed in about the same way as the general American white population. All of the parents (control, foster, and biological parents so far as was definitely known) were non-Jewish, white, American, British, or north European. All spoke English. None of the children had obvious biological defects, and they were homogeneous as to race and educational opportunity. Generalizations are therefore limited to this population. The foster children were legally adopted (not just cared for) and placed at a very early age ($\bar{X}$ = 3 months, 2 days). A great deal of evidence was generated to demonstrate that selective placement had not occurred. All children were, however, placed in "good" homes.

The home environments of all the children were carefully and extensively analyzed. Table 8-III shows the important correlations from this study.

Burks estimated a number of statistics from her data.[47] The most important were: (1) the proportional contribution of total home environments to the IQ variance; (2) the total contribution of heredity; (3) a numerical estimate of the potency of home environment to raise or depress IQ.

TABLE 8-III
CORRELATIONS BETWEEN CHILD'S IQ AND ENVIRONMENTAL AND HEREDITARY
FACTORS (ADAPTED FROM BURKS[47])

Correlate	Foster			Control		
	r_1	r_2	N	r_1	r_2	N
Father's M.A.	.07	.09	178	.45	.55	100
Mother's M.A.	.19	.23	204	.46	.57	105
Mid Parent M.A.	.20	-	174	.52	-	100
Father's Vocabulary	.13	.14	181	.47	.52	101
Mother's Vocabulary	.23	.25	202	.43	.48	104
Whittier Index	.21	.24	206	.42	.48	104
Culture Index	.25	.29	186	.44	.49	101
Income	.23	.26	181	.24	.26	99

r_1 = raw correlation r_2 = r_1 corrected for attenuation

Adapted from B. S. Burks, "The Relative Influence of Nature and Nurture upon Mental Development: A Comparative Study of Foster Parent-foster child Resemblance and True Parent-True Child Resemblance," *Yearbook of the National Society for the Study of Education*, 27:219 (1928).

We will examine each of these statistics in light of subsequent research and use them as a stepping-off place to discuss a number of methodological and conceptual issues.

Contribution of Home Environment to IQ Variance

The multiple correlation between father's mental age, father's vocabulary, mother's vocabulary, income, and child's IQ in the *Foster* Group corrected for attenuation was .42. The square of this correlation (.176) represents the proportion of the variance in children's IQ's accounted for by the home environments, which included the parents' IQ. This is probably an underestimate, and more precise environmental measures would most likely raise the multiple *r* somewhat. Nevertheless, this is the only study where environmental factors have been directly and carefully examined when *random assignment of cases to environments has been achieved,* i.e. genotypes and environments are completely crossed. Many researchers are still unaware of this methodological necessity and continue to study the effect of environmental variables within families. Bloom, for example, cites a study by Wolf where thirteen environmental variables yielded a multiple correlation of +.76 (variance accounted for = 58%) with children's IQ.[48] Unfortunately, Wolf's "environmental variables" are by no means pure, and in the within-family situation which he assessed they must reflect both genetic and environmental influences. As Jessen has pointed out:

Since no major study of the heritability of intelligence has attributed more than 25 percent of the total variance to environmental differences, some 33 percent of the variance in IQs that Wolf "accounted for" by environmental variables must actually have a genetic basis. This observation leads to interesting questions: To what degree does the parental genotype create the deprivation or enrichment of the environment in which the offspring are reared? And to what degree does the child's genotype shape its own environment?[49]

Burks had an additional crude check on the effect of environment within a part of her sample.[47] There were twenty-one cases where two unrelated foster children were reared in the same home. The correlation between these pairs was .23 ± .21. This correlation is a direct estimate of the percentage of variance accounted for by environmental factors.[50] Unfortunately, we do not know how representative these cases are. An examination of Table 8-II shows that the median correlation of five studies (which probably includes the one by Burks) is .24. Again, however, we do not know how representative the cases are.

The Total Contribution of Heredity for the Environment Sampled in This Study

Using a relatively complex scheme, Burks estimated from the control data that "close to 75 or 80 percent of IQ variance is due to innate and heritable causes." Unfortunately, as Wright pointed out, her method of analysis was erroneous. He reanalyzed her data and obtained a broad heritability coefficient of .81[51] This figure is in precise agreement with the outcome of literally dozens of studies that have calculated broad heritability in a number of different ways.[27] It also agrees closely with the independently derived figure of .17 as the amount of variance accounted for by the environment.

An Estimate of the Potency of the Home Environment to Raise or Depress IQ

Table 8-IV shows the effects of various environmental factors on the IQ of foster children. The table was constructed by multiplying the correlation between the factors in question and the child's IQ by the S.D. (15) of the IQs. For example, total environment correlated .42 (corrected for attenuation) with IQ, therefore 15 x .42 = 6.3 IQ points.

Burks felt this was the most significant table in her study and concluded from it that:

1. The total effect of environmental factors one standard deviation up or down the scale is only about 6 points, or, allowing for a maximal oscillation in the corrected multiple correlation (.42) of as much as .20, the maximal effect almost certainly lies between 3 and 9 points.

TABLE 8-IV
AVERAGE SHIFT, DUE TO ENVIRONMENT, IN POINTS OF I.Q. OF FOSTER CHILDREN, WHEN VARIOUS FACTORS ARE ONE S.D. ABOVE OR BELOW THE POPULATION MEAN

Factor	Measured	Actual
Foster father's mental age	1.0	1.4
Foster mother's mental age	2.9	3.5
Foster mid-parent mental age	3.0	-
Whittier rating of foster home	3.1	3.6
Culture rating of foster home	3.7	4.4
Total environment	5.3	6.3

From B. S. Burks, "The Relative Influence of Nature and Nurture upon Mental Development: A Comparative Study of Foster Parent-foster child Resemblance and True Parent-True Child Resemblance," *Yearbook of the National Society for the Study of Education,* 27:219 (1928).

2. Assuming the best possible environments to be three standard deviations above the mean of the population (which, if "environments" are distributed approximately according to the normal law, would only occur about once in a thousand cases) the excess in such a situation of a child's IQ over his inherited level would be between 9 and 27 points—or less if the relation of culture to IQ is curvilinear on the upper levels, as it well may be.[47]

These conclusions must be very carefully interpreted. They do not mean, as some have suggested, that everyone under the IQ distribution could be shifted up nine to twenty-seven IQ points by improving their environments. Nor do they mean that IQ cannot be depressed more than this amount. Burks' cases were randomly distributed across their environment. Normal populations are not. As we will show in the section on social class, in open societies social mobility leads to a better genotype-by-environment fit, and gene segregation and recombination lead to social mobility. The question is really one of fit. Consider the widely cited case of the monozygotic twins, Gladys (IQ 92) and Helen (IQ 116) in the study by Newman, Freeman, and Holzinger.[52] The twenty-four-point separation is the largest one found in the literature. These twins were separated at age eighteen months and were tested at age thirty-five. Helen had good health, obtained a B.A. degree, and taught school. Gladys, suffering from severe illnesses, only went to third grade and was brought up for much of her childhood in an isolated district in the Canadian Rockies. The twenty-four-point IQ difference can best be understood as a poor genotype-by-environment fit. Gladys' genotype probably would have permitted the development of an IQ of 116, but it was not allowed to develop because of a less-than-favorable environment.[53] Using Burks' figures of six IQ points as the effect of environmental factors one standard deviation up or down the scale, we would assume that Gladys had an environmental disadvantage of four standard deviations. This does not seem out of line with the difference between ill health and third grade education versus good health and a B.A. from a good school.

All of the twenty-four IQ point discrepancy need not, however, be ascribed to the psychosocial environment. Jensen has pointed out that early prenatal effects (as indexed by fingerprint ridge counts) could easily account for much of the difference.[54]

A second study of the IQ of foster children was that of Skodak and Skeels.[55] The important findings from this study for our purposes have been summarized by Honzik and are shown in Figure 8-4.[56]

The figures show that both the rate of increasing resemblance to true parents and the final level attained is the same regardless of whether children are reared by their true parents or not. This is extremely strong evidence in support of genetic control of intelligence. Interestingly enough, however, the Skodak and Skeels study also yielded evidence for the malleability of IQ.[55] The mean IQ of mothers of the adopted children was 85.7, while the IQ of the children was 106.

Such apparently contradictory findings always astound those who confuse genetic control or heritability with unmodifiability. Such confusion often results because of a serious limitation in correlational statistics. The Pearsonian coefficient of correlation is the mean of the cross products of standard scores. Since standard scores are used, the absolute mean difference between the related variables is ignored. Thus, the IQ of the adopted children in the Skodak and Skeels study could have moved up, moved down, or remained constant due to adoption; but as long as their relative standing remained the same, the correlation of their IQs with that of their true parents would remain unaffected.[57, 58] It is of some interest to note that genetic researchers tend to rely on correlational statistics while researchers interested in environmental effects tend to focus on mean differences.

COMPLICATIONS NECESSARY IN THE GENETIC MODEL AND ADDITIONAL CONCEPTS

The sample of studies we have reviewed up to this point constitutes an impressive array of evidence for the genetic control of IQ. The simple additive polygenic model does not, however, account for all the facts. As we have seen there is also environmental control of IQ. In this section we will discuss a conceptual scheme, reaction range, that succinctly synthesizes both the genetic and environmental points of view. This will be followed by a discussion of the widely misunderstood and misinterpreted concepts of interaction and heritability.

Reaction Range

Each genotype has a norm of reaction, a largely unspecifiable (for the individual) but limited set of outcomes that could oc-

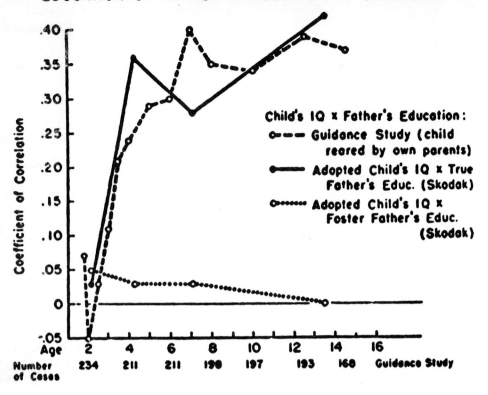
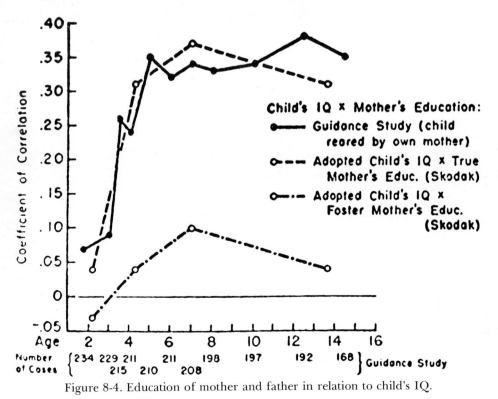

Figure 8-4. Education of mother and father in relation to child's IQ.

cur, given different environmental treatments. Different genotypes have different norms of reaction. Gottesman has depicted, in the form of a set of curves, the norm of reaction for a subset of genotypes that determine IQ.[25] His scheme is shown in Figure 8-5.

The curves are hypothetical but summarize our knowledge about the mutability of IQ very nicely. Curve A (mean phenotypic IQ = 25) represents the performance of individuals with severe mental deficiency known to be due to dominant or recessive gene inheritance or chromosomal aberrations. Curves B, C, and D represent three genotypes drawn from the normal distribution of IQ. In the natural habitat Curve B represents an $\bar{X}$ IQ of 70 (familial retardation). Curve C represents an average genotype, $\bar{X}$ IQ of 100, and Curve D represents a superior genotype, $\bar{X}$ IQ of 150. The potential range of scores available to each genotype across a wide range of environments is shown by the curves and indexed by the reaction ranges on the right side of graph.

The large difference in reaction range between Curves B and D suggests that there is an interaction between genotype and environment with D genotypes, demonstrating a greater capacity to utilize an enriched environment than B genotypes, or, conversely, a greater decrease in IQ for D genotypes than B under deprived conditions. Although interactions of this type have been demonstrated with animals, the evidence for humans is not very strong.[59, 60] Part of the difficulty is due to an inadequate sampling of extreme environments. There is simply no question that a D genotype raised in a seriously deprived environment would suffer more than a B genotype. The enhancing effects of superior environments is more open to question. The problem of interactions will be treated more fully below.

The Interaction of Environment and Heredity

No concept in genetics has caused as much confusion among social scientists as that of the interaction between heredity and environment. Social scientists use the term to mean that environmental stimuli have an important biological effect on the organism. Hunt says: "Ample evidence has now accumulated to show that the consequences of informational interaction with circumstances, through the ears and the eyes (and especially the latter for the evidence extant), is quite as biological in nature as the effects of nutrition or genetic constitution. Interaction through the eyes, especially early in life, has genuine neuroanatomical and neurochemical consequences."[34]

This is a perfectly legitimate way to define interaction and it is compatible with standard English usage. No geneticist would deny the existence of the importance of interactions of this type. They are necessary if life is to proceed, and much remains to be learned about them. For the sake of clarity, let us call the kinds of interactions described above *transactions*. Geneticists apply the term *interaction* to a somewhat different level of analysis, i.e. to mean that different genotypes respond differently across environmental conditions (nonadditivity). Thus, the relationship shown in Figure 8-6a would *not* be an interaction. Genotypes X and Y differ from each other under E_1, indicating that they transact differently with that environment (the same stimuli have a different effect). They also transact with E_2 in such a way as to improve their overall performance relative to E_1, but they maintain the same ranking under both environments. In analysis of variance terms, Figure 8-6a represents a main effect for genotypes and a main effect for environments, but not an

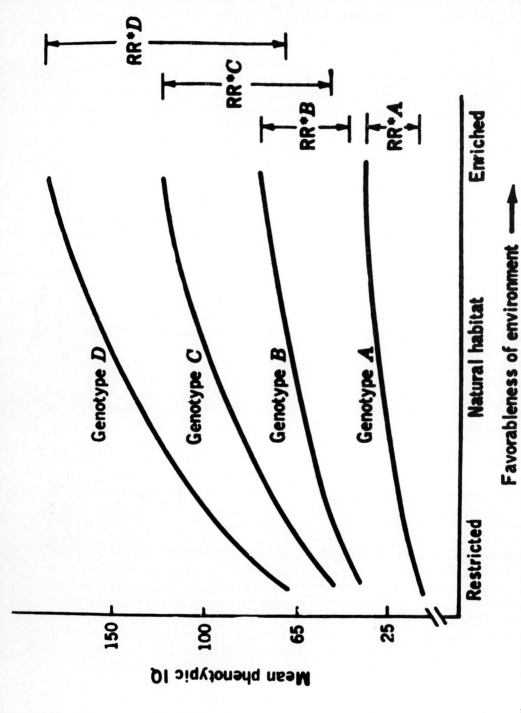

Figure 8-5. Scheme of the reaction range concept for four hypothesized genotypes. (From I.I. Gottesman, "Genetic Aspects of Intelligent Behavior," in N. Ellis (ed.), *Handbook of Mental Deficiency: Psychological Theory and Research* [New York, McGraw-Hill, 1963].) Note: Marked deviation from the natural habitat has a low probability of occurrence. R.R. signifies reaction range in phenotypic IQ.

interaction. Figure 8-6b is an example of an ordinal interaction. In this case genotype X responds much more favorably to E₂ than genotype Y. The interaction is ordinal because the rank ordering of genotypes does not change. Gottesman's reaction range curves are examples of this type of interaction.[25] Some geneticists are more restrictive with their use of the term and would require disordinality before defining an effect as an interaction. Figure 8-6c is an example of a disordinal interaction. It shows a reversal of rank under two different environments. The data in Figures 8-6b and 8-6c would yield interactions in the analysis of variance sense (Fig. 8-6c would yield only an interaction effect; there would be no main effects due to genotypes or environments). Later, we will use the term V_I to symbolize the interaction of heredity and environment in the analysis of variance sense described above.

It is important to recognize that, given a sufficiently sensitive experiment, at least *two* genotypes and *two* environments are necessary to detect an interaction. A single genotype and a single environment are sufficient for a transaction to occur. Note that each genotype-by-environment juncture plotted in Figure 8-6a, 8-6b, and 8-6c reflects a genotype-by-environment transaction. The kinds of transactions in which a set of genotypes can engage in under different environmental conditions determine which, if any, type of interaction will occur when a contrast is made.

As we pointed out earlier, the evidence for ordinal interaction for IQ is not very strong, although such a situation is very plausible when all the disparate data are assembled, as in Gottesmen's reaction range curves.[25]

The few relevant studies deal mostly with educational attainments and have been summarized by Weisman, who says: "Our own results fall into line with this general trend; brightness has higher correlations with environmental factors than does backwardness."[61] After summarizing his own data he concludes that the evidence points "to the virtual certainty that an adverse environment has its greatest educational effect on children of above average ability." Scarr-Salapatek has also reported data indicating that "white children with lower IQs are less susceptible to environmental differences between families than are children with higher IQs even in an advantaged population."[62] Jinks and Fulker,[63] however, in their reanalysis of twin data on the Mill Hill Vocabulary test collected by Shields,[64] found suggestive evidence for the conclusion that unfavorable genotypes were more influenced by environment than favorable ones. This issue deserves much more research.

There is no evidence to my knowledge for disordinal interactions between IQ and environmental treatments. The lack of evidence for disordinal interaction effects may come as a surprise to some psychologists, but not to those familiar with the intense, but as yet fruitless search for a similar phenomenon in psychology, namely aptitude by treatment interactions.[65]

An important problem often confused with interaction is that of correlated environments or covariance, symbolized as Z_{COVHE}. Covariance exists when genotypes are differentially distributed across environments. Covariance for intelligence is generally positive; good genotypes tend to be located where the environment is good. For example, intelligent children are likely to have intelligent parents who are above average in socioeconomic status. Innately intelligent children are also likely to both modify and select and attend to stimulating features of their environment. (See Bell[66] and Cattell[67] for a general treatment of this problem, and Hayes[68] for a unique

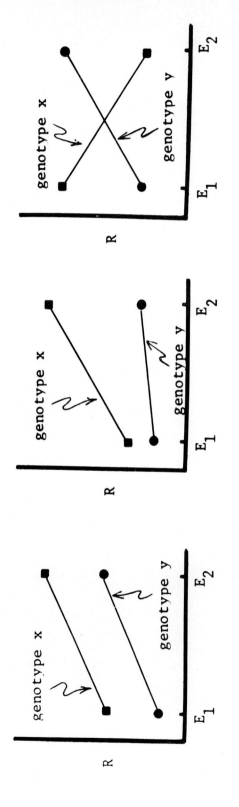

Figure 8-6. Three possible types of relationships between two genotypes and two environments.

point of view as to what is inherited with respect to intelligence.) The question of whether or not to include this source of variation with heredity or environment is an open one. Regarding the possible covariance for IQ described above, Jinks and Fulker ask:

But is not this a more or less inevitable result of genotype? To what extent could we ever get a dull person to select for himself an intellectually stimulating environment to the same extent as a bright person might? Even when these correlations exist because of the pressure of others on an individual, it is not clear to what extent the correlations can be manipulated. Perhaps it can to some extent by such drastic procedures as intensively coaching the dull, and drastically depriving the intelligent, but the effect on the correlation is still not entirely clear. We need much more evidence concerning the effects and causes of correlated environments in order to decide their importance.[63]

As will be shown in the next section, estimates of the magnitude of the contribution of V_I and Z_{COVHE} to total phenotypic variance in IQ under current environmental conditions are small. As Burt's adjusted data will also show, the percentage of total covariance due to Z_{COVHE} becomes very small as the assessment of IQ is improved. Nevertheless, we concur with Jinks and Fulker that much more research is needed before any conclusions can be reached regarding the importance of this source of variance.[63]

The issue of environment-by-heredity interactions involves numerous additional complexities which cannot be treated here. The reader is referred to Erlenmeyer-Kimling, Haldane, and Vale and Vale for additional discussion.[69, 70, 71]

Heritability

According to Falconer[72]: "The heritability of a metric characteristic is one of its most important properties."[72] Falconer is referring to traits of economic importance in animal work. In our opinion intelligence is of no less importance in human populations. Much controversy and misunderstanding surrounds the concept of heritability. Heritability is a population statistic describing an important property of a given trait in a given population at a given time. It cannot be used to characterize any individual or any characteristic in an individual. Like other population statistics (e.g. standard deviation) it takes on different values as the population changes. Unfortunately, it is subject to a variety of sources of perturbation and may fluctuate dramatically from time to time, even when measured on the same population.[73] Nevertheless, the heritability of intelligence has been assessed in many human populations of similar genetic background (white, northern European) and the findings have been very consistent.[27]

Technically, heritability (h^2) is defined as the ratio of additive genetic variance to phenotypic variance.[72] (In all of our equations we will omit variance due to errors of measurement for the sake of simplicity.)

$$h^2 = \frac{V_A}{V_P} \quad (1)$$

where V_A = additive variance component (breeding value) and V_P = total phenotypic variance of the trait measured in the population.

Heritability can be estimated in a number of ways, some of which are much more efficient than others.[72] The choice of method depends on the type of data at hand. One estimate of heritability is obtained by taking the correlation between the "midparent" value (mean of two parents) and "midchild" value (mean of all children), provided there is no nongenetic cause of resemblance between offspring and parents. The latter requirement suggests that many of the heritabilities of

IQ in the literature are overestimates because of the common environmental influences which tend to make relatives similar. In animal studies a good estimate comes from analysis of the correlations between fathers and children reared apart because common maternal circulation (an environmental effect) may inflate the correlation between mother and child. For IQ, h^2 generally falls between .50 and .60, depending on the degree of assortive mating, which increases additive variance.[74]

Heritability is typically used to predict response to artificial selection by way of the formula:

$$R = h^2 S \qquad (2)$$

where S = mean phenotypic value of those selected parents expressed as a deviation from the population mean, and R = mean deviation of their offspring from the population mean.

More complex formulas are used to predict the effects of natural selection because the character selected for is not IQ but fitness (total number of offspring). Falconer has outlined the appropriate formulas to use in those circumstances.[75]

The following is an example of how heritability can be used to predict response in IQ as a function of selection. The data are taken from Reed and Reed.[28] The average IQ of twenty-nine couples when the male had an IQ of 131 or above was 122.7 and that of their forty-three children for whom IQ scores were available was 117.7. On the assumption that $h^2 = .55$, the children's mean IQ should be 112.5. The deviation of this prediction from the data is -5.2 IQ points. The average IQ of twenty-seven couples when the average of the female was 131 or above was 119.4 and that of their forty-five children for whom IQs were available was 116.0. The expected value is 110.7, and is in error by -5.3 IQ points. The actual heritability in this population may be higher or lower and accurate knowledge as to its value would allow us to make a more accurate prediction of scores expected on a genetic basis for any subset of the population.

This set of predictions, crude and inaccurate as they are, should be contrasted with those which might be made from models that overemphasize the role of *current environments* in the determination of variance in IQ. Under the assumption that variance in IQ is determined almost entirely by environment, one would expect that the above parents could maintain their children's IQ at their own level, if not raise it, and certainly they would not be expected to drive it down. The term *current environments* is emphasized in order to differentiate them from potential future environments (both pre-and postnatal) that might be found to interact with different genotypes and account for a larger portion of phenotypic intelligence at that time. The possibility that certain kinds of environments (biochemical?) might simulate the effects of polygenes for intelligence also exists.

Formula (1) gives us an estimate of heritability in the narrow sense. That is, it tells us what is transmitted additively, but it does not index the total influence of heredity on phenotypic variance (degree of genetic determination or genetic variance).

The total influence of heredity on phenotypic variance is indexed by:

$$H^2 = \frac{(V_A + V_{AM}) + V_D + V_i}{V_P} \qquad (3)$$

The two additional sources of variance to be defined here are dominance (V_D) and epistasis (V_i).

V_D: consider a single locus with two alleles, A and a. As in our earlier discussion, assume that A has a value of 1 and that a has a value of 0. Then AA = 2, Aa = 1, and aa = 0. This is the simple additive case. If AA = 2, Aa = 2, and aa = 0, then A is said to be dominant. If AA = 2, Aa = 0, and aa = 0, then A is said to be recessive. If Aa takes a value between 1 and 2, A is said to be partially dominant, and if Aa takes a value between 0 and 1, A is partially recessive. If Aa takes a value over 2, A is said to be overdominant. Note that dominance indicates a lack of additivity and therefore can be considered a form of interaction.

V_I: epistasis is a measure of the interdependence of gene effects or nonallelic (interlocus) interactions. There is little reason to expect epistasis when a large number of additive genes control a trait, and it is typically pooled with dominance variance when one attempts to estimate these effects apart from additive variance.

Note that we have also partitioned the additive variance into two components: V_A, the additive random variance expected under random mating (panmixia), and V_{AM}, the additive variance due to assortive mating. As we will show below, assortive mating, when sufficiently strong, has considerable effect in increasing the genetic variance.

Estimation of Genetic and Environmental Components of Phenotypic Variance

Up to this point we have defined all the components of the total phenotypic variance of a trait except V_E which symbolizes a main effect due to environment. The full equation reads as follows:

$$V_P = V_A + V_{AM} + V_D + V_I + V_E + 2Cov_{HE} + V_I \quad (4)$$

What is the relative importance of each of these components? According to Eaves psychogenic studies of IQ have seldom used large enough samples to allow estimation of genetic and environmental parameters with reasonably narrow confidence limits.[76] The best of these studies was conducted by Burt,[41] and his results are shown in Table 8-V

The numbers in parentheses are adjusted estimates. The children's IQ scores in this sample were submitted to their teachers for criticism and, in those cases where it was decided that the scores were not a fair estimate of the child's ability, the child was retested and an adjusted score was obtained.

Jinks and Fulker[63] reanalyzed the Burt adjusted data as published in Burt,[77] using more powerful biometric techniques which allowed them to partition the variance somewhat differently and test for the significance of their parameter estimates as well as the overall fit of their model to the data. The data were partitioned into four components: G_1 = within family genetic component, G_2 = between family genetic component, E_1 = within family environmental component, E_2 = between family environmental component. The results are presented below. The significance level indicates a test against the null hypotheses that the parameter does not differ significantly from zero.

$$G_1 = 0.40 \pm 0.03 \qquad p<.001$$

TABLE 8-V
ANALYSIS OF VARIANCE OF INTELLIGENCE TEST SCORES

Source of Variance	Percent*	
Genetic:		
Genetic (additive)	40.5	(47.9)
Assortative Mating	19.9	(17.9)
Dominance & Epistasis	16.7	(21.7)
Environmental:		
Covariance of Heredity & Environment	10.6	(1.4)
Random Environmental Effects Including H x E Interaction (V_I)	5.9	(5.8)
Unreliability (test error)	6.4	(5.3)
Total	100.00	(100.0)

*Figures in parentheses are percentages for adjusted assessments. See text for explanation.
Data from C. Burt, "The Inheritance of Mental Ability," *American Psychology,* 13:1 (1958); table from A. R. Jensen, "How Much Can We Boost IQ and Scholastic Achievement?" *Harvard Educational Review,* 39:1 (1969).

$G_2 = 0.47 \pm 0.04 \quad p<.001$
$E_1 = 0.07 \pm 0.01 \quad p<.001$
$E_2 = 0.06 \pm 0.03 \quad p<.025$

The model as a whole fits the data extremely well as tested by appropriate statistics. A conservative test of dominance variance yielded a value of .149 (p<.05). The dominance ratio was .75 (0 = no dominance, 1 = complete dominance) indicating considerable dominance.

The best esitmate of h^2 was .706, but this figure is probably inflated due to common environments. The broad heritability was .863. There was a strong suggestion that there was no significant effect due to covariance of heredity and environment. The effect of heredity-by-environment interaction could not be tested.

INTELLIGENCE AND SOCIAL STRUCTURE

In the first part of this chapter we argued that intelligence was an important biological parameter, but that an adequate understanding of the concept required simultaneous consideration of both environmental and genetic influences. In the introduction we also suggested that intelligence was an important social parameter, particularly in highly stratified modern industrialized societies. Here we would like to argue that an adequate understanding of social structure also requires a simultaneous consideration of environmental and genetic factors. Our discussion applies only to white samples. Gottesman should be examined for a discussion of the race x IQ x social class issues.[26]

According to Tyler: "The relationship of measured intelligence to socio-economic status is one of the best documented findings in mental test history."[78] Typical correlations are .42 .45, and .57.[79, 80, 32] In most of these studies the subjects are still moving up the socioeconomic status (SES) ladder, and this could drive the correlation higher. Ball calculated the correlation between a measure of occupational status determined in 1937 on two groups of men measured on the Pressey Mental Survey test in 1918 and 1923.[81] For the 1923 group it was .57 and for the 1918 group it was .71. The increase with time is quite large. (See Waller[32] for a similar effect across generations.) Correlations in the .40 to .50 range thus appear to be low estimates.

There are two opposing explanations for the relationship between SES and IQ, the deprivation model and the genetic model.[82] Herrnstein has succinctly stated the genetic model in the following syllogism:[83]

1. If differences in mental abilities are inherited, and
2. if success requires those abilities, and
3. if earnings and prestige depend on success,
4. then social standing will be based to some extent on inherited differences among people.

We have presented considerable evidence for the first premise. The second, however, needs elaboration. Status achievement in advanced industrial societies is, to a large extent, mediated by education. It is necessary, therefore, to demonstrate that educational achievement is dependent on mental ability rather than only on level of social origins. We will do this below. Some of the tables presented also demonstrate that achieved status varies considerably within level of social origin as a function of IQ. This is a result difficult to account for under the assumption that environmental conditions are the sole determinants of IQ. Assortive mating is also an important determinant of social class differences in IQ because it increases the variance of the IQ distribution and reduces regression.

Table 8-VI presents data from a large sample of 20,700 men who "represent the approximately 45 million men 20-64 years old in the civilian, non-institutional population of the United States in March, 1962."[84] That table shows that the correlation between the occupational characteristics of both the wife's and husband's fathers are appreciable (∼.3). Thus, there is considerable assortment with respect to parental occupation status. More important, however, is the much higher correlation between years of school completed for each spouse. This correlation averages about .60. Such a high correlation suggests that more than social class origins are at work. Column 4 of Table 8-VI gives a picture of what happens to the correlation between mates' fathers' occupational status when the effect of education is parcelled out. The correlations shrink considerably but do not disappear, indicating that there is still some assortment related to background factors of fathers. The remaining variance is small, however, and it seems that assortive mating for education is relatively free of the influence of family background. The most likely explanation for the correlation between education of spouses is selection based on IQ, with the schools supplying the spatial context for meeting and subsequent mating.[82, 85] Evidence for the influence of IQ on educational achievement is supplied by Duncan.[79] He finds that 42 percent of the variance in educational attainment can be accounted for by the following four variables: (1) early IQ (2) number of siblings (negative weight), (3) father's education, (4) father's occupation. IQ alone accounts for 38 percent of the variance quite independently of other factors.

Thus, the achieved status of educational attainment is in large part independent of family status and due to IQ. The correlation between education and occupational status is about .60.[32, 79, 86] Thus, status must in part be due to IQ and, therefore, be influenced by genetic factors.

One of the consequences of the process of random gene segregation and recombination, as we pointed out earlier, is regression. This is illustrated for a large sample in Table 8-VII. The children's IQs fall roughly half-way between the adult mean and the population mean. The S.D. of the children within occupational class has gone down from 9.6 to 14.0, very close to the S.D. (15) for the entire population. Since the spread in IQ between various occupations has remained relatively stable,[87] some

TABLE 8-VI
SIMPLE AND PARTIAL CORRELATIONS BETWEEN SELECTED CHARACTERISTICS OF SPOUSES BY AGE OF WIFE

Age of Wife	Number of Couples (thousands) (1)	Years of School Completed (2)	Father's Occ. Status (3)	(3) Holding (2) Constant (4)
22-26	3,726	.63	.33	.16
27-31	4,077	.61	.27	.10
32-36	4,685	.59	.25	.10
37-41	4,907	.55	.31	.18
42-46	4,311	.62	.30	.18
47-51	3,785	.60	.29	.14
52-56	2,810	.60	.32	.21
57-61	1,846	.63	.37	.25

Adapted from P. Blau and O. D. Duncan, *The American Occupational Structure* (New York, John Wiley & Sons, Inc., 1967).

TABLE 8-VII
MEAN IQs OF ADULTS AND CHILDREN ACCORDING TO OCCUPATIONAL CLASS

Class	Adults	Children
Higher Professional	139.7	120.8
Lower Professional	130.6	114.7
Clerical	115.9	107.8
Skilled	108.2	104.6
Semiskilled	97.8	98.9
Unskilled	84.9	92.6

Children classified by occupational class of fathers. Adults classified by their own occupational status.
From C. Burt, "Intelligence and Social Mobility," *British Journal of Statistical Psychology*, 14:3 (1961).

children of each generation must move up and some must move down in order to reconstitute the adult distribution. If such redistribution did not occur, the difference between class means would disappear in about five generations.[88] This movement away from level of origin is called "basic mobility." From these and other data, Burt estimates that the total amount of intergenerational mobility in England is about 30 percent.[88]

Burt's conclusions are based on latitudinal data and are therefore not entirely satisfactory. Small longitudinal studies, however, support all his basic findings. Waller examined the relationship of difference in social position between fathers and sons (N = 146 pairs) to difference in IQ scores.[32] The correlation was + .37, and the relationship is illustrated in Figure 8-7. This figure shows clearly that the greater the difference in IQ scores between father and son, the greater the social mobility both in an upward and downward direction.

Waller's sample had poor representation in the upper two social classes.[32] This makes the findings all the more striking but leaves us with the question: How much social mobility occurs in that range?

Gibson has provided us with the necessary data.[89] He analyzed the relationship between IQ and social mobility in a sample of scientists whose IQ scores were similar in range to those expected of the higher 25 percent of a representative general population sample. There are a number of anomalies in Gibson's data which suggest that the samples are unrepresentative of the male children produced by the parents. This does not, however, influence the argument we are developing. The IQs of the scientists and their brothers and fathers, by social class, are presented in Table 8-VIII.

The relationship between socioeconomic status and IQ is significant even in this highly selected sample (regression coefficient = .34). Table 8-VIII also illustrates regression. The IQs of the sons, who were selected because they were high (extreme), are higher (except in Class I) than their fathers'. The IQs of the brothers must therefore be lower than their fathers', and they are (except in Class IIIM).

The relationship of IQ to social mobility by social class for twenty-two families in which the IQ of the father and at least two sons is known are shown in Table 8-IX.

There is no effect for Class I. The IQ of those sons who move down (almost all only moved down one class) does not differ from those who are stable. The remaining three classes demonstrate clear segregation up and down as a function of IQ. Thus, we have evidence from both latitudinal and longitudinal studies that IQ is heavily implicated in the social mobility which takes place between social classes.

Some researchers will admit to the exis-

TABLE 8-VIII
IQ OF SCIENTISTS, THEIR FATHERS AND SIBS IN DIFFERENT SOCIO-ECONOMIC CLASSES

Occupational class of father (scientist's and brothers initial class)	Scientists'			Fathers'		Scientists' Sibs		
	N	M	S.D.	M	S.D.	N	M	S.D.
Class I	18	129.7	4.24	130.16	8.00	19	125.84	9.67
Class II	36	128.6	6.25	122.6	10.90	23	120.83	9.94
Class III NM	12	125.66	5.97	121.58	9.08	4	113.25	10.44
Class III M	10	123.4	5.93	113.1	10.72	5	116.8	12.09

From J. B. Gibson, "Biological Aspects of a High Socio-economic Group: I. IQ, Education and Social Mobility," *Journal of Biosocial Science*, 2:1 (1970).

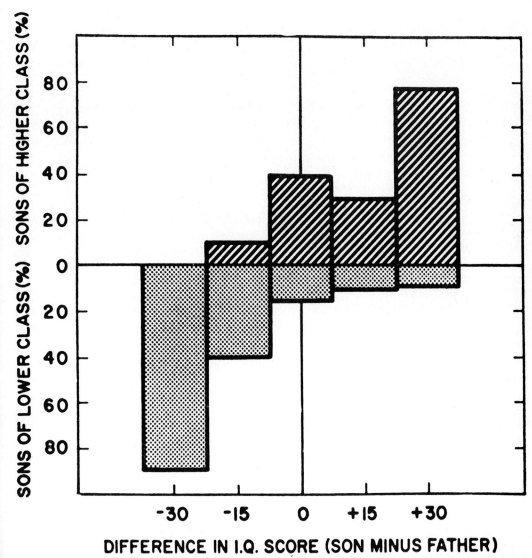

Figure 8-7. Relationship of difference in social position to difference in IQ scores. (From J.H. Waller, "Achievement and Social Mobility: Relationship among IQ Score, Education, and Occupation in Two Generations," *Social Biology*, 18:252 [1971].)

tence of large innate individual differences in IQ but still deny the possibility of genetic differences between social classes.[90, 91]

Phenotypic differences between classes are considered to have arisen largely from environment effects. Thoday and Gibson have pointed out that when genetic differences and environmental effects are confounded:

and both give rise to parent-offspring correlations—components of social heredity may spuriously inflate "heritability" estimates.

In such a situation the heritability of the variable may be shown to be high within families, or even within classes, but the question of the

TABLE 8-IX
IQ AND SOCIAL MOBILITY OF SCIENTISTS AND THEIR BROTHERS IN EACH OCCUPATIONAL CLASS

Occupational Class of Fathers	Fathers' Mean IQ	Sons' Mean IQ and Subsequent Mobility	
Class I	127·8	129·6	stable
		129·7	moved down
Class II	121·45	128·6	moved up
		120·8	stable
Class III NM	117·0	118·5	moved up
		-	stable
		99.0	moved down
Class III M	111·5	121·6	moved up
		101·0	stable
		-	moved down

From J. B. Gibson, "Biological Aspects of a High Socio-economic Group: I. IQ Education and Social Mobility," *Journal of Biosocial Science*, 2:1 (1970).

heritability of between-class or between-group differences remains completely open, because there is no warrant for extrapolating from within- to between-group heritabilities without satisfactory "transplant" experiments.[92]

In order to determine if genetic differences between groups will occur due to intergroup mobility on a variable, even under conditions where environment is a confounding factor, Thoday and Gibson carried out a model transplant experiment using the fruit fly, *Drosophila melanogaster*.[92] The experiment is elegant and highly instructive. It will, therefore, be described in some detail.

The trait studied was sternopleural bristle number (hairs on a certain part of the fly's body). It is a continuous trait, with incomplete heredity, and is systematically responsive to temperature (more bristles grow at lower temperatures.) Mobility was controlled by the experimenters and made to act in the same direction as environmental effects (much like IQ and social advantage). The experiment has the following structure: (1) two populations are established from a wild stock, one at 20°C (to produce a high bristle number) and one at 25°C (to produce a low bristle number); (2) in each generation ten virgin females and ten males are removed from each population. Of the twenty females the ten with the highest bristle numbers go to the 20°C environment (e.g. if seven originally come from the 20°C environment and three come from the 25°C environment, this constitutes 30% mobility for females). The remaining ten with the low bristle numbers go to the 25°C environment. The same sorting occurs for the males. The experiment can be diagrammed as shown in Figure 8-8.

The first generation demonstrates 10 percent mobility. In each environment, eighteen individuals remain in the environment of origin and two have immigrated in. This change constitutes assortive mating, since individuals are being matched on number of bristles.

Table 8-X gives the mean number of bristles for both environments for the base population and later generations as well as intergroup mobility. The groups have come to differ over and above the difference due to temperature (environment) alone. Mobility is less than 50 percent, indicating that the groups truly differ on the variable.

As in the real world, when we are dealing with variables like IQ, the causes of the differences between groups are confounded. Both genes and the environment are working together. In this case, how-

TABLE 8-X
MEAN NUMBERS OF BRISTLES OF TWO GROUPS OF FLIES, AND INTERGROUP MOBILITY

Generation	Group 20°C	25°C	Difference	Mobility %
0	19.8	18.9	0.9	-
6	19.5	17.4	2.1	20
7	19.9	17.7	2.2	20
8	19.9	17.4	2.5	30
9	19.8	17.4	2.4	20

From J. M. Thoday and J. B. Gibson, "Environmental and Genetical Contributions to Class Differences: A Model Experiment," *Science*, 167:990 (1970).

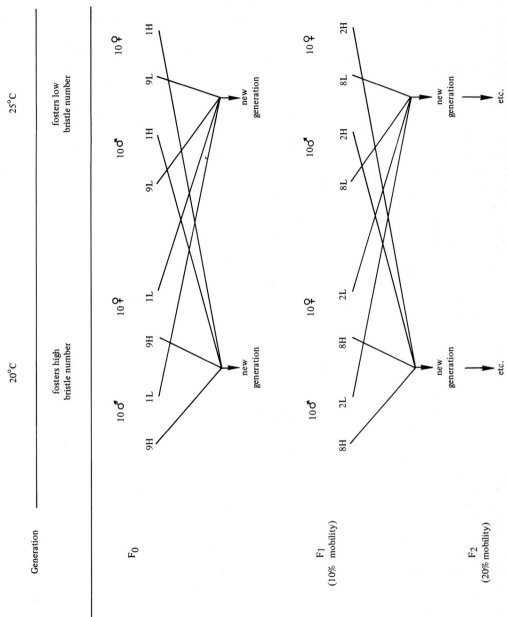

Figure 8-8. Schematic diagram of the Thoday and Gibson Model Experiment. See text for details.

TABLE 8-XI
MEAN NUMBERS OF BRISTLES OF FLIES OF GENERATION 8 RAISED IN TWO ENVIRONMENTAL CONDITIONS

Environment	Progeny of Flies with High Numbers of Bristles	Progeny of Flies with Low Numbers of Bristles
20°C	19.9	18.5
25°C	18.3	17.4

From J. M. Thoday and J. B. Gibson, "Environmental and Genetical Contributions to Class Differences: A Model Experiment," *Science,* 167:990 (1970).

ever, it is possible to conduct a transplant experiment. One sample of flies from the seventh generation, 20°C group, was grown at 25°C, and a second sample from the 25°C group was grown at 20°C. The progenies of these samples can be compared with generation 8 in Table 8-X in order to parcel out the effects of genes and environment. The appropriate data is shown in Table 8-XI.

Thoday Gibson described their results as follows.[92] "The difference between the means of the two groups when grown in their 'own' environmental conditions is 19.9 − 17.4, or 2.5 bristles per individual. When the environmental conditions are made the same, the difference is reduced to an average of 1.15 bristles per fly, giving us a between-group heritability of 1.15/2.5 = 0.42."* Additional tests showed that the within-groups heritability was 0.13. Thus, more between-group variance is genetic than within group variance in spite of the strong environmental differences between groups. After warning the reader not to attach any particular importance to any of the heritabilities or other statistics in their study as well as to be cautious about extrapolation, the authors conclude:

---

*In the original paper, Thoday and Gibson report 1.15/2.5 = 0.42. This must be a miscalculation. It should read 0.46. An analysis of variance of triplicate assays of generation 7, however, yields a between-group heritability of .42.

Nevertheless, we do feel that our experiment is relevant to the human situation in as much as it strengthens the expectation that social mobility related to a heritable variable will give rise to some genetic difference between class means despite strong parent-offspring environmental correlation. We, therefore, believe that our experimental results support those who hold the view that neither cultural nor genetic approaches alone are likely to lead to adequate explanations of social class phenomena.[92]

We conclude this section on IQ and social class by pointing out some of the social implications of the ideas we have developed. It has been argued that to the extent that our society (1) continues to follow the principles of a meritocracy, (2) manifests increasing assortive mating for intelligence, and (3) increases environmental and educational opportunity for everyone, the more our society will be stratified on the basis of biological differences.[26, 83,] The range of differences will be greater and the uniformity within families and classes greater. In Scarr-Salapatek's words, "The thesis of egalitarianism surely leads to its antithesis in a way that Karl Marx never anticipated."[93]

While the general thrust of the above argument is correct, and the trends described will probably develop to a considerable extent, like any other argument based on linear extrapolation, it can only be wrong if pushed too far. There are a number of flaws in the argument,[93] but the most serious one is the myopic focus on one trait.

In spite of the general law of positivity, i.e. most socially desirable traits including IQ tend to correlate positively,[94] IQ is not an overall measure of the social value of an individual. Individuals are multidimensional, and we have no idea what the contribution of particular combinations of traits, transmitted independently, is to social value. The focus on IQ tests is overly analytic and verbal. Terman's sample of gifted (defined by IQ> 160) children, for example, contained no gifted painters, sculptors, composers, or poets.[95] One might also ask about actors, musicians, dancers, etc. We are raising the issue of the creativity-IQ distinction,[96] on which we cannot elaborate here. But, even if the correlation between IQ and creativity is high, it certainly is not unity, and the thrust of our argument is not changed.

Selection for coadaptive gene complexes, as opposed to single traits, is the rule in nature[97, 98]; and our society is slowly coming to realize both the wastefulness and decreased diversity to which the overemphasis on IQ leads. Some scholars have recognized this and have attacked this misplaced emphasis in selection for college entrance.[99, 100] As our educational system opens itself to greater diversity of talent, total genotypic diversity should increase, not decrease. As our society becomes more diversified in the social, political, technological, educational, and cultural domains, it should require greater heterogeneity of talent, not less. Thus, there is likely to be a limit on biological stratification because of the processes of random gene segregation and recombination which influence the distribution of these multiple traits in a population. These considerations should also be kept in mind when eugenic proposals that favor the maximization of a single trait such as IQ are confronted. The reader is referred to Gottesman and Erlenmeyer-Kimling are for sensitive discussion of these issues with respect to population policies.[101]

THE FUTURE OF INTELLIGENCE

Differential reproduction lies at the heart of modern evolutionary theory. Natural and artificial selction work because they allow the carriers of some genotypes to survive and the carriers of others to be eliminated. Fitness in the Darwinian sense is measured only in terms of reproductive efficiency.

Thinking men have long been concerned with the effect of differential reproduction on the level of intelligence in a population. Galton condemned the peerage as a disastrous institution because of the breeding patterns it fostered.[102] More recently, concern has been expressed over the relationship between IQ and family size. This correlation has generally been reported to be of the order of −.20 to −.30.[103, 104, 105,] Such a high negative correlation should have led to a decline in the intelligence of the populations sampled. When follow-up studies were conducted, however, small improvements in mean scores were found, rather than the predicted declines.[106, 107] The failure of the decline to materialize came to be called Cattell's paradox.

The paradox was resolved by Higgins, Reed, and Reed who showed that the correlations existed because the studies omitted all those persons in each generation who did not reproduce at all.[108] When these childless members of a generation are included, the paradox is resolved. It necessarily follows that childless members of a generation must, on the average, have a low IQ; and, indeed in the Higgins, Reed, and Reed study, it was 80.46. This classic study has been replicated twice,[109, 110] and the combined results of all three studies are shown in Figure 8-9.

There had been some speculation about the importance of the slight bimodality in the Higgins, Reed, and Reed, and Bajema data.[26, 27] It could have indicated a degree of polarization at the high and low ends of the IQ scale with a marked loss of heterozygotes. The Waller data, however, suggest that the bimodality is due to sampling problems rather than a real effect.

Bajema pointed out that in order to estimate the direction and intensity of natural selection for IQ, one must take into account fertility, mortality, and generation length, simultaneously.[109] This can be done by using the statistic, intrinsic rate of natural increase. Table 8-XII shows the intrinsic rate of natural increase, average generation length, and fitness of both Waller's[110] and Bajema's[109] data. All fitness ratios are expressed as decimal fractions of the value for the highest group.

In both studies natural selection is positive and seems to indicate that parents with the highest IQ scores have a reproductive advantage. Although we should be cautious in generalizing from these data (due to the restricted nature of the samples), they do coincide with the findings of incomplete directional dominance found in the Burt data by Jinks and Fulker.[63] Such findings would be expected if natural selection for intelligence is the continuing process that Reed has suggested it should be.[112]

Like many of the parameters discussed in this chapter the relationship between

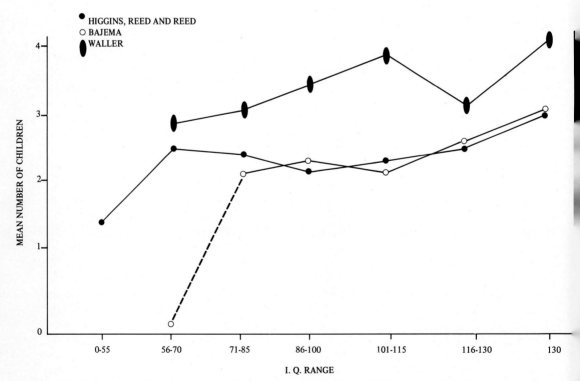

Figure 8-9. Differential fertility and intelligence. (From A. Falek, "Differential Fertility and Intelligence Current Status of the Problem," *Social Biology,* 18:550 1971].[111])

TABLE 8-XII
INTRINSIC RATE OF NATURAL INCREASE, AVERAGE GENERATION LENGTH, AND FITNESS ACCORDING TO PARENTAL IQ

Parental IQ Range	Intrinsic Rate of Natural Increase	Average Generation Length (Yrs.)	Fitness (Av. No. Offspring/Person)
Bajema (1963)			
-120	+0.0089	29.42	1.000
105-119	+0.0039	28.86	0.8614
98-104	+0.0003	28.41	0.7771
80-94	+0.0075	28.01	0.9484
69-79	-0.0100	28.76	0.5774
Total sample	+0.0039	28.49	---
Waller (1971)			
131-150	+0.0299	26.59	1.0000
116-130	+0.0198	27.74	0.8218
101-115	+0.0213	27.91	0.8597
86-100	+0.0205	28.26	0.8520
71-85	+0.0211	28.12	0.8067
56-70	+0.0188	27.58	0.8073
Total sample	+0.0212	27.97	0.8504

From A. Falek, "Differential Fertility and Intelligence: Current Status of the Problem," *Social Biology*, 18:550 (1971).

family size and IQ is not a constant. It will vary as a function of many other variables such as general economic conditions, differential availability of birth control information, welfare policies, and level of medical care. The studies discussed above were based on completed families that were raised many years ago. Conditions are probably very different today. Whether the same relationship continues to hold is an open question, one worthy of continual scrutiny.

SUMMARY

Human intelligence can be usefully construed as a single relatively coherent trait whose phenotypic variance is largely under genetic control. The relative influence of various genetic and environmental factors have been estimated, but only crudely.

There is probably a moderate ordinal interaction between genotypes for intelligence and environment with genotypes for higher intelligence exhibiting greater susceptibility to both environmental deprivation and enrichment. This interaction can be best understood as an example of the genetic concept of "norm of reaction" with favored genotypes manifesting a wider reaction range than the less favored.

There is increasing evidence that societies which reward on the basis of individual ability (meritocracies) are becoming stratified in such a way that class differences in intelligence have an appreciable genetic component. There are, however, reasons to expect that biological stratification on the basis of intelligence will be self-limiting.

Earlier studies of family size and IQ led to the prediction of a decline in the intelligence of the populations studied. These predictors were shown to be due to a methodological error, and natural selection for IQ was shown to be positive in the same populations. Nevertheless, the relationship is a dynamic one, varying with different social conditions, and it should be subject to continual scrutiny.

REFERENCES

1. Cancro, R. (Ed.): *Intelligence: Genetic and Environmental Influences.* New York, Grune, 1971.
2. Ehrman, L., and Omenn, G. (Eds.): *Genetic Endowment and Environment in the Determination of Behavior.* New York, Acad Pr, 1972.
3. Harvard Educational Review: *Environment, Heredity and Intelligence.* Reprint Series No. 2. Cambridge, Harvard U Pr, 1969.
4. Vandenberg, S.G.: The nature and nurture of intelligence. In Glass, D.C. (Ed.): *Genetics.* New York, Rockefeller, 1968.
5. Burt, C.: The evidence for the concept of intelligence. *Br J Educ Psychol, 25*:158, 1955.
6. Partridge, E.: *Origins: A Short Etymological Dictionary of Modern English.* New York, Macmillan, 1958.
7. Wechsler, D.: *The Measurement and Appraisal of Adult Intelligence.* Baltimore, Williams, 1958.
8. Bouchard, T.J., Jr.: Current conceptions of intelligence and their implications for assessment. In McReynolds, P. (Ed.): *Advances in Psychological Assessment.* Palo Alto, Science Behavior Books, 1968, vol. I.
9. McNemar, Q.: Lost: Our intelligence? Why? *Am Psychol, 19*:871, 1964.
10. Vernon, P.E.: Ability factors and environmental influences. *Am Psychol, 20*:723, 1965.
11. Cattell, R.B.: The structure of intelligence in relation to the nature-nurture controversy. In Cancro, R. (Ed.): *Intelligence: Genetic and Environmental Influences.* New York, Grune, 1971.
12. Horn, J.L.: Organization of abilities and the development of intelligence. *Psychol Rev, 75*:242, 1968.
13. Guilford, J.P.: *The Nature of Human Intelligence.* New York, McGraw, 1967.
14. Wolf, T.H.: The emergence of Binet's conceptions and measurement of intelligence: A case history of the creative process. *J His Behav Sci, 5*:113, 1969.
15. Wolf, T.H.: The emergence of Binet's conceptions and measurement of intelligence: II. A case history of the creative process. *J His Behav Sci, 5*:207, 1969.
16. Tuddenham, R.D.: The nature and measurement of intelligence. In Postman, L. (Ed.): *Psychology in the Making.* New York, Knopf, 1962.
17. Furth, H.G.: *Thinking Without Language: Psychological Implications of Deafness.* New York, Free Pr, 1966.
18. Furth, H.G.: Linguistic deficiency and thinking: Research with deaf subjects 1964-1969. *Psychol Bull, 76*:58, 1971.
19. Binet, A., and Simon, T.: New methods for the diagnosis of the intellectual level of subnormals. *Ann Psychol, 11*:191, 1905.
20. Ghiselli, E.: *The Validity of Occupational Aptitude Tests.* New York, Wiley, 1966.
21. Burt, C., and Howard, M.: The multifactorial theory of inheritance and its application to intelligence. *Br J Statist Psychol, 9*:95, 1956.
22. Burt, C.: Quantitative genetics in psychology. *Br J Math Statist Psychol, 24*:1, 1971.
23. Burt, C.: Inheritance of general intelligence. *Am Psychol, 27*:175, 1972.
24. Erlenmeyer-Kimling, L., and Jarvik, L.F.: Genetics and intelligence: A review. *Science, 142*:1477, 1963.
25. Gottesman, I.I.: Genetic aspects of intelligent behavior. In Ellis, N. (Ed.): *Handbook of Mental Deficiency: Psychological Theory and Research.* New York, McGraw, 1963.
26. Gottesman, I.I.: Biogenetics of race and class. In Duetsch, M., Katz, I., and Jensen, A.R. (Eds.): *Social Class, Race and Psychological Development.* New York, HR&W, 1968.
27. Jensen, S.R.: How much can we boost IQ and scholastic achievement? *Harvard Educ Rev, 39*:1, 1969.

28. Reed, E.W., and Reed, S.C.: *Mental Retardation: A Family Study.* Philadelphia, Saunders, 1965.
29. Li, C.C.: A tale of two thermos bottles: properties of a genetic model for human intelligence. In Cancro, R. (Ed.): *Intelligence: Genetic and Environmental Influences.* New York, Grune, 1971.
30. Vanderberg, S.G.: Human behavior genetics: Present status and suggestions for future research. *Merrill-Palmer Q, 15*:121, 1969.
31. Pickford, R.W.: The genetics of intelligence. *J Psychol, 28*:129, 1949.
32. Waller, J.H.: Achievement and social mobility: Relationship among IQ score, education, and occupation in two generations. *Soc Biol, 18*:252, 1971.
33. Li, C.C.: The correlation between parents and offspring in a random mating population. *Am J Hum Genet, 6*:383, 1954.
34. Hunt, J. McV.: Has compensatory education failed? Has it been attempted? In Harvard Educational Review: *Environment, Heredity and Intelligence.* Reprint Series No. 2. Cambridge, Harvard U Pr, 1969.
35. Lewis, D.G.: The normal distribution of intelligence: A critique. *Br J Psychol, 48*:98, 1957.
36. Burt, C.: The distribution of intelligence. *Br J Psychol, 48*:161, 1957.
37. Burt, C.: Is intelligence normally distributed? *Br J Statist Psychol, 16*:175, 1963.
38. Penrose, L.S.: *The Biology of Mental Defect.* London, Sidgwick & Jackson, 1963.
39. Zigler, E.: Familial mental retardation: A continuing dilemma. *Science, 155*:292, 1967.
40. Roberts, J.A.F.: The genetics of mental deficiency. *Eugen Rev, 44*:71, 1952.
41. Burt, C.: The inheritance of mental ability. *Am. Psychol, 13*:1, 1958.
42. Garrison, R.J., Anderson, V.E., and Reed, S.C.: Assortative marriage. *Eugen Q 15*:113, 1968.
43. Higgins, J.V.: *An Analysis of Intelligence of 1,016 Families.* Unpublished doctoral dissertation, U Minn, 1961.
44. Conrad, H.S., and Jones, H.E.: A second study of familial resemblance in intelligence. *Yearbook of the National Society for the Study of Education, 39*:97, 1940.
45. Rosenthal, D.: *Genetic Theory and Abnormal Behavior.* New York, McGraw, 1970.
46. Rosenthal, D.: A program of research on heredity in schizophrenia. *Behav Sci, 16*:191, 1971.
47. Burks, B.S.: The relative influence of nature and nurture upon mental development: A comparative study of foster parent-foster child resemblance and true parent-true child resemblance. *Yearbook of the National Society for the Study of Education, 27*:219, 1928.
48. Bloom, B.S.: *Stability and Change in Human Characteristics.* New York, Wiley, 1964.
49. Jensen, A.R.: Social class and verbal learning. In Deutsch, M., Katz, I., and Jensen, A.R. (Eds.): *Social Class, Race and Psychological Development.* New York, HR&W, 1968.
50. Jensen, A.R.: A note on why genetic correlations are not squared. *Psychol Bull, 75*:223, 1971.
51. Wright, S.: Statistical methods in biology. *J Am Statist Assoc, 26*:155, 1931.
52. Newman, H.H., Freeman, F.N., and Holzinger, K.J.: *Twins: A Study of Heredity and Environment.* Chicago, U of Chicago Pr, 1937.
53. Newman, H.H.: *Twins and Super-Twins.* London, Hutchison, 1942.
54. Jensen, A.R.: A reply to Gage: The causes of twin differences in IQ. *Phi Delta Kappan, 53*:420, 1972.
55. Skodak, M., and Skeels, H.M.: A final follow-up study of one hundred adopted children. *J Gen Psychol, 75*:85, 1949.
56. Honzik, M.P.: Developmental studies of parent-child resemblance in intelligence. *Child Dev, 28*:215, 1957.
57. McCall, R.B.: Intelligence quotient pattern over age: comparisons among

siblings and parent-child pairs. *Science, 170*:644, 1970.
58. Weizmann, F.: Correlational statistics and the nature-nurture problem. *Science, 171*:589, 1971.
59. Cooper, R., and Zubek, J.: Effects of enriched and restricted early environments on the learning ability of bright and dull rats. *Can J Psychol, 12*:159, 1958.
60. Henderson, N.D.: Genetic influences on the behavior of mice can be obscured by laboratory rearing. *J Comp Physiol Psychol, 72*:505, 1970.
61. Wiseman, S.: Environmental and innate factors and educational attainment. In Meade, J.E., and Parkes, A.S. (Eds.): *Genetic and Environmental Factors in Human Ability.* London, Oliver & Boyd, 1966.
62. Scarr-Salapatek, S.: Race, social class and IQ. *Science, 174*:1285, 1971.
63. Jinks, J.L., and Fulker, D.W.: Comparison of the biometrical, genetical, MAVA, and classical approaches to the analysis of human behavior. *Psychol Bull, 73*:311, 1970.
64. Shields, J.: *Monozygotic Twins Brought Up Apart and Brought Up Together.* London, Oxford U Pr, 1962.
65. Cronbach, L.J., and Snow, R.E.: Individual differences in learning ability as a function of instructional variables. Eric Document, ED-029001, 1969. Bethesda, Maryland, Eric Document Reproduction Services (Leasco Information Products, Inc., 4827 Rugby Ave, Bethesda, Maryland, 20014).
66. Bell, R.Q.: A reinterpretation of the direction of effects in studies of socialization. *Psychol Rev, 75*:81, 1968.
67. Cattell, R.B.: The interaction of heredity and environmental influences. *Br J Statist Psychol, 16*:191, 1963.
68. Hayes, K.J.: Genes, drives and intellect. *Psychol Rep, 10*:299, 1962.
69. Erlenmeyer-Kimling, L.: Gene-environment interactions and the variability of behavior. In Ehrman, L, and Omenn, G. (Ed.): *Genetic Endowment and Environment in the Determination of Behavior.* New York, Acad Pr, 1972.
70. Haldane, J.B.S.: The interaction of nature and nurture. *Ann Eugen, 13*:197, 1946.
71. Vale, J.R., and Vale, C.A.: Individual differences and general laws in psychology. *Am Psychol, 24*:1093, 1969.
72. Falconer, D.S.: *Introduction to Quantitative Genetics.* New York, Ronald, 1960.
73. Roberts, R.C.: Some concepts and methods in quantitative genetics. In Hirsch, J. (Ed.): *Behavior Genetic Analysis.* New York, McGraw, 1967.
74. Cavalli-Sforza, L.L., and Bodmer, W.F.: *The Genetics of Human Populations.* San Francisco, Freeman, 1971.
75. Falconer, D.S.: Genetic consequences of selection pressure. In Meade, J.E., and Parkes, D.S. (Eds.): *Genetic and Environmental Factors in Human Ability.* Edinburgh, Oliver & Boyd, 1966.
76. Eaves, L.J.: Computer simulation of sample size and experimental design in human psychogenics. *Psychol Bull, 77*:144, 1972.
77. Burt, C.: The genetic determination of differences in intelligence: a study of monozygotic twins reared together and apart. *Br J Psychol, 57*:137, 1966.
78. Tyler, L.: *The Psychology of Human Differences.* New York, Appleton, 1965.
79. Duncan, O.D.: Ability and achievement. *Eugen Q, 15*:1, 1968.
80. Stewart, N.: AGCT scores of Army personnel grouped by occupation. *Occupations, 26*:5, 1947.
81. Ball, R.S.: The predictability of occupational level from intelligence. *J Consult Psychol, 2*:184, 1938.
82. Eckland, B.K.: Social class structure and the genetic basis of intelligence. In Cancro, R. (Ed.): *Intelligence: Genetic and Environmental Influences.* New York, Grune, 1971.
83. Herrnstein, R.: I.Q. *The Atlantic, 228*:43, 1971.
84. Blau, P., and Duncan, O.D.: *The American*

Occupational Structure. New York, Wiley, 1967.
85. Eckland, B.K.: New mating boundaries in education. *Soc Biol, 17*:269, 1970.
86. Bajema, C.: Relation of fertility to occupational status, IQ, educational attainment, and size of family of origin: A follow-up study of a male Kalamazoo public school population. *Eugen Q, 15*:198, 1968.
87. Conway, J.: Class differences in general intelligence. II. *Br J Statist Psychol, 12*:5, 1959.
88. Burt, C.: Intelligence and social mobility. *Br J Statist Psychol, 14*:3, 1961.
89. Gibson, J.B.: Biological aspects of a high socio-economic group: I. IQ, education and social mobility. *J Biosoc Sci, 2*:1, 1970.
90. Floud, J.E., and Halsey, A.H.: Measured intelligence is largely an acquired characteristic. *Br J Educ Psychol, 28*:290, 1958.
91. Halsey, A.H.: Class differences in general intelligence: I. *Br J Statist Psychol, 12*:1, 1959.
92. Thoday, J.M., and Gibson, J.B.: Environmental and genetic contributions to class differences: a model experiment. *Science, 167*:990, 1970.
93. Scarr-Salapatek, S.: Unknowns in the IQ equation. *Science, 174*:1223, 1971.
94. Humphreys, L.G.: Theory of intelligence. In Cancro, R. (Ed.): *Intelligence: Genetic and Environmental Influences.* New York, Grune, 1971.
95. Terman, L.M.: Scientists and nonscientists in a group of 800 gifted men. *Psychol Monogr, 68*:No. 7, 1954.
96. Wallach, M.A.: *The Intelligence/Creativity Distinction.* New York, General Learning Press, 1971.
97. Caspari, E.: Introduction to Part I and remarks on evolutionary aspects of behavior. In Hirsch, J. (Ed.): *Behavior-Genetic Analysis.* New York, McGraw, 1967.
98. Lerner, I.M.: *Heredity Evolution and Society.* San Francisco, Freeman, 1968.
99. Gough, H.G.: Misplaced emphases in admissions. *J College Stu Pers, 6*:130, 1965.
100. MacKinnon, D.W.: The nature and nurture of creative talent. *Am Psychol, 17*:484, 1962.
101. Gottesman, I.I., and Erlenmeyer-Kimling, L.: Prologue: a foundation for informed eugenics. *Soc Biol, 18*:51, 1971.
102. Galton, F.: *Hereditary Genius.* Cleveland, World Publishing, 1962. (Originally published 1869.)
103. Cattell, R.B.: *The Fight for Our National Intelligence.* London, King, 1937.
104. Roberts, J.A.F.: The negative association between intelligence and fertility. *Hum Biol, 13*:410, 1941.
105. Burt, C.: *Intelligence and Fertility.* London, Hamilton, 1946.
106. Scottish Council for Research in Education: *The Trend of Scottish Intelligence: A Comparison of the 1947 and 1932 Surveys of the Intelligence of 11-Year Old Pupils.* London, U of Lond Pr, 1949.
107. Cattell, R.B.: The fate of national intelligence: Test of a thirteen-year prediction. *Eugen Rev, 42*:136, 1951.
108. Higgins, J.V., Reed, E.W., and Reed, S.C.: Intelligence and family size: A paradox resolved. *Eugen Q, 9*:84, 1962.
109. Bajema, C.: Estimation of the direction and intensity of natural selection in relation to human intelligence by means of the intrinsic rate of natural increase. *Eugen Q, 10*:175, 1963.
110. Waller, J.H.: Differential reproduction: Its relation to IQ test score, education, and occupation. *Soc Biol, 18*:122, 1971.
111. Falek A.: Differential fertility and intelligence: Current status of the problem. *Soc Biol, 18*:550, 1971.
112. Reed, E.W., and Reed, S.C.: The evolution of human intelligence. *Am Sci, 53*:317, 1965.

Chapter 9

GENETIC FACTORS IN PERSONALITY DEVELOPMENT

H. J. Eysenck

A CHAPTER DEALING WITH the genetics of personality should ideally consider in detail two points which in effect will only be discussed in a somewhat perfunctory manner. These two points are: (1) the nature of personality and personality measurement, and (2) the methodological and statistical problems involved in genetic analysis. The problems raised by the definition of the term personality and the many divergent views about the best ways of measuring the various aspects of personality which can be quantified are not amenable to anything but fairly dogmatic assertion within the limitations of a single chapter; readers who feel dissatisfied with the cursory treatment given these topics here may wish to consult more detailed discussions and presentations.[1, 2, 3] We shall here use the term to mean semipermanent behavior patterns characteristic of individuals and of social importance and relevance. The notion of individual differences is obviously central to the personality concept, but there are many accidental and unimportant differences between individuals which would not be considered part of personality by most psychologists. The most widely accepted conceptualization of these individual differences is in terms of traits and types, and there is now considerable agreement on some form of hierarchical scheme, in which a "type" concept, such as extraversion, is based on the observed correlations between traits such as sociability, impulsiveness, activity, liveliness, and excitability (see Fig. 9-1). Traits themselves are similarly based on observed correlations between different manifestations. Thus, the trait sociability is based on the observation that individuals who like to go to parties also like to have many friends, like to talk to strangers, and dislike reading in solitude. Type-concepts are not differentiated from trait-concepts in terms of distribution; just as sociability is fairly normally distributed, so is extraversion. The erroneous notion that type-concepts implied bimodal distributions or actual categorical divisions into qualitatively different types dates back to Galen and later medieval writers and is still present in Kant; it does not form part of the teaching of any of the modern typologists (e.g. Jung, Kretschmer, Sheldon). In order to obviate misconceptions of this sort, Eysenck suggested the conception of dimensions of personality.[4] This also makes clear the close relation between the hierarchical model and the mathematical

---

The preparation of this material was aided by a grant from the Colonial Research Fund.

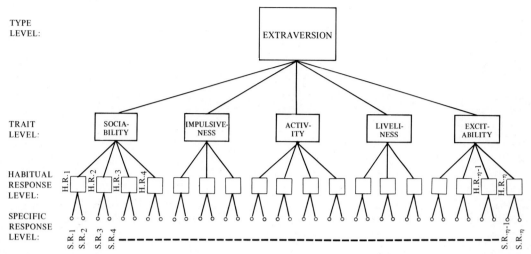

Figure 9-1. Types are superordinate concepts arising from the intercorrelation of traits (subordinate concepts). Extraversion is a type concept: sociability, activity, etc. are trait concepts.

technique of factor analysis which is fundamental to its empirical working out.[5]

At the type level, most investigators have discovered two major dimensions, extraversion-introversion (E) and neuroticism-stability (N). Cattell prefers the labels of exvia-invia and anxiety, and other writers, such as Burt, have chosen the term *emotionality* instead of neuroticism. Terminology apart, there is substantial agreement on the facts; Eysenck and Eysenck have surveyed the literature in some detail and demonstrated empirically the almost total agreement at the type level between the measuring instruments constructed by Cattell, Guilford, and Eysenck.[3] There is much less agreement at the trait (primary factor) level; the factors discovered by Cattell, Guilford, and Eysenck do not show much overlap, and with regard to Cattell's factors in particular, many large-scale replications using his own items have failed to produce factors at all similar to those claimed by him.[6,7,8,9] In evaluating papers on the genetic component in such nonreplicable factors, it must always be borne in mind that we may be dealing with statistical artifacts and that the reality and meaningfulness of these factors are very much less firmly established than that of type factors like E and N. It should also be borne in mind that often questionnaires and other personality measures are labeled *extraversion* or *neuroticism* but that, unless there is direct evidence (through factor analysis, through correlation with well-established measures of E or N, or by reference to clinical groups) of the correctness of the application, these terms may be misleading; examples are given by Eysenck.[2]

Many personality measures, particularly questionnaires, have face validity but have not been submitted to proper factor analytic study of item intercorrelation; this applies particularly to the orthodox scales of the MMPI, the California Psychological Inventory, the Edwards scales, and many other widely used tests. It cannot be stressed too forcefully that a scale does not measure ambition simply because it has been so labeled, nor does it necessarily measure any one thing or concept at all; without direct proof it may simply consist of a collection of items having nothing whatever in common. Psychologists have been too ready to accept personality measures as meaningful without looking carefully into their credentials; many of the contradictions in the literature, particularly that concerning genetic effects, might vanish if a more critical attitude were adopted. The same may be said about tests of the projective kind; these are largely unreliable, subjective, and invalid and require special precautions in their use if the results are to be taken seriously.[10]

Factor-analytic models are descriptive; science prefers models which contain causal contingencies. A causal model linking E and N with certain physiological mechanisms has been advanced by Eysenck;[11] it is suggested that introversion is produced essentially by high arousal levels in the cortex (which in turn are produced by an overly active ascending reticular activating system), while extraversion is characterized by low arousal levels. Studies of EEG records bear this out; introverts tend to show low amplitude, high frequency alpha, as compared with extraverts. High N is connected with excessive responsiveness of the autonomic system (particularly the sympathetic branch) and the visceral brain. These hypotheses link the study of personality with electrophysiological work and also with traditional experimental laboratory studies of conditioning, vigilance, sensory thresholds, etc., as these are known to be related to differences in arousal. Studies along these lines would seem to hold out most promise for genetic analysis; but, as we shall see, they have been pursued much less actively than those involving questionnaires. Possibly the ease with which questionnaire data can be gathered has been a factor in this; it is to be hoped that in the future more attention will be paid to these apparently more promising areas.

It will be seen that factor analysis is fundamental to modern conceptions of personality, and to the unraveling of the complex relations underlying even the descriptive model. There is, however, another reason why factor analysis is an indispensable tool in relation to genetic analysis in particular. Biometrical statistics work with scores of one kind or another, but a simple consideration will show that no interpretation is possible unless the variance can be partitioned into several different parts of the total score. Consider a simple questionnaire purporting to measure impulsiveness. The total variance on this test has at least four components, as follows:

$$s^2{}_T = s^2{}_{Imp.} + s^2{}_{Ex} + s^2{}_S + s^2{}_E. \quad (1)$$

In this formula, s^2 stands for the variance, T for the test score; the test variance is made up of variance due to the trait of impulsiveness, to extraversion, to specific factors affecting this test alone, and to error. Thus, T cannot be interpreted as a measure of impulsiveness alone; it is a compound measure. If we should discover by twin research that a specified degree of heritability attaches to T, we would still not know what it was that was being determined by heredity: it might be impulsiveness, it might be extraversion, or it might be the specific factors of the test; it might, of course, also be any combination of all three. Thus, our analysis should, by preference, be of factors and factor scores, not of tests and test scores; only in this way does any clearly specifiable psychological meaning attach to our genetic analysis. This point of view was first put forward by Eysenck and Prell,[12] but it does not seem to have been seriously considered by other workers in this field. No criticism has been made of the argument, but the suggested method has not been taken up. This is regrettable, as interpretation of published work is made more difficult, and perhaps ultimately impossible, by this neglect to partition the total score variance. In assessing the evidence, this point should certainly be borne in mind.

When we turn to the actual statistical methods used for determining heritability, dominance, and other genetic variables from psychological experimentation, we note that there are three distinct stages through which development has taken place (disregarding for this purpose Galton's original work which was, of course, of considerable historical importance, but not designed to give quantitative assessments of genetic variables). First, we have what might be called the classical approach through correlations between relatives (mostly MZ and DZ twin pairs), issuing in the estimation of various ratios of genetic and environmental influences on trait variation. Among the better known ones are Holzinger's H,[13] Neel and Schull's E,[14] and Nichols' HR.[15] These estimates are concerned with within-family environmental variation, not with between-family variation. Second, we have Cattell's more systematic and comprehensive Multiple Abstract Variance Analysis approach, which leads to both an estimation of nature/nurture ratios and an assessment of the importance of the correlation between genetic and environmental influences within the family as well as within the culture.[16,17] And, third, we have the biometrical genetical approach initiated originally by Fisher,[18] extended and applied by Mather,[19] and finally revised and used in connection with the genetic analysis of intelligence and personality by Jinks and Fulker.[20] This last reference is now the *loc. class.* for research in this field, and we will rely heavily both on the methodological discussion given there, and also on the results of their reanalysis of the Maudsley personality data.*

The biometrical genetical approach includes the first two approaches as special cases but goes farther by attempting an assessment of the kinds of gene action and mating system operating in the population. Although clearly much more powerful and inclusive and much less subject to unprovable assumptions, this new approach has been used almost exclusively in investigations with animals and not in the study of human populations; for this reason most of the data here discussed are derived from studies whose planning and analysis were suboptimal. Fortunately, it is possible to transfer some of the findings from the Jinks and Fulker study[20] to a discussion of the results of earlier work, thus demonstrating the applicability of assumptions which up to now seemed quite arbitrary but for which we now have some support.

Before turning to the results of the biometrical approach and before discussing the results of the classical approaches, it may be useful to discuss the question of assortative mating. Departures from panmixia are well documented in relation to such variables as age,[21,22] religion,[21,23] ethnic background,[24,25] physical traits,[26,27,28] intelligence,[29,30,31,32] and opinions and attitudes. In all these fields the calculation of heritability ratios must therefore take into account assortative mating. When we turn to personality we find that the situation is rather different. Correlations between married partners

---

*Svancara[53] has suggested that heritability estimates may change from one point of the life cycle to another; he puts forward various models, but his data are not sufficient to lend support to any of these. The point is an intriguing one and might repay detailed study. It would, of course, be necessary to equate reliabilities and validities of tests used at various ages, a task itself not without considerable difficulty!

tend to be quite low, and they tend to depart from chance in both directions; coefficients may be positive[23, 33, 34] or negative.[35, 36, 37] The latter findings are usually interpreted in terms of some form of "complementarity" hypothesis, the former in terms of "like marries like"; Tharp has reviewed the literature[38] (see also Schellenberg and Bee,[39] and Katz, et al.[40]). Before trying to explain an effect, however, it might be wise to consider whether anything substantial is, in effect, in need of explanation; the few statistically significant correlations emerge from a forest of small and insignificant ones and may be simply chance artifacts. When we consider the results obtained with the MMPI by Hill[33] and by Gottesman[34] on ninety-seven couples in Minnesota and sixty-six couples in Boston, respectively, we find that there are two and three significant correlations (out of eleven) in their tables, but these do not coincide. Thus, Hill finds a significant correlation for paranoia (.27) in his sample, but Gottesman finds an insignificant one of $-.11$ in his sample, differing in direction from that found by Hill. Hypochondriasis is significant for Gottesman's sample (.26), but not for Hill's ($-.02$). Kerckhoff and Davis have suggested a complex "filter" theory in which the first two filters eliminate too dissimilar candidates, while the last filter selects for complementariness of needs; but their evidence hardly suffices for so complex a theory.[41]

Cattell and Nesselroade raise the possibility that stable marriages may be characterized by homogamy, unstable ones by complementarity.[42] Their results on 102 stable and thirty-seven unstable marriages lend some weak support to this notion. Out of fifteen noncognitive traits measured by the 16 PF, fourteen are positively correlated for partners in stable marriages, while in the unstable marriages eight coefficients are negative, three significantly so.

Interpretation, however, is difficult because of the doubtful interpretability and meaningfulness of the 16 PF scales.[49] However, Pickford, et al. found results which are vaguely in the same direction.[44] Even if true, this rule does not concern geneticists interested in heritability unless unhappy marriages produced fewer children; it is the fact of child-producing marriage which determines heritability. Our conclusion must be that, so far, no good evidence has been produced to rule out panmixia and that, if assortative mating takes place, its effects are likely to be small, at best.

There is one possible exception to this rule; as Vandenberg points out, there is homogamy for psychopathology. Kallmann and Mickey support this view as far as schizophrenia is concerned.[43, 45] Kreitman provides further evidence based on both neurotics and psychotics.[46, 47] Whether this generalization is true or not is not very relevant to our discussion; we are dealing with normal subjects in this chapter, not with schizophrenics or neurotics. There may be one qualification to this view, however; in addition to E and N, Eysenck and Eysenck have recently suggested a third type factor in the normal personality field, called *psychoticism* (P).[48, 49] If there is homogamy for psychopathology, this factor might show it. Two separate (unpublished) studies from our own Department give figures on assortative mating for P, E, and N. Nias studied 700 pairs of parents who, between them, had 1,200 children.[50] Selection was reasonably random in the sense that whole classes of children were approached and tested first, and later their parents were brought into the investigation. In the other study, Insel made a thorough investigation of ninety-eight families, including 589 subjects in all, involving in every case three generations.[51] The correlations found by them were quite insignificant (and indeed opposite in sign) for E (.06 and −.10) and N (.12 and −.09). For P, however, both report significant, positive correlations (.22 and .36, respectively.) These findings are in support of our general conclusion of general panmixia for most personality traits but homogamy for psychopathology of subpsychotic type in normal persons. If this conclusion is right, then we need not worry about correcting heritability estimates arrived at on the basis of assumptions about nonassortative mating. It may be worthwhile to point out that these two studies are of particular relevance to our discussion because they: (1) contain an adequate number of married couples, unlike most of the studies quoted and (2) used personality dimensions for which, unlike the MMPI or the 16 PF, there exists solid empirical support.

We must now turn to a consideration of empirical studies directly concerned with the genetic basis of personality. Most of these, as already noted, made use of the classical method of design and analysis, which means in effect nothing more than that MZ and DZ twin pairs of the same sex are administered questionnaires (or, more rarely, given laboratory tests, or rated on some behavior checklists). The investigator then applies one of the formulae for the determination of heritability already noted and states his conclusion that the variable putatively measured by means of the test or questionnaire is determined to a given degree by hereditary factors. We have already noted some obvious criticisms of this method, in so far as the tests chosen and their interpretation are concerned; we will later on note some further criticisms concerned with the genetic analysis and the general experimental design. Before doing so, it may be worthwhile pointing out that our review of the literature quite naturally falls into two parts. The first part deals with the American literature, which falls into the pattern outlined above, with the possible exception of the work of Cattell. The second part deals with the English literature, particularly the work done at the Maudsley Hospital; this differs from the American work in a number of ways which will be detailed after we have described the American studies. These begin with the classical book by Newman, Freeman, and Holzinger, which is still widely quoted.[52] These writers compared fifty pairs of MZ and fifty pairs of DZ twins with respect to a number of physical measurements and mental and educational tests; they concluded that "the physical characteristics are least affected by the environment, that intelligence is affected more; educational achievement still more; and personality or temperament, if our tests can be relied upon, the most. This finding is significant, regardless of the absolute amount of the environmental influence." The lack of hereditary influence on personality suggested here is still assumed to be a true statement of fact in the accounts given by most textbooks of psychology and psychiatry, but there are reasons for regarding it with suspicion, not only in the light of more recent work, but also in view of the many criticisms to be made of this early work. These have been summarized by Eysenck[11] as follows:

> There are two main criticisms. In the first place, the measures used would not now be regarded as either reliable or valid. They included the Woodworth-Mathews Personality Inventory, the Kent-Rosanoff Scale, the Pressey Cross-Out Test, and the Downey-Will Temperament Test. It is doubtful whether any psychologist would nowadays wish to make very strong claims for these measures; even if they could be regarded as reliable and valid, the question would still have to be asked; valid for what? The Woodworth-Mathews Inventory is the only one for which detailed statistics are presented, and we discover that for identical twins the intraclass correlation is .562, for fra-

ternal twins it is .371, and for identical twins brought up in separation it is .583. If we regard, as the original authors certainly did, this questionnaire as an inventory of neurotic tendency, then we would here seem to have some mild indication of the importance of heredity, seeing that identical twins are distinctly superior in point of intraclass correlation to fraternals. Moreover, and this is a particularly interesting feature of this table, identical twins brought up in separation are more alike than are identical twins brought up together; this is the only test used, including physical measurements, where this is true. The authors comment that "the Woodworth-Mathews Test appears to show no very definite trend in correlations, possibly because of the nature of the trait and also because of the unreliability of the measure." It is not quite clear to the present writer why there is this denial of a definite trend; it seems fairly clear that identical twins, whether brought up in separation or together, are more alike than are fraternal twins and we shall see, later modern work has amply justified such a conclusion.

We must now turn to our second criticism, which has curiously enough not to our knowledge been made before. The personality tests used by Newman, Freeman, and Holzinger were essentially tests for adults; the Woodworth-Mathews Inventory, for instance, was constructed specifically for selection purposes in the army and in hospitals. It is quite inadmissible to use tests of this kind on children, and as is made clear on page 106 of the twin study, the average age of the whole group of identical and fraternal twins is only about thirteen years. No details are given, but it is clear that there must have been children as young as eight or even younger in this group, and it is doubtful whether a large proportion of the children were in a position to understand the terms used in the tests or to give meaningful replies to them. Our own work in questionnaire construction[54] has clearly shown the difficulties attending the construction of personality inventories for children, and the difficulties which children may encounter in answering questions even when these are specifically constructed for them. Taking together these two criticisms of the Newman, Freeman, and Holzinger study,[52] it is perhaps justifiable to say that the data do not support their conclusions. Identical twins whether brought up together or in separation are clearly more alike than are fraternal twins,

thus suggesting the importance of heredity in contributing to the temperamental differences studied; the poor reliability and inappropriate nature of the test used suggest that the differences found would have been larger and possibly much larger had more suitable tests been employed. We shall indeed find in our examination of the evidence that later studies using more appropriate methods of examination have resulted in much better discrimination between identical and fraternal twins. It must, of course, be remembered that the Newman, Freeman, and Holzinger study[52] was a pioneering venture and that at the time, few if any personality tests existed, particularly as far as children were concerned. More to blame perhaps are later writers who have cited their work as support for the proposition that heredity is a relatively unimportant factor in the causation of individual differences in temperament and personality, without a closer look at the details of the evidence offered.[11]

The Newman, Freeman, and Holzinger study thus emerges with a heritability of about 30 percent, which is not negligible when we consider the inappropriateness of the tests; more impressive perhaps is the fact that MZ twins brought up separately are if anything more alike than MZ twins brought up together. It is possible that this odd result caused the authors to be hypercautious in their conclusions; we will see that later studies have found similar results so that we may consider the finding as probably nonartifactual. (It should be added here that our strictures apply only to that portion of the Newman, Freeman, and Holzinger work concerned with personality; for the rest, their book fully deserves its high standing in the literature.[52] They, themselves, obviously did not feel happy with the section on personality, as their caveat, already quoted, makes clear.)

Carter's early work[55] gave similar results to that of Newman, Freeman, and Holzinger[52]; he used Bernreuter's Personality Inventory on a group including high school students as well as more mature subjects. Results for fifty-five pairs of identical twins and forty-three pairs of like-sex twins are available for the four traits (allegedly) measured by the Bernreuter test: neurotic tendency, introversion, self-sufficiency, and dominance. The intraclass correlations for identical and fraternal twins are, respectively: .63 and .32 for neuroticism; .50 and .40 for introversion; .44 and −.14 for self-sufficiency; and .71 and .34 for dominance. The interpretation of these results must be subject to the proviso that the scales are by no means independent, as indeed emerges quite clearly from a table of intercorrelations given by Carter himself. For various reasons it seems that Bernreuter's arbitrary and subjective naming of his scores is misleading; what he has called "self-sufficiency" is probably nearest to our concept of introversion; with neurotic tendency, introversion (so-called), and lack of dominance, all forming part of our conception of neuroticism. Whether this be the correct interpretation or not, Carter's results certainly support the view that heredity plays some part in the genesis of the individual differences measured by the four scales of the Bernreuter Personality Inventory.

Gottesman[56] studied thirty-four pairs each of identical and fraternal same-sex adolescent twins, using the Minnesota Multiphasic Personality Inventory and Cattell's High School Personality Questionnaire. He summarized his results as follows: "Within the limits of the assumptions, the attempt at quantification of the proportion of scale variance accounted for by heredity gave positive results for six of the HSPQ factors. Factors E, Submissiveness versus Dominance; H, Shy, Sensitive versus Adventurous; and J, Liking Group Action versus Fastidiously Individualistic, showed appreciable variance accounted for by heredity but with environment predominating. Factors F, Q2 and O, Confident Adequacy versus Guilt Proneness, showed about equal contributions of heredity and environment. The same kind of analysis of the MMPI gave positive results for five of the ten scales. Scales 7 (Psychasthenia) and 8 (Schizophrenia) showed appreciable variance accounted for by heredity but with environment predominating. Scales 2 (Depression) and 4 (Psychopathic Deviate) showed about equal contributions of heredity and environment. Scale 0 (Social Introversion) showed a predominance of variance (H = .71) accounted for by heredity. The value H for the Otis IQ in this study was .62." Table 9-I shows the results in some detail.

In a second study, Gottesman[34] reports on results from another twin experiment using eighty-two pairs of MZ twins and sixty-eight pairs of DZ twins. The detailed figures of his analysis are given in Table 9-II. It will be seen in both Tables 9-I and 9-II that the highest heritability values relate to extraversion-introversion variables (social introversion, psychopathic deviate, psychasthenia, and to a lesser degree, hysteria). The psychotic scales (schizophrenia and depression) also reach satisfactory levels. It is unfortunate that Gottesman used the MMPI scales as they stood; it is well known that these scales are far from univocal.[2] A factor-analytic study might have revealed much more clear-cut results by relating herita-

TABLE 9-I
MINNESOTA STUDY MMPI SCALE HERITABILITY INDICES
FOR TOTAL GROUP, FEMALES AND MALES

Scale		Total Group H	F^a	Females H	F^a	Males H	F^a
1	Hypochondriasis	.16	1.19	.25	1.33	.01	1.01
2	Depression	.45	1.81	.22	1.28	.65	2.83
3	Hysteria	.00	.86	.00	.56	.43	1.74
4	Psychopathic Deviate	.50	2.01	.37	1.60	.77	4.35
5	Masculinity-Femininity	.15	1.18	.00	.99	.45	1.83
6	Paranoia	.05	1.05	.00	.70	.52	2.09
7	Psychasthenia	.37	1.58	.47	1.89	.24	1.31
8	Schizophrenia	.42	1.71	.36	1.56	.50	2.00
9	Hypomania	.24	1.32	.33	1.50	.00	.81
10	Social Introversion	.71	3.42	.60	2.49	.84	6.14

[a]The three values of F required for significance at the .05 level are 1.78, 2.04, and 2.72 for the total group, the females, and males.

Quoted with permission from I. I. Gottesman, "Personality and Natural Selection," in S. G. Vandenberg (ed.), *Methods and Goals in Human Behavior Genetics* (New York, Academic Press, Inc., 1965).

TABLE 9-II
HARVARD TWIN STUDY MMPI SCALE HERITABILITY INDICES
FOR TOTAL GROUP, FEMALES, AND MALES

Scale		Total Group		Females		Males	
		H	F^a	H	F^a	H	F^a
1	Hypochondriasis	.01	1.01	.00	.89	.09	1.10
2	Depression	.45	1.82	.42	1.91	.37	1.58
3	Hysteria	.30	1.43	.44	1.79	.11	1.12
4	Psychopathic Deviate	.39	1.63	.46	1.47	.46	1.84
5	Masculinity-Femininity	.29	1.41	.30	1.43	.29	1.42
6	Paranoia	.38	1.61	.40	1.67	.40	1.67
7	Psychasthenia	.31	1.46	.60	2.53	.00	.99
8	Schizophrenia	.33	1.49	.37	1.58	.29	1.41
9	Hypomania	.13	1.15	.19	1.23	.04	1.04
10	Social Introversion	.33	1.49	.35	1.53	.29	1.40

[a]The three values of F required for significance at the .05 level are 1.47, 1.66, and 1.78 (preliminary analysis).

Quoted with permission from I. I. Gottesman, "Personality and Natural Selection," in S. G. Vandenberg (ed.), *Methods and Goals in Human Behavior Genetics* (New York, Academic Press, Inc., 1965).

TABLE 9-III
THE F RATIOS FOR FRATERNAL AND IDENTICAL WITHIN-PAIR VARIANCES
ON 11 SCORES OF HIGH SCHOOL PERSONALITY QUESTIONNAIRE

	Score	Cattell 1965	Vandenberg 1962	Gottesman 1963
1	Tender-minded vs. tough-minded	1.47	0.97*	1.07
Q4	Nervous tension vs. autonomic relaxation	1.56*	2.08†	0.53
C	General neuroticism vs. ego strength	1.60*	3.20†	1.03
Q3	Will control	1.08	1.87*	1.53
D	Impatient dominance	1.35	0.93	0.62
A	Cyclothymia vs. schizothymia	1.08	1.30	1.11
H	Adventurous cyclothymia vs. withdrawn schizothymia	1.34	0.93	1.62
K	Socialized morale vs. boorishness (education-minded)	1.39	1.06	-
E	Dominance vs. submissiveness	0.90	0.97	1.44
J	Energetic conformity vs. quiet egocentricity	1.57*	1.56	1.41
F	Surgency vs. desurgency	1.47	1.45	2.29†
	Number of DZ pairs	32	37	34
	Number of MZ pairs	52	45	34

*p 0.05
†p 0.01

bility to more meaningful dimensions of personality. Even as they stand, however, the figures are of some interest. (Another small sample of twins has been analyzed for MMPI scores by Reznikoff and Honeyman.[57]

Another questionnaire on which there are available several different samples of twins whose scores were analyzed is Cattell's High School Personality questionnaire. Results are shown in Table 9-III for three different studies. (The Gottesman study has already been referred to above.) Factors C (neuroticism) and F (extraversion) reinforce the general picture of genetic determination for these two dimensions of personality; in the Cattell scales we deal, of course, with primary factors, but these two are the basic traits at the primary level on which the second-order traits of E and N are based. The numbers of children involved are not large, and the size of F ratios for different scales is very variable across investigators; this is not an uncommon feature of work in this field. When the three studies using the MMPI, mentioned above, are compared in a similar manner, similarities in size of F ratios for identical scales are also not very apparent.[58]

Vandenberg administered the Thurstone Temperament Schedule and obtained results summarized in Table 9-IV.[59] There is good evidence for hereditary determination of behavior subsumed under the headings "active" and "vigorous," and slightly weaker evidence for "impulsive" and "sociable." Again, the number of twin pairs is not very large, and the relative size of the F ratios cannot be taken too seriously. Note the very low DZ correlations and the fact that in three cases they are actually negative (although not significantly so). This point may be important, and we will return to it again later on.

Vandenberg also administered, on another occasion, the Myers-Briggs Type Indicator; results are shown in Table 9-V.[60] Only the value for extraversion-introversion is significant, and that is difficult to interpret in view of the esoteric way in which the inventory was constructed and the lack of independent proof for his validity. No such proof is available either for the Comrey Personality and Attitude Factor Scales, on which Vandenberg has reported more recently and for which Table 9-VI shows the results.[61]

In a later study, Gottesman used the California Psychological Inventory on adolescent twins, seventy-nine pairs being MZ and sixty-eight pairs DZ.[62] Intraclass correlations for the eighteen scales of the inventory ranged from .29 to .60 for the MZ

TABLE 9-IV
MZ AND DZ INTRACLASS CORRELATIONS AND H AND F VALUES FOR THURSTONE TEMPERAMENT SCHEDULE

Scale	rMZ	rDZ	H	F
Active	0.55	−0.06	0.67	3.01†
Vigorous	0.58	0.00	0.59	2.43†
Impulsive	0.44	−0.12	0.46	1.84*
Dominant	0.61	0.23	0.20	1.25
Stable	0.10	0.08	0.31	1.45
Sociable	0.50	−0.06	0.47	1.90*
Reflective	0.55	0.28	0.06	1.06

Number of MZ pairs 45
Number of DZ pairs 35

*p 0.05
†p 0.01
From S. G. Vandenberg, "The Hereditary Abilities Study in Hereditary Components in a Psychological Test Battery," *American Journal of Human Genetics*, 14:220 (1962).

TABLE 9-V
THE F RATIOS FOR FRATERNAL AND IDENTICAL WITHIN-PAIR VARIANCES AND VALUES OF H ON FOUR SCALES OF MYERS-BRIGGS TYPE INDICATOR FOR 27 LIKE-SEXED DZ and 40 MZ TWIN PAIRS

Name of Scale	F	H
Extroversion-introversion	1.84*	0.46
Thinking-feeling	0.80	0.00
Judgment-perception	0.76	0.00
Sensing-intuition	0.70	0.00

*p 0.05

TABLE 9-VI
F RATIOS BETWEEN FRATERNAL (DZ) AND IDENTICAL (MZ) WITHIN-PAIR VARIANCES OF 12 SCORES ON THE COMREY PERSONALITY AND ATTITUDE FACTOR SCALES

Name of Scale	Boys	Girls	All Cases
Empathy	1.46	1.18	1.28
Neuroticism	.77	1.32	1.23
Welfare state attitude	1.71	1.04	1.27
Achievement need	1.91*	2.10†	2.20†
Dependence	1.25	1.15	1.15
Compulsion	1.49	1.50*	1.50*
Self-control	1.24	1.24	1.28
Religious attitudes	.84	1.86†	1.49*
Hostility	.93	.72	.82
Punitive attitudes	.89	1.61*	1.27
Shyness	2.83†	1.51*	1.94†
Ascendance	.50	.89	.80

Number of pairs	DZ	27	63	90
	MZ	52	59	111

*p .05
†p .01

twins, all being statistically significant for the MZ twins, but only half for the DZ twins. Seven traits had approximately one third or more of the within-family variance significantly associated with genetic factors: Sociability, Dominance, Self-Acceptance, Social Presence, Socialization, Good Impression, and Psychological Mindedness. A further five traits had H values greater than 25 percent. Gottesman gives clusters of traits which have been shown by factor analysis to belong together; the first groups together extraverted scales (sociability, self-acceptance, social presence, dominance), all with H values between .35 and .49, averaging over 40 percent. A second, less highly heritable cluster is identified as "dependability—undependability," and may be similar to neuroticism; heritability values average just below 30 percent. Similar data, on much larger samples of questionnaire-diagnosed MZ and DZ twins, are reported by Nichols[63] and Schoenfeldt[120]; the findings from these two studies, plus those of Gottesman[62] and Vandenberg, et al.,[64] have been averaged by Lindzey, et al.[65] and give MZ correlations of .45 and .46 for males and females, and of .26 and .29 for DZ males and females.

Lindzey, et al. summarize results and derive certain "fairly clear conclusions" as follows:[65] (1) There is ample evidence that MZ twins are more alike with respect to personality than are DZ like-sexed twins. (2) Correlations are lower, in both groups, for personality traits than they are for IQ test scores; even when corrected for reliability (personality tests are usually less reliable); correlations only come up to .61 and .37 respectively, assuming reliabilities of .75. (3) Heritability is considerably lower than in the case of IQ tests; even when corrected for attenuation, coefficients would be below 50 percent. (4) The fourth point is perhaps the most interesting. It appears from some calculations made by Thompson and Wilde[66] that when different studies are compared for the order of heritability coefficients for component scales, no similarity whatever appears; in other words, scale A may have a higher heritability than scale B in one study, but a lower one in another. Lindzey, et al. conclude that "for the present, the most economical interpretation of the data would seem to be that while the genotype may have an appreciable effect on personality, the network of causal pathways between genotype and phenotype is so complex in this realm that the effect of the genotype is spread almost evenly across the broad phenotypic measures that personality and interest questionnaires provide."[65] An alternative explanation, and one which is preferred by some, is that the "spreading" is due to the failure to use univocal trait measures, by failing to employ factor analysis in the purification of the scales in question. Where scales are based on relatively arbitrary collocation of items, nothing else can really be expected.

The same comment must apply to a Finnish study, which may be grouped with the American ones for convenience.

Partanen, Bruun and Markkanen studied 902 male twins of twenty-eight to thirty-seven years of age from all over Finland;[67] among the instruments used by them was a set of questionnaires covering sociability, need of achievement, neuroticism and aggressiveness; of these, unfortunately, only one (sociability) would appear to be sufficiently reliable and valid to give worthwhile data; the others are too short, arbitrarily selected and unvalidated to an extent which makes it impossible to draw any serious conclusions from. Sociability "shows evidence of the presence of hereditary influences;" a canonical variate abstracted from the questionnaire and representing largely this variable, showed an index of heritability of .47. This canonical variate may be regarded as an approximation to extraversion, which thus again shows itself to be determined to a marked extent by hereditary influences.

The studies reviewed so far were carried out on adults and adolescents. Do young children give results dissimilar to those reported on older groups? Scarr studied twenty-four MZ and twenty-eight DZ pairs of girl twins of elementary school age, using ratings and observations in standardized situations;[68, 69, 70] mothers' ratings on the Adjective Check List were also used. Median intraclass correlations for the former type of observation were .39 and .23, and for the mothers' ACL ratings they were .40 and .11; again, there is overall evidence for greater personality similarity of MZ twins as compared with DZ twins. The work of Koch lends some mild support to this conclusion.[71] For even younger twins, Brown, Stafford and Vandenberg assessed eight variables, ratings being based on interviews with the mother.[73] MZ twins were more alike on seven of these variables, impressively so on feeding and sleeping problems. Freedman had films of a small group of twins rated for behavior in standard situations; smiling and fear of strangers showed the most marked MZ similarities.[74] As Lindzey, et al. point out, "these data on younger twins, taken together, tend to support the common observation that the greater resemblance of identical twins has early roots, but the data are insufficient to cast much light on the differential influence of heredity and environment on different traits, and thus to assist much in the interpretation of findings at later ages. A really large study of twins followed through the early years of life would be most welcome."[65] This conclusion is supported by the work of Juel-Nielsen, who presents detailed case history type information of twelve Danish pairs of MZ twins reared apart from early infancy, or early childhood in some cases.[75] Unfortunately, he relied on projective tests, notoriously unreliable and lacking in validity (Zubin, et al.[10]) and subjective interviews which are difficult to quantify. A proper study of this kind would be extremely interesting and important.

We must now turn to the English work which, as already mentioned, shows some important differences from the American work reviewed so far. (No chauvinistic overtones are intended here in making this comparison; the distinction made is a real one, and for historical reasons is linked with nationality.) This difference can perhaps best be characterized in terms of what the respective authors are trying to accomplish. As far as the American studies are concerned, one feels that the only aim of the work is to demonstrate that heredity plays some part in the genesis of the type of behavior under investigation; the writers are satisfied if they can show that some arbitrarily selected aspect of personality has a significant H. Hence, the choice of questionnaire used is almost perfunctory; no reasons are given why the MMPI or the CPI is administered rather than some other inventory. When a reason is given for the choice, the argument is extremely weak. Gottesman for instance, tries to justify the selection of the MMPI by stating that the scales are put together by choosing items which differentiated between normal and psychiatric criterion groups.[34] "This is one reason why the MMPI is different from personality tests derived by factor analysis." It is also one reason why the test is almost useless for his purposes, and why the results are quite impossible to interpret. The method of construction used assumes the validity and reliability of psychiatric diagnosis; the empirical evidence is overwhelmingly opposed to this assumption. Reliabilities are known to be very poor, and few psychiatrists would make much of an argument for the validity of what is at best a compromise type of nosology.[76] Recent studies of the UK-US Diagnostic Unit have shown that identical patients would have a 500 percent higher likelihood of being classified "depressive" or "manic-depressive" by an English than by an American psychiatrist.[77] Clearly, the nationality of the psychiatrist is far more important for the diagnosis than the personality of the patient. On the psychometric side, Comrey has shown in a series of papers that when each scale of the MMPI is taken and factor analyzed, the items are quite heterogeneous and do not define a single factor, as they are intended to; factor analysis of the whole scales gives rise to two major factors which can be identified as estraversion and neurotism. (Eysenck and Eysenck review some of this literature.[3]) Last, but not least, the questions in the MMPI have a heavy pathological content and are quite unsuitable for normal subjects. None of these obvious criticisms are anticipated by Gottesman or the other writers who have used this test.

The English work was based on the views that test selection: (1) should be based on the best available knowledge of the major dimensions of personality and (2) should be directed at the discovery of the basic facts relating to the genetic system involved. This means that attempts should be made to deal with such issues as assortative mating, dominance, and interaction, in addition to simple heritability. Hence, the English work presents a more unified aspect; for some twenty years research has been progressing along certain lines in an attempt to improve both experimental and statistical methods in the furtherance of this goal. The first study in this campaign was undertaken by Eysenck and Prell;[12] and this made two new contributions.

In the first place, they argue that objective tests of behavior are superior to personality questionnaires, particularly when used on children, and are in any case less liable to faking. In the second place, they argue that conceptions such as neuroticism are essentially based on the notion of intercorrelated traits and measurements and that twin studies carried out on single measures confound the issue by mixing up variance due to the trait under investigation and specific variance relative to the test in question. They suggest, therefore, that a whole battery of tests should be given and factor analyzed and that a score based on a combination of tests having the highest saturations with the factor in question should be used. In addition they suggest that these factor scores should be validated against some form of external control; in their own work they have done so by comparing an experimental group of children under treatment at a child guidance clinic with normal children in school, demonstrating significant differences in "neuroticism" between the two groups of children.[78] Using this approach, Eysenck and Prell found that the factor score derived from all the tests gave an intraclass correlation of .851 for identical twins and of .217 for fraternal twins;[12] this factor score showed a greater difference between identical and fraternal twins than any single constituent test, thus suggesting that it was indeed a general factor of neuroticism which was an inherited rather than a specific variance for any single test. The Holzinger h^2 coefficient showed a hereditary determination of .810 if we are willing to assume that this coefficient can indeed be used to measure hereditary determination in this manner.

In another similar study, Eysenck has reported results on a battery of tests of intelligence, extraversion, and autonomic activity.[79] Again, intraclass correlations for identical and fraternal twins were calculated for three factors corresponding to these concepts rather than individual tests.

For extraversion the correlation for identical twins was .50, for fraternal ones −.33; for the autonomic factor the two correlations were respectively .93 and .72. For intelligence the values were .82 and .38. Holzinger's h^2 statistic was calculated for all three factors, giving very similar results in the neighborhood of .7 for all three. The appearance of a negative intraclass correlation for the fraternal twins is unusual, and Eysenck concluded that it seems "likely that this value represents a chance deviation from a true correlation of zero, or of some slight positive value, an assumption strengthened by the fact that a correlation of the observed size is not statistically significant. Under the circumstances, however, we cannot regard the h^2 statistic derived for the factor of extraversion as having very much meaning ... much more reliance fortunately can be placed on the significance of the differences between identical and fraternal twins for this factor which ... is fully significant."

The results of the Eysenck studies give values indicating a greater influence of hereditary causes than would be true of the other studies quoted so far. Apart from the possibility of chance deviations, it may be suggested that the following causes have possibly been operative: (1) behavioral tests are more likely than questionnaires to reveal deep-seated constitutional features of the personality, (2) factor scores are more reliable and valid than single tests, (3) the measures selected have been chosen on the basis of a theory of personality which has perhaps more experimental and theoretical backing than that which gave rise to the measures used by the earlier workers. It is possible that any or all of these causes may have been operative, and it must be left to future investigation to discover to what extent these hypotheses can be upheld.

Eysenck's work was followed by that of Wilde,[72] who

administered a personality questionnaire to eighty-eight monozygotic and forty-two dizygotic twin pairs. He obtained scores for the following variables: (a) neurotic instability or neuroticism (N); (b) neurotic instability as manifested by the presence of functional bodily complaints (NS); (c) extraversion score (E); (d) lie score (L); (e) masculinity/femininity scale (MF). Wilde[72] divided his twin pairs into those who had been living separately for five years or more and those who had been living in the same home or who had been separated for less than five years. His results are reproduced in Table 9-VII. It will be seen that as far as neuroticism is concerned, we again find little in the way of difference between monozygotic twins brought up together or brought up in separation, but a great difference between monozygotic twins and dizygotic twins. Findings for the NS score, i.e. neurotic instability as shown by functional bodily complaints, show the same pattern; but here we find that both monozygotic and dizygotic twins brought up in separation have much greater intraclass correlations than do twins brought up together. The pattern for extraversion is rather curious. Monozygotic and dizygotic twins brought up together show the usual greater intraclass correlation among the monozygotic twins as compared to the dizygotic twins. When we turn, however, to twins brought up in separation, we find that dizygotic twins show greater intraclass correlations. The L score (lie score) shows no evidence of hereditary determination, but the masculinity/femininity scale does so very clearly. We may perhaps summarize Wilde's results by saying that except for the curious finding relation to extraversion scores in twins brought up in separation, the results are very similar to those discussed previously.

Of somewhat lesser importance is a more recent study by Young, Fenton, and Lader, in which they used the Middlesex Hospital Questionnaire, as well as Eysenck's E and N scales.[80] The small number of MZ and DZ twins (seventeen and fifteen pairs, respectively) means that little accuracy attaches to the h^2 values calculated; these center on .5 for the various scales, with N and E having values of .48

TABLE 9-VII
INTRAPAIR CORRELATIONS OF 88 MZ AND 42 DZ TWIN PAIRS

		N	NS	E	L	MF
1 MZ-cohab.	n = 50	0.55	0.46	0.58	0.48	0.45
2 MZ-separ.	n = 38	0.52	0.75	0.19	0.46	0.44
3 MZ-all	n = 88	0.53	0.67	0.37	0.46	0.44
4 DZ-cohab.	n = 21	−0.14	−0.05	0.19	0.33	−0.34
5 DZ-separ.	n = 21	0.28	0.64	0.36	0.49	0.30
6 DZ-all	n = 42	0.11	0.34	0.35	0.54	0.02

Double underlining refers to a significance of 01; single underlining refers to a significance of 05; assuming normality of distributions.
Quoted with permission from G. J. S. Wilde, "Inheritance of Personality Traits," *Acta Psychologica; European Journal of Psychology (Amsterdam)*, 22:37 (1964).

and .43, respectively. Phobic anxiety has the highest value (.64), depression the lowest (.07). Hysterical personality (.56) and Free floating anxiety (.50) are somewhat higher than Somatic concomitants of anxiety (.47) and Obsessional traits (.43). However, there is no guarantee that on another sample the rank order of these values would not be inverted. None of the DZ intraclass correlations were significant, and five out of eight were negative; this must make one rather suspicious of the meaningfulness of the h^2 values reported. Taking as an example the "Obsessional traits and symptoms" scale of the Middlesex Hospital Questionnaire, we find r_{MZ} = .22 (which is insignificant), and r_{DZ} = −.38 (which is also insignificant). The h^2 calculated is .43, which looks quite respectable, but whose meaning must remain doubtful in view of the large negative r for DZ twins and the lack of significance for both correlations. It is probably rather pointless to carry out and report studies with such small numbers; the probable errors involved in the statistics are too large to allow any useful conclusions.

Much more important is a recent monograph by Shields, in which he used a self-rating questionnaire devised by the writer which is very similar to the MPI, sharing a number of items in common with it; this questionnaire provides scores for extraversion and neuroticism.[81] Shields applied it and two tests of intelligence to a collection of pairs of twins who had been separated from one another in childhood; he also had a control group of twins who had been brought up together. There were forty-four separated MZ, forty-four nonseparated MZ control pairs, and thirty-two pairs of DZ twins of which eleven had been brought up apart. His findings with the questionnaire are given in Table 9-VIII. It will be seen that identical twins are much more alike than fraternal twins, regardless of whether they are brought up together or in separation; it will also be seen that in each case the twins

TABLE 9-VIII
INTRACLASS CORRELATION COEFFICIENTS

	C	S	DZ
Height	+0.94	+0.82	+0.44
Weight	+0.81	+0.37	+0.56
Dominoes	+0.71	+0.76	−0.05
Mill Hill	+0.74	+0.74	+0.38
Combined intelligence	+0.76	+0.77	+0.51
Extraversion	+0.42	+0.61	−0.17
Neuroticism	+0.38	+0.53	+0.11

Intraclass correlation coefficients for MZ twins brought up together (C), MZ twins brought up apart (S), and DZ twins brought up together (DZ).
Quoted with permission from J. Shields, *Monozygotic Twins Brought up Apart and Brought up Together* (London, Oxford University Press, 1962).

brought up separately are more alike than twins brought up together. These results are, therefore, in good accord with those originally reported by Newman, Freeman, and Holzinger[52] as well as with those of later writers. They are rather more clear-cut, perhaps due to two possible causes: (1) the subjects of the experiment were adults rather than children, and consequently the questionnaires applied to them much more readily than they would apply to children; and (2) the questionnaires used had been elaborated for many years on the basis of factor analytic studies of personality and were consequently perhaps more reliable and valid than those used earlier. It is of some interest that in the above table the fraternal twins have a negative intraclass correlation for extraversion, very much as in the study by Eysenck cited above.[79] Again, this intercorrelation is not significantly different from zero, but the coincidence is certainly striking, although the writer cannot present any reasonable hypothesis which would account for such a negative correlation.

So far, we have been concerned with the field of classical "heritability." Now, we turn to the reanalysis of Shields' data by Jinks and Fulker.[20] They are concerned with the construction of genotype-environment models, including interaction variables, and in doing so they go far beyond the simplistic classical models. As they point out, even in the absence of genotype-environment interaction, partitioning the total variation between two components, the genetic (G) and the environmental (E), must lead to a partitioning of these two components into within-family and between-family (G_1 and G_2, E_1 and E_2). Classical formulae ignore explicitly important sources of variation (Holzinger's H ignores G_2 and E_2, for example). Nor do classical formulae enable us to test for interaction or for dominance or for assortative mating. Jinks and Fulker develop a model which does enable us to do these things,[20] and then apply this model to the data collected by Shields,[81] described above very briefly. What is the outcome of their analysis?

The model requires, first of all, that the four groups used (male and female MZ and DZ twins) should be homogeneous with respect to N (the first variable to be discussed); the data pass this test. The second assumption to test is the possible importance of genotype-environment interaction. There is no evidence of GE_1 or GE_2, nor are correlated environments found to be a complication in these data:[20] "Thus on the basis of these tests we are justified in fitting the simple G and E model to the data." When the calculations are done (they are given in detail in the original paper), G and E_1 are clearly significantly greater than O, while E_2 is not. There is some slight indication of assortative mating, but the figures are not significant. "At the same time, the absence of dominant gene action is clearly indicated by the fact that G_1 is not greater than G_2. The absence of dominant gene action strongly suggests that an intermediate level of neuroticism has been favored by natural selection and constitutes the population optimum for this personality trait." Further tests confirm the adequacy of the simple model, and "this simplified model, which fits the data extremely well, . . . may now be used to calculate heritabilities with some degree of confidence." The broad heritability = narrow heritability = .54, which is fully significant. Cattell's nurture:nature ratios were also calculated and indicated that, although environment is more important than genotype in producing differences among siblings, the differences in neuroticism observed between families is entirely genotypic in origin: "Evidently cultural and class differences have no effect on this major personality dimension."[20]

We now turn to the analysis of extraversion scores.[20] Here, too, "we conclude that the types of family represent reasonably adequate samples from the same population." Testing for interaction, it is found that "there is evidence of a certain amount of GE_1, but not GE_2 ... Introvert genotypes are more susceptible to environmental influences than extravert genotypes, the latter's being relatively impervious. This finding is, of course, fully consistent with Eysenck's (1967) theory that the introvert is more conditionable than the extravert..."[11] Broad heritability was calculated as accounting for 67 percent of variation; but it should be added that this figure depends on certain assumptions which are required to be made because the simple model does not fit the data as well as it does in the case of neuroticism. Jinks and Fulker raise the possibility that failure may be due to "competition" (intrauterine or later), particularly between DZ twins;[20] and Eysenck's data, giving a negative intraclass correlation on E for DZ twins, support this

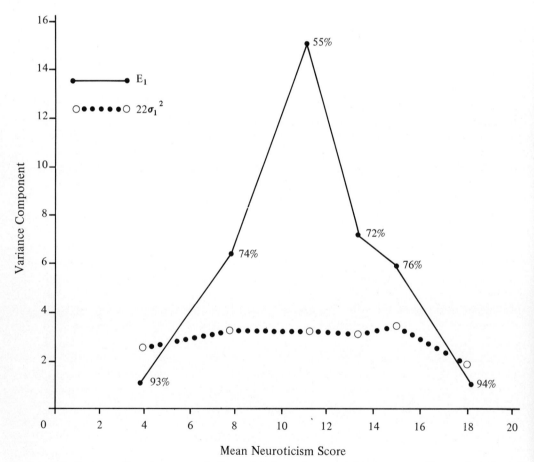

Figure 9-2. Degree of environmental determination of neuroticism as a curvilinear function of degree of neuroticism.

view.[79] Jinks and Fulker discuss the reasons for the failure of the simple model in some detail; it would take us too far to follow them in this.[20]

An interesting further analysis of some of these data was undertaken by Fulker and Eaves (unpublished). Using only MZ twin pairs, they plotted mean scores for each pair against differences, arguing that mean scores were the best available estimate of the degree of extraversion (or neuroticism) of the pair, while the difference would indicate the action of environment (and errors of measurement).

These plots are shown in Figures 9-2 and 9-3; they are clearly bowed, rather than linear, and indicate very clearly that environmental factors are more important in the middle range of scores than at the extremes. In part this must be a statistical artifact; very high or low mean scores are incompatible with large within-pair differences. However, this is clearly not the whole story; when the same analysis was done on IQ data the plotted data were almost entirely linear, with a very weak quadratic component. The data suggest, then, that hereditary influences are

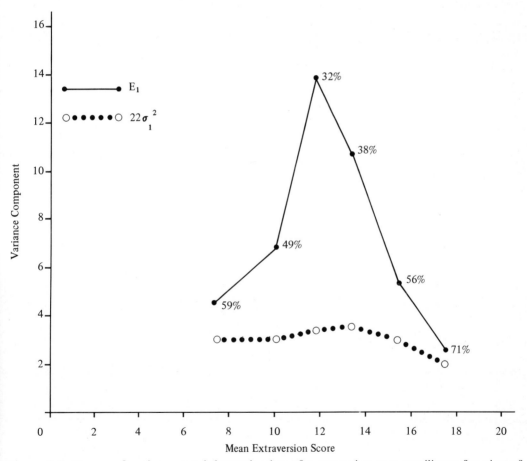

Figure 9-3. Degree of environmental determination of extraversion as a curvilinear function of degree of extraversion.

stronger for E and N when these personality traits are either very strong or very weak; ambiverts are more determined in their conduct by environmental determinants. This finding may be worthy of being followed up in future work.

Another interesting and novel idea for analyzing twin data is suggested by Loehlin, who divided personality inventory items into those showing high heritability and those showing low heritability and then proceeded to factor analyze items of either kind separately in a sample of normal subjects.[82] He found similarities between the factors in the two sets. In particular, there as an extraversion and an adjustment (neuroticism) factor in each.

But there also seems to be a difference in the flavor of corresponding factors from the two analyses: factors in the first set seem more focused on the individual himself; the ones in the second set, on his reaction to his environment. The extraversion factor in the first set stresses what the individual *brings* to group activities; the one in the second shifts slightly toward emphasizing the rewards and punishments the individual *derives* from such activities. The emotionality factor in the first table is centered on impulses and their control; the one in the second on response to environmental restriction and threat.

Leohlin suggests the possibility that personality dimensions may tend, in general, to be two-sided, with one aspect reflecting certain (probably complex) heredity tendencies, and the other the social institutions, and interpersonal pressures which grow up around these tendencies. If this is indeed the case, it could lead to much confusion in nature-nurture work in the area of personality.[82]

This study could bear repetition with a better selected set of items.

Our discussion has centered upon questionnaire studies or studies using factor scores derived from objective laboratory tests and inventories (e.g. Eysenck and Prell[12]). Reasons have already been given why studies using projective techniques along traditional lines are not acceptable in a survey of scientific data; the reliability and validity of such scores are simply not high enough to provide data upon which replicable conclusions are based. Given such low reliability, one might predict that twin studies using such tests as the Rorschach or the Szondi would find insignificant heritabilities; this is indeed so. Reports have been published of the Rorschach test by Bleuler,[83] Elmgren,[84] Schachter,[85] Troup,[86] Kiev,[87] Kerr,[88] and Vacca and Ciolfi[89]; and of the Szondi test by Rabin[90] and Nolan.[91,92] All of these were essentially negative. Vandenberg argues "that one is tempted to conclude that the Rorschach measures largely environmental factors to the extent that it provided reliable measures."[60] The sting, of course, is in the tail; it is known that the measures provided are not in fact reliable, so that the Rorschach is in fact not measuring anything at all. It is curious to note that even now such tests as the Rorschach or the Szondi are spoken of in a hushed and reverent manner in spite of demonstrated faults which would have led to the summary execution of any other test. The interpretation of the results quoted above which will be offered here differs from Vandenberg's only in doubting if the data provided have any meaning; they cannot, in our opinion, be used to support any kind of environmentalist hypothesis about the genesis of personality traits. The only reason for discussing these studies here is that in the past they have been quoted as lending support to such an hypothesis.

The same arguments may be applied to another type of measure that has frequently been used in this connection: handwriting analysis (graphology). This type of projective or "expressive" analysis used to be very popular on the European continent, and a number of twin studies have been reported in the literature.[52,93,94,95,96,97,98] Results are somewhat conflicting (which is not surprising, as no agreed methods of scoring exist), but on the whole there is little evidence for heritability to be found in these studies. Again, we must conclude, however, that these results tell us nothing about the heritability of personality; the reliability and validity of graphological analysis is too low to make possible any substantive conclusions. It might have been more advisable to have used direct measures of expressive movement, like those pioneered by Allport and Vernon;[99] results might have been more promising.

Studies using laboratory measures of learning, conditioning, perception, etc., raise other problems, particularly when these tests are used singly, rather than as part of a factorial battery. Consider an experiment by A. Jensen (unpublished) in which he administered the pursuit rotor learning task to thirty-five pairs of MZ and thirty-four pairs of DZ twins; each of 30 ten-second trials was followed by a ten-minute rest period which in turn was followed by twenty more ten-second trials. The reminiscence score (mean of the first two postrest trials minus the mean of the last two prerest trials) showed strong evidence of heritability ($h^2 = 2(r_{MZ} - r_{DZ}) = .86$). It is possible to regard this simply as evidence of heritability of pursuit rotor reminiscence; however, such reminiscence is known to correlate with extraversion, and thus to be a measure of this personality trait.[100] Clearly, we may, with equal justice, claim that we have some evidence of hereditary determination of E; the problem, as indicated in the formula given near the beginning of this chapter, lies in the proper apportioning of the variance. Substitute "Re." for "Imp." in that formula, and the exact nature of the problem will be apparent: our data do not tell us whether part or all of the heritability discovered in relation to the reminiscence score is related to reminiscence, to extraversion, or is specific to the test itself, i.e. would not be replicated if a different test of reminiscence were to be used. Without the benefit of a factorial model such single determinations of heritability, using tests which are known to be also measures of a given personality variable are inherently ambiguous and cannot be properly interpreted.

Other examples abound; only a few will be cited. Vandenberg has reported that Critical Flicker Fusion scores have a heritability of .44;[59] C.F.F. is known to be a measure of cortical arousal, and of extraversion.[11] Size constancy ($h^2 = .47$) and perception of the upright ($h^2 = .40$) are cited in the same paper; these, too, are known to be related to personality.[11] Mirror drawing ($h^2 = .70$), also quoted by Vandenberg, is known to be a measure of neuroticism.[59] Visual illusions may also be mentioned,[101,102] and Fuller and Thompson give many other examples.[103] To these might be added EEG patterns, which are known to be much more similar for MZ pairs than for DZ pairs.[104,105,106,107] Here, too, it remains to be determined what (if any) part of the total variance which shows heritability is due to personality factors. One suspects that the data do indicate some such personality heritability, but the proof lacks conviction. One can only conclude that from the point of view of personality research such single studies are useless; a factorial design is absolutely essential if far-reaching conclusions about heritability of personality factors are to be sustained.

We must now turn to an important area of criticism. Most of the evidence we have reviewed was collected with reference to twins, and twin research has been seriously criticized on several grounds. A detailed discussion of these criticisms is given by Thompson and Wilde;[66] many of these criticisms are not applicable to personality research and will not be considered here. Some criticisms clearly show that heritability estimates are likely to be underestimates; thus failures in correct assignment of twins to the MZ or DZ category must lower the calculated heritabilities below the true value. Similarly, the existence of extra-chromosomal inheritance would lower the genetic similarity of MZ twins from the 100 percent value assumed in most formulae,[108] and thus decrease estimated heritability values below the true value. There is, however, one frequently made criticism which must be discussed in some detail.

Jones, Smith, and Scarr have shown that MZ cotwins are more similar in the treatment they receive from their parents than are DZ cotwins,[109, 110, 111] and it seems fairly well established that the home environment of MZ twins is in general more similar as far as the two children are concerned than is the home environment of DZ twins.* Such findings have sometimes been interpreted as throwing doubt on the results of twin work, because it is alleged that the excess MZ intraclass correlation so frequently found may be due to more similar environmental treatment of MZ twins rather than to hereditary causes. Such criticism, of course, fails to touch such studies as those of Shields, which made use of twins brought up in separation;[81] but it may be considered quite cogent in relation to the majority of other studies quoted so far or to be mentioned later.†

Scarr has pointed out that the facts are not in dispute but that their interpretation is by no means obvious.[69] When parents are correct about their twins' zygosity, two important factors are confounded: (1) the greater genetic differences of DZ cotwins, with accompanying physical, intellectual, and behavioral differences; and (2) the greater differences of parental treatment of DZ pairs, which might create additional intrapair dissimilarities.

If parents are simply reacting to the existing differences between their DZ twins' behavior, then no bias is introduced into twin studies. But, if they effectively train differences, then these environmentally determined differences would bias the comparisons of intraclass correlations in favour of genetic hypotheses, by reducing the possible similarities of DZ cotwins. By the same token, the parents of MZ twins who know their twins are identical may react to existing similarities or seek to train greater similarities than would otherwise exist. When parents are correct about their twins' zygosity, it is impossible to distinguish between parental behavior that is a reaction to the phenotypic behavior of their twins and parental treatment that seeks to train greater differences or similarities.

Scarr goes on to point out that fortunately not all parents are correct about their twins' zygosity and that these parents offer a critical test of environmental bias in twin studies. "By examining the cases of parents who are wrong about their twins' zygosity, it is possible to separate parental reactions to similarities and differences based on genetic relatedness from parental behaviors which arise from their belief that

---

*Takuma has published some data to show that in Japan twins (both MZ and DZ) are treated very differently, depending on which is regarded as the older, which as the younger.[112] This is in line with Japanese feudal tradition, and he publishes some evidence which gives substance to his conclusion that this difference in treatment produces differences in late behavior. Unfortunately, most of this evidence is anecdotal and purely descriptive; few empirical studies using proper tests seem to have been done in Japan. From many points of view, such studies would be particularly interesting in view of this great diversity of treatment.

---

†The evidence certainly does not suggest that MZ twin pairs are more alike if they are kept together longer; Vandenberg and Johnson have shown that, if anything, earlier separation is associated with greater similarity, at least as far as IQ is concerned.[113]

their twins should or should not be similar."[69]

She found that, in fact, something like 20 percent of twins in a group of 164 twin pairs were misclassified by their mothers; twelve MZ pairs and twenty-one DZ pairs were misclassified. Using various rating scales, checklists, and other devices, she came to the conclusion that, "... differences in the parental treatment that twins receive are much more a function of the degree of their genetic relatedness than of parental beliefs about 'identicalness' and 'fraternalness' ... The comparisons of parental behavior for correctly and incorrectly classified pairs suggest that environmental determinants of similarities and difference between MZ and DZ co-twins are not as potent as the critics charge." The number of twin pairs about whom mothers were mistaken is, of course, not very large, and the study should be repeated with larger numbers; but the direction of the results is quite definitely contrary to that expected from the point of view of the environmentalistic hypothesis. The data thus support the results from studies carried out on MZ twins brought up in isolation from each other. If this criticism is to be taken seriously, then it will have to be supported by stronger evidence than seems at hand at the moment.*

When considering the evidence regarding the adequacy of the simple genetic model for extraversion and neuroticism, no mention was made of one rather curious fact which seems too well documented to be entirely due to chance, namely the closer correspondence in personality of mothers and children as compared with the correlations reported for fathers and children. The first indication of such a relation appeared in the work of Crook,[114] who used the Bernreuter Personality Inventory; he found near zero correlations between father and son on the scales of neuroticism and dominance, and slightly higher, but still insignificant, correlations between father and daughter. In contrast, mother-daughter correlations were as high

---

*An interesting finding relevant to this discussion is reported by Claridge, Canter, and Hume in an as yet unpublished book entitled *Personality Differences and Biological Variations: A Study of Twins*, (Oxford, Pergamon Press). They divided their sample of forty-four MZ and fifty-one DZ twin pairs into separated and unseparated, making five years the separation line. Using the Eysenck Personality Inventory they found, with respect to extraversion-introversion, that both types of twins were more alike when brought up apart than when brought up together. "This is particularly true of MZ twins where for both the EPI and 16 PF measures, the correlations were low and nonsignificant in 'unseparated pairs,' but highly significantly positive in 'separated pairs.' For DZ twins, results were less consistent across the two personality tests." The correlations in question are .67 as opposed to .10 (for MZ twins) and .36 as opposed to .22 (for DZ twins); these figures are for the EPI. For the 16 PF test, they are .85 versus .29, and .50 versus −.65. As the authors point out, "the results suggest that the effect of the interaction between members of a twin pair in close contact with each other is to make them less alike in extraversion, perhaps because they tend to adopt complementary roles. It is only after separation, when the modifying influences arising from the twin situation are eliminated or reduced, that the genotypic similarities can be fully expressed at the phenotypic level." (As far as neuroticism is concerned, exactly the opposite is true; "twin pairs, whether MZ or DZ, were more alike when together or recently separated than when they had been separated for more than five years.")

The authors separated out the "sociability" and "impulsiveness" components of extraversion, finding similar effects on both when separation was considered. For sociability, the intraclass correlations for MZ twins were .51 and .91; for DZ twins they were .25 and .25. The resulting F values were 1.41 (N.S.) and 9.50 (p < .01). For impulsivity the correlations are much smaller, probably because the number of items in the EPI relevant to this trait are smaller, and hence this subscale much less reliable. This technique of comparing twin pairs in terms of length of separation seems a promising one; more data are urgently needed to replicate, if possible, these results.

as .57, and mother-son correlations .30 (both fully significant). Coppen, Cowie and Slater used the Maudsley Personality Inventory to investigate family similarities among neurotic patients and their relatives.[115] With a sample of 224 patients and 735 relatives, they found statistically significant correlations only between mother-offspring and between siblings; no significant correlations were found between father and other relatives in this study. Both Crook[114] and Coppen, et al.[115] suggested that the affective relationship between the mother and her offspring could account for their results, thus implying an environmental determination. In order to test this hypothesis, Insel tested three generations (grandparents, parents, children) on the PEN scale;[51] this as yet unpublished measure gives scores for psychoticism, extraversion, and neuroticism, and also contains a lie (dissimulation) scale.[48,49] He also administered the Wechsler Intelligence Test, and the Wilson and Patterson Conservative Scale.[116]

Correlations between different relations are shown for these variables in Tables 9-IX through 9-XIV. For C and for IQ (Tables 9-XIV and 9-XIII, respectively) there are no obvious large differences between the correlations of mothers and fathers and their offspring, nor between the different grandparents and their offspring and grandchildren. The position is clearly different when we turn to the personality variables; here again, mothers show much higher correlations with their children as compared with fathers. So far, results are similar to those already discussed, and similar hypotheses are tenable regarding the origin of these correlations. Insel inquires,

> What correlations should be expected between grandparents and child? Surely the father's relationship with his children would not be more remote in contact and intensity than the grandparents' relationship. And therefore one would expect father-child correlations to be as high or higher than grandparent-child correlations. Surprisingly, this is not what one finds with psychoticism, extraversion or neuroticism. Both grandparents on both sides of the family show consistently higher correlations with their grandchildren than do fathers with their children. This is especially true of the maternal grandmother . . . For example, on psychoticism maternal grandmother-eldest son correlate .301, while father-eldest son correlate .006; maternal grandmother-eldest daughter correlate .109 while father-eldest daughter correlate .065. On extraversion maternal grandmother-eldest son correlate .277 while father-eldest son correlate −.031; maternal grandmother-eldest daughter correlate .212 while father-eldest daughter correlate .061. On neuroticism maternal grandmother-eldest son correlate .253 while father-eldest son correlate −.070; maternal grandmother-eldest daughter correlate .296 while father-eldest daughter correlate .061. Since the pattern of family correlations,—i.e. the exceptionally high correlations between mother-offspring, and near zero correlations between father-offspring, and the intermediate correlations between grandparents-grandchildren—is not consistent with what is known about environmental effects . . . , it is suggested that an environmental hypothesis cannot account for the personality results, and a genetic model would seem more appropriate.[51]

Insel discovered five genetic models which might be considered: sex-linked, sex-limited, sex-influenced, sex-modified, and cytoplasmic inheritance.[51] It would take us too far afield to follow him into these complex issues; what seems clear is that at the moment there are not sufficient data to arrive at any sort of conclusion. Replication of the study is urgently needed, and so is an extension, e.g. to foster children, and the correlations between their P, E, and N scores and those of their true and foster mothers and fathers. Unless replication studies fail to provide support for the main findings in Insel's study,[51] however, it is clear that we have a feature which is liable to complicate con-

TABLE 9-IX
PSYCHOTICISM
CORRELATIONS BETWEEN PARENTS, GRANDPARENTS, AND CHILDREN FOR PSYCHOTICISM

	N	Mother	Maternal Grandmother	Paternal Grandmother	Father	Maternal Grandfather	Paternal Grandfather
Eldest son	70	.60++	.30	.03	.01	−.05	.13
Second son	41	.54	.25	.00	.01	.01	.09
Third son	10	.55	.25	.10	.01	.00	.09
Eldest daughter	83	.19	.11	.00	.06	.10	−.13
Second daughter	39	.16	.04	.10	.09	.05	.07
Third daughter	8	.10	−.08	.01	.09	−.06	−.10

TABLE 9-X
EXTRAVERSION
CORRELATIONS BETWEEN PARENTS, GRANDPARENTS, AND CHILDREN FOR EXTRAVERSION

	N	Mother	Maternal Grandmother	Paternal Grandmother	Father	Maternal Grandfather	Paternal Grandfather
Eldest son	70	.32++	.28	.18	−.03	−.16	−.13
Second son	41	.38++	.18	.11	.02	.00	.10
Third son	10	.49	.25	.30	.01	.06	.57
Eldest daughter	83	.55++	.21	.18	.06	.03	.07
Second daugher	39	.40+	.25	.10	.09	.11	.06
Third daughter	8	.50	.44	.09	.11	.13	.14

TABLE 9-XI
NEUROTICISM
CORRELATIONS BETWEEN PARENTS, GRANDPARENTS, AND CHILDREN FOR NEUROTICISM

	N	Mother	Maternal Grandmother	Paternal Grandmother	Father	Maternal Grandfather	Paternal Grandfather
Eldest son	70	.32++	.25	.39+	−.07	.13	.10
Second son	41	.52++	.28	.00	.01	−.15	.10
Third son	10	.50	.20	.01	−.12	.13	−.10
Eldest daughter	83	.55++	.30+	.16	.06	.10	.13
Second daughter	30	.48++	.20	−.17	.04	.15	.05
Third daughter	8	.40	.28	.06	.09	.10	.18

TABLE 9-XII
LIE SCALE
CORRELATIONS BETWEEN PARENTS, GRANDPARENTS, AND CHILDREN FOR LIE SCALE

	N	Mother	Maternal Grandmother	Paternal Grandmother	Father	Maternal Grandfather	Paternal Grandfather
Eldest son	70	.38++	.15	.09	−.14	.17	.30
Second son	41	.32+	−.20	.13	−.08	.00	.29
Third son	10	.35	.18	.23	.20	.14	.42
Eldest daughter	83	.41++	.20	.23	.61++	.24	.26
Second daughter	39	.45++	.22	.31	.57++	.31	.29
Third daughter	8	.35	−.19	.21	.49	.02	.37

TABLE 9-XIII
IQ
CORRELATIONS BETWEEN PARENTS, GRANDPARENTS, AND CHILDREN FOR IQ

	N	Mother	Maternal Grandmother	Paternal Grandmother	Father	Maternal Grandfather	Paternal Grandfather
Eldest son	29	.53++	.29	.20	.41+	.25	.24
Second son	12	.56	.30	.23	.38	.22	.28
Third son	-	-	-	-	-	-	-
Eldest daughter	28	.58++	.32	.20	.34	.24	.29
Second daughter	14	.50	.28	.11	.30	.10	.02
Third daughter	-	-	-	-	-	-	-

TABLE 9-XIV
CONSERVATISM
CORRELATIONS BETWEEN PARENTS, GRANDPARENTS, AND CHILDREN FOR CONSERVATISM

	N	Mother	Maternal Grandmother	Paternal Grandmother	Father	Maternal Grandfather	Paternal Grandfather
Eldest son	70	.61++	.39++	.39+	.62++	.21	.28
Second son	41	.55++	.33+	.35+	.56++	.28	-.17
Third son	10	.57+	.36	.28	.40	.22	.13
Eldest daughter	83	.70++	.40++	.30	.41++	.31	.20
Second daughter	39	.63++	.40++	.32	.44++	.37+	.05
Third daughter	8	.60	.50	.38	.40	.40	.20

siderably the simple model which would otherwise satisfy most of our tests. It is clearly too early to give up that model, but it seemed worthwhile mentioning this possible danger to it at this point.

We are now in a position to state certain conclusions which would seem to follow from the evidence reviewed in this chapter. Some of these conclusions, particularly those regarding the Jinks-Fulker type model,[20] and their use of biometrical genetical methods, as well as the writer's insistence on the need for factorial methods in the assessment of heritability, may be regarded as controversial; arguments regarding these issues have been presented in the text and will not be repeated. In the absence of valid arguments against the positions adopted on these points, it seemed justified to include them in our conclusions, but it seemed fair to warn the reader that not all experts might be willing to accept them as such.

1. There is overwhelming evidence that questionnaire measures of personality, embodying many diverse traits, show evidence of heritability when tested on samples of MZ and DZ twins; the former are invariably more alike than the latter.
2. The heritability of different traits cannot be established with any great degree of accuracy at the moment, different studies giving different rankings of heritability estimates for identical sets of traits.
3. This failure may in part be due to

TABLE 9-XV
CORRELATIONS BETWEEN MID-PARENT AND MID-CHILD ON PERSONALITY AND SOCIAL ATTITUDE VARIABLES (N = 94).

P:	E:	N:	L:	C-Scale
.23+	.63++	.65++	.26+	.69++

From P. M. Insel, *Family Similarities in Personality, Intelligence, and Social Attitudes,* unpublished doctoral dissertation (London, Institute of Psychiatry, 1971).

the smallness of the samples characteristically employed in these studies; standard errors are too high to make the estimates very meaningful. Other causes may be the low reliabilities of the measures used and their lack of factorial homogeneity.

4. Results from investigations making use of projective techniques (e.g. Rorschach test, Szondi test) cannot be interpreted in any meaningful sense because of the lack of reliability and validity characteristic of these measures. Graphological techniques suffer from the same debility.

5. Results from investigations using single laboratory measures of conditioning, C.F.F., EEG, or perceptual variables cannot be interpreted in any meaningful sense in spite of high reliability and evidence of validity (correlation with personality) because there is no way of allocating genetic variance to personality, to the perceptual, motor or other variable measures, and the specific components of the test.

6. Heritability studies using the classical approach make assumptions regarding such issues as assortative mating, between-family environmental influences, interaction; in the absence of direct evidence on these points or the existence of a genetic model enabling tests to be carried out on the applicability of these assumptions, even the heritability estimates so widely published carry little conviction.

7. A model of the kind mentioned in the last paragraph does now exist and has been used in connection with two higher-order personality factors, extraversion and neuroticism; with minor complications the model has been found to fit the experimental data reasonably well.

8. With respect to these two major dimensions of personality, there is little, if any, evidence of assortative mating; there appears to be no evidence of simple dominance; interaction effects are minimal or nonexistent; between-family environmental effects are minimal or nonexistent. For extraversion, there is some evidence of GE_1, supporting the writter's theory that introverts condition/learn better than do extraverts.

9. It seems likely that higher heritabilities are found at extreme ends of the E and N continua, although certain possible artifacts may make the estimates depart from the true values.

10. Higher heritabilities than for simple traits measured by questionnaires are found for higher-order factors, like E and N, particularly when these are measured by means of multiple measures, and when factor scores are used. This superiority may be due to any or all of the following causes: (a) the use of personality dimensions having a long historical development and good theoretical backing; (b) the use of different types of measures (questionnaires, experimental laboratory measures, psychophysiological tests); and (c) the employment of factor analysis in delineating the major dimensions involved and the use of factor scores.

11. There is suggestive evidence that inheritance of personality traits may be linked with sex in an ill-understood manner; inheritance seems to proceed through the

mother rather than through the father. No acceptable theory exists at the moment to account for this phenomenon, and the possibility remains that it may be purely environmental.

12. Criticisms of the twin method, while plausible, have not given rise to empirical evidence which would discredit this method. Studies of twins separated early in life and study of twins whose parents were mistaken concerning their status (MZ or DZ) are among those which suggest that the twin method is more reliable than was thought at one time.

It may be useful to suggest certain ways in which future research endeavors can be improved, basing these suggestions on the facts and theories surveyed in this chapter. These suggestions are, of course, to some extent subjective, embodying as they do the views of the writer about more general problems than can easily be settled by specific experiments. Nevertheless, they may be of some use, if only as the basis for argument and discussion. (1) It would seem that investigations using relatively small numbers of twin pairs (fifty pairs or less in each of the MZ and DZ categories) are only justified if there is a very specific aim which does not necessitate small standard errors in the final estimates of heritability; for most purposes it would seem that fifty pairs in each category is an absolute minimum, with one hundred pairs a much more desirable number. Even this does not enable us to estimate population parameters with any great accuracy. (2) In addition to MZ and DZ twins brought up together, a well-designed study should, wherever possible, contain MZ twins brought up separately; and it seems equally desirable that various degrees of relationship within the family should be studied. It is now possible to state with considerable accuracy what particular types of data give the optimum return for expenditure of energy in selection of cases[117, 118, 119] and simple imitation of time-honored procedures, such as comparison of MZ and DZ twin groups, does not any longer constitute the optimum research strategy. (3) Simple statements of heritability, obtained along the classical lines and using ther classical formulae, are not longer sufficient although they are, of course, better than nothing. What is required now are explicit models of a genetic kind (of the kind pioneered by Jinks and Fulker[20]), together with tests of the applicability of the data of the model. (4) On the personality side, the importance of using factorially meaningful measures must be stressed; by preference these should be multimeasure, i.e. contain questionaires, laboratory tests, psychophysiological measures in as diverse a form as possible. By including maximum diversity of measuring instrument in the battery, it becomes possible to eliminate instrument factors and reduce the influence of specific factors to a minimum. It seems likely that by following these four simple rules, the genetic study of personality will make very rapid advances in the next few years.

REFERENCES

1. Hall, C.S., and Lindzey, G.: *Theories of Personality*, 2nd ed. New York, Wiley, 1968.
2. Eysenck, H.J. (Ed.): *The Structure of Human Personality*, 3rd ed. London, Methuen, 1970.
3. Eysenck, H.J., and Eysenck, S.B.G.: *The Structure and Measurement of Personality*. San Diego, Knapp, 1969.
4. Eysenck, H.J.: *Dimensions of Personality*. London, Routledge and Kegan Paul, 1947.
5. Thurstone, L.L.: *Multiple Factor Analysis*. Chicago, U of Chicago Pr, 1947.
6. Howarth, E., and Browne, A.: An item-factor-analysis of the 16 PF. *Personality: An International Journal*, 2:117, 1971.
7. Sells, S.B., Damaree, R.G., and Will, D.P.: *A Taxonomic Investigation of Personality*. Dallas, Texas Christian Inst Behavioral Res, 1968.
8. Greif, S.: Untersuchungen zur deutschen Übersetzung der 16 PF Fragebogen. *Psychol Beiträge*, 12:186, 1970.
9. Eysenck, H.J.: Primary and second-order factors in a critical consideration of Cattell's 16 PF battery. *Br J Soc Clin Psychol*, 11:265, 1972.
10. Zubin, J., Eron, L.D., and Schumer, F.: *An Experimental Approach to Projective Techniques*. London, Wiley, 1965.
11. Eysenck, H.J.: *The Biological Basis of Personality*. Springfield, Thomas, 1967.
12. Eysenck, H.J., and Prell, D.: The inheritance of neuroticism: an experimental study. *J Ment Sci*, 97:441, 1951.
13. Holzinger, K.J.: The relative effect of nature and nurture influences on twin differences. *J Educ Psychol*, 20:245, 1929.
14. Neel, J.V., and Schull, W.J.: *Human Heredity*. Chicago, U of Chicago Pr, 1954.
15. Nichols, R.C.: The National Merit Twins study. In Vandenberg, S.G. (Ed.): *Methods and Goals in Human Behavior Genetics*. New York, Acad Pr, 1965.
16. Cattell, R.B.: The multiple abstract variance analysis. Equations and solutions for nature-nurture research on continuous variables. *Psychol Res*, 67:353, 1960.
17. Cattell, R.B.: Methodological and conceptual advances in evaluating hereditary and environmental influences and their interaction. In Vandenberg, S.G. (Ed.): *Methods and Goals in Human Behavior Genetics*. New York, Acad Pr, 1965.
18. Fisher, R.A.: The correlation between relatives on the supposition of Mendelian inheritance. *Trans Roy Soc*, 52:399, 1918.
19. Mather, K.: *Biometrical Genetics: The Study of Continuous Variation*. London, Methuen, 1949.
20. Jinks, J.L., and Fulker, D.W.: Comparison of the biometrical, genetical, MAVA and classical approaches to the analysis of human behavior. *Psychol Bull*, 73:311, 1970.
21. Hollingshead, A.B.: Cultural factors in the selection of marriage mates. *Am Sociol Rev*, 15:619, 1950.
22. Lutz, F.: Assortative mating in man. *Science*, 16:249, 1918.
23. Burgess, E.W., and Wallin, P.: Homogamy in personality characteristics. *J Abnorm Psychol*, 39:475, 1944.
24. Kennedy, R.J.R.: Single or triple melting pot? Intermarriage trends in New Haven, 1870-1940. *Am J Sociol*, 49:331, 1944.
25. Morton, N.E., Ching, S.C., and Ming, M.P.: *Genetics of Interracial Crosses in Hawaii*. Basel, Karger, 1967.
26. Pearson, K., and Lee, A.: On the laws of inheritance in man. *Biometrika*, 2:372, 1903.
27. Harris, J.A.: Assortative mating in man. *Popular Sci Monthly*, 80:476, 1912.
28. Pearson, K.: Reply to certain criticisms of Mr. G.U. Yule. *Biometrika*, 5:470, 1906.
29. Willoughby, R.R.: Family similarities in

30. Willoughby, R.R.: Family similarities in mental test abilities. *Genet Psychol Monogr, 2*:235, 1927.
30. Willoughby, R.R.: Family similarities in mental test abilities. *27th Yearbook Natl Soc Study Educ, Part I*:55, 1928.
31. Jones, H.E.: A first study of parent-child resemblance. *27th Yearbook Natl Soc Study Educ, Part I*:61, 1928.
32. Burks, B.S.: The relative influence of nature and nurture upon mental development: A comparative study of foster parent-foster child resemblance. *27th Yearbook Nat Soc Study Educ, Part I*:219, 1928.
33. Hill, M.: *Familial Trends in Personality Traits as Measured by the MMPI: Prospective Assortative Mating Patterns and Pedigree Analysis.* Unpublished doctoral dissertation, University of Minnesota, 1968.
34. Gottesman, I.I.: Personality and natural selection. In Vandenberg, S.G. (Ed.): *Methods and Goals in Human Behavior Genetics.* New York, Acad Pr, 1965.
35. Gray, H.: Psychological types in married people. *J Soc Psychol, 29*:189, 1949.
36. Winch, R.F.: *Mate Selection, A Study of Complementary Needs.* New York, Harper, 1958.
37. Roos, D.E.: *Complementary Needs in Mate Selection: A Study Based on R Type Factor Analysis.* Unpublished doctoral dissertation, Northwestern U, 1956.
38. Tharp, R.G.: Psychological patterning in marriage. *Psychol Bull, 60*:97, 1963.
39. Schellenberg, J.A., and Bee, L.S.: A reexamination of the theory of complementary needs in mate selection. *Marr Fam Liv, 22*:227, 1960.
40. Katz, I., Glucksberg, S., and Kraus, R.: Need satisfaction and Edwards PPS scores in married couples. *J Consult Clin Psychol, 24*:203, 1960.
41. Kerckhoff, A.C., and Davis, K.E.: Value consensus and need complementarily in mate selection. *Am Sociol Rev, 27*:295, 1962.
42. Cattell, R.B., and Nesselroade, J.R.: Likeness and completeness theories examined by sixteen personality factor measures on stably and unstably married couples. *J Pers Soc Psychol, 7*:351, 1967.
43. Eysenck, H.J.: An experimental and genetic model of schizophrenia. In Kaplan, A.R. (Ed.): *Genetic Factors in Schizophrenia.* Springfield, Thomas, 1972.
44. Pickford, J.H., Signori, E.I., and Rempel, H.: Husband-wife difference in personality traits as a factor in marital happiness. *Psychol Rep, 20*:1087, 1967.
45. Kallmann, F.J., and Mickey, J.S.: Genetic concepts and *Folie à Deux. J Hered, 37*:298, 1946.
46. Kreitman, N.: The patient's spouse. *Br J Psychiatr, 110*:159, 1964.
47. Kreitman, N.: Married couples admitted to mental hospitals. *Br J Psychiatr, 114*:699, 1968.
48. Eysenck, S.B.G., and Eysenck, H.J.: The measurement of psychoticism: a study of factor stability and reliability. *Br J Soc Clin Psychol, 7*:286, 1968.
49. Eysenck, S.B.G., and Eysenck, H.J.: Scores on three personality variables as a function of age, sex and social class. *Br J Soc Clin Psychol, 8*:69, 1969.
50. Nias, D.: *Personality Factors of Children and Adults.* Unpublished doctoral dissertation, Univ London, 1975.
51. Insel, P.M.: *Family Similarities in Personality, Intelligence, and Social Attitudes.* Unpublished doctoral dissertation, Inst Psychiatry, London, 1971.
52. Newman, H.H., Freeman, F.N., and Holzinger, K.J.: *Twins: A Study of Heredity and Environment.* Chicago, U of Chicago Pr, 1937.
53. Svancara, J.: Variability of psychological results in twins as a starting point for developmental hypotheses. *Psychologia Apato Psychologia Dietata, 6*:89, 1971.
54. Eysenck, S.B.G.: *Manual of the Junior Eysenck Personality Inventory.* London, U of London Pr, 1965.
55. Carter, H.D.: Twin similarities in personality traits. *Char Person, 4*:61, 1933.

56. Gottesman, I.I.: Heritability of personality: A demonstration. *Psychol Monogr,* 77:1, 1963.
57. Reznikoff, M., and Honeyman, M.G.: MMPI profiles of monozygotic and dizygotic twin pairs. *J Consult Clin Psychol, 31*:100, 1966.
58. Vandenberg, S.G.: Hereditary factors in normal personality traits (as measured by inventories). In Wortis, J. (Ed.): *Recent Advances in Biological Psychiatry.* New York, Plenum Pr Plenum Pub, 1967, vol. IX.
59. Vandenberg, S.G.: The hereditary abilities study in hereditary components in a psychological test battery. *Am J Hum Genet, 14*:220, 1962.
60. Vandenberg, S.G.: Contributions of twin research to psychology. *Psychol Bull, 66*:327, 1966.
61. Vandenberg, S.G. (Ed.): *Progress in Human Behavior Genetics.* Baltimore, Johns Hopkins, 1968.
62. Gottesman, I.I.: Genetic variance in adaptive personality traits. *J Child Psychol Psychiatry, 7*:199, 1966.
63. Nichols, R.C.: The resemblance of twins in personality and interests. *National Merit Scholarship Corp Res Rep, 8*:1966.
64. Vandenberg, S.G., Comrey, A.L., and Stafford, R.E.: *Hereditary Factors in Personality and Attitude Scales in Twin Studies.* Res Rep No. 16, Louisville Twin Study, Louisville, U of Louisville, 1967.
65. Lindzey, G., Loehlin, J., Manosevitz, M., and Thiessen, D.: Behavioral Genetics. In *Annu Rev Psychol, 22*:39, 1971.
66. Thompson, W.R., and Wilde, G.J.S.: Behavior genetics. In Wolman, S. (Ed.): *Handbook of Psychology.* New York, McGraw, 1971.
67. Partanen, J., Bruun, K., and Markkanen, T.: *Inheritance of Drinking Behavior, A Study on Intelligence, Personality and Use of Alcohol of Adult Twins.* Helsinki, Finnish Foundation for Alcohol Studies, 1966.
68. Scarr, S.: Genetic factors in activity motivation. *Child Dev, 37*:663, 1966.
69. Scarr, S.: The origin of individual differences in adjective check list scores. *J Consult Clin Psychol, 30*:354, 1966.
70. Scarr, S.: Social introversion-extraversion as a heritable response. *Child Dev, 40*:823, 1969.
71. Koch, H.T.: *Twins and Twin Relations.* Chicago, U of Chicago Pr, 1966.
72. Wilde, G.J.S.: Inheritance of personality traits. *Acta Psychol (Amst) 22*:37, 1964.
73. Brown, A.M., Stafford, R.E., and Vandenberg, S.G.: Twins: behavioral differences. *Child Dev, 38*:1035, 1967.
74. Freedman, D.: An ethological approach to the genetical study of human behavior. In Vandenberg, S.G. (Ed.): *Methods and Goals in Human Behavior Genetics.* New York, Acad Pr, 1965.
75. Juel-Nielsen, N.: *Individual and Environment.* Copenhagen, Munksgaard, 1965.
76. Eysenck, H.J. (Ed.): *Handbook of Abnormal Psychology,* 2nd ed. New York, Basic, 1973.
77. Cooper, J.E., Kendell, R.E., Gurland, B.J., Sartorius, N., and Farkas, T.: Cross-national study of diagnosis of the mental disorders: some results from the first comparative investigation. *Am J Psychiatry, 125*:21, 1969.
78. Eysenck, H.J., and Prell, D.: A note on the differentiation of normal and neurotic children by means of objective tests. *J Clin Psychol, 8*:202, 1952.
79. Eysenck, H.J.: The inheritance of extraversion-introversion. *Acta Psychol, 12*:95, 1956.
80. Young, J.P.R., Fenton, G.W., and Lader, M.H.: The inheritance of neurotic traits in a twin study of the Middlesex Hospital Questionnaire. *Br J Psychiatry, 119*:393, 1971.
81. Shields, J.: *Monozygotic Twins Brought Up Apart and Brought Up Together.* London, Oxford U Pr, 1962.
82. Loehlin, J.C.: A heredity-environment analysis of personality inventory data. In Vandenberg, S.G. (Ed.): *Methods and Goals in Human Behavior Genetics.* New York, Acad Pr, 1965.

83. Bleuler, M.: Die Rorschachsche Formdentung; Versuch bei Geschwistern. *Z Neurol, 188*:366, 1929.
84. Elmgren, J.: Le test de Rorschach au point de vue constitutionel. *Ann Psychol, 51*:593, 1951.
85. Schachter, M.: Contribution à l'étude du psychodiagnostic de Rorschach chez les jumeux. *Encephale, 41*:23, 1952.
86. Troup, E.A.: A comparative study by means of the Rorschach method of personality development in twenty pairs of identical twins. *Genet Psychol Mongr, 20*:461, 1938.
87. Kiev, M.: Temperamental differences in twins. *Br J Psychol, 27*:51, 1936.
88. Kerr, M.: Temperamental differences in twins. *Br J Psychol, 27*:51, 1936.
89. Vacca, E., and Ciolfi, F.: Le test de Rorschach appliqué à des jumeux italiens. Prem Recontre Rorschach, Paris, 1949.
90. Rabin, A.J.: Genetic factors in the selection and rejection of Szondi's pictures: A study of twins. *Am J Orthopsychiatry, 22*:551, 1952.
91. Nolan, E.G.: *Uniqueness in Monozygotic Twins.* Unpublished doctoral dissertation, Princeton, 1959.
92. Nolan, E.G.: Szondi test protocols of monozygotic and dizygotic twin populations. *J Project Technol, 25*:471, 1961.
93. Seeman, E., and Saudek, R.: The self-expression of identical twins in handwriting and drawing. *Char Person, 1*:91, 1932.
94. Nicolay, E.: Messungen an Handschrift proben von Zerwillingspaaren uber 14 Jahren. *Arch Ges Psychol, 105*:275, 1939.
95. Dennemark, H.G.F.: *Die Handschriften von Zwillingen.* Jena, Thomas and Hubert, 1940.
96. Orlyngen, E.: On hereditary and environmental determination of the variability in handwriting. *Acta Psychiatr, Scand 20*:75, 1945.
97. Roman-Goldzieher, K.: Untersuchung der Schrift und des Schreibens von 283 Zwillingen. *Graphologia, 1*:29, 1945.
98. Hoyler, A.: Messungen an Zwillingsschriften. *Z Diag Psychol Persönlichkeits Forsch, 6*:39, 1958.
99. Allport, G., and Vernon, P.E.: *Studies in Expressive Movement.* New York, Macmillan, 1933.
100. Eysenck, H.J.: Reminiscence, drive and personality—revision and extension of a theory. *Br J Soc Clin Psychol, 1*:127, 1962.
101. Smith, G.: *Psychological Studies in Twin Differences.* Lund, Gleerup, 1949.
102. Matheny, A.P.: Genetic determinants of the Panza illusion. *Psychonom Sci, 24*:155, 1971.
103. Fuller, J.L., and Thompson, W.R.: *Behaviour Genetics,* 2nd ed. London, Wiley, 1966.
104. Lennox, W.G., Gibbs, E.C., and Gibbs, F.A.: The brain-wave patterns, and hereditary traits. *J Hered, 36*:233, 1945.
105. Raney, E.: Brain potentials and lateral dominance in identical twins. *J Exp Psychol, 24*:21, 1939.
106. Vogel, F.: *Über die Erblichkeit des normalen Elektroenzephalogramms.* Stuttgart, Thieme, 1958.
107. Young, J.P.R., and Fenton, G.W.: An investigation of the genetic aspects of the alpha attenuation response. *Psychol Med, 1*:365, 1971.
108. Jinks, J.L.: *Extra-Chromosomal Inheritance.* New York, P-H, 1964.
109. Jones, H.E.: Perceived differences among twins. *Eugen Q, 5*:98, 1955.
110. Smith, R.T.: A comparison of socioenvironmental factors in monozygotic and dizygotic twins, testing an assumption. In Vandenberg, S.G. (Ed.): *Methods and Goals in Human Behavior Genetics.* New York, Acad Pr, 1965.
111. Scarr, S.: *Genetic and Human Motivation.* Unpublished doctoral dissertation, Harvard Univ, Harvard, 1964.
112. Takuma, T.: Einige japanische Untersuchungen zur Zwillingsforschung. *Z*

Exp Angew Psychol, 18:670, 1971.
113. Vandenberg, S.G., and Johnson, R.C.: Further evidence on the relation between age of separation and similarity in IQ among pairs of separated identical twins. In Vandenberg, S.G. (Ed.): *Progress in Human Behavior Genetics*. Baltimore, Johns Hopkins, 1968.
114. Crook, M.N.: Intra-family relationships in personality test performance. *Psychol Rec, 1*:479, 1937.
115. Coppen, A., Cowie, V., and Slater, E.: Familial aspects of 'neuroticism' and 'extraversion.' *Br J Psychiatry, 111*:70, 1965.
116. Wilson, G.D., and Patterson, J.C.: *Manual for the Conservatism Scale*. Windsor, Berks, Natl Found Educ Res, 1970.
117. Eaves, L.J.: The genetic analysis of continuous variation: a comparison of experimental designs applicable to human data. *Br J Math Statist Psychol, 22*:131, 1969.
118. Eaves, L.J.: The genetic analysis of continuous variation: a comparison of experimental designs applicable to human data. II. Estimation of heritability and comparison of environmental components. *Br J Math Statist Psychol, 23*:189, 1970.
119. Eaves, L.J.: Computer simulation of sample size and experimental design in human psychogenics. *Psychol Bull, 77*:144, 1972.
120. Schoenfeldt, L.F.: The hereditary components of the project TALENT two-day test battery. *Meas Eval Guid, 1*:130, 1968.

Chapter 10

PHYSIQUE AND PERSONALITY

Detlev von Zerssen

INTRODUCTORY REMARKS

The inclusion of a chapter on physique and personality in a textbook on human behavior genetics has, to a certain extent, historical reasons. The study of individual differences in both physical appearance and behavioral traits and the relationships between them is a traditional subject of constitutional research.[1,2,3,4] Since constitution has been identified with an individual's genotype by several authors,[5] the term *constitutional* has gained the connotation of "genetically determined." This, however, is entirely misleading.[6,7] With very few exceptions (e.g. the classical blood groups) even the most constant physical characteristics[8,9,10,11,12,13] and, still more so, all the habitual mental traits [14,15,16,17] are not entirely determined by the individual's genetic endowment. To a varying degree they are molded by environmental influences on growth processes on the one hand, and personality development on the other.

It is now generally acknowledged that one's bodily makeup can be modified considerably by variations in physical conditions during the fetal period and early childhood.[18] It is also well-known that personality development is, to a large extent, under the control of environmental stimuli.[14,17] The importance of experiences during the first years of life is particularly well established in this respect.[19,20] Early childhood experiences may also influence general growth patterns,[21,22,23] although most likely in a rather indirect way.[24] Moreover, physical conditions in the pre- and postnatal life-period are relevant not only for the physical but likewise for the mental development.[14]

As with variations in the physical and mental components of the entire constitution, their interrelationships are caused by genetic as well as by nongenetic factors. This is convincingly demonstrated by abnormal states such as Down's syndrome and endemic cretinism.[25] In Down's syndrome,[26,27] the concomitant abnormalities in the physical and mental spheres are caused by the individual's genotype, in most cases characterized by a trisomy of chromosome 21. In endemic cretinism,[28] the corresponding relationship is usually of external origin, the lack of iodine in the fetal organism.

There are several additional examples of concomitant abnormalities in physical and mental development.[29] These are partly due to chromosomal[30] or other deficiencies of the genotype[31] and partly to external conditions of fetal or early postnatal life (e.g. malnutrition,[32,33] toxoplasmosis, and others[25]). The resulting positive correlation between mental retardation and physical dysplasia,[14] established in investigations of adult psychiatric populations,[34, 35] is, therefore, only to a certain extent determined by genetic factors.[14] The interrelationships among physical abnormalities, gender role, sexual orientation, and personality traits in the various forms and degrees of intersexuality are of a particular complexity and cannot simply be related to either genetic or environmental factors.[36, 37, 38, 39] Very subtle investigations are necessary to distinguish the different, genetic, as well as nongenetic, sources of concomitance between physical and mental characteristics. The same is true for psychophysical relationships within the normal range of variation which are often thought to indicate mainly the influence of polyphenic genes.[2, 40, 41, 42, 43] However, a tendency of extraverted individuals toward higher weight, found in male samples from the average population,[44, 45] was also shown to exist in pairs of monozygotic twins.[16] It can, therefore, not entirely be traced back to the genetic basis of constitutional differences.[46]

Even if certain differences in physique are almost exclusively determined by the individuals' genotypes, mental concomitants of these differences may result mainly from external influences. An example is the reaction of the human environment to visible differences in the individuals' appearance. Certainly, this is partially true in the cases of sexual and racial differences.[14, 17] Therefore, the relationship between physique and personality is usually studied within groups of the same sex and the same (main) race. Since most of these investigations were restricted to the male sex and the white (Europid) race, it seems doubtful how far their results can be generalized to either the female sex or other races.

Age is another factor which must be considered when studying psychomorphological relations. During the period of maturation, a close but trivial relationship exists between physical and mental development, which is largely age-dependent. Correlations between tests of the developmental stage in the physical and mental spheres usually drop considerably if the age factor is eliminated.[47] Also during adulthood, there are changes in physique[5] as well as in personality[14, 17] which must be considered with respect to psychophysical relationships; to overlook them may have serious consequences for the conclusions drawn from investigations of such relationships.

There are, of course, variations in the physical and mental spheres of human constitution which are not due to differences in sex, race, or age. These variations and their interrelationships are the main concern of constitutional research[45] and will, therefore, be emphasized in our report on physique and personality. Only investigations of the external aspect of physique (morphological habitus) and its relation to personality features (intelligence and temperament) will be reviewed, because, historically, they have thus far predominated in the entire field of constitutional research.[2] Definitions of terms as they are used in this review are given in the glossary (see: *constitution, dimension, factor, habitus, intelligence, personality, temperament, type*).

VARIATIONS IN PHYSIQUE

The Typological Approach

The typological approach[7] to the analysis of variations in physique dates back to ancient Greek medicine. Valuable information on the historical development of this branch of human biology can be found in a number of surveys.[1, 2, 4, 48] We have restricted ourselves to a brief outline of Kretschmer's typological system.[40, 49] His typology of body build is in basic accordance with previous ones, but it has become more widely accepted in various fields of scientific work (medicine, anthropology, psychology, and criminology) all over the world for many years or even decades. This is mainly due to Kretschmer's vivid and detailed description of the physical types and to his stimulating notion of their affinity for types of temperament, types of functional psychoses, and other abnormalities (e.g. physical diseases and types of criminality).

According to Kretschmer, the indefinite number of individual differences in physical habitus can be related to some basic types or patterns which are thought to reflect the main differences among people. By comparing the individual pattern with those basic patterns, a "type diagnosis" can be achieved. Diagnosing a case means, in this connection, assigning him to the reference pattern (type) with which he shows the highest degree of concordance. The typing procedure is thus a kind of pattern recognition.[50] If the case in question resembles more than one type, he may be designated a "mixed type" (a combination of more than one reference pattern). If he is intermediate among all types under consideration, he is termed "uncharacteristic" from a typological point of view.

In systems other than Kretschmer's, this last diagnosis may be a positive one of allocating the case to the "average type" of the population in question. This usually leads to a much higher incidence of average cases than in Kretschmer's approach of forced assignment to types which deviate from the average.[34]

Kretschmer has described three normal deviations from the average physical habitus which he assumes to be typical. This implies that the characteristics of each type are positively intercorrelated. Therefore, Kretschmer defines types as *Schnittpunkte von Korrelationsmaxima* by which he apparently means central points of clusters of correlated traits.[7] In addition to these normal types, Kretschmer distinguishes a group of abnormal constitutions, the so-called "dysplastic types." They are either quantitatively abnormal, i.e. with regard to the degree of deviation from the average; or they are qualitatively abnormal, presenting a disharmonious mixture of features of different types or exhibiting peculiar features not characteristic for any of the normal types. The latter are described as follows:

1. The pyknic type is generally stout with a relatively long trunk and short limbs and a broad and particularly deep chest. In a normal nutrition state he tends to develop a large amount of subcutaneous fat mainly concentrated in the neck and abdominal regions. Depending on the amount of food intake, this type easily gains or loses weight.

2. The leptosome type was originally called *asthenic*, a term which subsequently was only used for an underdeveloped variant of this type. Contrary to the pyknic type, he is generally slender with relatively long limbs, a narrow flat chest, and poor development of subcutaneous fat. He, therefore, constantly falls below the normal average in weight relative to height, even when well nourished.

3. The athletic type is not, as assumed by some authors, merely intermediate between both other types of physique, but deviates from the average in both muscular and bone development. Consequently, the shoulders are broad, hands and feet are large, and the body surface is structured by muscles. The development

of fatty tissue is relatively low and the weight-curve over time is uncharacteristic.

There is no standardized procedure for making an individual diagnosis in relation to these types. General impression, special morphognostic features, and various anthropometric measures and indices all enter into the final judgment in an undefined way. This results in gross discrepancies in statements on the incidence of morphological types in comparable samples drawn from the average population or from diagnostic subgroups of psychiatric inpatients, etc.[51] Therefore, several authors have attempted to define the types operationally on the basis of anthropometric indices alone.[34, 52, 53, 54] Relative weight or relative chest measurements were mainly used for discriminating the pyknic and the leptosome type,[53, 54] whereas the athletic type was usually characterized by shoulder breadth and other skeletal measures.[34, 52, 54] Because even the joint distributions of these measures and indices[55] are continuous and unimodal,[2] the anthropometric definition of physical types necessarily leads to a dimensional approach of physique.[56]

The Dimensional Approach

General Remarks

The dimensional approach to individual differences in physique seeks to discover fundamental traits which are assumed to account for the correlations among a large number of different physical characteristics.[57, 58] Those traits are usually called dimensions or factors in the realm of factor-analytic theory.[59, 60] They can be represented by single or combined measures which are most closely related to the entire group of correlating characteristics. Historically, the measures were first chosen on the basis of inspection or on theoretical grounds, and, only afterwards, were they tested by a correlation with other variables.[41, 61, 62, 63] When factor analysis was introduced into the investigation of morphological variations, however, some authors derived criteria for complex variations of physique primarily from statistical data analysis.[64, 65, 66]

Pre-factor-analytic Studies

Conrad, one of Kretschmer's pupils, was among the authors who started their analyses of morphological variations from inspection as well as from theoretical considerations on the nature of these variations.[52] He revised his teacher's system by converting it from a typological to a dimensional one.[49] Pyknics and leptosomes were interpreted as representing deviations from the (metromorphic) average within a variation of proportional growth between a conservative (pyknomorphic) and a propulsive (leptomorphic) extreme. This means that pyknomorphic build is looked upon as relatively childlike in proportions as compared with the leptomorphic build which is considered markedly more adult. This variation of physique is quantitatively expressed by a modification of Strömgren's index.[53]

A combination of hand breadth and shoulder breadth is used by Conrad as a measurable criterion for another variation in physique opposing the hyperplastic athletic habitus to the hypoplastic asthenic habitus. Thus (in accordance with Schlegel[41] but at variance with Kretschmer[40]), asthenic physique is not regarded as an underdeveloped variant of leptomorphic habitus; rather, it is attributed to a deviation from the normoplastic mean of habitus variation in direction to the (according to Conrad) abnormal extremes of general hypo- versus hyperplasia.

In 1940 the American psychologist, Sheldon, described a dimensional system of morphological variations which was partially different from those described by European authors.[42] Methodologically, it was based on the comparison of standard photos of a large number of individuals of the same sex and age level (4000 male college students). This comparison resulted in a plain triangular schema, the *somatotype distribution chart,* in which the individuals were arranged according to their resemblance with each other. The average type held the central position, while the most markedly extreme variants were located at the three corners of the schema, as shown in Figure 10-1. Sheldon called these extremes the endomorphs, mesomorphs, and ectomorphs, respectively, because he considered them related to the dominance of one of the germinal layers in the individual's growth process.

According to this theoretical assumption, the endomorphic habitus is dominated by derivatives of the endoderm. In fact, however, the typical aspect of endomorphs is mainly due to overdevelopment of subcutaneous fat, a tissue component which stems from the mesoderm.[61, 67, 68, 69, 70, 71, 72] Consequently, relatively pure endomorphs display a rounded appearance, with the trunk dominating over the limbs. In contradiction to Sheldon's theoretical assumption, the mass of visceral organs is, on the whole, not positively correlated with a stout and obese physique.[73]

The mesomorphic habitus is determined by a marked development of bones and muscles which, in accordance with the basic theoretical assumption, are derived from the mesoderm.[71] This habitus appears rectangular and broad (in Kretschmer's terminology like the athletic type, but with pyknic tendencies).

The ectomorphic habitus is characterized by slenderness due to the poor development of bone, muscle, and fatty tissue in relation to length. It therefore resembles Kretschmer's asthenic type. Body surface (epidermis) and cerebral tissue, each stemming from the ectoderm, seem overdeveloped in ectomorphs compared with the mass of tissues of mesodermic origin. However, there is no empirical basis for the assumption that people of ectomorphic (linear) physique have a larger amount of cerebral tissue and a greater body surface than other extreme variants of physique or individuals of average physical appearance. For this reason, the term ectomorphy is rather misleading, as endomorphy is in another respect.

Indeed, Sheldon's somatotypology can be traced back to variations in derivatives of the mesoderm alone: relative overdevelopment of fatty tissue (endomorphy), of bone and muscle tissue (mesomorphy), and relative underdevelopment of all mesomorphic tissues (ectomorphy).[63, 74, 75] On a purely descriptive level, this modification of Sheldon's concept does not affect his attempt to arrange human individuals according to global resemblance in physical habitus; neither does it necessarily influence the rating based on it.

Sheldon determines an individual's position in his somatotype distribution chart by photoscopic comparison with reference cases within the same range of relative body weight which he had rated in the original process of standardizing his schema.[76] The rating is expressed by three numerals, each of them arranged along a seven-point scale. The three scale values are reciprocal to the distances of an individual's habitus from each of the three corners of the distribution chart.

Yet it is quite obvious that any point in a

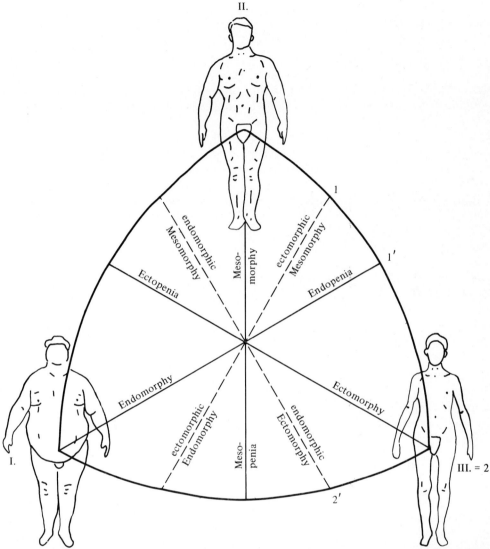

Figure 10-1. Sheldon's somatotype distribution chart (simplified) with examples of three extreme types of physique and with two additional axes of reference (1 and 2'.) Compare with Figure 10-3 and 10-4.

plane (any individual in the somatotype distribution chart) can be adequately described by two numerals only. Sheldon's description of the distribution is redundant because he projects the points on three oblique axes of reference representing the degree of endomorphy, mesomorphy, and ectomorphy.[57, 77] A simpler and more economic procedure would be the projection of points on two orthogonal axes[74, 78] (e.g. 1 = ectomorphic mesomorphy versus ectomorphic endomorphy, and 2 = ectomorphy versus ectopenia; or 1' endomorphy versus endopenia, and 2'

=endomorphic mesomorphy versus endomorphic ectomorphy;[79] see Fig. 10-1).

The British physician, Parnell,[80] and the American anthropologist, Hooton,[70] independently related endomorphy and mesomorphy to variations in the tissue components *fat* and *muscularity*. Neither author tried to derive ectomorphy from these components (as a lack in fat and muscularity).[71,74] Instead, each expressed ectomorphy (as linearity or attenuation, respectively) by the (inverse) ponderal index (height/$\sqrt[3]{\text{weight}}$). This index is, of course, mainly determined by variations in fat and muscularity.[63] Parnell assessed these components somatometrically, using the total score of skinfold thickness measured at three different sites as the criterion for fat (=endomorphy) and a combined score of extremital girth (corrected for fat) and of condylar breaths as the criterion for muscularity (=mesomorphy). A similar approach was employed by B.H. Heath.[69]

Meanwhile, Sheldon realized that his original system of variations in body build was redundant and that the three interdependent dimensions should be substituted by two independent ones to achieve a more expedient description of the variations under consideration.[81] To make this description as objective as possible, he defined the two dimensions anthropometrically. His operational definitions were not based on tissue components (fat and muscularity) but rather related to bodily proportions (as in his photoscopic approach). In a manner similar to that of Parnell and Hooton he assessed ectomorphy by using the ponderal index (taken at the time of an individual's maximum weight and corrected for age). For the assessment of an independent dimension of habitus variation, he combined two measures of mesomorphy and endomorphy, respectively, in only one measure. This so-called *trunk index* is a rather complicated photometric index which relates the upper part of the trunk (dominating in mesomorphs) to its lower region (dominating in endomorphs). According to Sheldon, this index remains constant throughout life in both sexes, at least from the third year onwards; it is considered almost unaffected by the nutritional state. The correlation with the ponderal index is said to be practically zero.

By adding the body's "extension into space" (size) to the two shape dimensions of body build, Sheldon created a three-dimensional system. The criterion for the third dimension was stature. Thus, variations in physique are now described by Sheldon along three independent dimensions. One represents variations in general growth (extension into space), while the other two represent independent aspects of variations in proportional growth (development of volume in relation to height and development of the upper relative to the lower part of the trunk).

In his original version of somatotypology based on somatoscopic assessment,[42] Sheldon also considered dysplasia, but defined it in a more restricted sense than Kretschmer.[40] In Sheldon's typology, this term refers only to discrepancies in the ratings of endomorphy, mesomorphy, and ectomorphy in different parts of the body. According to Kretschmer, however, extreme endomorphs and ectomorphs would also be designated "dysplastic" (obese-dysplastic or asthenic-hypoplastic, respectively;[82] Fig. 10-3).

The American anthropologist, Seltzer, has quantitatively expressed disharmonious growth of different body regions in terms of deviations from the normal range of variation in certain anthropometric indices.[83] In comparing the percentages of bodily disproportions in Sheldon's somatotypes, he found them to be positively related to ectomorphy. Apparently, this is because only deviations in one direction were taken into account, and these were always toward ectomorphy versus mesomorphy and/or endomorphy (e.g. extremely thin calves in relation to shoulder breadth and the like). Thus, it seems doubtful whether Seltzer's approach has indeed led to a biometric delineation of disharmonious growth or whether it has only given some additional information on ectomorphic physique.

As a special aspect of dysplasia, Sheldon considered the relative degree to which secondary sex characteristics of the opposite sex were developed in the male.[42] This *gynandromorphy* was rated via standard photos according to a series of morphognostic features assumed to have a high discriminating power in an anthroposcopic comparison of both sexes. A similar approach was later pursued by Draper[84] and by others.[85, 86] Some authors have investigated morphological sex differences through statistical analyses of anthropometric data. For the most part, scores and indices expressing the relation of shoulder breadth to hip breadth,[87] and that of muscle and bone development to that of subcutaneous fat,[63, 88, 89, 90] were elaborated as criteria of *androgyny*, the variation of secondary sex characteristics within one sex.

Tanner described the androgyny score (3 times biacromial diameter minus biiliac diameter minus 82) which he had derived from a discriminant analysis of anthropometric measures of male and female samples.[87] Wilde, together with the present author, has shown, however, that the androgyny score correlates as high as .96 with biacromial diameter.[91] Therefore, it adds hardly any further information to this singular measure.[63] Moreover, neither measure combined in the androgyny score has a high discriminating power for sex differences: hip breadth, in absolute values,[61, 92] is rather worthless in this respect,[5, 90] and shoulder breadth related to height, is, too, almost equal in both sexes;[92] in absolute values it is, as a sex-discriminating parameter, inferior compared with other somatometric measures like weight and, even more so, hand girth.[5, 90] Nonetheless, the score has received a great deal of attention in the study of psychomorphological relationships.[93, 94, 95, 96, 97]

A better discrimination between the sexes is achieved by combining X-ray measures of bone breadth (tibial breadth measured on standard radiograms of the calf) and/or muscle breadth with subcutaneous fat (measured on the same X-ray photographs). The respective scores were described by Reynolds[88] and by Tanner.[89] Consequently, it can be assumed that the mesomorphy/endomorphy-ratio within one sex is closely related to the dimension of androgyny.[63]

This hypothesis can be tested by the correlation between Sheldon's index for gynandromorphy and the relation between his ratings of mesomorphy and endomorphy taken from a sample of 134 male individuals portrayed in the *Atlas of Men*.[76] In these cases, Sheldon provided information not only on the somatotype, but also on his ratings of gynandromorphy. In an investigation of the present author, the correlation between both ratings (gynandromorphy and mesomorphy/endomorphy-ratio) turned out to be as high as $-.64$.[45] A similar value ($r = .66$) was found when comparing Sheldon's rating of gynandromorphy (g-index) with a photometric criterion for androgyny. This criterion had been developed by the present author before Sheldon published the similar, although much more complicated, trunk index (upper trunk volume divided by lower trunk volume). It is defined as the difference between Sheldon's measures of upper trunk breadth (TB_1) and lower trunk breadth (TB$_3$). The score reflects the relative weight of the soft tissue components, muscle and fat, as well as that of the skeletal width of shoulder and hip. Some authors might prefer Sills' more neutral term *omomorphy* for the body build variation described by the trunk-breadths-difference.[72] However, the linear relationship of the score values to Sheldon's g-index, as demonstrated in Figure 10-2, leaves no doubt that it is, indeed, a criterion for androgyny.

As the interrater reliability of a global photoscopic assessment of androgyny was found by the present author to be .67,[90] it seems doubtful whether the subjective rating of androgyny adds much to what can be expressed in terms of the mesomorphy/endormorphy ratio, objectively defined by Sheldon's trunk index. Further evidence supporting this argument is found in the result of a factor analysis of data from the sample of 134 individuals from the *Atlas of Men* (table 19). This will be discussed below.

Nevertheless, the question remains

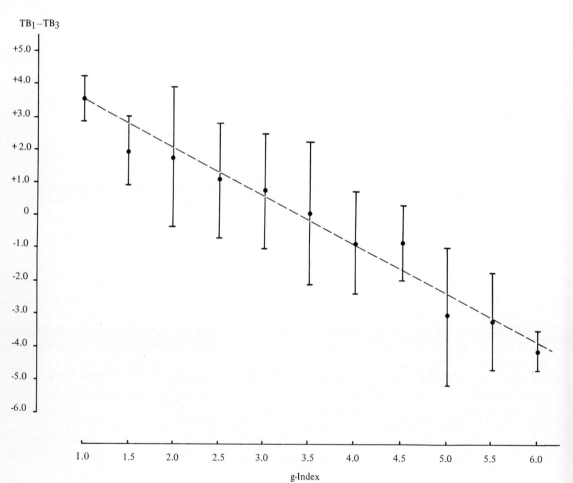

Figure 10-2. Means and standard deviations of the photometric index of androgyny (TB$_1$–TB$_3$) within the classes of Sheldon's g-index (n = 134).

whether it is more useful to describe variations in physique, disregarding height, on the basis of differences in shape (e.g. slenderness versus stoutness and masculinity versus femininity of build), or on the basis of differences in tissue components (e.g. fat and muscularity). The description according to shape has three advantages:
1. It is in basic agreement with the older typological approach.
2. The proportions under consideration can be objectively defined by anthropometric indices, even when derived from standard photos.
3. They show definite relationships to such independent variables as sex and age.

Ad 1: The extreme ectomorph with low relative weight (high value of ponderal index) is practically identical with the asthenic habitus, described by Kretschmer as the extreme variant of leptosomatic physique. Extremely ectopenic individuals with high relative weight (mesomorphic endomorphs) correspond to Kretschmer's pyknics.[79] These types have been described under various headings since the period of Hippocrates.[2] Moreover, individuals with high degrees of mesomorphy in relation to endormorphy can easily be identified with the athletic type outlined by Kretschmer;[82] Figure 10-3.

Ad 2: Precise quantitative measures of variations in physical proportions can be taken not only from the individual himself but also from standard photos.[42, 90, 98, 99] It is not so evident that measurable criteria for tissue components can be obtained photometrically.

Ad 3: Stoutness (which corresponds to ectopenia versus ectomorphy) varies with age. From adolescence to the beginning of senility there is a progressive tendency toward pyknomorphy (ectopenia).[100, 101, 102] This has also been demonstrated biometrically in an investigation of groups of psychotics matched according to sex, age, and verbal intelligence.[79, 103] Significant differences in a series of somatometric criteria for stoutness which, according to the literature on physical type and psychosis, should discriminate between manic-depressives and schizophrenics, were found only when comparing the older with the younger age group. This comparison was made irrespective of the diagnosis. The type of psychosis, however, did not show any definite relationship to body build.

Obviously, differences in stoutness within one age group cannot be explained by the age factor. Nevertheless, the relationship to different age groups may very well be interpreted as indicative of stoutness as a biologically meaningful dimension of physical variation. The same is valid for the mesomorphy/endomorphy-ratio with respect to sex differences and androgyny.

Consequently, the hypothesis seems to be plausible that these dimensions of bodily proportions may show specific correlations to personality traits.[63] On the other hand, there might be a more specific connection between mental traits and tissue components such as fat and muscularity.[104]

As previously mentioned, the variations in physique described by Sheldon can be interpreted as resulting from variations in tissue components. Some authors, in analyzing interindividual differences in physical habitus, paid particular attention to such variations.[105] A prominent representative of this group of research workers is the Swedish physician, Lindegård.[61, 62] He differentiated four components which, according to him, vary rather independently of each other: two components represent the development of fat and muscle, respectively (as in Parnell's and Hooton's modification of Sheldon's somatotypology[70, 80]); the others refer to two different aspects of bone development, length of extremeties, and skeletal "sturdiness", the latter mainly depending on the development of joints.

The assessment of these components is objective, though only in the case of skeletal measures (tibia length and femoral condylar breadth, respectively) does it seem to be truly valid. Muscle development is rather indirectly estimated from the dynamometric measure of handgrip. The fat measure is usually derived from body weight and from the physiometrically recorded measure of muscle development by means of a regression equation.

Factor-analytic Studies

The typological and dimensional systems referred to thus far were based mainly on clinical observation and/or biological speculation. Only afterwards were they objectified by measurements and statistical data analyses. The factor-analytic approach has partly reversed the sequence by starting from measurement and achieving a structure through a statistical procedure.[57,59,60] Certainly, speculation usually enters into this approach when the primarily formal, factorial structure is interpreted semantically.

Expressed in the language of matrix algebra, factor analysis consists of the transformation of a $n \times n$ matrix of correlations between n observed variables into an $n \times m$ matrix of the n variables and a smaller number ($m < n$) of m hypothetical variables, the factors or dimensions. The factors are supposed to be more fundamental and general in nature than the observed variables. With this assumption, the latter can be replaced by the factors when describing the variations within the subject under investigation.

The factor-analytic structure of a set of variables greatly depends on the number and nature of the variables, the specific procedures applied for data collection (sampling, measurement), and on data analysis (type of correlation, extraction of factors, and rotation of factors). Therefore, one cannot always expect to find the same factors in different analyses within the same field of research, e.g. variations in body build.

The first factor analyses of anthropometric measurements were performed by British psychologists.[2] In several of these analyses two factors were extracted from the matrix of intercorrelations but not rotated. This was partly for historical reasons and partly because the first two unrotated factors, together accounting usually for 50 percent of the variance or more, could easily be interpreted in a biologically meaningful way as representing variations in general size and relative stoutness, respectively. Both factors had remarkably high weights for all variables included in the analyses. The loadings on the first general factor were positive, indicating a tendency of all variables (absolute measures of weight, length, breadth, depth, and circumference) toward common variation in the same direction. This can be interpreted as an expression of general growth which influences extension of all body parts in the same basic way. The second factor, however, was a bipolar one, length measures having loadings opposite to those of all other measures. Therefore, this factor can be defined as a factor of proportional growth, opposing stoutness to slenderness. It is apparently identical with the leptopyknomorphic dimension of morphological differences described by numerous authors before factor-analyses were used to investigate variations in body build.

By rotating these factors according to simple structure criteria, an equivalent solution can be achieved. In this instance, the common variance is related to two group factors, one with weight only for length measures, the other for body depth, breadth, etc.[106, 107, 108] For this reason, the rotated factors may be termed length factor and cross sectional factor, respectively.

However, different solutions were frequently reported in the literature.[2, 5] This was partially due to differences in the investigated samples which varied in sex and age composition, social status, and the like. Even more influential was the selection of other measures[109] and the application of other data processing techniques.[108, 110] For example, if several head measures were included in the analyses, and more than two factors were extracted and then rotated according to simple structure criteria, a head factor usually emerged.[106, 108, 111] If a greater number of various measures were included, the length and the breadth factor were broken down to more specific dimensions of habitus variation (e.g., limb length,[6, 110] limb girth,[112] size of joints,[112] trunk thickness,[111] etc.).

In an unpublished investigation which the present author conducted with Hüneke, the different results previously published by other authors could be reproduced by reanalyzing data from the same sample in the following manners:

1. selecting the respective variables,
2. extracting the respective number of factors,
3. rotating them or not as the other authors had done.

Thus, a general size and a bipolar shape factor were elicited in a simulation of an analysis published by Rees and Eysenck.[66] In a simulation of an analysis performed by Hofstätter,[106] we, too, found a length, a cross sectional, and a head factor (though only in a four-factor solution instead of Hofstätter's use of three factors).

From the standpoint of psychomorphological relationships, primarily two kinds of factor analytic studies of physique are relevant:

1. those where the results can be paralleled to the findings of previous investigations relating physical variations to personality variations, and,
2. those which themselves are concerned with such relationships.

Regarding previous investigations, the first factor analytic studies of physique by British authors were completely in accord with the concept of a leptopyknomorphic variation in bodily proportions.[2] Nevertheless, the literature contains only a few reports on a factor which may be interpreted as representing androgyny.[45] A probable explanation is that a factor can emerge only if proper criteria have been included in the set of variables under investigation, and this was usually not the case with respect to androgyny. On the other hand, a fat and/or muscle factor could be isolated in analyses using measures of fat,[6, 72, 107] and/or muscle development[6] and/or global ratings of endomorphy and mesomorphy;[72] these ratings also can be regarded as representing variations in fat and muscle development, respectively.

The fat factor had high loadings on the following measures: direct measures of fat (e.g. skinfold thickness or corresponding X-ray parameters of subcutaneous fat), weight, circumference of trunk and limbs (at sites where subcutaneous fat shows considerable variation), some related diameters like chest depth, and ratings of endomorphy. Measures of skeletal development and direct measures of muscle development usually show no relationship to this factor. The muscle factor, on the other hand, has no weight for direct measures of fatty tissue. Rather, it relates to direct (X-ray) measures as well as to indirect measures of muscle development.[6] The latter are several measures of limb circumference, measure of physical strength (e.g. dynamometrically recorded handgrip[72]), and ratings of mesomorphy.[72] Moreover, measures of skeletal sturdiness and some criteria of physical stoutness were shown to be related to this factor. Thus, the term *physical robustness* was chosen by the present author instead of muscle or muscularity.[45,46]

According to what had been stated above on the two dimensions of bodily proportions (leptopyknomorphy and androgyny), it should be possible to relate them to variations in fat and robustness. Within the framework of the factor-analytic model, this implies the following possibilities:

1. to reproduce the distribution of individuals in Sheldon's somatotype distribution chart on the basis of factor scores for fat ($\approx$ endomorphy versus endopenia) and physical robustness ($\approx$ endomorphic mesomorphy versus endomorphic ectomorphy);
2. to rotate this structure to an equivalent one where the two reference axes represent stoutness versus slenderness (=ectopenia versus ectomorphy = high degree of fat and robustness versus low degree of both components) and the mesomorphy/endormorphy-ratio (= variation between the extremes of a combination of a high degree of robustness with a low degree of fat and vice-versa);
3. to identify the latter dimension with variations in androgyny.

If the first solution leads to the factorial structure under 2, then it logically follows that a rotation of this structure would elicit the results referred to under 1.

In an attempt to test this hypothesis, the present author has, in collaboration with Gammel and Koeller, computed data from Sheldon's *Atlas of Men*.[76] All 134 cases, on whom Sheldon's ratings of androgyny were given (as g-index), were selected from the total of 1145 individuals, on whom detailed somatotypological information was presented. For this sample, Pearson's correlation coefficient among six variables was calculated. The variables were the g-index (g), ratings of endomorphy (I), mesomorphy (II), and ectomorphy (III), the mesomorphy/endomorphy ratio (II/I), and the ponderal index (PI = height/$\sqrt[3]{\text{weight}}$). A standardized program for principle component analysis was applied to the matrix of intercorrelations. Only factors with eigenvalues greater than one were rotated to achieve a simple structure solution (using varimax criteria).

The resulting factor matrix, with two factors fulfilling the criteria for rotation, is reproduced in Table 10-I. These two factors (accounting for 89% of the total variance) are in complete accordance with our expectation. The first factor can easily be interpreted as one of androgyny (high loading of g), contrasting mesomorphy ($\approx$ robustness) with endomorphy ($\approx$ fat; high loadings with opposite signs for II and I). This relation is also expressed by the mesomorphy/endomorphy-ratio (high loading of II/I in opposite direction to g).

The structure of the second factor is, likewise, self-evident. It opposes ectomorphy (III) and its somatometric criterion (PI) to endomorphy (I) and, to a lesser degree, mesomorphy (II). The specific criteria of androgyny (g and II/I) do not show a definite relationship to this factor. The first factor, on the other hand, is practically unrelated to the classical criteria of slenderness (III and PI). It can be inferred that slenderness versus stoutness (= ectomorphy versus ectopenia = leptomorphy versus pyknomorphy) is independent of secondary sex characteristics,[63,90] a conclusion which is at variance with the opinion of several authors.[5]

TABLE 10-I

FACTORIAL STRUCTURE OF TWO PRINCIPLE COMPONENTS AFTER VARIMAX ROTATION (LEFT COLUMNS 1, 2) AND AFTER FURTHER GRAPHICAL ROTATION (RIGHT COLUMNS 1' 2')

(n = 134)

variable				factor				
kind	no	designation	symbol	1	2	1'	2'	h²
som.-metric index	1	ponderal index	PI	.05	.97	.70	.68	.95
	2	ectomorphy	III	-.14	.96	.55	.80	.94
photo-scopic rating	3	mesomorphy	II	.88	-.35	.41	-.85	.89
	4	endomorphy	I	-.72	-.67	-.98	.00	.96
	5	meso-/endomorphy	II/I	.89	.11	.73	-.53	.81
	6	g-index	g	-.88	-.04	-.67	.57	.77
		% total variance		48.0	40.8	48.3	40.5	88.8

Factor Loadings ≥ .30 (in absolute values) in heavier print.

If the two factors represented in Table 10-I are orthogonally rotated so that the loading of endomorphy (I) is maximized, the other factor obtains almost equal weights with opposite signs for mesomorphy and ectomorphy. This solution equals the description of morphological variations along the dimensions of fat (endomorphy versus endopenia) and robustness (mesomorphy versus ectomorphy or, more precisely, endomorphic mesomorphy versus endomorphic ectomorphy). Figures 10-3 and 10-4 make the relationships between the factorial structures and the reference axes in Sheldon's somatotype distribution chart as shown in Figure 10-1 even more evident than does the factor matrix in Table 10-I.

Depending on the variables included in a factor analysis, the rotation, according to purely formal criteria of simple structure, may lead primarily to a solution of this kind. A structure comparable to our varimax factors in Table 10-I could then be achieved by a rotational procedure placing the axes of reference intermediate between the criteria of fat and robustness. A factor analysis (according to the maximum likelihood method) of fifteen anthropometric measures (including X-ray measures of fat and muscle) which was performed by Tanner, together with Healy and Whitehouse, could serve as an example.[6] The first three of five rotated factors clearly represent the development of fat, muscle, and skeletal size. This factor solution is cross-validated by independent analyses of data from both a male and female sample, comprising 125 and 166 individuals, respectively.

If the first two factors (fat and muscle) were rotated approximately 45°, another structure would emerge, representing the aspect of body shape dimensions rather than that of tissue components: one vector would change to a factor of androgyny, opposing fat (at the female pole) to muscle (at the male pole). The other vector would then represent stoutness versus slenderness, contrasting a high development of fat

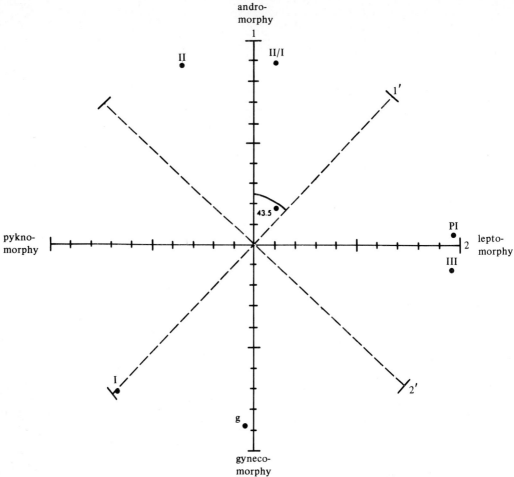

Figure 10-3. Graphical representation of factorial structures presented in Table 10-I. Varimax factors 1 and 2, and graphically rotated factors 1' and 2' (broken lines). Varimax factor 1 is represented by the ordinate and varimax factor 2 by the abscissa to make the relationship to Sheldon's somatotype distribution chart (see Fig. 10-1) more obvious.

and muscle (at the pyknomorphic pole) to a low development of both components (at the leptomorphic pole).

Within the framework of this geometric representation of the factorial structure of variations in physique, the factor of general skeletal size is to be considered an axis vertical to the plane determined by these two factors. If the three orthogonal axes represent factor scores, according to which each individual in a given sample receives a position within the frame of the factorial structure (described independently by Tanner and by the present author), the resulting structure will be quite similar to the revised version of Sheldon's system of somatotypes.

Virtually the same structure was obtained by the present author in a factor analytic approach guided by hypotheses

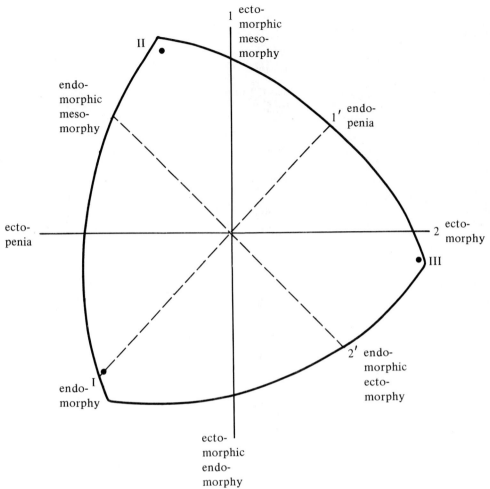

Figure 10-4. Projection of the outline of Sheldon's somatotype distribution chart (fig. 10-1) on the factorial structure of Figure 10-3.

derived from the classical literature on dimensions of morphological habitus variations.[90] The data were selected from an investigation performed with Wilde on variations and covariations of body build and personality in a sample of healthy young men drawn from the general population.[91] We intercorrelated twenty-seven independently defined photometric[42, 99] and somatometric measures,[113] a photoscopic rating of hirsutism,[42] and six special criteria of two hypothetical dimensions of variations in bodily proportions (leptopyknomorphy and androgyny). There were three criteria for each of the two dimensions in question: one somatometric index, one photometric index, and a photoscopic rating score as shown in Table 10-II.

Seven factors were extracted from the correlation matrix by the centroid method. Since the third unrotated factor was quite

TABLE 10-II
FACTORIAL STRUCTURE OF THE FIRST THREE OF SEVEN CENTROID FACTORS AFTER GRAPHICAL ROTATION OF FACTOR 1 AND 2 (LEFT COLUMNS) AND AFTER FURTHER GRAPHICAL ROTATION OF THE FIRST WITH THE THIRD FACTOR (1' 3'; RIGHT COLUMNS).
(p = 115)

kind		no	designation	1	2	3	1'	2	3'	h^2
single measures	photometric (ph.)	1	head height	.18	.06	-.09	.12	.06	.16	.04
		2	neck length	-.51	.33	.25	-.35	.33	-.45	.43
		3	trunk length	.11	.67	.16	.17	.67	-.09	.49
		4	leg length	-.30	.76	.27	-.15	.76	-.38	.75
		5	arm length (exl. hand length)	-.28	.68	.24	-.14	.68	-.34	.60
		6	hand length	-.28	.68	.01	-.25	.68	-.13	.54
		7	foot length	-.26	.76	-.05	-.26	.76	-.07	.65
		8	facial-breadth-two (FB$_2$*)	.57	.26	-.16	.44	.26	.40	.42
		9	neck-thickness-transverse (NTt*)	.56	.28	-.21	.41	.28	.44	.44
		10	neck-thickness-anteroposterior (NTap*)	.55	.45	.25	.60	.45	.02	.56
		11	forearm thickness	.56	.49	-.31	.36	.49	.53	.65
		12	wrist thickness	.25	.54	-.13	.17	.54	.23	.37
		13	knee-joint-thickness	.29	.64	-.03	.25	.64	.15	.50
		14	leg-thickness-lower-two (LTL$_2$*)	.46	.35	-.06	.39	.35	.26	.34
	somatometric (s.)	15	chest depth	.61	.30	-.07	.51	.30	.34	.47
		16	chest breadth	.55	.45	-.33	.35	.45	.54	.62
		17	biacromial distance	.11	.64	-.22	.00	.64	.25	.47
		18	bi-iliac distance	.19	.73	.16	.24	.73	-.06	.59
		19	condular breadth, humerus	-.04	.69	-.33	-.18	.69	.28	.59
		20	condular breadth, femur	.11	.76	-.07	.07	.76	.11	.60
		21	hand girth, left	.04	.65	-.51	-.19	.65	.47	.68
		22	upper arm girth	.76	.33	-.24	.57	.33	.55	.74
		23	calf girth	.60	.47	-.07	.51	.47	.33	.59
		24	weight	.56	.75	.08	.54	.75	.18	.89
		25	fat over triceps	.68	.12	.36	.77	.12	.02	.61
		26	fat scapular	.80	.06	.32	.86	.06	.07	.75
		27	fat iliacal	.76	.11	.43	.87	.11	-.05	.77
indices	s.	28	F/M-ratio**	.38	.17	.80	.70	.17	-.55	.82
		29	type index (Rees-Eysenck)	-.67	-.03	.50	-.38	-.03	-.75	.71
	ph.	30	neck thickness index (NTt*/NTap*)	-.14	-.22	-.42	-.31	-.22	.32	.25
		31	rel. neck-depth (NTap*/neck length)	.67	-.17	-.17	.52	-.17	.45	.50
photo-scopic scales		32	androgyny scale***	.30	.08	.59	.53	.08	-.39	.44
		33	leptopyknomorphy scale***	.84	-.03	-.23	.65	-.03	.58	.76
		34	hirsutism scale***	.33	-.06	-.11	.25	-.06	.25	.13
			% total variance	19.0	23.0	13.1	19.7	23.0	12.4	55.1

*according to Sheldon et al. (1940)
**F and M according to Parnell (1958)
***global rating according to v. Zerssen (1966) [4] [5]

Factors 1' and 3' with coverted signs
Factor loadings ≥ .30 (in absolute values) in heavier print

in accordance with the assumption of an androgynic factor, only the first and second factors were graphically rotated so that the photoscopic and somatometric criterion for leptopyknomorphy (the latter defined as Rees-Eysenck's body build index: stature times one hundred divided by transverse chest diameter times six)[66] reached maximum values on one of them, as indicated in the left columns of Table 10-II. All other loadings on this factor corresponded to the hypothesis of a factor of stoutness versus slenderness. The loadings on the other rotated factor fulfilled our prediction of a factor of general skeletal development having relatively high positive loadings on all length and skeletal breadth measures and low loadings on head, fat, and shape measures as well as on the ratings of hirsutism and shape dimensions.

A rank correlation of variables from this matrix with the same variables, ordered according to sex differences, was also in complete agreement with the expectation from our hypothesis: for the factor of leptopyknomorphy, the correlation was practically zero ($\rho = .12$), for the androgynic factor it was extremely high ($\rho = .90$), and for the factor of general skeletal development it was intermediate ($\rho = -.45$). There is only one finding which seems to contradict our hypothesis regarding the relationships between body shape dimensions and secondary sex characteristics, as shown in the left columns of Table 10-II: the hirsutism rating obtained the highest loading on the leptopyknomorphy factor rather than on that of androgyny.

This finding is, however, in accordance with that of Lindegård and co-workers in a similar sample of young male adults.[114] These authors correlated specific male sex characteristics, testis size, penis size, and body hair (hirsutism) with each other, with traits of body build (length, sturdiness, muscle, fat, and others), and also with urinary steroid metabolites presumed partially to reflect testicular endocrine activity. Whereas testis size and penis size both correlated positively with length and sturdiness, and penis size, in addition, with muscle, to a small but significant degree ($r \approx .2$), body hair showed a significant association only with fat ($r \approx .3$). It can be concluded that penis size, and less so, testis size, are related to general skeletal size (length and sturdiness) as well as to the male components of variations in secondary sex characteristics of body build (muscle and sturdiness). In contrast to this, hirsutism seems related to a component common to pyknomorphy and gynecomorphy (fat). Neither of the specific sex characters were significantly correlated with urinary steroid compounds in the Swedish investigation.

A slight association of body hair to physical robustness (comprising muscle and sturdiness), besides its relationship to fat, is indicated by the equal loadings of hirsutism rating on factors 1' and 3', as shown in the right columns of Figure 10-2. These factors were achieved by a graphical rotation of the two vectors representing shape (left columns, 1 = slenderness and 3 = androgyny). The rotation was so performed that one vector cut the center of the fat cluster in the respective plane of the common factor space. This resulted in a structure which represented a fat factor (1') and a factor of physical robustness (3') in addition to the factor of skeletal size, as shown in the right columns of Table 10-II.

According to these findings, the main variations of body build can be described optimally by means of three independent dimensions: One represents general skeletal size, the others represent fat and robustness, or (in an equivalent solution) the shape factor of slenderness versus stoutness as well as that of androgyny.

Criteria for these dimensions should be so defined that they are as specific as possible.

The fourth factor of our analysis, not represented in Table 10-II, was related to head size. As previously mentioned, a similar factor was usually found, if several head measures had been included in the analysis. There are other dimensions of morphological variations (e.g. trunk thickness, size of extremities), but only those described here in detail seem relevant in relation to variations in personality. Moreover, physical disproportions must be considered in this respect.

PSYCHOMORPHOLOGICAL CORRELATIONS

Physique and Intelligence

In several investigations, tests of general intelligence were correlated with height and other criteria of general size. These correlations tended to be low positive, the coefficients usually varying between values of .1 to .2.[14,115] Similar values were found for correlations of intelligence with head size and with criteria for the body shape dimensions, leptopyknomorphy, and androgyny.[2,14]

Regarding the shape dimensions, mainly young male individuals of leptomorphic build[66,116,117] and those with low scores of physical masculinity[86] were found intellectually slightly superior to those with contrasting types of physique.[2] However, this is valid only with respect to verbal intelligence.[118] The relationship can be explained on the basis of findings by the Swedish authors, Lindegård and Nyman,[104] and by the present author.[45] It was shown that muscularity (or robustness), which is low in both leptomorphic and gynecomorphic individuals, correlates negatively with verbal intelligence; fat, however, which is low in leptomorphs but high in gynecomorphs (see discussions above and Tables 10-I, 10-II), varies independently of intelligence. This will be discussed later in relation to the etiology of psychomorphological correlations.

It is not yet conclusive to what extent the correlations of intelligence to physical dimensions are linear. Particularly with respect to general size and to head size, one might assume that mainly low scores of physical dimensions and intelligence are intercorrelated.[3,2,1] This may also be true for the relationship between dysplastic build and low intelligence inferred from clinical investigations in the physical habitus of oligophrenics.

Physique and Temperament

General Remarks

The majority of investigations on psychomorphological relationships are concerned with correlations between body build and temperament.[2] They were usually guided by the hypothesis that people who look different should also behave differently, while those who are more alike in appearance should also be similar in their behavior.[119] Therefore, relatively specific correlations between types of physique and types of personality were assumed by most authors working in this field. Curiously enough, differences and similarities in general size were often neglected in this respect, while body shape was predominantly taken into account.

If the same authors rated both the physical and temperamental features themselves, investigators' bias may have entered into their judgment on the relationships between those features. Our survey, therefore, will emphasize findings that have been repeatedly elicited on the basis of independent ratings of physique and personality, particularly when obtained by means of modern biometric techniques in assessing physique[113, 120] and personality features[57, 58, 121, 122] and by statistical data analysis.[60, 123] In order to eliminate the influence of maturation on the psychomorphological correlations, mainly investigations of adults will be reviewed in this context.

The Typological Approach

Kretschmer was one of the few constitutionalists who considered the mental correlates of abnormal (dysplastic) forms of physique (e.g. neurotic traits as correlates of partial retardation in physical development).[40]

His main concern, however, was to relate the normal types of physique—pyknic, leptosome, and athletic—to types of temperament which he and some of his co-workers had partly derived from the symptomatology of functional psychoses and personality features in chronic epileptics.[49] This is due to Kretschmer's assumption of a very close affinity between physical types and types of abnormal behavior (affective psychoses, schizophrenic psychoses, and epilepsy). Following Tiling,[124] he conceptionalized mental disorders as exaggerated forms of normal behavior. Kretschmer concluded that the association between abnormal behavioral and physical types was rooted in a more fundamental affinity between physical types and types of normal temperament, and that the latter were caricatured in the abnormal behavior of psychotics and epileptics. Consequently, he called the temperament associated with leptosome physique *schizothymic*. This term indicates the similarity of the temperamental type to the schizophrenic symptom pattern. The typical temperament of pyknics was labeled *cyclothymic* because of an assumed relationship to manic-depressive mood swings. The temperament of the athletic type[125] was called *viscose* because of its tenacity, which seemed similar to that found in chronic epileptics.[126, 127] Due to a tendency toward explosive impulsiveness, Kretschmer related the viscose athletic type also to the catatonic type of schizophrenia as well as to criminally assaultive behavior.[128]

It is striking that this basically speculative system received so much empirical support, not only from the work of Kretschmer and his pupils, but also from clinical investigations by many other authors.[40] This continued until proper scientific criteria were applied to the problem. Then the following results were achieved:

1. The association between physical types and types of psychoses is mainly caused by the influence of age on each of them.[2, 14, 34, 79, 103]
2. In many cases, particularly those with endogenous periodic depressions,[129, 130, 131, 132, 133, 134] the psychotic symptomatology does not show any similarity to premorbid personality traits.[103]
3. Psychomorphological correlations are usually much lower than expected from Kretschmer's writings, when based upon independent assessment of physical and temperamental traits.[2, 14, 17, 55]
4. Judging from the intercorrelation of questionnaire parameters of temperamental types, the viscose temperament and schizothymia must be regarded as closely related to each other and to introversion, and opposed to cyclothymia and extraversion.[55] This, however, inevitably

leads to a dimensional concept of variations in temperament, which is discussed in the next section.

The Dimensional Approach

PREFACTOR-ANALYTIC STUDIES: Within the framework of a dimensional concept of body build variations derived by Conrad[49, 52] and by others from Kretschmer's typology, several authors reported significant correlations of ratings and self-ratings of extraversion versus introversion, as well as of emotional and autonomic balance versus imbalance (neuroticism) with measurable criteria for pyknomorphy versus leptomorphy (stoutness versus slenderness).[2, 57] The association, as such, is well-established, but it does not seem to be a very close one since correlation coefficients usually vary between values of approximately .2 to .3.

In quality, though not in quantity, this correlation is in basic agreement with Sheldon's concept.[43, 135, 136] He described the temperament of the (leptomorphic) ectomorphs as *sociophobic,* thus emphasizing the social aspect of introversion. This was contrasted with two different forms of extraversion which he conceived as associated with the fat endomorphs and the robust mesomorphs, respectively. Both forms are well in agreement with cyclothymia. The one related to mesomorphy, however, is almost contrary to Kretschmer's concept of the viscose temperament in the athletic type, even though the latter presents predominantly mesomorphic features.[75] This cannot really be explained on the basis of the assumption that Sheldon employed a more empirical approach to the problem. His thinking, too, was greatly influenced by speculations on the nature of psychomorphological relationships. As previously mentioned, Sheldon related constitutional differences to differences in the relative development of derivatives of the germinal layers in the process of ontogenesis. Within this hypothetical framework he conceptionalized his findings regarding psychomorphological relationships in the following way:

A preponderance of the endoderm leads to a fleshy (endomorphic) physique with strongly developed viscera and to a *viscerotonic* temperament governed by the desire for visceral activity in eating. This goes along with love of comfort and *sociophilia* (viscerotonic extraversion).

A preponderance of the mesoderm leads to a robust, muscular (mesomorphic) physique, the motoric system (bones and muscles) being accentuated, and to a *somatotonic* temperament. Such individuals are motivated for high motoric activity, rivalry in sports, and leadership (somatotonic extraversion). In addition, there may be a tendency toward aggressive and even criminal behavior.[82]

A preponderance of the ectoderm leads to a linear ectomorphic physique with a strong development of cerebral tissue relative to other tissues, and of body surface relative to body mass. This constellation includes a predisposition to a *cerebrotonic* temperament characterized by a high degree of sensitivity toward stimuli from the environment and a resulting sociophobic attitude (sociophobia = social introversion).

The correlation between the physical dimensions and the corresponding temperamental traits is reported to be as high as .8, which would indicate an almost complete concordance of variations in body build and temperament.

Approximately half a century before Sheldon, Huter,[137] a German layman physiognomist, similarly described and explained constitutional differences.[35] The basic agreement between Huter's and

Sheldon's concepts is particularly striking with respect to the mentally robust muscular type, hypothetically related to the embryonic mesoderm by each author. Huter also assumed that a disharmonious mixture of components of different types in an individual predisposed him to criminal behavior. This is reminiscent of Lombroso's well-known concept of physical abnormalities in criminals.[138] Sheldon does not offer detailed information on mental correlates of dysplasia. However, he does report a very close relationship between physical and mental traits of androgyny which was not considered in Huter's system.

One must question to what degree the highly speculative concepts of psychomorphological relationships, developed by Huter and by Sheldon, are supported by the results of objective biometric investigations. It is evident that the embryological basis of these concepts is open to severe criticism. Nonetheless, on a descriptive level they may have advantages over Kretschmer's concept.

As mentioned above, the finding of correlations between criteria for extraversion versus introversion, and those for stoutness versus slenderness, is qualitatively suited to both concepts; in quantity the correlation is at variance with both of them. But are there any positive findings of two different kinds of extraversion being related to two different kinds of variation in body mass, caused either by fat (endomorphy) or by robustness (mesomorphy)?

Do dysplastic variants indeed tend to have disharmonious minds, and are physical and mental criteria of androgyny interrelated as Sheldon assumes?

Biometric investigations conducted to test Sheldon's hypotheses on psychomorphological correlations usually failed to confirm the high coefficients reported by him. Lubin, in reanalyzing Sheldon's own data on temperamental traits, could show that some computational error must have entered the table of correlations because it contained values which were mathematically impossible.[139] Some investigators, using psychometric tests not constructed to assess the personality dimensions described by Sheldon, found mainly insignificant correlations between test-scores and Sheldon's dimensions of physique.[116, 140, 141] Positive findings by other authors could not be cross-validated.[44, 117, 142, 143] However, if self-rating questionnaires were specifically designed for testing Sheldon's hypotheses, not only was the well-established correlation between ectomorphy (slenderness) and cerebrotonia (social introversion) confirmed in principle, but also correlations of mesomorphy (muscularity) and endomorphy (fat) with specific aspects of extraversion (somatotonia or viscerotonia, respectively).[144, 145]

The results of one of these studies[144] could be cross-validated by the present author[75] in collaboration with Wilde.[91] However, this was only true with regard to a reverse correlation of mesomorphy (measured as muscularity according to Parnell[80]) and ectomorphy (measured as linearity by the Ponderal-Index according to the same author) to the scale values of somatotonia as opposed to cerebrotonia as shown in Table 10-III. The result is well in accord with findings in two samples of American junior high school boys.[146]

Endomorphy (measured as fat according to Parnell) and the scale values of viscerotonia were not significantly correlated with each other in our sample; there was merely a slight negative correlation between them. Each variable was significantly correlated only with other variables of the same (morphological or temperamental) sphere: a high negative correlation was

TABLE 10-III
CORRELATIONS AMONG VARIABLES REPRESENTING DIMENSIONS OF PHYSIQUE AND PERSONALITY ACCORDING TO SHELDON

(Data from D. von Zerssen, "Eine biometrische überprüfung der Theorien von Sheldon über Zusammenhänge zwischen Körperbau und Temperament," *Zeitschrift für experimentelle und agewandte Psychologie*, 12:521 [1965])
(n = 123)

		I	II	III	V	S
I	fat					
II	muscularity	−.08				
III	linearity	−.68***	−.54***			
V	viscerotonia	−.12	.15(*)	−.03		
S	somatotonia	−.01	.31***	−.23**	.33***	
C	cerebrotonia	.07	−.20*	.10	−.46***	−.75***

(*) p <0.10   * p <0.05   ** p <0.01   *** p <0.001

I, II, III		dimensions of physique according to Parnell
I	endomorphy	skinfold thickness at three different sites
II	mesomorphy	combined measure of extremital girth (corrected for fat) and condylar breadth
III	ectomorphy	Ponderal index = $\dfrac{\text{height}}{\sqrt[3]{\text{weight}}}$
V, S, C		test-scores derived from Child's self-rating questionnaire

found between endomorphy and ectomorphy but not between endomorphy and mesomorphy; viscerotonia was correlated positively with somatotonia and negatively with cerebrotonia. (It may be added that viscerotonia also showed a tendency toward correlation with mesomorphy.) The highest negative correlation was found between somatotonia and cerebrotonia. Mesomorphy and ectomorphy were also negatively correlated with each other to a marked extent.

It was concluded that the findings lend support to a psychomorphological correlation between only one dimension of physique (muscularity versus linearity) and one dimension of temperament (somatotonia versus cerebrotonia). The latter dimension was interpreted as an aspect of extraversion/introversion which was correlated with physique. Viscerotonia was regarded as another aspect of extraversion, which was not related to physique to the same extent. Finally, it was assumed that the fat component of body mass (endomorphy versus ectomorphy) was not related to temperamental traits.

Our conclusions were, of course, in contradiction to Sheldon's assumption that there existed specific relationships between different aspects of physique and those of extraversion/introversion. Still, his assumption of such specific relationships was, in a sense, substantiated by Lindegård's and Nyman's work.[104] They found correlations between muscle (measured as muscular strength) and two dimensions of personality:[147, 148, 149] one (validity according to Sjöbring) resembling somatotonia, the other (stability according to Sjöbring) resembling cerebrotonia, and a further correlation of the latter dimension to fat (measured by skinfold thickness and by a complicated weight index). It may be worth mentioning that Sjöbring himself had assumed a basically similar negative relationship of stability to pyknomorphic build and a positive one of validity to a strong, muscular type of physique.[149]

If one assumes that stability opposes

cerebrotonia (as social introversion) mainly to the viscerotonic aspect of extraversion (sociability) and only to a lesser degree to somatotonic extraversion (impulsiveness),[150, 151] then the correlations are in accordance with Sheldon's and, thus, with Huter's, typology. The findings of a relatively high negative correlation (–.44) of stability ($\approx$ cerebrotonia) with relative weight (a criterion for ectomorphy) would also agree with this concept. Skeletal sturdiness (measured as condylar breadth) and height were not significantly correlated with personality dimensions. This could be explained by the independence of variations in general skeletal size and temperamental traits. Since the assessment of temperament was based on clinical ratings, the findings were not protected against investigator bias. Therefore, the results are suggestive rather than conclusive.

In an investigation carried out in England (with self-ratings of similar aspects of personality and with measurements of physique, which included the components fat and muscle as defined by Lindegård), the correlations found in the Swedish investigation could not be replicated.[96] This casts some doubt on their validity. On the other hand, the results of the British study were, in part, at variance with findings of several previous investigations.[91, 93] For this reason, they also are not quite convincing. Thus, the question whether there are specific correlations between different kinds of extraversion and different aspects of variations in body mass cannot yet be answered with any certainty. We will later examine this problem within the context of factor-analytic studies of psychomorphological relationships.

This problem is closely connected with that of a psychomorphological variation of bisexual characteristics. One can ascertain from the intercorrelations among the mental and somatic variables rated by Sheldon[43] that mental androgyny (gynandrophrenia) is correlated with somatotonia as high as –.69 and that these two aspects of temperament show virtually the same correlations to other variables. Considering the rather low reliability of personality ratings,[58] it seems plausible that gynandrophrenia is just the reverse of somatotonia. Its relationship to viscerotonia is not so clear, but the relationship of mental androgyny (gynandrophrenia) and the second dimension in Sheldon's system of varieties of temperament is as striking as in the case of morphological androgyny and the second dimension in his system of varieties of human physique.

Judging from other authors' investigations in androgyny,[91, 116, 152] the psychomorphological correlations seem to be much lower than originally assumed by Sheldon. Yet a positive relationship of extraversion to relative maleness of temperament, implicitly pointed out by the high correlation between gynandrophrenia and somatotonia, is in accordance with some unbiased investigations; comparing samples of both sexes in self-rating scales of extraversion, males usually obtain higher values than females.[153] Moreover, male students with a relatively feminine physique were found by Seltzer to be more introverted than those with a pronounced masculinity of build.[86] Further, they showed a greater degree of emotional lability and increased signs of autonomic imbalance. A similar difference is usually found when comparing females with males.[153]

This positive correlation of introversion, emotional lability, and autonomic imbalance with somatic femininity, may, however, not be a specific one since it is well established that the same or similar mental traits are correlated with slenderness versus stoutness as well.

It should be noted that the leptopyknomorphic variation is a sex-independent dimension of physical variation. Furthermore, Seltzer found in another investigation a combination of introverted traits, emotional lability, and signs of autonomic imbalance in male individuals with a high degree of body disproportions, which he had defined operationally through deviations in certain anthropometric indices.[83] Seltzer's findings suggest a rather unspecific relationship between physical and temperamental traits in sharp contrast to the assumption of relatively specific correlations of the kind depicted in Structure 10-I.

The hypothesis of an unspecific psychomorphological relationship is illustrated in the Structure 10-II.

Seltzer's findings on mental correlates of bodily disproportions may be explained by the fact that he, in defining disproportions operationally, has only considered deviations in the direction toward slenderness.[83] Thus in dealing with personality characteristics of physically disproportioned individuals, he may have simply described the mental correlates of slenderness. However, a similar explanation for his findings regarding mental correlates of somatic androgyny seems unlikely because the latter dimension of physique varies independently of slenderness versus stoutness.[86] Instead, it opposes endomorphy (fat) to mesomorphy (muscularity). Here the question arises whether the mental correlates of this endomorphy/mesomorphy ratio are indeed correlates of the ratio as such, or of only one of its constituents. Our findings favor a relationship between personality

STRUCTURE 10-I

Morphological Variations	Temperamental Correlates
slenderness *vs.* stoutness = leptomorphy *vs.* pyknomorphy = ectomorphy *vs.* ectopenia	social introversion *vs.* social extraversion (sociability) = schizothymia *vs.* cyclothymia = cerebrotonia *vs.* cerebropenia
somatic androgyny (gynandromorphy) = gynecomorphy *vs.* andromorphy = ectomorphic andomorphy *vs.* ectomorphic mesomorphy	mental androgyny (gynandrophrenia) = gynecothymia *vs.* androthymia = somatopenia *vs.* somatotonia (impulsiveness)
bodily disproportions	disharmonious (neurotic or criminal) personality

STRUCTURE 10-II

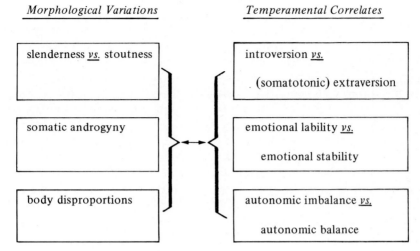

and only one constituent of the ratio, e.g. muscularity as shown in Table 10-III.[75]

Since muscularity is composed of a muscle and sturdiness component, a further question which arises is whether each or only one of them is related to temperament. Moreover, general skeletal size has to be considered with regard to variations in temperamental traits. Here the results are particularly divergent.[64, 91, 96, 97, 104, 154] These issues will be more thoroughly discussed in the next section on factor-analytic investigations of psychomorphological relationships.

FACTOR-ANALYTIC STUDIES: Factor-analytic investigations have, in part, clarified the complex interrelations within the entire spectrum of morphological and temperamental variations.[57, 58] However, the majority of studies are concerned with either the morphological or the temperamental aspect of constitution alone. There are only a few factor-analytic investigations in which morphological and psychological variables are computed simultaneously. This is quite striking since some psychologists (e.g., Thurstone, Eysenck[57, 58, 136]) have done a great deal of independent work in factorizing correlations within the physical and personality spheres. The usual result in combined studies was that, after rotation according to simple structure criteria, only dimensions representing either the somatic or the mental sphere emerged.[155]

If physiological measures were included in the analyses, they, too, were separated by the rotational procedures from the morphological, as well as the behavioral, aspects of the whole constitution. This was also the result of an animal study in which a large number of morphological, physiological, and behavioral measures were factor-analyzed.[156] However, in an investigation of the interrelationship between physiological and personality variables in human individuals, where the primary factors were only represented by either variations within the physiological or the mental spheres, slight correlations between physiological and mental factors appeared if the rotation was not restricted to orthogonality of vectors.[157] In higher order analyses, this association led to the formation of a global factor with weight for the otherwise differentiated physiological and personality dimensions. Consequently, this indicates some common variance for the somatic and mental spheres of the constitution.

The common variance of mental and somatic measures may, in addition, be reflected by their loadings on the first unrotated factor. Once rotational procedures were introduced into factor analysis, however, the unrotated factors no longer received much attention by most authors. In an unpublished analysis performed by Othmer in collaboration with the author,[45] the following measures obtained loadings (in absolute values) on the first unrotated factor >.35 in one direction: mental test scores representing vitality, extraversion, masculinity of temperament or cyclothymia, respectively, handgrip,[72, 104, 158, 159] and somatometric criteria of physical robustness (upper-arm girth, chest breadth, shoulder breadth, weight, and Parnell's muscularity score [80]); in the opposite direction: mental test scores representing emotional and autonomic imbalance, anxiety, femininity of temperament or schizothymia, respectively, as well as the Rees-Eysenck index of body type.[66] Measures of length and fat, an intelligence test score, scores of self-rating scales for rigidity,[160] social desirability,[161] and interests, as well as objective physiological measures of autonomic functions, were unrelated to it.

After varimax-rotation, the morphological and temperamental measures contained in the first unrotated factor were also isolated from each other. Criteria for physical robustness then constituted one of three morphological factors (the other two representing length and fat). The loadings of mental tests, previously related to the first factor, were mainly distributed on two varimax-factors: one representing emotional and autonomic imbalance (neuroticism), the other representing certain aspects of extraversion versus introversion. A test score for femininity of temperament ($\approx$ gynandrophrenia) received a markedly negative loading on one of them, thereby reflecting one aspect of introversion. The other varimax-factors (comprising further aspects of personality, interests, motoric strength, and various autonomic functions) need not be mentioned in this context because only variables with loadings on neuroticism and extraversion/introversion factors showed more frequent and higher correlations to morphological variables than expected by chance. The morphological correlates of these temperamental traits had comparatively high loadings on the varimax factor of physical robustness. This indicates a relationship between only one dimension of morphological variation (robustness $\approx$ muscularity) and temperament (extraversion versus introversion and neuroticism, respectively).

It is feasible to objectify this relationship quantitatively: if the twenty-four morphological variables computed in this investigation were arranged according to the number and degree of their statistically significant correlations with personality traits, these rankings could be compared with their sequence according to the loadings they obtained (in absolute values) on the factor of physical robustness. Using Spearman's rank coefficient as a measure of concordance between the two rankings, we found a value of .43. This indicates that they were not completely independent of each other.

The hypothesis of the relationship between only the robustness factor and temperament was verified in a similar way. The present author utilized information from three independent investigations which he had conducted in collaboration with Netter-Munkelt, Othmer, and Wilde, respectively. Anthropometric and psychometric data from three samples, altogether comprising 331 young male adults (not including 61 females) had been

computed. Eighteen morphological variables (single measures and indices), which had been considered in all three studies, were arranged in the following manner:
1. according to the number and degree of their significant correlations with measures of temperamental traits,
2. according to their loadings on the three morphological factors found in that investigation where the greatest amount of information on physical habitus had been available (including photometric measures and indices and photoscopic ratings as shown in the right columns of Table 10-II).

The comparison of the ranking, according to psychomorphological correlations of variables, with the rankings according to their factor loadings resulted in the following values of Spearman's rank coefficient: with physical robustness .74, with fat $-.18$, and with skeletal size $-.17$. It can be concluded that only variations in physical robustness (factor 3′) are definitely related to temperament; variations in fat (factor 1′) and skeletal size (factor 2′), however, seem to be unrelated to temperamental traits.

The coefficient of .74 with respect to psychomorphological correlations and factor loadings on physical robustness does not, of course, reflect the degree of correlation between physical robustness and temperamental traits. It merely indicates that such a correlation exists. To evaluate the extent of this relationship, the present author, together with Rey, intercorrelated factor scores for physical robustness, fat, and skeletal size, as well as for a temperamental factor (mental vitality), thought to represent the mental correlates of robustness. The usual low value of .2 was found for the expected psychomorphological correlation among the variables under investigation as indicated in Table 10-IV.

The temperamental factor considered in this analysis appeared as the first unrotated factor in an analysis of data from the same sample of 115 young male adults from which the morphological factors in Table 10-II had been derived. This analysis included thirty-seven scores for mental test scales and individual test items. The first unrotated factor had the highest weight ($\geq .4$) in the positive direction for the test scores of extraversion (from Eysenck's MPI[162]), male temperament (according to the present author), and the scores of some test items indicative of either cyclothymia (according to Kretschmer[40]), viscerotonia, or somatotonia (according to Sheldon[43] and Child[144]). In the negative direction, the highest weight was for test scores of neuroticism (from Eysenck's MPI[162] and for a list of complaints constructed by the present author[163]), schizothymia (accord-

TABLE 10-IV
CORRELATIONS OF FACTOR SCORES REPRESENTING A TEMPERAMENTAL FACTOR (MENTAL VITALITY) AND THREE MORPHOLOGICAL FACTORS
(n = 115)

		vitality	1′	2
1′	fat	$-.09$		
2	skeletal size	$-.04$	.00	
3′	robustness	.20*	.00	.00

* $p < 0.02$

ing to Kretschmer[40]), and for scores of test items indicative of cerebrotonia (according to Sheldon[43] and Child[144]). All of these scores correlated in the same direction as their factor-loadings with criteria of physical robustness. However, they did not show a systematic association with criteria of general skeletal size or of the development of subcutaneous fat.

The result is, therefore, that psychomorphological correlations are not as specific as formulated in Structure 10-I. Instead, they are of the more unspecific variety outlined in the schema discussed earlier. One argument for the validity of this conclusion is that our investigations started with the hypothesis which assumed relatively specific relationships.[45, 63, 91]

It is evident that the mental correlates of physical robustness resemble the mental correlates of eurymorphy (= pyknomorphy = endopenia) as described by Rees and Eysenck[66] and by Sanford and his coworkers.[164] They are also similar to the mental correlates of maleness of build described by Seltzer.[86] This is not surprising because robustness is reasonably high in the pyknomorphic as well as the andromorphic physique. However, both physical types differ markedly in the development of subcutaneous fat, which is high in pyknomorphs and low in andromorphs, as shown in the left columns in Table 10-II. As fat appears unrelated to temperament, the latter will show principally the same relationship to both types of physique (pyknomorphs and andromorphs).

The schema of unspecific relationships between temperament and types of physique as outlined previously (in Structure 10-II) can, therefore, be further simplified on the left (somatic) side (direction of variations converted on both sides) (Structure 10-III).

Further simplification is achieved by conceiving the mental correlates of physical robustness as belonging to a higher order factor opposing extraversion to introversion and emotional imbalance. It is reflected in the first unrotated factor of mental test scores and items mentioned above.

We labeled this dimension of temperament mental vitality because in one of our analyses it received the highest weight on self-rating scales constructed to investigate the concept of vitality.[45, 103, 165] The scale values of these tests were also correlated significantly (up to .3) with criteria for physical robustness.

Basically, the psychomorphological correlations were concordant in the three male samples which we investigated. In a small female sample comprised of sixty-one individuals, the results were not as distinct in this respect, a finding which is in accordance with reports from the literature.[65, 166] On the other hand, the factorial structures of variations in physique and personality were, on the whole, similar in the female and in a male sample of equal

STRUCTURE 10-III

Morphological Variation | Temperamental Correlates

- physical robustness
- ≈ muscularity
- ≃ endomorphic mesomorphy

↔

- (somatotonic) extraversion
- emotional stability
- autonomic balance

size and same age composition. There were, of course, some differences, one of which may be worth mentioning: In the female sample, the Rees-Eysenck index of body type[66] and its one component, chest breadth, constituted a factor separated from that of leptopyknomorphy. In contrast to this, both variables were closely associated with the leptopyknormorphic dimension in the male sample. Here they also showed significant correlations to mental test scores and test items, a finding which corresponded to those in the two other male samples.

In all three male samples, upper-arm girth, chest breadth, and the index of body build had the highest rankings among eighteen morphological variables arranged according to the number and degree of their correlations with mental traits. In this respect they also exceeded photoscopic ratings of leptopyknomorphy (slenderness versus stoutness) and andogyny. It is, therefore, unlikely that psychomorphological correlations are insufficiently reflected by the values .2 to .3 which are usually found in a combined anthropometric and psychometric approach to the problem of physique and temperament.

Moreover, it can be inferred from unpublished graphical analyses of correlations between mental test scores and criteria of physical robustness performed by the present author that most probably the association between mental vitality and physical robustness does not increase toward the extremes of variations in physique. In our analyses it proved even more pronounced within the medium range of robustness. This trend may be due, though, to the higher number of individuals within this range, which made the means of their test scores more reliable compared to those of the relatively small groups with extreme values of physical measures.

Physique, Personality, and Abnormal Behavior

Several forms of abnormal behavior have been related to variations in body build and their mental correlates. Some of the relationships described in the literature are predominantly based on speculations, while others can be ascribed to the influence of age. Even when elaborate statistical procedures are applied to a large set of exact measures, spurious correlations of this kind may emerge if age is not considered.[111] In this case, manic-depressives will appear more pyknic, compared with schizophrenics, who, on the average, are ten to twenty years younger in most samples investigated.[2] For the same reason, paranoid schizophrenics will exhibit more pyknic features than nonparanoid schizophrenics.[111] Hebephrenics, the youngest diagnostic subgroup among schizophrenics, will look comparatively more leptosomatic ($\approx$ ectomorphic) than any other subgroup.[167] As psychoses of early onset usually have a particularly unfavorable prognosis,[168] it follows that the course of the disease seems to be influenced by the patients' physical makeup. Therefore, it is questionable whether a leptosomatic physique is still indicative of a bad prognosis once the age factor is controlled. After all, some findings on the relationship between physical measures and the prognosis of psychotic disorders contradict the assumption of the prognostic significance of leptosomia.[169, 170]

Correlations between physique and type or subtype of psychosis tend to be negligibly low when samples of equal age are compared with each other. Usually, statistical significance is only reached on the basis of large samples (comprised of more than 100 cases[34, 97, 171,] and is, therefore, of no practical value for diagnostic purposes.[103] The correlations must be definitely below the figure for the age correlation of leptopyknomorphy,[79, 103] which can be estimated as .2 to .3 within the age range of about twenty to sixty years.[66] In our survey on the relationships between psychomorphological constitution and deviant forms of behavior, we will preferably review those investigations in which the age factor was considered.

A positive correlation between oligophrenia and physical dysplasia (as defined by Kretschmer) is the most obvious relationship of this kind.[34, 35, 172, 173] In addition, general growth seems diminished in a comparatively high proportion of the mentally retarded.[174] Their reduced general skeletal size, as well as their dysplastic features, are in accordance with the slight positive correlation of intelligence with stature and the slight negative one with dysplasias found in the general population.[14] There are, of course, even dysplastic dwarfs who are very intelligent[21] (e.g. among those with achondroplasia[29]); in addition, there are tall and well-set individuals among the mentally retarded. In some syndromes, various degrees of mental backwardness tend to be associated with high stature, particularly in the XXY- and, more pronounced, in the XYY-syndromes.[175, 176, 177] On the whole, however, shortness and dysplasias are more typical for oligophrenics.

Epileptics, too, display a relatively large proportion of dysplastic features.[34, 35, 126, 127] Moreover, a comparatively large number of them were found to be of athletic build.[34, 35, 126, 127] Athletic and dysplastic features may not be equally distributed in epileptics with different forms of the disorder.[178] In an unpublished anthropometric and psychometric investigation performed by Leder in collaboration with the present author, two groups, each comprised of eight male epileptics, and a group of eight male patients suffering from various diseases of the peripheral nervous system were compared with each other. The epileptics had been assigned to the subsamples according to the type of seizure;[179, 180] one exhibiting grand mal on awakening, combined with pyknolepsy, the other having night epilepsy, or diffuse epilepsy with psychomotoric fits. The three samples were exactly matched according to age and verbal intelligence. Despite the small number of cases, the two epileptic groups deviated significantly from each other and from the control group in several physical measures: patients with grand mal on awakening exceeded the other epileptics as well as the controls in criteria for physical robustness, particularly in extremital girths (see Fig. 10-5). The other epileptics tended to be more leptomorphic, mainly in thoracic measures. Furthermore, they exhibited an incongruency between upper arm girth (which was below the average; see Fig. 10-5), and calf girth (which was above the average). Thus they may be designated leptosome-dysplastic in Kretschmer's terminology. On a self-rating list of complaints,[163] they scored significantly higher than either the epileptics with grand mal on awakening or the controls.

In schizophrenics the percentage of dysplasias may not be so high as originally postulated by Kretschmer[40] and by others.[34, 35, 181, 182] Perhaps they are more pronounced in hebephrenics[34] and/or in chronic cases; but even then, the question arises whether this is due to an unfavorable course of the disease in patients having dysplastic physique,[40] or to the influence of hospital life (including insulin treatment) on physical habitus.[183]

According to some investigators, schizophrenics have a reduced general size.[2, 97, 184] Moreover, their physical habitus seems to be slightly less robust compared with controls of the same

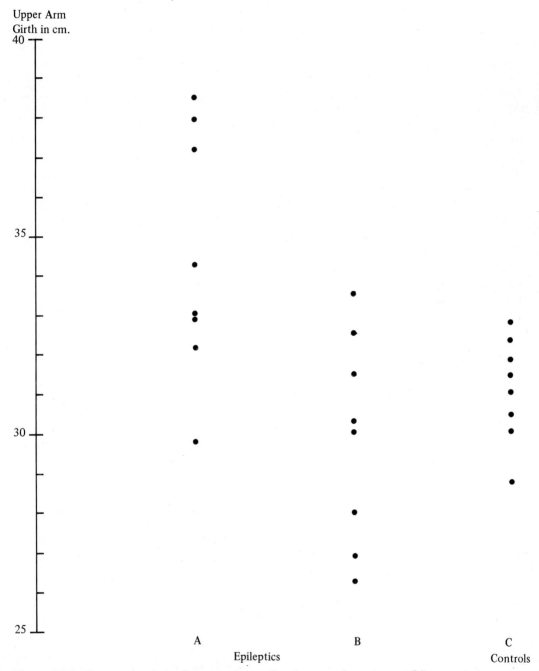

Figure 10-5. Upper arm girth of three groups of male outpatients (n_{total}= 24).
(A) Grand mal on awakening and pyknolepsy
(B) Noctnural or "diffuse" grand mal and psychomotor attacks
(C) Peripheral neurological disorders

age.[2, 95, 97] Both findings, however, are not unequivocal.[2, 34] Sagittal chest diameter was found by some investigators to be most characteristic of the schizophrenics' body build.[2, 171] It may indicate a dysplastic feature rather than a tendency toward leptosomatic build because other criteria for this type of physique do not differ from the normal means.[171, 185] In Conrad's terminology, the typical habitus of schizophrenics may be designated hypoplastic rather than leptomorphic.[52]

The paranoid subgroup of schizophrenics was shown to differ from other schizophrenics in physical appearance,[171] although this difference was not marked when the samples were properly matched for age.[97, 185] Most likely, the paranoid group does not show any morphological deviations from the average population of comparable age.

Regarding premorbid personality traits of schizophrenics,[168] the present author, together with Rudersdorf,[186] and later with Fritsch,[187] could prove their tendency toward schizoid features through statistical analyses of questionnaire data from patients' relatives. There is some doubt as to whether these traits correlate with those physical traits considered typical for the habitus of schizophrenics. It is worth mentioning in this context that a tendency toward reduced physical robustness, as well as to more pronounced schizoid features and, in addition, a lower intelligence level was found in the index cases of a group of monozygotic twins discordant for schizophrenia.[188] Furthermore, in an investigation of the offspring of schizophrenics,[189] it was found that the children with leptosomatic physique had a higher risk of becoming psychotic;[190] but this could not be corroborated by anthropometric comparison of psychiatric patients (including schizophrenics) with their siblings.[191]

Reports on physical and premorbid mental characteristics of patients with affective psychoses are almost as contradictory as corresponding statements on the psychomorphological constitution of schizophrenics.[2, 95] This may partially result from the fact that patients with genetically different forms of affective psychoses, monopolar endogenous depression, as well as bipolar manic-depressive psychosis, are usually included in the same sample.[192, 193] It is doubtful whether the affective group as a whole is physically different from the average population[97] (more pyknic and less dysplastic[34, 40]) as many authors presume.[2] From clinical impression, it was suggested that patients with monopolar depressive psychosis may be leptosomatic more often than those with bipolar affective psychoses who tended to be more pyknic in build.[192] However, anthropometrically, no difference between monopolar and bipolar affective psychotics could be demonstrated.[194]

On the other hand, there are differences in premorbid personality traits between these groups.[131, 195] It is, however, not known whether each of them differs from the average population in this respect. It is only empirically well-founded that patients with monopolar depressive psychoses deviate from the average population premorbidly as well as in their lucid intervals between psychotic episodes.[134] The typical personality of these cases may be described as orderly, rigid, and emotionally dependent rather than cyclothymic or schizothymic, extraverted or introverted, and the like.[196] Their personality features were first described by psychoanalysts.[197] Some authors[198, 199, 200] regard these features most characteristic of patients with involutional melancholia, a disorder which is now considered a late manifestation of monopolar depression.[192, 193] Bipolar psychotics may show certain premorbid characteristics which are less obvious and less homogenous than those of monopolar depressives.[201] However, the premorbid personality traits of all kinds of patients with affective disorders seem unrelated to body build.

Another situation is found with respect to the psychomorphological constitution of neurotics. Here, deviations in physique and correlated temperamental traits have been shown to exist.[2] The finding of a reduced general size is thus far restricted to a sample of southern Chinese patients and may not be generalized to other populations.[97] A more typical finding, however, is one expected from psychomorphological correlations within the average male population, where neurotic tendencies are associated with reduced physical robustness and (toward the extreme of neuroticism)[162] with introversion to a slight, but significant degree. The same physical characteristic was proven to dominate in male students with emotional disorders when comparing them with their emotionally better adjusted colleagues.[202] This finding, which is corroborated in other studies (e.g. Winter[203]), may be more pronounced in patients with mood disturbances and autonomic dysfunctions (Eysenck's[57] dysthymics) who were found to have, comparatively, the highest values in the body type index (height times one hundred divided by transverse chest diameter times six).[66] As one can see from the right columns of Table 10-II, the index is markedly related to the factor of physical robustness (3').

In an investigation of personality characteristics of neurotic inpatients performed by the present author together with Krauss,[204] neurotics, in general, were found to be habitually more sensitive, uncooperative, and inhibited long before the onset of actual clinical symptomatology. This tendency was most marked in patients with depressive symptoms and those with predominantly autonomic dysfunctions. Our findings favor Slater's[205] and Eysenck's[57] concepts of constitutional predisposition to neurotic reactions in stressful life situations.

According to Coppen,[94] male homosexuals seem to resemble neurotics in their physical makeup by reduction in breadth measures. In the fat/bone index, they deviate from neither neurotics nor normal controls.[89] Consequently, their physical habitus is characterized by a lack in robustness rather than by femininity.

Findings in the physique and personality of delinquent youths, as well as adult criminals, are in a way contrary to those in the psychomorphological constitution of neurotics. Hooton's finding of reduced general size in prisoners[206] was not reproduced by subsequent investigators. Rather, the physical habitus of delinquents and criminals was repeatedly described as markedly more robust than the average.[80, 82, 207, 208, 209] According to observations of the Gluecks and of others, this corresponds to the high degree of impulsiveness which they tend to exhibit from childhood.[210, 211]

Probably impulsiveness, as a temperamental correlate of physical robustness, predisposes a person only to certain types of offenses (e.g. assaults[206]). As with neurotics, there seem to be physical differences among heterogenous groups of delinquents.[128, 206, 212] However, the general contrast between the psychomorphological constitution of neurotics on the one hand and delinquents/criminals on the other, agrees with the incidence of neurotic disorders and criminal offenses in both sexes: women, who tend to be physically less robust and emotionally less stable, as well as less impulsive than men, show a higher frequency of neurotic reactions,[213, 214, 215,] whereas criminal acts occur much more frequently in the male sex.[216]

The hypothesis that a disharmonious mixture of different types predisposes an individual to neurotic disorders or to criminal behavior is empirically not well-founded,[128] though in one investigation neurotics were found to be more often dysplastic than emotionally healthy individuals.[203] It must be confessed that the methodology of assessing physical dysplasias is still in its infancy. Here is a vast field for new approaches to an old problem.

Etiological Considerations

In several independent investigations (most of them using the comparative twin method) it was demonstrated that the main dimensions of variations in the morphological, as well as in the mental, phenotype are largely determined by the genotype and, to varying degrees, by environmental factors. The influence of these factors is relatively low in respect to the dimensions of skeletal variations (e.g. general skeletal and head size).[9, 11, 12, 13] It is higher for body shape dimensions,[12, 217] higher still for the muscle,[9, 12, 218, 219] and highest for the fat component.[9, 12] Regarding personality dimensions, the relative weight of external influences is apparently lower in general intelligence[220] than it is in extraversion[221] and neuroticism.[15, 16, 222, 223, 224, 225]

The exact nature of psychomorphological relationships within the normal range has, however, not yet been convincingly elucidated by empirical findings. It is likely to be as manifold as that of abnormal variations of psychomorphological constitution. This is highly probable for the correlations of intelligence with both general size and head size. They may be explained in the same way as with the high incidence of dysplastic features in the body build of oligophrenics and epileptics: deviations in genes, whole chromosomes, or environmental conditions of early life interfere with normal growth processes as well as with normal development of brain functions.[18, 29] This is clearly demonstrated in Klinefelter's syndrome: The higher the degree of numerical aberration of the karyotype, the greater the incidence of dysplastic features,[226] as well as the degree of mental retardation (see Fig. 10-6).

The psychomorphological differences in monozygotic twins, discordant for schizophrenia, cannot be explained genetically but rather on the basis of data referring to physical conditions in their early life period.[188] Another intrapair difference in the discordant twin group was the higher frequency of lower birth weight, birth complications, neonatal asphyxia, and CNS illness as a child in the prospective psychotics. In the light of these[188] and related findings,[227, 228] noxious stimuli which affect the developing brain seem to be responsible for reduced physical robustness, lowered intelligence level, and accentuated schizoid features as well as for the enhanced predisposition toward psychotic breakdown in the index cases.

With respect to the apparently linear correlation between physical robustness and mental vitality within the average population, an endocrinological explanation seems more plausible.[96] The influence of steroid compounds on bone and muscle,[104, 229] as well as personality development,[104, 230] is fairly well established. Hormonal action during the fetal period of life is of particular interest in this respect because it may lead to certain behavioral manifestations much later in life.[231] An example of such a delayed mental effect of the hormonal state in the fetus is the tomboyish behavior of girls whose mothers during pregnancy had been treated with synthetic progestagenes (derived from testosterone) because of

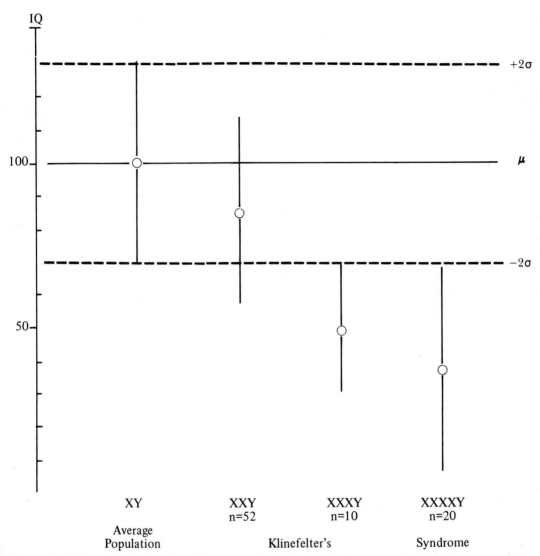

Figure 10-6. Intelligence level and karyotype in Klinefelter's Syndrome. IQ of forty cases with karyotype XXY measured by the present author and the others derived from the literature (see W. Doering, *Intelligenz, Verhalten und Persönlichkeit chromatinpositiver Männer*, unpublished doctoral dissertation [Munich, University of Munich, 1971].)

habitual abortion.[230] Physically, these girls showed enlargement of the clitoris at birth. (This was corrected surgically in early childhood.) Another example is given in a report on reduced masculine behavior of boys whose diabetic mothers had received estrogens as well as progesterone during pregnancy.[232] Physically, these boys were also shorter than control subjects and tended to have a relatively shorter upper body half. Two of them had hypospadias.

Endocrinological investigations in adults

may be inadequate for the detection of differences in those hormonal states responsible for the differences in physique and personality as well as for their interrelationships. However, at least skeletal and muscular development in the adult male is correlated with the excretion of steroid metabolites to a certain extent.[104, 229] In addition, findings regarding correlations between temperamental traits and steroid metabolites were reported in the literature.[96, 104] More extensive research is needed in this field to validate these findings and to detect further psychoendocrinological relationships.

The endocrinological hypothesis of correlations between physical robustness (which includes muscular development) and mental vitality (which includes tomboyishness) does not explain why physical robustness is negatively correlated with verbal intelligence. On the contrary, according to clinical experience with early androgenic and/or progestagenic influence on intellectual functioning manifested later in life,[29, 230, 233] a positive correlation would have been expected. Therefore, a psychogenic origin of this correlation seems more probable: tomboys are usually more interested in sports than in intellectual activities. The development of verbal intelligence, however, will to a certain extent depend on such motivational factors.

A similar explanation must also be considered with respect to the relationship between mental vitality and physical robustness, taking into account that this correlation is more marked in males than in females: boys with a robust physique may prove superior to others in physical combat, thereby developing a self-confident attitude toward life and extraverted traits, combined with the emotional and autonomic stability of assertive personalities. On the other hand, boys lacking in physical robustness will more often lose than win in physical combat, and therefore feel inferior to others; this can easily lead to general insecurity, introversion, and emotional and autonomic imbalance. It seems natural that these boys develop more interest in intellectual activities than in sports, because in physical pursuits their feelings of inferiority are only reinforced.[6, 14, 17, 75, 146]

A further explanation of the psychomorphological relationships between verbal intelligence and mental vitality, on the one hand, and robustness, on the other, lies in the assumption that those who are more active, extraverted, and stable are less interested in intellectual learning. Instead, their motivation is directed more toward motor activities, thereby stimulating muscular development rather than that of intellectual functioning.[14, 75] The fact that robustness also includes chest breadth and other measures which cannot be directly related to muscular development but correlate with mental test scores, casts some doubt on this explanation.

Yet another interpretation of the psychomorphological relationships requires consideration.[14, 17] As already mentioned in the introduction, visible differences among people may influence the reactions of others toward them and thereby shape certain modes of their behavior. People may even behave in a manner expected by others or at least conceive of themselves as behaving according to these expectations. This would, of course, also influence the result of self-ratings so often used in psychomorphological investigations for the assessment of temperamental traits. It has been argued that psychomorphological correlations might be due to mechanisms of this kind because they were in accordance with popular concepts about such relationships.[234, 235]

This argument, however, is insufficiently founded. As outlined in this survey, the self-ratings of people with a markedly different physique are to a large extent at variance with both classical concepts of constitutional research and popular beliefs: the closest correlation between physical and mental traits is not found at the extremes of the variations in physique, as is to be expected from most existing type concepts. Rather, the correlation has been shown to exist within the medium range of physical variations where differences in body build are most likely not marked enough to become social signals. Moreover, the correlation is elicited with respect to a dimension of physique which is less obvious than variations in general size and subcutaneous fat. This dimension, physical robustness, is an aspect of physique which has not even been detected as such in many investigations, including those employing factor analysis.

Therefore, the hypothesis of an endocrinological causation of the correlation between physical robustness and mental vitality and the hypothesis of an influence of early experience of physical superiority versus inferiority on personality development are the most convincing ones. They do not exclude each other nor do they exclude the influence of other factors not considered within this context.[236]

The higher incidence of dysthymics (introverted neurotics) in male individuals lacking in physical robustness, together with the higher frequency of delinquent youths and adult criminals among the physically robust, can be explained on the basis of differences in mental vitality as a correlate of robustness: mentally vital individuals, self-assertive and aggressive as they are, will react to frustration extrapunitively rather than intropunitively. The socially less adapted among them may then even offend the law by an unrestricted acting-out of aggressive impulses.

Mentally less vital individuals, on the other hand, will tend toward more intropunitive reactions and, by repressing aggressive impulses, more often develop emotional and autonomic disturbances. From this point of view dysthymic neurosis could be regarded as "the negative of criminality,"[79] which is a basic modification of Freud's concept of neurosis as "the negative of perversion."[237]

Our etiological considerations are, of course, highly speculative and can only serve to stimulate further research in this field.

CONCLUDING REMARKS

Constitutional research, rooted in ancient medicine, is directed toward the individual as a whole by analyzing the various relatively stable aspects of the phenotype and relating them to each other as well as to determining factors. Body build and personality offer themselves as starting points for this kind of research because they represent two basically different aspects of individuality which can be observed readily in daily life.[2] This, however, has also been a frequent source of bias which led to a marked overrating of psychomorphological correlations. When more objective and more reliable techniques for the assessment of physical and personality characteristics and for multivariate statistical data analysis were introduced into constitutional research, many illusions about the specificity and the degree of such correlations, including their clinical relevance, were destroyed. The consequence was a growing disappointment among scientists and a loss of interest in the subject. Yet the investigation of the whole individual still is, and will remain, a legitimate task of anthropological research.

A broader and more sophisticated approach to the problem is apparently needed. Physique must be analyzed more thoroughly regarding variations in both material composition[238] and inner structure[73] from the macroscopic down to the molecular level.[239] Moreover, individual differences in functions, not only in permanent material structures, must be considered. This refers to differences in biochemical and physiological processes,[240] in immunological reactions, as well as in reactions to drugs.[241, 242] Finally, variations in the processes of growth, differentiation, and involution should be regarded.[89, 238] This developmental aspect is of similar importance in studying personality.[19, 243]

Longitudinal studies, which include the above mentioned aspects of constitution, are therefore urgently required. The design for such investigations must take into account genetic and nongenetic factors which are expected to contribute to the relationships under consideration. A large-scale cohort study of children of schizophrenic mothers in comparison with a control group may be considered a model for investigations of this kind.[228, 244] The follow-up of monozygotic twins would also be of particular value for etiological research on individual differences.[12] The introduction of modern methods of planning and organization,[245] as well as of information handling,[246] including computerization,[247] into investigations of this kind will help preserve the holistic approach of constitutional research. Otherwise, it has no chance of surviving in our era of highly specialized scientific work.

REFERENCES

1. Conrad, K.: Konstitution. In Gruhle, H.W., Jung, R., Mayer-Gross, W., and Müller, M. (Eds.): *Psychiatrie der Gegenwart.* Berlin, Springer, 1967, vol. III/1.
2. Rees, L.: Constitutional factors and abnormal behaviour. In Eysenck, H.J. (Ed.): *Handbook of Abnormal Psychology*, 2nd ed. London, Pitman, 1973.
3. Schwidetzky, I.: Konstitution. In Heberer, G., Schwidetzky, I., and Walter, H. (Eds.): *Anthropologie*, 2nd ed. Frankfurt, Fischer, 1970.
4. Tucker, W.B., and Lessa, W.A.: Man: A constitutional investigation. *Q Rev Biol, 15*:265, 411, 1940.
5. Knussmann, R.: Entwicklung, Konstitution, Geschlecht. In Becker, P.E. (Ed.): *Humangenetik.* Stuttgart, Thieme, 1968, vol. I/1.
6. Tanner, J.M.: Human growth and constitution. In Harrison, G.A., Weiner, J.S., Tanner, J.M., and Barnicot, N.A. (Eds.): *Human Biology.* Oxford, Clarendon, 1964.
7. Von Zerssen, D.: Methoden der Konstitutions– und Typenforschung. In Thiel, M. (Ed.): *Enzyklopädie der geisteswissenschaftlichen Arbeitsmethoden,* No. 9: Methoden der Anthropolgie. Munich, Oldenbourg, 1973.
8. Hierneaux, J.: Heredity and environment: Their influence on human morphology. *Am J Phys Anthropol, 21*:575, 1963.
9. Osborne, R.H., and DeGeorge, F.V.: *Genetic Basis of Morphological Variation.* Cambridge, Harvard U Pr, 1959.
10. Ostertag, W.: Genetic and environmental factors influencing certain anthropometric traits. *Proc Indiana Acad Sci, 68*:59, 1958.
11. Von Verschuer, O.: Beiträge zum Konstitutionsproblem aus den

Ergebnissen der Zwillingsforschung. *Z Menschl Vererb-Konstit-Lehre, 30*:646, 1952.
12. Von Verschuer, O.: *Wirksame Faktoren im Leben des Menschen.* Wiesbaden, Steiner, 1954.
13. Vogel, F., and Wendt, G.G.: Zwillingsuntersuchung über die Erblichkeit einiger anthropologischer Masse und Konstitutionsindices. *Z Menschl Vererb-Konstit-Lehre, 33*:425, 1956.
14. Anastasi, A.: *Differential Psychology*, 3rd ed. New York, Macmillan, 1965.
15. Eysenck, H.J.: *The Biological Basis of Personality.* Springfield, Thomas, 1967.
16. Shields, J.: *Monozygotic Twins Brought Up Apart and Brought Up Together.* London, Oxford U Pr, 1962.
17. Tyler, L.E.: *The Psychology of Human Differences*, 3rd ed. New York, Appleton, 1965.
18. Sobel, E.H.: Organic disorders interfering with somatic growth. In Gardner, L.E. (Ed.): *Endocrine and Genetic Diseases of Childhood.* Philadelphia, Saunders, 1969.
19. Schaefer, E.S.: An analysis of consensus in longitudinal research on personality consistency and change: Discussion of papers by Bayley, Macfarlane, Moss and Kagan, and Murphy. *Vita Hum (Basel), 7*:143, 1964.
20. Thompson, W.R., and Schaefer, T., Jr.: Early environmental stimulation. In Fiske, D.W., and Maddi, S. (Eds.): *Functions of Varied Experience.* Homewood, Dorsey, 1961.
21. Drash, P.W., Greenberg, N.E., and Money, J.: Intelligence and personality in four syndromes of dwarfism. In Cheek, D.B. (Ed.): *Human Growth.* Philadelphia, Lea & Febiger, 1968.
22. Gardner, L.I.: Deprivation dwarfism. *Sci Am, 227*:76, 1972.
23. Patton, R.G., and Gardner, L.I.: Short stature associated with maternal deprivation syndrome: Disordered family environment as cause of so-called idiopathic hypopituitarism. In Gardner, L.I. (Ed.): *Endocrine and Genetic Diseases of Childhood.* Philadelphia, Saunders, 1969.
24. Rutter, M.: Maternal deprivation reconsidered. *J Psychosom Res, 16*:241, 1972.
25. Eastham, R.D., and Jancar, J.: *Clinical Pathology in Mental Retardation.* Bristol, Wright, 1968.
26. Benda, C.E.: *Down's Syndrome.* New York, Grune, 1969.
27. Lenz, W.: Anomalien der Autosomen unter besonderer Berücksichtigung des Schwachsinns. In Becker, P.E. (Ed.): *Humangenetik.* Stuttgart, Thieme, 1967, vol. V/2.
28. Lehmann, W.: Krankheit der Drüsen mit innerer Sekretion. In Becker, P.E. (Ed.): *Humangenetik.* Stuttgart, Thieme, 1964, vol. III/1.
29. Money, J.: Intellectual functioning in childhood endocrinopathies and related cytogenetic disorders. In Gardner, L.I. (Ed.): *Endocrine and Genetic Diseases of Childhood.* Philadelphia, Saunders, 1969.
30. Pallister, P.D., Kaveggia, E.G., Inhorn, S.L., Meisner, L., and Opitz, J.M.: Chromosome studies in malformation-retardation syndromes. In Primrose, D.A.A. (Ed.): *Proceedings of the Second Congress of the International Association for the Scientific Study of Mental Deficiency.* Warsaw, Polish Medical Publ, 1971.
31. Carter, C.H. (Ed.): *Medical Aspects of Mental Retardation.* Springfield, Thomas, 1965.
32. Cravioto, J.: Protein-caloric undernutrition and mental retardation. In Jervis, G.A. (Ed.): *Expanding Concepts in Mental Retardation.* Springfield, Thomas, 1968.
33. Widdowson, E.M.: Early nutrition and later development. In Wolstenholme, G.E.W., and O'Connor, M. (Eds.): *Diet and Bodily Constitution.* London, Churchill, 1964.
34. Elsässer, G.: Körperbauuntersuchungen bei endogen Geisteskranken, sonstigen Anstaltsinsassen und Durch-

schnittspersonen. *Z Menschl Vererb-Konstit-Lehre, 30*:307, 1951.
35. Von Rohden, F.: Konstitutionelle Körperbau-Untersuchungen an Gesunden und Kranken. *Arch Psychiatr Nervenkr, 79*:786, 1927.
36. Bräutigam, W.: Körperliche, seelische und soziale Einflüsse auf die Geschlechtszugehörigkeit des Menschen. *Internist (Berlin), 4*:171, 1964.
37. Doering, W.: *Intelligenz, Verhalten und Persönlichkeit chromatinpositiver Männer.* Unpublished doctoral dissertation, Munich, Germany, Univ Munich, 1971.
38. Meyer, A.E., and Von Zerssen, D.: A psychological investigation of women with so-called idiopathic hirsutism. *J Psychosom Res, 4*:206, 1960.
39. Ounsted, C., and Taylor, D.C. (Eds.): *Gender Differences.* London, Churchill, 1972.
40. Kretschmer, E.: *Körperbau und Charakter,* 25th ed. Berlin, Springer, 1967.
41. Schlegel, W.S.: *Körper und Seele.* Stuttgart, Enke, 1957.
42. Sheldon, W.H., with the collaboration of Stevens, S.S., and Tucker, W.B.: *The Varieties of Human Physique: An Introduction to Constitutional Psychology.* New York, Harper, 1940.
43. Sheldon, W.H., with the collaboration of Stevens, S.S.: *The Varieties of Temperament: A Psychology of Constitutional Differences.* New York, Haprer, 1942.
44. Hood, A.B.: A study of the relationship between physique and personality variables measured by the MMPI. *J Pers, 31*:97, 1963.
45. Von Zerssen, D.: *Körperbau, Persönlichkeit und seelisches Kranksein.* Med. Habilitationsschrift, Heidelberg, Germany, Univ Heidelberg, 1966.
46. Von Zerssen, D.: Methoden und Ergebnisse der biometrischen Konstitutionsforschung. In Keiter, F. (Ed.): *Verhaltensforschung im Rahmen der Wissenschaften vom Menschen.* Göttingen, Musterschmidt, 1969.
47. Undeutsch, U.: Das Verhältnis von körperlicher und seelischer Entwicklung. In Lersch, P., Sander, F., Thomae, H., and Wilde, K. (Eds.): *Handbuch der Psychologie,* 2nd ed. Göttingen, Hogrefe, 1959, vol. III.
48. Ciocco, A.: The historical background of the modern study of constitution. *Bull Inst Hist Med Johns Hopkins Univ, 4*:23, 1936.
49. Jaspers, K.: *General Psychopathology.* Chicago, U of Chicago Pr, 1963.
50. Grüsser, O.J., and Klinke, R. (Eds.): *Pattern Recognition in Biological and Technical Systems.* Berlin, Springer, 1971.
51. Von Rohden, F.: Methoden der konstitutionellen Körperbauforschung. In Abderhalden, E. (Ed.): *Handbuch der biologischen Arbeitsmethoden.* Berlin, Urban & Schwarzenberg, 1937, vol. IX/3.
52. Conrad, K.: *Der Konstitutionstypus,* 2nd ed. Berlin, Springer, 1963.
53. Strömgren, E.: Über anthropometrische Indices zur Unterscheidung von Körperbautypen. *Z ges Neurol Psychiatr, 159*:75, 1937.
54. Wertheimer, F.I., and Hesketh, F.E.A.: A minimum scheme for the study of the morphological constitution in psychiatry, with remarks on anthropometric technique. *Arch Neurol Psychiatry, 17*:93, 1925.
55. Von Zerssen, D.: Biometrische Studien über "Körperbau und Charakter." *Fortschr Neurol Psychiatr, 33*:455, 1965.
56. Ekman, G.: On typological and dimensional systems of reference in describing personality. *Acta Psychol (Amst), 8*:1, 1951.
57. Eysenck, H.J.: *The Structure of Human Personality,* 2nd ed. London, Methuen, 1965.
58. Guilford, J.P.: *Personality.* New York, McGraw, 1959.
59. Harman, H.H.: *Modern Factor Analysis,* 2nd ed. Chicago, U of Chicago Pr, 1967.

60. Overall, J.E., and Klett, C.J.: *Applied Multivariate Analysis.* New York, McGraw, 1972.
61. Lindegård, B.: *Variations in Human Body-Build.* Copenhagen, Munksgaard, 1953.
62. Lindegård, B.: Differential somatology, In Lindegård, B. (Ed.): *Body-Build, Body-Function, and Personality.* Lund, Gleerup, 1956.
63. Von Zerssen, D.: Dimensionen der morphologischen Habitusvariationen und ihre biometrische Erfassung. *Z Menschl Vererb-Konstit-Lehre, 37*:611, 1964.
64. Rees, L.: Body size, personality and neurosis. *J Ment Sci, 96*:168, 1950.
65. Rees, L.: Body build, personality and neurosis in women. *J Ment Sci, 96*:426, 1950.
66. Rees, L., and Eysenck, H.J.: A factorial study of some morphological and psychological aspects of human constitution. *J Ment Sci, 91*:8, 1945.
67. Brožek, H., and Keys, A.: Body build and body composition. *Science, 116*:140, 1952.
68. Dupertuis, C.W., Pitts, G.C., Osserman, E.F., Welham, W.C., and Behnke, A.R.: Relation of specific gravity to body build in a group of healthy men. *J Appl Physiol, 3*:676, 1951.
69. Heath, B.H., and Carter, J.E.L.: A modified somatotype method. *Am J Phys Anthropol, 27*:57, 1967.
70. Hooton, E.A., cit. after Hunt, E.E., Jr.: Human constitution: an appraisal. *Am J Phys Anthropol, 10*:55, 1952.
71. Reynolds, E.L., and Asakawa, T.: A comparison of certain aspects of body structure and body shape in 200 adults. *Am J Phys Anthropol, 8*:343, 1950.
72. Sills, F.D.: A factor analysis of somatotypes and of their relationship to achievement in motor skills. *Res Q Am Assoc Health Phys Educ, 21*:424, 1950.
73. Im Obersteg, J.: Über Beziehungen des Körperbautypus zu Gewicht und Maß innerer Organe. *Acta Genet, 3*:193, 1952.
74. Ekman, G.: On the number and definition of dimensions in Kretschmer's and Sheldon's constitutional systems. In Ekman, G., Husén, T., Johansson, G., and Sandström, C.I. (Eds.): *Essays in Psychology Dedicated to David Katz.* Uppsala, Almqvist & Wiksell, 1951.
75. Von Zerssen, D.: Eine biometrische Überprüfung der Theorien von Sheldon über Zusammenhänge zwischen Körperbau und Temperament. *Z Exp Angew Psychol, 12*:521, 1965.
76. Sheldon, W.H., with the collaboration of Dupertuis, C.W., and McDermott, E.: *Atlas of Men: A Guide for Somatotyping the Adult Male at All Ages.* New York, Harper, 1954.
77. Humphreys, L.G.: Characteristics of type concepts with special reference to Sheldon's typology. *Psychol Bull, 54*:218, 1957.
78. Lorr, M., and Fields, V.: A factorial study of body types. *J Clin Psychol, 10*:182, 1954.
79. Von Zerssen, D.: Comparative studies in the psychomorphological constitution of schizophrenics and other groups. In Sankar, D.V.S. (Ed.): *Schizophrenia: Current Concepts and Research.* Hicksville, PJD Publ, 1969.
80. Parnell, R.W.: *Behaviour and Physique: An Introduction to Practical and Applied Somatometry.* London, Arnold, 1958.
81. Sheldon, W., Lewis, N.D.C., and Tenney, A.M.: Psychotic patterns and physical constitution: A thirty-year follow-up of 3800 psychiatric patients in New York state. In Sankar, D.V.S. (Ed.): *Schizophrenia: Current Concepts and Research.* Hicksville, PJD Publ, 1969.
82. Sheldon, W.H., with the collaboration of Hartl, E.M., and McDermott, E.: *Varieties of Delinquent Youth: An Introduction to Constitutional Psychiatry.* New York, Harper, 1949.
83. Seltzer, C.C.: Body disproportions and dominant personality traits. *Psychosom*

Med, 8:75, 1946.
84. Draper, G.: The mosaic of androgyny maleness within female and femaleness within male. *N Engl J Med, 255*:393, 1941.
85. Bayley, N., and Bayer, L.M.: The assessment of somatic androgyny. *Am J Phys Anthropol, 4*:433, 1946.
86. Seltzer, C.C.: The relationship between the masculine component and personality. *Am J Phys Anthropol, 3*:33, 1945.
87. Tanner, J.M.: Current advances in the study of physique. *Lancet,* 574, 1951/I.
88. Reynolds, E.L.: The fat/bone index as a sex-differentiating character in man. *Hum Biol, 21*:199, 1949.
89. Tanner, J.M.: *Growth at Adolescence,* 2nd ed. Oxford, Blackwell, 1962.
90. Von Zerssen, D.: Habitus und Geschlecht. *Homo, 19*:1, 1968.
91. Wilde, K.: *Eine Überprüfung verschiedener Konstitutionstypologien mit objektiven anthropometrischen und psychometrischen Methoden.* Unpublished doctoral dissertation, Hamburg, Univ Hamburg, 1964.
92. Carpenter, A.: An anthropometric study of masculinity and femininity of body build. *Res Q, 12*:714, 1941.
93. Bridges, P.K., and Jones, M.T.: Personality, physique and the adrenocortical response to a psychological stress. *Br J Psychiatry, 113*:601, 1967.
94. Coppen, A.J.: Body-build of male homosexuals. In Jores, A., and Freyberger, H. (Eds.): *Advances in Psychosomatic Medicine.* New York, Brunner, 1961.
95. Coppen, A., Julian, T., Fry, D.E., and Marks, V.: Body build and urinary steroid excretion in mental illness. *Br J Psychiatry, 113*:269, 1967.
96. Segraves, R.T.: Personality, body build and adrenocortical activity. *Br J Psychiatry, 117*:405, 1970.
97. Singer, K., Chang, P.T., and Hsu, G.L.K.: Physique, personality and mental illness in the Southern Chinese. *Br J Psychiatry, 121*:315, 1972.
98. Dupertuis, C.W.: Anthropometry of extreme somatotypes. *Am J Phys Anthropol, 8*:367, 1950.
99. Tanner, J.M., and Weiner, J.S.: The reliability of the photogrammetric method of anthropometry, with a description of a miniature camera technique. *Am J Phys Anthropol, 7*:145, 1949.
100. Brožek, J., Hunt, E.E., and Škerlj, B.: Subcutaneous fat and age changes in body build and body form in women. *Am J Phys Anthropol, 11*:577, 1953.
101. Kemsley, W.F.F.: Weight and height of a population in 1943. *Ann Eugen, 15*:161, 1950.
102. Newman, R.W.: Age changes in body build. *Am J Phys Anthropol, 10*:75, 1952.
103. Von Zerssen, D.: Körperbau, Psychose und Persönlichkeit. *Nervenarzt, 37*:52, 1966.
104. Lindegård, B., and Nyman, G.E.: Interrelations between psychologic, somatologic, and endocrine dimensions. In Lindegård, B. (Ed.): *Body-Build, Body-Function, and Personality.* Lund, Gleerup, 1956.
105. Brožek, J.: Quantitative description of body composition: physical anthropology's "fourth" dimension. *Curr Anthropol, 4*:3, 1963.
106. Hofstätter, P.R.: Die Faktorenanalyse somatischer Merkmale. *Z Menschl Vererb-Konstit-Lehre, 27*:579, 1944.
107. McCloy, C.H.: An analysis for multiple factors of physical growth at different age levels. *Child Dev, 11*:249, 1940.
108. Thurstone, L.L.: Factor analysis and body types. *Psychometrika, 11*:15, 1946.
109. Howells, W.W.: Factors of human physique. *Am J Phys Anthropol, 9*:159, 1951.
110. Thurstone, L.L.: Factorial analysis of body measurements. *Am J Phys Anthropol, 5*:15, 1947.
111. Moore, T.V., and Hsu, E.H.: Factorial analysis of anthropological measurements in psychotic patients. *Hum Biol, 18*:133, 1946.
112. Heath, H.: A factor analysis of women's

measurements taken for garment and pattern construction. *Psychometrika, 17*:87, 1952.
113. Martin, R.: *Lehrbuch der Anthropologie*, 3rd ed. Stuggart, Fischer, 1957, vol. I.
114. Lindegård, B., Morsing, C., and Nyman, G.E.: Male sex characters in relation to body-build, endocrine activity, and personality. In Lindegård, B. (Ed.): *Body-Build, Body-Function, and Personality*. Lund, Gleerup, 1956.
115. Paterson, D.G.: *Physique and Intellect*. New York, Appleton, 1930.
116. Child, I.L., and Sheldon, W.H.: The correlation between components of physique and scores on certain psychological tests. *Char Person, 10*:23, 1941.
117. Smith, H.C.: Psychometric checks on hypotheses derived from Sheldon's work on physique and temperament. *J Pers, 17*:310, 1949.
118. Heidbreder, E.: Intelligence and the height-weight ratio. *J Appl Psychol, 10*:52, 1926.
119. Krisch, H.: Woher stammt die populäre Überzeugung, daβ eine Relation zwischen somatischem und psychischem Habitus besteht? *Arch Psychiatr Nervenkr, 79*:489, 1927.
120. Howells, W.W.: Measurement and analysis in anthropology. In Whitla, D.K. (Ed.): *Handbook of Measurement and Assessment in Behavioral Sciences*. Reading, Addison-Wesley, 1968.
121. Cattell, R.B.: *The Scientific Analysis of Personality*. Chicago, Harmondsworth, 1965.
122. Fiske, D.W.: *Measuring the Concepts of Personality*. Chicago, Aldine, 1971.
123. Campbell, R.C.: *Statistics for Biologists*. London, Cambridge U Pr, 1967.
124. Tiling, T.: Individuelle Geistesartung und Geistesstörung. *Grenzfr Nervenleb, 27*:1904.
125. Kretschmer, E., and Enke, W.: *Die Persönlichkeit der Athletiker*. Leipzig, Thieme, 1936.
126. Delbrück, H.: Über die körperliche Konstitution bei der genuinen Epilepsie. *Arch Psychiatr Nervenkr, 77*:555, 1926.
127. Westphal, K.: Körperbau und Charakter der Epileptiker. *Nervenarzt, 4*:96, 1931.
128. Von Rohden, F.: Körperbauuntersuchungen an geisteskranken und gesunden Verbrechern. *Arch Psychiatr Nervenkr, 77*:151, 1926.
129. Becker, J., Spielberger, C.D., and Parker, J.B.: Value achievement and authoritarian attitudes in psychiatric patients. *J Clin Psychol, 19*:57, 1963.
130. Gneist, J.: Religiöse Prägung und religiöses Verhalten der depressiven Primärpersönlichkeit. *Confin Psychiatr, 12*:164, 1969.
131. Perris, C.: A study of bipolar (manic-depressive) and unipolar recurrent depressive psychoses. IV: A multidimensional study of personality traits. *Acta Psychiatr Scand, 42 (Suppl. 194)*:68, 1966.
132. Spielberger, C.D., Parker, J.B., and Bekker, J.: Conformity and achievement in remitted manic-depressive patients. *J Nerv Ment Dis, 137*:162, 1963.
133. Von Zerssen, D., with the collaboration of Koeller, D.-M., and Rey, E.-R.: Objektivierende Untersuchungen zur prämorbiden Persönlichkeit endogen Depressiver, Methodik und vorläufige Ergebnisse. In Hippius, H., and Selbach, H. (Eds.): *Das depressive Syndrom*. Munich, Urban & Schwarzenberg, 1969.
134. Von Zerssen, D., Koeller, D.-M., and Rey, E.-R.: Die prämorbide Persönlichkeit von endogen Depressiven. Eine Kreuzvalidierung früherer Untersuchungsergebnisse. *Confin Psychiatr, 13*:156, 1970.
135. Diamonds, S.: *Personality and Temperament*. New York, Harper, 1957.
136. Hall, C.S., and Lindzey, G.: *Theories of Personality*. New York, Wiley, 1957.
137. Huter, C.: *Illustriertes Handbuch der praktischen Menschenkenntnis*, 5th ed. Schwaig bei Nürnberg, Kupfer Nachf, 1952.

138. Lombroso, C.: *Crime, It's Causes and Remedies.* Boston, Little, 1911.
139. Lubin, A.: A note on Sheldon's table of correlations between temperamental traits. *Br J Psychol* (Stat. Sect.) *3*:186, 1950.
140. Bull, K.R.: An investigation into the relationship between physique, motor capacity and certain temperamental traits. *J Educ Psychol, 28*:149, 1958.
141. Janoff, I.Z., Beck, L.H., and Child, I.L.: The relation of somatotype to reaction time, resistance to pain, and expressive movement. *J Pers, 18*:454, 1950.
142. Coffin, T.E.: A three-component theory of leadership. *J Abnorm Soc Psychol, 39*:63, 1944.
143. Smith, D.W.: The relation between ratio indices of physique and selected scales of the Minnesota Multiphasic Personality Inventory. *J Psychol, 43*:325, 1957.
144. Child, I.L.: The relation of somatotype to self-ratings on Sheldon's temperamental traits. *J Pers, 18*:440, 1950.
145. Cortés, J.B., and Gatti, F.M.: Physique and self-description of temperament. *J Consult Psychol, 29*:432, 1965.
146. Hanley, C.: Physique and reputation of junior high school boys. *Child Dev, 22*:247, 1951.
147. Coppen, A.: The Marke-Nyman temperament scale: An English translation. *Br J Med Psychol, 39*:55, 1966.
148. Nyman, G.E.: Variations in personality. *Acta Psychiatr Scand, 31* [Suppl]:*107,*1956.
149. Sjöbring, H.: Personality structure and development. *Acta Psychiatr Scand* [Suppl] *244*: 1973.
150. Carrigan, P.M.: Extraversion-introversion as a dimension of personality: A reappraisal. *Psychol Bull, 57*:329, 1960.
151. Eysenck, S.B.G., and Eysenck, H.J.: On the dual nature of extraversion. *Br J Soc Clin Psychol, 2*:46, 1963.
152. Terman, L.M., and Miles, C.C.: *Sex and Personality.* New York, Russell, 1968.
153. Eysenck, H.J., and Eysenck, S.B.G.: *Personality Structure and Measurement.* London, Routledge & Paul, 1969.
154. Burt, C.: Subdivided factors. *Br J Psychol (Stat. Sect.), 2*:41, 1949.
155. Royce, J.R.: Concepts generated in comparative and physiological psychological observations. In Cattell, R.B. (Ed.): *Handbook of Multivariate Experimental Psychology.* Chicago, Rand 1966.
156. Brace, G.L.: *Physique, Physiology and Behavior: An Attempt to Analyze a Part of Their Roles in the Canine Biogram.* Unpublished doctoral dissertation, Harvard, Harvard Univ. 1962. (Cit. by Royce, 1966.)
157. Fahrenberg, J., and Myrtek, M.: Ein kritischer Beitrag zur psychophysiologischen Persönlichkeitsforschung. *Z Exp Angew Psychol, 13*:222, 1966.
158. Jones, H.E.: The relationship of strength to physique. *Am J Phys Anthropol, 5*:29, 1947.
159. Seltzer, C.C., and Brouha, L.: The "masculine" component and physical fitness. *Am J Phys Anthropol, 1*:95, 1943.
160. Brengelmann, J.C., and Brengelmann, L.: Deutsche Validierung von Fragebogen der Extraversion, neurotischen Tendez und Rigidität. *Z Exp Angew Psychol, 7*:291, 1960.
161. Edwards, A.L.: Social desirability and personality test construction. In Bass, B.M., and Berg, I.A. (Eds.): *Objective Approaches to Personality Assessment.* Princeton, Van N-Rein, 1959.
162. Eysenck, H.J.: Manual of the Maudsley Personality Inventory. London, U of London Pr, 1958.
163. Von Zerssen, D.: Die Beschwerden-Liste als Test. *Therapiewoche, 21*:1908, 1971.
164. Sanford, R.N., Adkins, M.M., Miller, R.B., and Cobb, E.A.: Physique, personality and scholarship. *Monogr Soc Res Child Dev, 7*:73, 1943.
165. Von Zerssen, D.: Vitalität. In Müller, C. (Ed.): *Lexikon der Psychiatrie.* Berlin, Springer,1973.

166. Knussmann, R.: Konstitution und Partnerwahl. *Homo, 11*:133, 1960.
167. Kline, N.S., and Oppenheim, A.N.: Constitutional factors in the prognosis of schizophrenia: Further observations. *Am J Psychiatry, 108*:909, 1952.
168. Lehmann, H.E.: Schizophrenia. IV: Clinical features. In Freedman, A.M., and Kaplan, H.I. (Eds.): *Comprehensive Textbook of Psychiatry*. Baltimore, Williams, 1967.
169. Lodge Patch, I.C., Post, F., and Slater, P.: Constitution and the psychiatry of old age. *Br J Psychiatry, 111*:405, 1965.
170. Sherman, L.J., Mosely, E.C., Point, P., Ging, R., and Bookbinder, L.J.: Prognosis in schizophrenia. *Arch Gen Psychiatry, 10*:123, 1964.
171. Pivnicki, D., and Christie, R.G.: Body build characteristics in psychotics. *Compr Psychiatry, 9*:574, 1968.
172. Gellis, S.S., and Finegold, M.: *Atlas of Mental Retardation Syndromes*. Washington, D.C., U S Govt Printing Office, 1968.
173. Holmes, L.B., Moser, H.W., Halldórsson, S., Mack, C., Pant, S.S., and Matzilewich, B.: *Mental Retardation. An Atlas of Diseases with Associated Physical Abnormalities*. New York, Macmillan, 1972.
174. Wheeler, L.R.: A comparative study of physical growth of dull children. *J Educ Res, 20*:273, 1929.
175. Murken, J.D.: *The XYY-Syndrome and Klinefelter's Syndrome*. Stuttgart, Thieme, 1973.
176. Nielsen, J.: Klinefelter's syndrome and the XYY syndrome. *Acta Psychiatr Scand, [Suppl] 45*: 209, 1969.
177. Owen, D.R.: The 47,XYY male: A review. *Psychol Bull, 78*:209, 1972.
178. Mauz, F.: *Die Veranlagung zu Krampfanfällen*. Leipzig, Thieme, 1937.
179. Janz, D.: The grand mal epilepsies and the sleeping-waking cycle. *Epilepsia, 3*:69, 1962.
180. Janz, D.: *Die Epilepsien*. Stuttgart, Thieme, 1969.
181. Beringer, K., and Düser, G.: Über Schizophrenie und Körperbau. *Z Ges Neurol Psychiat, 69*:12, 1921.
182. Stertz, G.: Psychiatrie und innere Sekretion. *Z Ges Neurol Psychiatr, 53*:39, 1920.
183. Häfner, H.: Änderung von Konstitutionsmerkmalen nach Schock- und Krampftherapie. *Arch Psychiatr Nervenkr, 184*:493, 1950.
184. Brooksbank, B.W.L., MacSweeney, D.A., Johnson, A.L., Cunningham, A.E., Wilson, D.A., and Coppen, A.: Androgen excretion and physique in schizophrenics. *Br J Psychiatry, 117*:413, 1970.
185. Polednak, A.P.: Body build of paranoid and non-paranoid schizophrenic males. *Br J Psychiatry, 119*:191, 1971.
186. Rudersdorf, M.: *Objektivierende Untersuchungen über fremdanamnestische Angaben zur prämorbiden Persönlichkeit Schizophrener*. Unpublished doctoral dissertation, Univ Heidelberg, Heidelberg, Germany, 1968.
187. Fritsch, W.: *Objektivierende Untersuchungen zur prämorbiden Persönlichkeit Schizophrener*. Unpublished doctoral dissertation, Univ Heidelberg, Heidelberg, Germany, 1972.
188. Pollin, W.: A new approach to the use of twin study data in studies of the pathogenesis of schizophrenia and neurosis. In Kaplan, A.R. (Ed.): *Genetic Factors in "Schizophrenia."* Springfield, Thomas, 1972.
189. Elsässer, G.: *Die Nachkommen geisteskranker Elternpaare*. Stuttgart, Thieme, 1952.
190. Kallmann, F.J.: The Genetics of Psychoses: An Analysis of 1232 Twin Index Families. Congrès International de Psychiatrie. Paris, Hermann, 1950.
191. Price, J.: An anthropometric comparison of psychiatric patients and their siblings. *Br J Psychiatry, 115*:435, 1962.
192. Angst, J.: *Zur Ätiologie und Nosologie endogener depressiver Psychosen*. Berlin, Springer, 1966.
193. Perris, C.: A study of bipolar (manic-depressive) and unipolar recurrent depressive psychoses. I: Genetic in-

vestigation. *Acta Psychiatr Scand 42* [*Suppl*] *194:* 15, 1966
194. Perris, C.: A study of bipolar (manic-depressive) and unipolar recurrent depressive psychoses. V: Body-build. *Acta Psychiatr Scand 42* [*Suppl*] *194:* 83, 1966.
195. Hofmann, G.: *Vergleichende Untersuchungen zur prämorbiden Persönlichkeit von Patienten mit bipolaren (manish-depressiven) und solchen mit monopolar depressiven Psychosen.* Unpublished doctoral dissertation, Univ Munich, Munich, Germany, 1973.
196. Tellenbach, H.: *Melancholie,* 2nd ed. Berlin, Springer, 1964.
197. Mendelson, M.: *Psychoanalytic Concepts of Depression.* Springfield, Thomas, 1960.
198. Malamud, W., Sands, S.L., and Malamud, I.: The involutional psychoses: A socio-psychiatric study. *Psychosom Med, 3*:410, 1941.
199. Noyes, A.: *Modern Clinical Psychiatry,* 3rd ed. Philadelphia, Saunders, 1954.
200. Titley, W.B.: Prepsychotic personality of patients with involutional melancholia. *Arch Neurol Psychiatry, 36*:19, 1936.
201. Tellenbach, R.: *Untersuchungen zur prämörbiden Persönlichkeit von Psychotikern.* Unpublished doctoral dissertation, Techn Univ Munich, Munich, Germany, 1974.
202. Parnell, R.W., Davidson, M.A., Lee, D., and Spencer, S.J.G.: The detection of psychological vulnerability in students. *J Ment Sci, 101*:810, 1955.
203. Winter, E.: Über die Häufigkeit neurotischer Symptome bei "Gesunden." *Z Psychosom Med, 5*:153, 1958-59.
204. Krauss, W.: *Objektivierende Untersuchungen zur prämorbiden Persönlichkeit von Neurotikern.* Unpublished doctoral dissertation, Univ Munich, Munich, Germany, 1972.
205. Slater, E., and Slater, P.: A heuristic theory of neurosis. *J Neurol Neurosurg Psychiatry, 7*:49, 1944.
206. Hooton, E.A.: *The American Criminal.* Cambridge, Harvard U Pr, 1939, vol. I.
207. Eysenck, H.J.: *Crime and Personality.* London, Routledge & Paul, 1964.
208. Gibbens, T.C.N.: *Psychiatric Studies of Borstal Lads.* New York, Oxford U Pr, 1962.
209. Seltzer, C.C.: A comparative study of the morphological characteristics of delinquents and non-delinquents. In Glueck, S., and Glueck, E. (Eds.): *Unraveling Juvenile Delinquency.* Cambrige, Harvard U Pr, 1964.
210. Glueck, S., and Glueck, E.: *Unraveling Juvenile Delinquency.* Cambridge, Harvard U Pr, 1964.
211. Miller, M.C.: Follow-up studies of introverted children. III: Relative incidence of criminal behaviour. In Eysenck, H.J. (Ed.): *Readings in Extraversion-Introversion. I: Theoretical and Methodological Issues.* London, Staples, 1970.
212. Glueck, S., and Glueck, E.: *Physique and Delinquency.* New York, Harper, 1956.
213. Baumeyer, F.: Erfahrungen über die Behandlung psychogener Erkrankungen in Berlin. Fünfzehn Jahre Zentralinstitut für psychogene Erkrankungen. *Z Psychosom Med, 8*:167, 1961-62.
214. Cooper, B.: Psychiatric disorder in hospital and general practice. *Soc Psychiatry, 1*:7, 1966.
215. Hagnell, O.: *A Prospective Study of the Incidence of Mental Disorders.* Stockholm, Svenska Bokförlaget, Norstedts-Bonniers, 1966.
216. Scheinfeld, A.: *Women and Men.* London, Chatto & Windus, 1943.
217. Lasker, G.W.: The effects of partial starvation on somatotype: An analysis of material from the Minnesota Starvation Experiment. *Am J Phys Anthropol, 5*:323, 1947.
218. Lindegård, B.: Body-build and physical activity. In Lindegård, B. (Ed.): *Body-Build, Body-Function, and Person-*

ality. Lund, Gleerup, 1956.
219. Tanner, J.M.: The effect of weight-training on physique. *Am J Phys Anthropol, 10*:427, 1952.
220. Burt, C.: The genetic determination of differences in intelligence: A study of monozygotic twins reared together and apart. *Br J Psychol, 57*:137, 1966.
221. Eysenck, H.J.: The inheritance of extraversion-introversion. In Eysenck, H.J. (Ed.): *Readings in Extraversion-Introversion, 1: Theoretical and Methodological Issues.* London, Staples, 1970.
222. Eysenck, H.J., and Prell, D.B.: The inheritance of neuroticism: An experimental study. *J Ment Sci, 97*:441, 1951.
223. Young, J.P.R., Fenton, G.W., and Lader, M.H.: The inheritance of neurotic traits: A twin study of the Middlesex Hospital Questionnaire. *Br J Psychiatry, 119*:393, 1971.
224. Fuller, J.L., and Thompson, W.R.: *Behavior Genetics.* New York, Wiley, 1964.
225. Slater, E., and Cowie, V.: *The Genetics of Mental Disorders.* London, Oxford U Pr, 1971.
226. Hambert, G.: *Males with Positive Sex Chromatin.* Göteborg, Akademiförlaget, 1966.
227. Mednick, S.A., and Schulsinger, F.: Some pre-morbid characteristics related to break-down in children with schizophrenic mothers. In Rosenthal, D., and Kety, S.S. (Eds.): *The Transmission of Schizophrenia.* Oxford, Pergamon, 1968.
228. Mednick, S.A., and Schulsinger, F.: Studies of children at high risk for schizophrenia. In Dean, S.R. (Ed.): *Schizophrenia. The First Ten Dean Award Lectures.* New York, MSS Inform Corp, 1973.
229. Tanner, J.M., Healy, M.J.R., Whitehouse, R.H., and Edgson, A.C.: The relation of body build to the excretion of 17-ketosteroids and 17-ketogenic steroids in healthy young men. *J Endocrinol, 19*:87, 1959.
230. Money, J., and Ehrhardt, A.A.: Prenatal hormonal exposure: possible effects on behaviour in man. In Michael, R.P. (Ed.): *Endocrinology and Human Behaviour.* London, Oxford U Pr, 1968.
231. Goy, R.W.: Organizing effects of androgen on the behaviour of rhesus monkeys. In Michael, R.P. (Ed.): *Endocrinology and Human Behaviour.* London, Oxford U Pr, 1968.
232. Yalom, I.D., Green, R., and Fisk, N.: Prenatal exposure to female hormones. Effect on psychosexual development in boys. *Arch Gen Psychiatry, 28*:554, 1973.
233. Dalton, K.: Ante-natal progesterone and intelligence. *Br J Psychiatry, 114*:1377, 1968.
234. Staffieri, J.R.: Body build and behavioral expectancies in young females. *Dev Psychol, 6*:125, 1972.
235. Wells, W.D., and Siegel, B.: Stereotyped somatotypes. *Psychol Rep, 8*:77, 1961.
236. Barker, R.G.: *Adjustment of Physical Handicap and Illness: A Survey of the Social Psychology of Physique and Disability,* 2nd ed. New York, Social Science Research Council, 1953.
237. Freud, S.: Three contributions to the theory of sex. In Brill, A.A. (Ed.): *The Basic Writings of Sigmund Freud.* New York, Modern Lib, 1938.
238. Cheek, D.B. (Ed.): *Human Growth. Body Composition, Cell Growth, Energy, and Intelligence.* Philadelphia, Lea & Febiger, 1968.
239. Williams, R.J.: *Biochemical Individuality.* New York, Wiley, 1956.
240. Fahrenberg, J.: *Psychophysiologische Persönlichkeitsforschung.* Göttingen, Hogrefe, 1967.
241. Klerman, G.L., Dimascio, A., Greenblatt, M., and Rinkel, M.: The influence of specific personality patterns on the reactions to phrenotropic agents. In Masserman, J.H. (Ed.): *Biological Psychiatry.* New York, Grune, 1959.
242. Sheard, M.H.: Effect of lithium on human aggression. *Nature, (Lond) 230*:113, 1971.

243. Meili, R., and Meili-Dworetzki, G.: *Grundlagen individueller Persönlichkeitsunterschiede*. Bern, Huber, 1972.
244. Mednick, S.A., and McNeil, T.F.: Current methodology in research on the etiology of schizophrenia. *Psychol Bull, 70*:681, 1968.
245. MacMahon, B., and Pugh, T.F.: *Epidemiology: Principles and Methods.* Boston, Little, 1970.
246. Bourne, C.P.: *Methods of Information Handling.* New York, Wiley, 1965.
247. Kretschmann, H.-J., and Wingert, F.: *Computeranwendungen bei Wachstumsproblemen in Biologie und Medizin.* Berlin, Springer, 1971.

Chapter 11

GENETIC AND ENVIRONMENTAL DETERMINANTS OF NEUROSIS

William Pollin

THE NEUROTIC REACTIONS constitute a group of disorders which, though in many respects poorly defined and delimited, are nonetheless of great psychiatric significance. This is true because of both theoretical and medical considerations. From the theoretical point of view, they are of particular interest because they represent that borderline sector of behavior that links normal unhappiness on the one hand and severely disorganized behavior, sufficiently chaotic so as to be considered psychotic, on the other. From the medical and public health points of view, the neuroses, though usually considerably less severe than psychotic reactions, show a substantially greater incidence,[1] and therefore constitute the most common group of psychiatric disorders which can properly be included within that aspect of the human condition called illness. In the difficult and, to date, still quite unsuccessful attempt to achieve agreement on the definition of this category of psychiatric states, we are therefore challenged to clarify our concepts concerning behavioral mechanisms, behavior pathology, and the limits of normality. We must be clear in our minds how we distinguish between discomfort, deviance, abnormality, and disease, before we can look forward to a clear understanding of the pathogenesis of neurosis and the distinctive roles played by genetic and environmental factors.

Inability to define with any precision the limits of a given condition or phenomenon obviously makes the task of teasing apart the role of genetic and environmental determinants of such a phenomenon most difficult, if not impossible. Nevertheless, a considerable body of work has attempted such an analysis of neurotic conditions. Some of this work has been motivated by varied theoretical interests relevant to pathogenic or genetic issues. Other such studies, however, have resulted from the assumption that clarification of genetic and environmental interaction is required in order to indicate the possible areas and limits of optimal therapeutic intervention. There exists a widespread belief in psychiatry and in the whole area of behavior genetics that environmental and genetic determinants have that type of reciprocal relationship to each other which implies that, to the extent to which certain pathologies are genetically determined, they are relatively nonresponsive to therapeutic intervention; and conversely that, to the extent which environmental determinants predominate, the potential for therapeutic results is considerably enhanced. We shall subsequently return to further consideration and evaluation of this assumption and consider evidence which raises considerable doubts about it. Critical evaluation of this naive dichotomization may help us escape from an unnecessarily nihilistic therapeutic dead end. It may also help move us away from the idealogically toned confrontation between psychodynamically oriented environmentalists and genetically oriented organicists which has often generated considerable heat but questionable light in this field of inquiry.

It is important to keep in mind that historically and operationally the concept of the neuroses is a medical one,[2] springing initially from the experiences of medical practitioners called upon to attempt to treat, comfort, and help patients complaining of a variety of subjective discomforts and impediments to effective functioning. The fact that the category has been derived in this manner has had important implications and consequences in shaping our current efforts to define, use, and study it. Had the category been derived, instead, from the psychological laboratory, our current thinking might be quite different.

In view of the clinical background of the concept, our best source for its current operational definition, as used today in psychiatric practice, is the so-called DSM II (Diagnostic and Statistical Manual—Mental Disorders, second edition), a manual prepared by the Committee on Nomenclature and Statistics of the American Psychiatric Association.[3] This manual is used throughout the United States as the source for statistical reporting and recording of psychiatric conditions and can be considered, in effect, the current official yardstick of clinical definition and recording. Therefore, it constitutes, in all probability, the source of the most widely accepted definition and description of all clinical psychopathologies, including the neuroses. It is relevant not only for American Psychiatry, but also to international standards. The second edition of the "Official Manual of Approved Diagnostic Terms" (refer to the Foreward in DSM II, page 9) was based on the eighth revision of the "International Classification of Diseases" (ICD-8) produced under the auspices of the World Health Organization.[4] Thus, we are approaching a set of clinical definitions which, though still very loose and, in specific instances, very difficult to apply, nevertheless do represent the beginning of a worldwide international consensus of terminology and diagnostic points of view. However, these very definitions, arrived at after widespread international effort, demonstrate the difficulties encountered in attempting to define neurosis. One may note that the DSM definition begins with an immediate focus on one symptom (*anxiety*) and then proceeds to a discussion of psychodynamic mechanisms which represents the hypotheses of only one school of psychiatric thought about the relationship between anxiety and other symptoms. Then there follows a set of comparative statements indicating which symptoms present in other major categories of psychopathology are absent in the neuroses. No positive general statement of definition is anywhere present. The full DSM definition is as follows:

300: Neuroses. Anxiety is the chief characteristic of the neuroses. It may be felt and expressed directly or it may be controlled unconsciously and automatically by conversion, displacement, and various other psychological mechanisms. Generally these mechanisms produce symptoms experienced as subjective distress from which the patient desires relief. The neuroses, as contrasted to the psychoses, manifest neither gross distortion or misinterpretation of external reality nor gross personality disorganization. A possible exception to this is hysterical neurosis which some believe may occasionally be accompanied by hallucinations and other symptoms encountered in psychoses.

Traditionally, neurotic patients, however severely handicapped by their symptoms, are not classified as psychotic because they are aware of their mental functioning is distrubed.

Eight subtypes of neurosis are recognized.* These are:
300.0 Anxiety neurosis
300.1 Hysterical neurosis
      300.13 conversion type
      300.14 dissociative type
300.2 Phobic neurosis
300.3 Obsessive-compulsive neurosis
300.4 Depressive neurosis
300.5 Neurasthenic neurosis
300.6 Depersonalization neurosis
300.7 Hypochondrical neurosis

A more complete and vivid discussion of neurosis appears in the recent text by Redlich and Freedman.[5] They define neurotic behavior as:

... consisting of acts that can be characterized as inappropriate, inadequate, unadaptive, and infantile, resulting in a subjective and objective discrepancy between psychological potential and actual performance. For psychoanalysis, "neurosis" refers to core conflicts around which symptoms are built ... the variety of neurotic symptoms and character traits is extraordinarily large ... everybody commits a neurotic act at times.

Neurotic persons act inappropriately. Their social, sexual and occupational performance in general terms is below the level of their potential, and they suffer and/or impose misery on others ... they do not learn from their mistakes. The inappropriate behavior of neurotic persons is rigid and repetitive ... neurotic misjudge the inner world more than their outer world. Either they are ignorant of their infantile wishes, or suspending judgment, they attempt to satisfy them by inadequate or forbidden means. Neurotics do not clearly distinguish between wish and act; they feel guilty over fantasies and even unconscious wishes and invoke punishment for acts they did not commit. The overall result is frustration, shame, despair and fear of punishment. When they are loved they don't know it, and they seem to be incapable of loving in a mature way. They feel inferior over their incompetence, or they lose sight of their limitations and invite ridicule or abuse, which further lowers self-esteem and enhances isolation. The viscious circle of punishing and being punished and of destruction and self-destruction is one of the cardinal and most troublesome facets of neurotic behavior.

Redlich and Freedman emphasize that, although it is quite possible to convey the quality and characteristics of neurotic behavior in neurotic individuals, there are major problems attendant to the definition and distinction of normal from abnormal behavior.[5] They add significantly to our awareness of the problems that must be dealt with in these areas by pointing out that these concepts of normal and abnormal are more complex in psychiatry than in general medicine; and, particularly, by distinguishing between the three related and overlapping constructs: "A neurosis," "neurotic behavior," and "the sick role." One of the major difficulties in attempting to compare and assess the results of different research investigations involving neurosis is that most investigators have failed explicitly to consider these three distinct approaches to an operational definition and limitation of the phenomenon they are studying. We are therefore uncertain when incidence figures are quoted or results stated in clinical studies as to what

---

*To help visualize the boundaries of this category of behavior one may also cite the essential core of the DSM II definitions for "Psychosis," and for "Personality Disorder," the two major types of psychopathology from which neurosis must often be distinguished.

295-298: Psychosis. Patients are described as psychotic when their mental functioning is sufficiently impaired to interfere grossly with their capability to meet the ordinary demands of life. The impairment may result from a serious distortion in their capacity to recognize reality.

301: Personality Disorders. This group of disorders is characterized by deeply engrained, maladaptive patterns of behavior that are perceptibility different in quality from psychotic and neurotic symptoms. Generally these are lifelong patterns often recognizable by the time of adolescence or earlier.

extent the results refer to the appearance of some unspecified degree of neurotic behavior; the decision that the individual in question exhibits behaviors to such an extent that he can be considered a neurotic, or that his difficulties can be characterized as constituting a neurosis; or to what extent he demonstrates a sufficiently severe degree of neurosis so that the particular social stratum which he inhabits has, with his cooperation, decided that he should be considered as filling the sick role:

> In summary, the concept of health and normality is still quite muddled with empirical research mostly lacking. Only gross deviations are clearly recognized and agreed upon in all civilized societies; borderlines of normal and abnormal behavior are fuzzy and overlapping. Cultural relativism with respect to milder disorders is the rule. The judgments of psychiatrists cannot in reality be far removed from those of the common man of the societies and cultures in which the psychiatrists and patients live. At present we cannot make precise statements about normal and abnormal. We can, however, define to a certain extent the psychiatric sick role ... (which) legitimately excuses (the patient) from work and social obligations and granting him the right of special attention while requiring him to do his best to recover.[5]

This role is more complex in psychiatry than in general medicine and can well be divided into two basic types, dependent upon whether one is dealing with a patient with severe behavior disorders such as psychosis or severe neurosis or that of a patient with mild to moderate disorders. The patients with severe behavior disorders are easily accepted as filling the sick role in all technologically advanced societies, whether cooperative or not; are excused from work and other social obligations; and are not held accountable for antisocial acts. On the other hand, patients suffering from mild to moderate disorders, which would include the bulk of neuroses, cannot be readily differentiated from the normal general population.

Due to the marked limitations in our ability to clearly distinguish neurotic behavior from either normal or psychotic behavior, estimates of the prevalence of neurosis in medical practice are of little value. They range very widely, from approximately 5 to 75 percent. One of the most extensive and well-conducted case-finding studies of a substantial, complete urban population was the "Midtown" study undertaken by Srole, et al. in an area of East Manhattan.[6] This oft-quoted study reported 36 percent of their urban population with mild disturbances and 22 percent with mildly severe disturbance. However it is not clear what portion of either of these categories can properly be called neurotic and what portions demonstrate other behavior pathologies.

Pollin, et al. reported that among 31,818 individual male twins selected for health in that both members of the pair were admitted to the United States Armed Forces, 1,695 were diagnosed neurotic during a period of medical follow-up that average eighteen years, an incidence of 5.3 percent. Rosenthal quotes a study by Hagnell[7] of Swedish cohort of 2,400 people, all of whom were psychiatrically examined over ten years. He found that approximately one of every ten men, and one of every five women became neurotic in their lifetime. Two careful studies of first referrals to psychiatric outpatient clinics reported that 27 percent of these referrals were diagnosed neurotic in Denmark[8] and 43 percent at the Phipps Outpatient Psychiatric Clinic in Baltimore.[9] Rosenthal concludes, after surveying a number of additional such studies, that "psychoneurosis is a very common disease, that the reported rates vary with the methodology used, and ... the disorder is more common in women than men."[10]

REVIEW OF EMPIRICAL DATA

Three types of studies constitute the major sources of currently available empirical data that casts light on the relative role of genetic and environmental factors in the pathogenesis of neurosis. There are (1) twin studies, (2) family studies based on clinical diagnosis, and (3) family studies based on psychological test measures of neuroticism. Some of the major studies in each of these areas are reviewed below, and attempts are made to determine what type of overall conclusions, if any, can be tentatively reached based on this work. Before considering the data itself, however, some attention to methodologic rationale and problems is in order.

Twin Study Methods

The classical twin study method, in which the degree of concordance or similarity for a given diagnosis or trait is compared in a series of one-egg and two-egg twin pairs, has probably been the source of most data concerning the issue of genetic and environmental determinants in psychology and psychiatry. Its basic rationale, first propounded by Galton in 1875, assumes that monozygotic twins are completely identical, genetically speaking, and that all differences between them, therefore, must be due to environmental factors: some are prenatal, nongenetic but nonetheless constitutional; and others are postnatal and therefore experiential. On the other hand, differences between dizygotic twins may be due either to environmental influences or to genetic influences since such pairs are apt to be no more genetically alike than same-sex siblings. Therefore, in this method the extent to which identical twins are more alike, more concordant, than fraternal twins is considered a measure of the role played by genetic factors in the development of that particular diagnosis. Eysenck presents a useful discussion of this basic rationale.[11] Varied statistical techniques have been devised, beginning with Holzinger's (H2) which attempted to quantify the proportion of the variance contributed by heredity; Jinks and Fulker have recently reviewed the relationship of a number of these.[12] May has discussed some of the uncertain assumptions upon which H2 depends.[13] Serious questions have been raised concerning two major aspects of classical twin method assumptions: (1) the degree of genetic similarity in one-egg twin pairs, and (2) the extent to which nongenetic factors can bring about a higher level of concordance in MZ than in DZ twin pairs. Eysenck and Darlington have summarized several potential sources and types of genetic, organic differences between two monozygotic twins;[11,14] and Darlington, on the basis of such observations, has written, "We have to admit the comparisons of one-egg and two-egg twins do not give us the uncontaminated separation of heredity and environment which Galton and his successors had hoped for. The measure they give is in estimate: it must always be an underestimate of the effects of heredity."[15] Other workers, however, have pointed to the possibility of the opposite type of distortion: that such comparisons over estimate rather than underestimate genetic differences. Kety and Jackson, among others, have described a number of factors which could work in this opposite direction with regard to behavioral and psychiatric phenomena in particular.[16,17] Classical twin studies focus on the assumption that the major difference between monozygotic and dizygotic twin pairs is the degree of genetic similarity or difference they manifest. However, psychological and psychoanalytic studies of twin pairs have demonstrated many profoundly important, nongenetic developmental differences between identical and fraternal twin pairs. These include such factors as MZ twins spending more time together, being misidentified for each other by important figures and being more closely identified with each other. Such greater similarity of environmental pressures in identical pairs can lead to greater similarity of outcome variables for many behavioral characteristics on a nongentic basis. Scarr, however, has presented evidence to the contrary.[18]

Twin Studies Data

Ten published studies have been identified in which *clinical* diagnosis of neurosis were made in different series of twins. They are summarized in Table 11-I.

Pollin, et al. summarized the results of their own and the nine other studies which compared concordance rates in MZ and DZ pairs and permitted the computation of MZ/DZ ratios.[1, 19, 20, 21, 22, 23, 24, 25, 26, 27] The total number of pairs reported on was 1,264, 560 being MZ and 704 DZ pairs. The studies, which covered a time period of more than three decades, were undertaken in a number of different countries and obviously employed substantially different sample populations and diagnostic-criteria. Thus, it is perhaps not too surprising, though nonetheless disconcerting, that the percent of MZ concordance varied between 9 and 90 percent, and DZ concordance between 0 and 50 percent. Conversion of the MZ and DZ concordance rates into an MZ/DZ concordance ratio, however, tended to equalize and/or compensate for this variability to some extent. It was noted (Table 11-I), in the face of such great variation in basic concordance rates, that the MZ/DZ concordance ratios for all of these studies varied only between 1.3 and 2.1 This was in marked contrast to the range observed when ten similar twin studies in schizophrenia were similarly treated, and the MZ/DZ concordance ratios ranged instead between 3.5 and 6.1. Thus (with the single exception of the initial Tienari report,[25] subsequently modified in his later reports) no overlap was found in the MZ/DZ ratios for the two diagnostic categories, neurosis *versus* schizophrenic concordance (see Fig. 11-1.)

On the basis of the comparison discussed above, Pollin and associates concluded that there clearly was considerably stronger evidence for the existence of a genetic factor in the pathogenesis of schizophrenia

TABLE 11-I
CONCORDANCE RATES OF NEUROSIS IN TWINS

Investigator	No. of MZ Pairs	% MZ Concordance	No. of DZ Pairs	% DZ Concordance	MZ/DZ Ratio
Stumpful, 1937*	7	43	9	0	-
Slater, 1953†	8	25	40	13	2.0
Shields, 1954	23	74	18	50	1.5
Slater, 1961*	12	42	12	33	1.3
Ihda, 1961	20	50	5	40	1.3
Braconi, 1961	20	90	30	43	2.1
Tienari, 1963	21	57			
Inouye, 1965	21	48	5	40	1.2
Parker, 1966	9	67	11	36	1.8
Pollin, et al.,	419	9	577	7	1.3

*Hysteria only, but concordance based on cotwin's having any neurotic diagnosis.
†From group where index twin is neurotic and/or psychopathic; but concordance is based on cases where cotwin has diagnosis of neurosis.

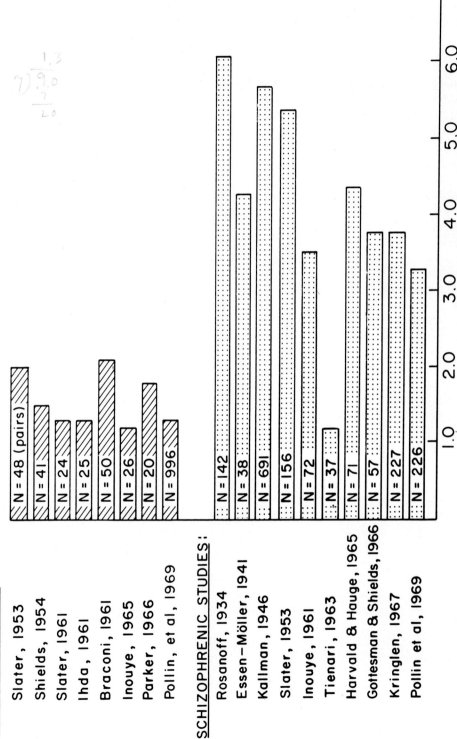

Figure 11-1.

than was true for psychoneurosis.[1] The original data in this study were derived from a large NAS series of 15,909 pairs of veteran twins whose medical follow-up averaged eighteen years.[28] The diagnoses were made not by the investigators but by independent treating physicians in the United States Armed Forces or the Veterans Administration. A total number of 1,695 individuals had the diagnosis of psychoneurosis. The MZ/DZ ratio for psychoneurosis in this study (1.5) was the same as that for fracture (N = 2,642 individuals) and slightly less than that for bacterial pneumonia. By this measure, it would appear that there was little, if any, evidence for the presence of a genetic component in psychoneurosis. When, however, H2 was computed, the value for psychoneurosis (0.069), though still quite low, as were all values in this population selected for health, was in the same order of magnitude as that for diabetes (0.066) and greater than that of eight of eleven other medical diagnosis which were investigated in this study. It was approximately 50 percent greater than that for fracture (0.048).

Rosenthal grouped the clinical twin studies of neurosis that appeared before the large NAS sample was analyzed (and excluded Shields' study of normative twins).[10] He reported a 53 percent concordance rate for MZ pairs, 40 percent for DZ pairs, and concluded, ". . . the difference is not great and leaves little doubt about the importance of environmental factors in the development of clinical neurosis." The fact that the MZ/DZ ratio of this earlier selection of twin studies (1.32) was essentially the same as that found in the much larger NAS sample despite the occurrence of much lower initial concordance rates in the NAS panel and its selection for health suggests that this figure may be considered reasonably valid.

Family Studies

Clinical Family Studies

Thirty years ago Brown reported the results of his study of 104 neurotic patients entitled "Heredity in the Psychoneuroses."[29] His report was notable for its data and also for Brown's clear enunciation of the difficulties of definition that still obscure this area. Brown noted, ". . . the chief difficulty is to define the condition the heredity of which one is attempting to trace," and suggested that the difficulties encountered in exploring this question for neurosis were similar to those one would encounter in trying to investigate the heredity of a symptom such as a cough. His point was well taken then, and we have made little progress in dealing with the issue since then. Brown treated most of his 104 neurotic probands himself (sixty-three anxiety states, twenty-one hysterics, and twenty obsessionals). He then obtained detailed family histories from both the patient and at least one other relative; in this connection, seeing some four hundred relatives and obtaining information about 2,288 relatives. Having obtained these case histories, he classified all the relatives seen according to their psychiatric abnormalities. He attempted a similar investigation of thirty-one nonpsychiatric medical patients and their families, pointing out, however, that in these instances family histories were somewhat less complete and satisfactory. He states that he undertook the investigation with some bias against the significance of heredity in the psychoneuroses but interpreted his results as showing a significant genetic role in the development of psychoneuroses. Whereas 16.4 percent of first-degree relatives of his probands were considered neurotic, only 1.1 percent of first-degree relatives of his control group were similarly diagnosed by him. The percent of first-degree relatives of the experimental group who showed a diagnosis similar to the proband varied considerably with neurotic subtype: 15.1 percent for anxiety states, 11.2 percent for hysteria, and 6.9 percent for obsessional states. Brown was aware of the methodological deficiencies in his approach, and he emphasized that they should not be seen as detracting from the importance of environmental factors or therapy.

Stenstedt studied the genetics of neurotic depression in a population of 176 probands, their parents, and siblings.[30] Including the probands, a total of 1,418 individuals received a thorough psychiatric examination and usually a personal follow-up examination. The results were somewhat inconclusive. The incidence of all mental disorders among the relatives did not differ from that in the general population in Stenstedt's view. There did, however, seem to be a higher incidence among the relatives of depressed probands than in the general population. Genetic factors seemed to be of greatest importance in manic-depressive psychosis, of less importance in involutional melancholia, and of least importance in neurotic depression. Stenstedt therefore emphasized the role of environmental factors in neurotic depression.

These two studies exemplify a group of family morbidity studies, which are summarized in Table 11-II.[29, 30, 31, 32, 33, 34, 35] These studies, which, as the above descriptions make clear, often involve a prodigious labor on the part of the investigator (note the huge number of subjects psychiatrically evaluated by Brown and by Stenstedt), suffer from serious methodological and conceptual difficulties.[29, 30] Methodologically, the essential data concern the relationship of frequency of diagnosis in probands and relatives in an experimental group compared to a control group (i.e. primary versus secondary relatives) or to general population incidence. Few, if any, of these investigators, however, are able to achieve an adequate degree of equivalent diagnostic contact with the several different subject groups involved. Thus, in Brown's study, which is not atypical, he was the proband's therapist; he talked directly with relatives, diagnosed others secondhand, and he pointed out that relevant behavioral and personal data were much less available to him from the normal control subjects and their relatives than from the probands and their relatives. Pollin has emphasized the major multiple distortions which can result from such substantial differences in the degree of need and in intensity of contact between the investigator and his experimental versus control groups.[36] An additional problem with studies such as these is the impossibility of evaluating the degree to which investigator bias may skew results. Diagnostic decisions are frequently difficult and ambiguous ones. To make these when the investigator is not blind to the implications of each diagnostic decision for the particular hypothesis he is investigating imposes a methodological burden which is very heavy indeed.

A more serious problem with studies of this type, however, rests upon a conceptual flaw. The basic rationale of these family morbidity studies rests on the assumption that the significant variable distinguishing first and second order relatives is the greater degree of genetic similarity. Therefore, if a correlation is found between the degree of consanguinity and the prevalence of neurotic illness, this is taken to be an indication of a genetic factor's playing a significant role in the pathogenesis of the given neurotic condition. However, it is clear that psychological

TABLE 11-II

FAMILY MORBIDITY STUDIES -- CLINICAL DIAGNOSES

INVESTIGATOR	PATIENTS & DIAGNOSES	RELATIVES	RESULTS	Probable range of neurosis:	CONTROL RESULTS
1. Brown--1942	Hospital Outpatients: anxiety state: 63 hysterics: 21 obsessives: 20	1°: 573 2°: 1247	Neurotic = 16.4% + "anxious personality" Neurotic = 2.8% + "anxious personality"	16.8% = 16.4 to 33.2% 12.3% = 2.8 to 15.1%	189 1° relatives 1.1--11.2% of 39 controls 1.1--11.2%
2. Rüdin--1953	130 obsessive-compulsives		Compulsive neurosis rate: parents........ sibs.......... children......	4.6% 2.3% 1.3%	
3. Stenstedt--1966	"neurotic depression" 176	1°: 1242	Depression rate:..................... (higher than estimated population rate)	4.8%	
4. Oki--1967	"early childhood neurosis" (age 4-6) 52	Parents Sibs 2°	Neurosis & "nervousness"............. " " " "	45% 33% 18%	52 matched controls 4.2% 0 % 0 %
5. Guze--1967	hysteria: 35 females	35 females	Hysteria: 14% of "possible" 10%........	24%	167 pregnant females = 1.8% hysterics
6. Sakai--1967	65 obsessive-compulsives	65 families	(Results stated as % of families with one or more members showing diagnosis of)	Neurosis: 40% (schizophrenia 22%; psychopathy 40%; depression 11%)	
7. Tsuda--1967	69 "depersonalization neurosis"		(Results stated as % of families with one or more members showing diagnosis of)	Neurosis: 38.5% (schizophrenia 21%; psychopathy 29%; depression 4%)	

and familial factors as well may be as closely tied to the degree of consanguinity as are genetic factors, and that an interaction of environmental and genetic variables, or the operation of experiential facors alone, may be involved when a correlation between prevalence of neurotic illness and consanguinity is demonstrated. Studies of the type described above and summarized in Table 11-II provide very little leverage for separating these two sets of factors; thus it is not certain how much significant light they can cast upon the existence of, and the mechanism of, possible genetic pathogenic factors.

Paralleling these studies of family distribution of psychopathology undertaken by genetically oriented investigators is a series of investigations somewhat similar in design, but with a quite different goal. These are studies which attempt to relate psychiatric morbidity in the parents, particularly the mothers, to the psychopathology of the patient on the basis of interpersonal and experiential factors rather than genetic ones. A good recent example is the work of R. S. Britton.[37] He studied 100 children (64 males, 36 females) with neurotic behavior disorders who were consecutive referrals to an overseas Army child guidance clinic. He then examined the occurrence of psychiatric morbidity in the mothers of these children. In this careful investigation, Britton allocated the children into three major groups: *neurotic* (symptoms including anxiety, phobias, separation anxiety, compulsions, and mood disturbance), *conduct disorders* (lying, stealing, running away), and *mixed* (symptoms in both categories). They were also categorized in terms of severity of disorder and individual symptoms. Mothers were placed into three categories with respect to presence or absence of psychiatric symptoms: (1) negative maternal history, (2) history of psychiatric symptoms treated by general practitioner, and (3) maternal history of psychiatric referral. There was relatively little positive psychiatric history in the fathers, in part because of the screening-out process that occurs in the Armed Forces. The thirty mothers in group 3 (psychiatric referral) fell into three categories: a depressive group (11 neurotic and 1 psychotic depression), a neurotic group (9), and a personality disorder group (9). There was no significant relationship between which diagnostic category the child was in and the presence or degree of psychopathology in the mothers as reflected by their being in group 1, 2, or 3. Only one symptom in the children, separation anxiety, showed a statistically significant relationship ($p < 0.02$) to severity of maternal psychopathology, being significantly more frequent in children with mothers in group 3. The significant qualitative relationship between child and maternal diagnostic categories was as follows: the children of mothers of personality disorder suffered from conduct disorders or mixed disorders, while the children of neurotic and depressive mothers were preponderantly neurotic or showed combined neurotic and conduct symptoms. Significant anxiety was recorded in sixteen of twenty-one children in neurotic and depressive mothers but in only one of nine children of mothers with personality disorders, and there was a similar difference for phobias (ratio of 10 to 1). Conversely, aggressiveness and fire-setting were much more present in children of personality-disorder mothers. Mood disorder was significantly more common in children of depressive mothers (9 out of 12) as compared with the other groups combined (6 out of 21). Britton concludes that the presence of a history of psychiatric disturbance in the mothers of sixty-seven of one hundred consecutive referrals to a child guidance clinic supports the hypothesis that maternal psychological adjustment is the crucial pathogenic variable in childhood neurosis:[37]

> ... this study would seem to suggest that where other factors such as gross social disorder, paternal psychiatric illness, and anomalous parenthood are reduced, maternal psychiatric ill health plays an even greater part in determining childhood disorder, both quantitatively and qualitatively. The distribution of the disorders between predominantly female sufferers in one generation, and male in the next, make a genetic explanation for this unlikely. Perhaps it conforms more to the social heredity described by Hilgaard.[38]

It seems very likely that the same data as that obtained by Britton could be interpreted by genetically oriented investigators so as to emerge with a very different set of conclusions. There was a correlation between occurrence of overanxious, neurotic children and neurotic mothers, whereas conduct disorders occurred in children of chronically unstable mothers with personality disorders rather than neurotic disorders. These observations and the similar relationship between individual symptoms in the child and the mother (i.e. the relationship between child phobia and anxiety to maternal neurosis and depression versus child aggressiveness and fire-setting related to maternal immature personality), could be interpreted to emphasize genetic determination regarding form of psychopathology. It would appear that a research design which permits two contending major hypotheses to use the same data in order to support contending theoretical points of view presents serious methodological deficiencies.

Psychological Family Studies

The twin and family morbidity studies described above have been limited to those employing *clinical* diagnoses of neurosis or other related psychopathology in probands and their relatives. Similar study designs have been based on the use of psychological tests rather than clinical diagnosis. Using either the twin or family morbidity protocol, different investigators have attempted to measure psychological variables relevant to neurosis and then look for patterns of distribution which might clarify the causal role of genetic and environmental factors. Some such psychological tests attempt to measure the relative presence or absence of specific symptoms or personality dimensions relevant to neurosis; others use factor analysis or varied theoretical constructs to define a central variable of neuroticism itself. The use of tests increases the possibility of obtaining meaningful quantitative rather than qualitative data and allows investigators from disciplines other than psychiatry to study and contribute to these issues. On the other hand, no test is universally accepted as a valid measure of the clinical category being studied, and when contradictory results are reported, the extent to which this is a result of test inadequacy is never clear.

Family studies using psychological measures can be usefully grouped into the same two categories as the clinical studies reviewed above. One of these is the group of twin studies where the major datum sought is the relative degree of concordance in MZ as compared to DZ twin pairs. The other category is that in which the psychological test data is used to compare presumed incidence of neuroticism, or personality traits assumed to be closely related to neuroticism, in different groups of nontwin relatives.

Table 11-III, "Psychological Tests of Neuroticism in Twins," summarizes the results obtained in six studies in which investigators using twin populations have at-

TABLE 11-III

PSYCHOLOGICAL TESTS OF NEUROTICISM, IN TWINS

	INVESTIGATOR	YEAR	INSTRUMENT	VARIABLES	POP. (N)	STATISTICS	COMMENTS
1.	Carter	1933	Bernreuter	"Neurotic tendency" (+ 3 others)	MZ: 55 DZ: 43	intraclass r: .63 " r: .32	
2.	Newman, Freeman, Holzinger	1937	Woodworth-Matthews Personality Inventory (+ 3 others)	"Neurotic tendency"	MZ: 50 DZ: 50	" r: .56 " r: .37	separated MZ: r = .58 authors question positive trend; Eysenck (11) reinterpreted data
3.	Shields	1962	Questionnaire similar to Maudsley Personality Inventory	"Neurotic tendency" (+ 4 others)	MZ separated 44 MZ together 44 DZ separated 21) DZ together 21)	" r: .53 " r: .58 " r: .11	essentially the same whether used separately or together
4.	Gottesman	1963	a) MMPI	10 standard scales + Block (? general neuroticism)	MZ: 34 DZ: 34	Heritability: 4/10 scales (Social Introversion + 3 others), + Block 25% or <.05 b) 6/14: H = 25%	
			b) Cattel	14 factors			Factor C, general neuroticism; negative
5.	Gottesman	1965	MMPI	10 MMPI Scales	MZ: 83	5/10(depression - introversion) <.05	
6.	Wilde	1964	Personality Questionnaire	"Neurotic tendency" (+ 4 others)	MZ: 88 DZ: 42	Intraclass r: .53 " r: .11	Essentially the same whether used separately or together

tempted to determine the degree of genetic control of a factor loosely termed neuroticism as measured by a variety of psychological test instruments.[39, 40, 41, 42, 43, 44] (Eysenck and Prell's study, one of the more sophisticated of such twin studies, using a battery of performance tests from which to abstract a neuroticism factor, with quite significant positive results, is not included.[45] Rosenthal reports that Blewett was unable to replicate these results.[10] He was not able to identify any factor clearly indicating neuroticism, as Eysenck and Prell had, and found that his tests did not intercorrelate in the predicted manner.)

Since the samples, the psychological instruments, and the statistics employed vary from each other significantly, great caution must be observed in any effort to summarize or combine these results. In addition, the qualifications described above that pertain to clinical twin studies are relevant here, as well. With such cautions kept in mind, one may observe that in each of these studies, whether measured by the intraclass correlation coefficient or by a heritability index, there is a greater correspondence for psychologically tested and measured neuroticism in MZ than in DZ pairs. There are probably very few variables, however, which do not show greater concordance in MZ than in DZ pairs.

Slater and Cowie recently evaluated the statistical significance of the MZ:DZ differences for neuroticism and related traits (neuroticism, nervous tension, psychoneurotic complaints, psychosomatic complaints).[46] Basing their analysis on recent summaries by Vandenberg[47] and Shields,[48] they concluded that only one half of such reported differences between MZ and DZ concordances are statistically significant at the .05 level or better. (Significance of increased MZ concordance: 0.01: 3 − 0.05: 1 − n.s: 4). It should be noted, however, in Table 11-III that in each of the twin studies where MZ pairs raised separately were looked upon as a separate subgroup, there was no essential difference between them and MZ pairs raised together whereas DZ pairs showed substantially lower values.

There has been relatively little work using psychological tests of neuroticism in other relatives than pairs of twins. Coppen, et al. used the Maudsley Personality Inventory to study 266 patients and 735 of their first-degree relatives and spouses.[49] They derived ratings of neuroticism from this inventory. Contrary to their expectations, the neuroticism scores of the patients' relatives were not uniformly elevated, and correlation coefficients between many classes of relatives were, on the whole, low. Significant positive correlations of this kind, however, were found in certain pairings: between male patients and their male siblings, and between male patients' mothers and their children (+0.4). Fuller and Thompson describe results obtained with the Bernreuter Personality Inventory to measure neuroticism in varying intrafamilial groupings.[50] In contrast to the above-mentioned report of Coppen, Cowie, and Slater, where significant positive correlations were only found in the families of male patients, the positive correlations in this study were primarily between mother and daughter (+0.6) and to a lesser extent between sister-sister pairs (+0.36).[49] Mothers and sons showed a 0.32 correlation, and as in other studies, the fathers' correlation with offspring were substantially lower (0.24 with daughters and only 0.06 with sons). Eysenck, quoting the work of Lienert and Reisse,[51] describes a similar difference between the correlations between mother and child, on the one hand (0.31), and father and child, on the other (0.13).[11]

EVALUATION AND CONCLUSIONS

Rosenthal's summary of the significance of the body of empirical work attempting to clarify the role of genetic determinants in neurosis is a sound and well reasoned one, and worth quoting.

By and large, we are limited in our conclusions about the heredity issue in neurosis because of the sparseness of studies, the relative lack of variety, their failure to take various diagnostic precautions, and the difficulty involved in assessing the role of environmental factors. However, the overall evidence points to the likelihood that heredity plays a role in the development of psychoneurotic symptoms, but we can say very little about the genetics involved except that various polygenic systems may be involved in a more or less low-keyed way . . . (Twin study results leave) little doubt about the importance of environmental factors in the development of clinical neurosis. (They also indicate) with equal strength that an inherited diathesis contributes significantly to the likelihood the clinical neurosis of more or less degree will occur.[10]

It is important to keep in mind, however, that there have been repeatedly individual studies in which, for a given entity or test trait evaluated in a particular subject group, results have indicated little, if any, genetic contribution to the variance.

Despite multiple difficulties which have impeded attempts to structure tight empirical studies that might yield conclusive data, there is some emerging tendency toward a theoretical consensus among workers that represents a wide diversity of points of view. This consensus postulates a varying degree of genetic contribution to a significant number of individual personality variables, all of which are also subject to greater or lesser, and in some instances major, influence by environmental variables as well. Such genetically influenced variation—along a quantitative continuum of factors which influence levels of arousal, attentiveness, ability to mobilize focal attention, emotionality, and the like—would thereby significantly influence several consequent, and related phenomena.

1. The development of *personality traits* such as passivity or hyperactivity and extroversion and introversion.
2. The capacity to cope with stress, and thus the potential *breakdown point*, for any given individual.
3. The determination or selection of those sectors of the personality which, in the face of stress beyond the individual's ability to cope, would be likely to break down, and thus the type of *symptoms* and individual is predisposed to have.

Thus, the overall tendency to break down could be influenced both by either the patterning or the quantitative level of certain genetically controlled variables or the interaction thereof. That is, an extreme variation from the mean, along the dimension of one or a few especially salient variables, might have equivalent potential for disposition to neurosis as a lesser degree of variability or deviance from the mean for a larger number of genetically influenced variables. Slater and Slater initially suggested such a polygenic model, which assumes "plural dimensional variation of a quantitative kind in the predisposition to neurotic disorders dependent upon a multifactorial genetic basis" in 1944;[52] and Slater and Cowie have recently described a more contemporaneous version.[46] In the more recent statement, they point out that this model will have greater or lesser applicability in different neurotic conditions, varying with the degree of correlation of specific personality variables with clinical symptoms.[46]

A very wide span of theoretical orientation is represented by investigators who reached conclusions essentially similar to those discussed above, from Freud[53] at one end of the psychodynamic continuum to investigators such as Slater whose interest and involvement are more heavily concentrated in genetic rather than psychodynamic issues. Although Redlich and Freeman have concluded that the "importance of inherited characteristics in neurosis and sociopathies is no longer asserted except by Hans J. Eysenck and D. B. Prell,"[5] most investigators in this area appear to believe that genetic factors are relevant either to the predisposition to develop some types of neurosis or to influence the pattern of symptomatology which appears.

Despite this apparent increasing area of widespread agreement, however, the empirical data available suggests that, in some respects, work in this area is approaching a dead end. One reason for this concerns the methodological deficiencies in terms of diagnostic difficulties, and investigator bias, referred to above. Certain additional conceptual problems, which are now becoming clear, however, are of equal, if not of greater, importance.

Early conceptualization and studies in this area tended to pose the basic question as: What share of the determinants of a given behavior or a given psychopathology are genetic in origin, and what portion are (instead) environmental? In principle, this basic issue, the distinction between genetic and environmental determinants of neurosis, seemed like a simple but necessary one, with potentially profound implications for understanding pathogenesis and also for therapy. We might draw the analogy to a tape recorder that we are using in order to record some music. If what emerges on the first playback is at many points extremely distorted and occasionally painful to hear, these poor results could be due to two very different processes. Either the input we were recording was very distorted, or the input was good but there was something wrong with some inner part of the machine. If we should correct the problem and emerge with a high-quality recording, it would obviously be essential to be certain where the fault lies and what needs correcting. If the inputs (one's environment and experience) are basically at fault, no matter how we improve the mechanism itself, we will continue to get very poor quality tape. Conversely, if there is a crucial defective part within the mechanism, no matter how fine the input, the product will still be unsatisfactory.

Posing the issue in this way was historically a necessary and, in its time, a productive phase of our intellectual development. However, increasing knowledge and awareness suggests that such formulations may be approaching a point of diminishing returns and has the further incidental disadvantage of tending to polarize points of view and approaches. Ernst Caspari referred to this point in summarizing the presentations at the 1964 Social Science Research Council Symposium on "Genetic Diversity and Human Behavior."[54] He pointed out that the "problem of whether a certain behavioral character is determined by genes or by the environment, or even a partitioning of the variance into genetic and environmental variance, is therefore misleading from the point of view of understanding behavior." Rather, "the question, particularly with respect to human psychology, is how genetic individuality, which is a demonstrated fact, will express itself under the influence of diverse social and environmental conditions."

Two additional phenomena which have been demonstrated in various animal studies seriously complicate the problem of separating genetic and environmental factors in the manner sought by the early models. One derives from the surprisingly powerful adult consequences of relatively minimal experiential differences in infancy; the other, the fact that genetic and environmental influences sometimes work synergistically rather than antagonistically.

Fuller studied the interaction of genetic and experiential variables on the development of socialization in dogs.[55] In one such study, individual animals of two different breeds of dogs (terriers and beagles) were completely isolated from three to fifteen weeks of age. Such isolation was followed by substantial differences in behavior with other dogs and persons and in the frequent appearance of bizarre posture and episodes of approach avoidance, all of which taken together constituted a "post-isolation syndrome." The genotype had substantial impact upon the direction of some of these changes. In addition, however, it was noted that "relatively brief breaks in isolation counteract the effects of experiential deprivation to a high degree." Animals "given fifteen minutes handling per week in two equal periods are more like pet-reared than isolated animals." That is the presence or absence of a change in experience which occupied less than 0.15 percent of each week was able, in these two dog breeds, to move the animals' behavior out of the experimental and into the control group category. Rosensweig and his co-workers, Denenberg, and others, have also described series of studies in which handling for similarly brief periods of time in infancy can evoke major differences in adult behavior in laboratory animals.[56,57] If analogous processes occur in human development, such profound consequences for adult behavior of quantitatively minimal differences in early experience must be kept clearly in mind. They substantially increase the difficulty to be anticipated in systematically teasing out and evaluating the importance of early environmental factors and experiential influences which may occupy one of these remarkably small factions of daily or weekly experience and yet have major consequences in terms of resultant behavioral differences.

Of even greater potential significance was Fuller's observations in the same study concerning the relationship between genetic and environmental factors in their ability to modify postisolation behavior.[50] It was observed that of seven quantified variables of persistent post isolation behavior, it was those variables which showed the strongest breed differences (i.e. the clearest genetic loading) that were also the ones most significantly affected by varying the number of breaks in isolation. Conversely, the four variables which did not show a breed difference were, in general, significantly less affected by changes in the amount of experience. Fuller concluded:

> The lesson to be learned is that behavior cannot be divided into two exclusive classes, one genetically programmed and insensitive to environmental manipulation; the other experientially programmed and insensitive to changes in the genotype. Actually it seems that those behavioral characteristics which are most affected by genotypic differences are the very ones which are most easily shifted by differences in the amount of experience, and those characteristics which are unrelated to genotypic differences are less sensitive to changes in experience. It is suggested here that there is synergism rather than antagonism between genetic and environmental effects upon behavior.

Fuller also reports evidence that the same environmental experiences can have different, or even opposite, effects, depending on individual differences in genetic structure. As a result of these two observations, he concludes, "It is time we rid ourselves forever of the heredity and environment dichotomy and begin to study gene-experience synergisms."

Additional impediments to clarification in this area have been the ideological intensity and fervor of contrasting points of view. Genetically oriented workers, some of them clinicians and some not, have tended to be very skeptical of family dynamic research and assume that such formulations as social heredity, which are based on nonquantitative clinical material, cannot be taken too seriously. The repeated history of the development of provocative family interaction formulations, such as the "double bind" hypothesis, which are never definitively established by hard empirical data, has led naturally to such caution. However, there is much recent work which uses more precise and quantitative methodologies to amplify and sometimes to support certain family-interaction formulations which were initially perhaps too global when first presented. Examples include Reiss's quantitative analyses of family interaction patterns and the studies by Wynne and Singer of intrafamilial communication deviances.[58, 59, 60] A description and evaluation of the possible significance of studies such as these needs to become part of the frame of reference of genetically oriented investigators.

Conversely, there are continuing major obstacles which stand in the way of a broad readiness to evaluate impartially the possible role of genetic factors in the pathogenesis of psychopathology. The powerful current of subliminal opposition to the acceptance of the role of genetic determinants in American psychiatry and psychology has frequently been commented upon, but not substantially reduced. Part of this stems from the impact of contemporary events that have involved all of us as citizens of the world, as well as scientists. The impact of the Nazi use of the concepts of inborn racial differences, and contemporaneously the angry heat engendered in the controversy over the propositions put forward by Jensen, are examples of this source of opposition. At a more basic level, however, there is the continuing conflict between notions of genetic determinism and a basic aspect of the American culture and style. Hall and Lindzey have made this point well:

One important by-product of American democracy, the Protestant ethic, and the dogma of the self-made man has been the rejection of formulations implying that behavior may be innately conditioned, immutable, a "given." Because it is commonly accepted that physical characteristics are linked closely to genetic factors, the suggestion that physical and psychological characteristics are intimately related seems to imply a championing of genetic determinism. It is not surprising that such conception has been unable to muster much support in the face of the buoyant environmentalism of American psychology.[61]

The fact that many psychodynamically oriented investigators seem to experience genetic studies, both the concepts and the comparative quantitative rigor of some of the work, as intrinsically alien and threatening is in some ways surprising. Skepticism concerning the role of innate factors is not historically an aspect of psychodynamic or analytic theory. Freud emphasized the existence of a complementary set of determinants in the pathogenesis of the neuroses that notably included innate as well as learned mechanisms.[53]

At another level, there is an obvious need in practically all clinical studies in these areas to pay more attention to problems of methodologic and diagnostic rigor, spelling out precisely the criteria used. Equally important is the avoidance of diagnostic contamination by investigator bias through the use of additional independent raters who are blind to facts of family relationship, and if possible, to study hypotheses. Hopkinson and Ley, in a study of manic-depressive psychosis and endogenous depression, demonstrate one manner in which clarification of genetic and environmental interactions may result from increased methodological rigor.[62] They studied 200 consecutive admissions with these diagnoses, and then all first-degree relatives were evaluated by separate family history. Among the 105 secondary cases found in these relatives, the morbidity risk was significantly lower in the relatives of late-onset probands than those of early-onset probands. This fall in morbidity risk was not a gradual one but quite sudden and highly significant. This finding, plus their replication of a bimodal onset age-rate curve, strongly suggests two different predominant types of pathogenesis. The cases with earlier onset seem to demonstrate significantly more genetic components than those with late onset. This finding now opens the way to similar analysis of other psychiatric entities. There is reason to believe that similar clarifications can arise from further additions of precise and reliable data, as was obtained in this study, in future studies of neurosis and personality deviations.

No matter how great the increase in methodological rigor, however, it seems likely that new approaches will be required before the issues and problems in this area can be substantially clarified. Some of these new approaches and requirements can be suggested at this point. One will be the development of new "hard" indicators of neurosis, or new psychological or psychophysiological measures more clearly and convincingly related to the clinical questions at hand. Buchsbaum, et al., at the NIMH, are currently studying average evoked responses (a measure which, by computer averaging techniques, can tease out the ongoing EEG activity, the EEG response to a precisely timed stimulus) in a series of twin subjects. Preliminary evidence is at hand, suggesting that this measure may reflect certain fundamental dimensions of psychological functioning and cognitive style. Wynne and his coinvestigators have suggested that measures such as these and the variations in cognitive style they reflect can be usefully conceptualized as response predispositions which may show a relationship to the likelihood of development of neurosis or psychosis. The measurement of such well-defined psychophysiological variables, the determination of the degree of genetic control they manifest, and their relationship to the appearance of clinical psychopathology are approaches which may bring new movement in this area.

Another approach is one which uses identical twins as a subject population in which to study environmental rather than genetic factors. This design—which perceives monozygotic twins as uniquely valuable and rare, well-controlled, natural experiments, focuses on the search for nongenetic determinants of significant behavior or personality discordance between them—is exemplified by the NIMH studies of twin pairs who are discordant for schizophrenia.[63] A more recent study from the same group, using a similar basic design, is studying personality development in a longitudinal design.[64] Families enter the study before the birth of twins, at that point where obstetricians predict a multiple birth. Those families in which the twins, observed at delivery, are found to be monozygotic, are followed longitudinally. The focus is on defining the earliest persistent differences in behavior patterns between the developing infant twins and explaining these differences on the basis of ongoing observations of birth differences, life history differences, and differences in patterns of intrafamilial perceptions of and relationships to the two members of the twin pair. This design may help to substantially clarify the origins of certain specific behavioral features such as anxiety proneness and passivity and early expressions of symptoms such as compulsive tendencies which have a relationship to, or constitute part of, the pattern of subsequent neurotic behavior.

Finally, however, it seems likely that whatever new, improved clinical and research methods can and will be devised, significant basic science progress—developments of fundamental knowledge which go beyond symptom description and instead relate neurosis to newly discovered aspects of CNS and endocrine function—will be required before truly definitive findings become available concerning the interrelationship of genes and environment in the pathogenesis of neurosis.

REFERENCES

1. Pollin, W., Allen, M., Hoffer, A., Stabenau, J., and Hrubec, Z.: Psychopathology in 15,909 pairs of veteran twins. *Am J Psychiatry, 126*:597, 1969.
2. Cullen, W.: *First Lines of the Practice of the Physic,* 3rd ed. Edinburgh, Creech, 1781.
3. American Psychiatric Association: *Diagnostic and Statistical Manual of Mental Disorders,* 2nd ed. Washington, D. C., American Psychiatric Association, 1968.
4. U. S. Department of Health, Education and Welfare: *International Classification of Diseases,* Adapted for Use in the United States. Public Health Service Publication No. 1963, 8th Revision. Washington, D. C., 1967.
5. Redlich, F., and Freedman, D.: *The Theory and Practice of Psychiatry.* New York, Basic, 1966.
6. Srole, L., Langner, T.S., Michael, S.T., Opler, M.K., and Rennie, T.A.C. (Eds.): *Mental Health in the Metropolis.* New York, McGraw, 1962, vol. I.
7. Hagnell, O.: *A Prospective Study of the Incidence of Mental Disorder.* Stockholm, Svenska Boknorlaget Norstedts-Bonniers, 1966.
8. Nielsen, J., Wilsnack, W., and Strömgren, E.: Some aspects of community psychiatry. *Br J Prev Soc Med, 19*:85, 1965.
9. Rosenthal, D., and Frank, J.: The fate of psychiatric clinic outpatients assigned to psychotherapy. *J Nerv Ment Dis, 127*:330, 1958.
10. Rosenthal, D.: *Genetic Theory and Abnormal Behavior.* New York, McGraw, 1970.
11. Eysenck, H.J.: *The Biological Basis of Personality.* Springfield, Thomas, 1967.
12. Jinks, J.L., and Fulker, D.W.: Comparison of the biometrical, genetical, MAVA and classical approaches to the analysis of human behavior. *Psychol Bull, 73*:311, 1970.
13. May, J.: Note on the assumptions underlying Holzinger's h^2 statistic. *J Ment Sci, 67*:466, 1951.
14. Darlington, C.: *Heredity and Environment.* Proc 9th Internatl Congr Genetics, Caryologia, 1954.
15. Darlington, C.: Psychology, genetics and the process of history. *Br J Psychol, 54*:292, 1963.
16. Kety, S.S.: Biochemical theories of schizophrenia. *Science, 129*:1528 and 1590, 1959.
17. Jackson, D.D.: A critique of the literature on the genetics of schizophrenia. In Jackson, D.D. (Ed.): *The Study of Schizophrenia.* New York, Basic, 1959.
18. Scarr, S.: Environmental bias in twin studies. *Eugen Q 15*:34, 1968.
19. Stumpfl, F.: Untersuchungen an psychopathischen Zwillingen. *Z Ges Neurol Psychiatry, 158*:480, 1937.
20. Slater, E.: *Psychotic and Neurotic Illnesses in Twins.* London, HM Stationery Office, 1953.
21. Shields, J.: Personality differences and neurotic traits in normal school children. *Eugen Rev, 45*:213, 1954.

22. Slater, E.: Hysteria 311. *J Ment Sci,* 107:359, 1961.
23. Ihda, S.: A study of neurosis by twin method. *Psychiatr-Neurol Jap,* 63:861, 1961.
24. Braconi, Z.: Le psiconeurosi e le psicosi nei gemelli. *Acta Genet Med Gemellol,* 10:100, 1961.
25. Tienari, P.: Psychiatric illnesses in identical twins. *Acta Psychiatr Scand,* 39: Suppl 393, 1963.
26. Inouye, E.: Similarity and dissimilarity of schizophrenia in twins. *Proc 3rd World Congr Psychiatry,* 1:524, 1961.
27. Parker, N.: Twin relationships and concordance for neurosis. *Proc 4th World Congr Psychiatry,* 2:1112, 1966.
28. Hoffer, A., and Pollin, W.: Schizophrenia in the NAS-NRC panel of 15,909 veteran twin pairs. *Arch Gen Psychiatry* 23:469, 1970.
29. Brown, F.W.: Heredity in the psychoneuroses (Summary). *Proc Roy Soc Med,* 35:785, 1942.
30. Stenstedt, A.: Genetics of neurotic depression. *Acta Psychiatr Scand,* 42:398, 1966.
31. Rüdin, E.: Ein Beitrag zur Frage der Zwangskrankheit, insbesondere ihre Hereditären beziehungen. *Arch Psychiatr Nervenkr,* 191:14, 1953.
32. Oki, T.: A psychological study of early childhood neuroses. In Mitsuda, H. (Ed.): *Clinical Genetics in Psychiatry.* Tokyo, Igaku Shoin, 1967.
33. Guze, S.B.: The diagnosis of hysteria: What are we trying to do? *Am J Psychiatry,* 124:491, 1967.
34. Sakai, T.: Clinico-genetic study on obsessive-compulsive neurosis. In Mitsuda, H. (Ed.): *Clinical Genetics in Psychiatry.* Tokyo, Igaku Shoin, 1967.
35. Tsuda, J.: Clinico-genetic study of depersonalization neurosis. In Mitsuda, H. (Ed.): *Clinical Genetics in Psychiatry.* Tokyo, Igaku Shoin, 1967.
36. Pollin, W.: Control and artifact in psychophysiological research. In Roessler, R., and Greenfield, N. (Eds.): *Physiological Correlates of Psychological Disorder.* Madison, U of Wis Pr, 1962.
37. Britton, R.S.: Psychiatric disorders in the mothers of disturbed children. *J Child Psychol Psychiatry,* 10:245, 1969.
38. Hilgard, J.R.: Sibling rivalry and social heredity. *Psychiatry,* 14:375, 1951.
39. Carter, H.D.: Twin similarities in personality traits. *J Genet Psychol,* 43:312, 1933.
40. Newman, H.H., Freeman, F.N., and Holzinger, K.J.: *Twins, A Study of Heredity and Environment.* Chicago, U of Chicago Pr, 1937.
41. Shields, J.: *Monozygotic Twins Brought Up Apart and Brought Up Together.* London, Oxford U Pr, 1962.
42. Gottesman, I.I.: Heritability of personality: A demonstration. *Psychol Monogr,* 77:1, 1963.
43. Gottesman, I.I.: Genetic variance in adaptive personality traits. Paper read at annual meeting of American Psychological Assn, Division of Developmental Psychology, Chicago, 1965.
44. Wilde, G.J.S.: Inheritance of personality traits. *Acta Psychol, (Amst),* 22:37, 1964.
45. Eysenck, H.J., and Prell, D.: A note on the differentiation of normal and neurotic children by means of objective tests. *J Clin Psychol,* 8:202, 1952.
46. Slater, E., and Cowie, V.: *The Genetics of Mental Disorders.* London, Oxford U Pr, 1971.
47. Vandenberg, S.G.: Hereditary factors in normal personality traits (as measured by inventories). In Wortis, J. (Ed.): *Recent Advances in Biological Psychiatry.* New York, Plenum Pr Plenum Pub, 1967, vol. IX.
48. Shields, J.: Heredity and psychological abnormality. In Eysenck, H.J. (Ed.): *Handbook of Abnormal Psychology,* 2nd ed. London, Basic Books, 1971.
49. Coppen, A., Cowie, V., and Slater, E.: Familial aspects of "neuroticism" and "extraversion." *Br J Psychiatry,* 111:70, 1965.
50. Fuller, J.L., and Thompson, W.R.: *Behavior Genetics.* London, Wiley, 1960.
51 Lienert, G.A., and Reisse, H.: Ein korrelationsanalytischer beitrag zur genetischen determination des

neurotizismus. *Psychol Beitr,* 7:121, 1961.
52. Slater, E., and Slater, P.: A heuristic theory of neurosis. *J Neurol Psychiatry,* 7:49, 1944.
53. Freud, S.: Introductory lectures on psychoanalysis, I and II. *Standard Edition.* London, Hogarth Press, 1961, vol. XV and XVI.
54. Caspari, E.: Behavioral consequences of genetic differences in man: a summary. In Spuhler, J.S. (Ed.): *Genetic Diversity and Human Behavior.* Chicago, Aldine, 1967.
55. Fuller, J.L.: Genetic influences on socialization. In Hoppe, R.A., Milton, A., and Simmel, E. (Eds.): *Early Experiences and Processes of Socialization.* New York Acad Pr, 1970.
56. Rosenzweig, M.R.: Role of experience in development of neurophysiological regulatory mechanisms and in organization of the brain. In Walcher, D., and Peters, D. (Eds.): *Early Childhood: The Development of Self-Regulatory Mechanisms.* New York, Acad Pr, 1971.
57. Denenberg, V.H., and Zarrow, M.X.: Effects of handling in infancy upon adult behavior and adrenocortical activity: Suggestions for a neuroendocrine mechanism. In Walcher, D., and Peters, D. (Eds.): *Early Childhood—The Development of Self-regulatory Mechanisms.* New York, Acad Pr, 1971.
58. Reiss, D.: Individual thinking and family interaction: II. A study of pattern recognition and hypothesis testing in families of normals, character disorders and schizophrenics. *J Psychiatr Res,* 5:193, 1967.
59. Reiss, D.: Varieties of consensual experience: Contrast between families of normals, delinquents and schizophrenics. *J Nerv Ment Dis, 152*:73, 1971.
60. Singer, M.T., and Wynne, L.C.: Family transactions and schizophrenia. In Romano, J. (Ed.): *The Origins of Schizophrenia.* Proc of the First Rochester International Conference. Amsterdam, Excerpta Medica, 1967.
61. Hall, C.S., and Lindzey, G.: *Theories of Personality,* New York, Wiley, 1957.
62. Hopkinson, G., and Ley, P: A genetic study of affective disorder. *Br J Psychiatry, 115*:917, 1969.
63. Pollin, W., and Stabenau, J.: Biological, psychological and historical differences in a series of monozygotic twins discordant for schizophrenia. In Rosenthal, D., and Kety, S.S. (Eds.): *The Transmission of Schizophrenia.* New York, Pergamon, 1968.
64. Cohen, D., Allen, M., Pollin, W., Inoff, G., Werner, M., and Dibble, E.: Personality development in twins: Competence in the newborn and preschool periods. *J Am Acad Child Psychiatry* in press.

Chapter 12

GENETICS AND HOMOSEXUALITY

John D. Rainer

Homosexuality, in both sexes, currently receives a large amount of attention as a social, legal, and psychiatric problem. Both in the medical literature and in the lay press, statements by those who treat homosexuals, as well as by homosexuals themselves, offer many theories of explanation, often manifestly oversimplified. It is not the task of this chapter to set forth another neat, incomplete, and unproved hypothesis but rather to explore the evidence for the existence of interacting genetic components and the possible ways in which they may play their part.

PREVALENCE AND CLASSIFICATION

Historically, overt homosexual behavior has existed in many societies with various degrees of social acceptance, but the extent is not accurately known. More recently, prior to the publications by Kinsey and his associates, it was generally believed that homosexuality in Western societies was confined to under 10 percent of the population.

After interviewing 5300 males and 5940 females, the Kinsey group arrived at incidence figures which were considerably higher.[1,2] Their findings, presented in 1948 and 1953, deserve scrutiny not only because of the social significance of their demographic data, but because of the implications of their definitions, assumptions, and etiological conclusions.

It will be remembered that the procedure used by this team was based on group samples rather than probability samples of the entire population. An attempt was made to obtain the cooperation of groups of individuals and then interview as many members of each group as possible. The interview rate within each group varied from 100 percent in one quarter or less of the groups to "something between 50 and 90 percent" in a "considerable proportion of the rest of the sample."

Kinsey and his co-workers obtained for each sex two sets of figures in contrast to the categories used in earlier estimates of homosexual behavior: one for homosexual response and one for homosexual experience. The former category included "erotic arousal," or "psychic response," to homosexual stimuli; the latter, actual contacts, whether leading to orgasm or not. In the case of psychic response, there was an accumulative incidence by the age of 45 of 50 percent in men and 28 percent in women.

Parts of this chapter are from J. D. Rainer, "Genetische Gesichtspunkte bei der Homosexualität," *Deutsche Medizinische Wochenschrift*, 87:377 (1962).

Rating Scale

Of particular interest was the design of a heterosexual-homosexual rating scale based both on psychological reactions and amount of overt experience. In this scale, 0 = entirely heterosexual; 1 = largely heterosexual, but with incidental homosexual history; 2 = largely heterosexual, but with a distinct homosexual history; 3 = equally heterosexual and homosexual; 4 = largely homosexual, but with a distinct heterosexual history; 5 = largely homosexual, but with incidental heterosexual history; 6 = entirely homosexual.

While 90 percent or more of married persons fell into category 0, the distribution of unmarried individuals at age twenty and at age thirty-five is shown in Table 12-I.

As to actual homosexual contact, the investigators reported that of individuals aged forty-five years and older, 37 percent of males and 13 percent of females had at least some homosexual experience. The incidence was greater in the case of unmarried persons (male, 50%; female, 26%). Apart from reversing the ratio of male to female homosexuality which was previously widely held, Kinsey and his group found homosexual choice in both sexes to be much more widespread than formerly believed.

The basic presuppositions underlying the collection of these data are of interest. The assumption was made that there is in each individual a heterosexual-homosexual balance with continuity from exclusive homosexuality to exclusive heterosexuality. With this balance to be measured by means of the seven-category rating scale, psychosexual response and overt experience were combined in the scale because they were assumed to be somewhat interchangeable.

Kinsey Hypothesis

The theories advanced by the Kinsey group on the etiology of homosexuality in both sexes were formulated in the second volume, dealing with female sex behavior. For want of evidence, hormonal, instinctive, hereditary, and psychoanalytic hypotheses were discounted and homosexual behavior attributed to the following: (1) the basic physiologic capacity of any mammal to respond to any sufficient sexual stimulus, (2) the conditioning effect of the first accidental sexual encounter with an individual of the same sex, and (3) the subsequent reinforcing or extinguishing effect of other persons' opinions and social codes.

To account for the fact that despite this equipotentiality, heterosexual contacts

TABLE 12-I
DISTRIBUTION OF RESPONSES BY SINGLE INDIVIDUALS
ACCORDING TO HETEROSEXUAL-HOMOSEXUAL SCALE (0-10).

Scale Rating	Male		Female	
	Age 20	Age 35	Age 20	Age 35
X (no sociosexual response)	3	4	14	19
0	53	78	61	72
1—6	18	42	11	20
2—6	13	38	6	14
3—6	9	32	4	11
4—6	7	26	3	8
5—6	5	22	2	6
6	3	16	1	3

tend to occur more frequently in man and all other mammalian species than homosexual ones, the investigators listed four additional conditions facilitating heterosexuality. First, the greater aggressiveness of the male was seen as a "prime factor in determining the roles" played by each sex. As a rule, similar levels of aggressiveness between individuals of the same sex were assumed to preclude homosexual mounting in animals.

Secondly, there was the greater ease of intromission in heterosexual contact with more of the "satisfactions which intromission may bring in a heterosexual relationship." The third was thought to be the presence of "olfactory and other anatomic and physiologic characteristics" differentiating the sexes, and fourth, the psychological conditioning effects of the "more frequently successful heterosexual contacts." While presented as incidental, these given aspects of sexual behavior may be of considerable importance and will be referred to later.

Kinsey and his associates concluded that the sexual behavior of any given human being could be described as a conditioning process, although they were unable to determine why it is that every individual does not engage in every type of sexual activity.

ETIOLOGICAL FORMULATIONS

The basis for differences in human sexual behavior has been formulated in the frameworks of a quasi-accidental (sociogenic) conditioning theory, of a perinatal hormonal conditioning theory, of a strictly genetic theory, and of particular psychodynamic theories. The eventual etiological synthesis will have great value for prevention, treatment, and sociolegal approaches. First, a brief review will be presented of the different theories of homosexuality from a biological or biosocial viewpoint.

Theories of Equipotentiality with Accidental and Sociogenic Conditioning

Based on statistical studies of human societies and the observation of animal behavior, the theory of equipotentiality implies that all individuals have a basic capacity for both heterosexual and homosexual responses; that most individuals in their psychological leanings and behavioral trends fall somewhere on a continuum between the two extremes; and that limitation of sexual response is determined by a combination of cultural conditioning and individual experience. Homosexual behavior is found in all human societies, in those that disapprove of it as well as those that sanction it as one of their social institutions. An assumed analog is observed in animals (though often as part of a dominance-submission pattern).[3]

Attempting to place this theory on a biological foundation, its proponents point to the anatomical identity of the two sexes in early embryological stages and to the homologies seen between the sexual apparatus in adulthood (if rudimentary portions are included). This bisexuality is then equated with equipotentiality for sexual behavior in early childhood (polymorphous perverse stage); and, by way of an analogy, the ultimate form of psychosexual development is attributed to a balance of conditioning forces. Since the aim of sexual behavior is seen only in achieving orgasm, this theory in its simplest form minimizes biological differences in the methods used for this purpose, placing the emphasis on early imprinting experiences which are usually accidental.

Theories of Prenatal and Neonatal Hormonal Influence

There is a body of experimental evidence indicating that androgenic hormone (testosterone) administered to female guinea pigs or rats in the perinatal period predisposes to masculine behavior at maturity, while male rats exposed to antiandrogenic drugs prenatally or castrated at birth show a reduced organization of the subsequent male response.[4] As stated by Harris and Levine:

It would seem that during maturation, the central nervous system passes through a critical period in which organization of the neural mechanism concerned with sexual behavior patterns in the adult is determined. During this period, which would seem to involve fetal life in the guinea pig and early neonatal existence in the rat, the administration of testosterone to a genetically constituted female leads to disorganization of the nervous elements underlying the expression of female behavior and to a differentiation or an increased sensitization of the elements underlying the expression of male behavior. ... It seems likely that some neural structure in the male becomes differentiated and fixed in its function under the influence of endogenous androgens during critical periods in development. It would appear that the effect of neonatal estrogen treatment (to the genotypic male) is to produce a functional castration by some direct action of the neonatal testis.[5]

Dörner and his co-workers extrapolated similar observations to humans and suggested that absolute or relative androgen deficiency during the critical period of fetal life in the genotypic male when presumed anatomically or functionally distinct hypothalamic mating centers develop would lead to a preponderant development of the "female" center.[6,7] There would subsequently result a predominantly female-like expression of the sex drive since, at puberty, the normally produced androgen acts upon both centers as a "sex-unspecific" activator. A similar hypothesis was formulated for the female—androgen excess in fetal life favoring the male center and predisposing to later male-like behavior. As Dörner puts it: "In the normal development of man, testosterone represents the mediator between the sex chromosomes and the neural differentiation of sexual behavior."

As for the source of the androgen imbalance in fetal life, Ward proposed an environmental stimulus.[8] Male rats were exposed to prenatal stress by severe restraint of their mothers during gestation; later, as adults, these rats showed reduction of male-like and enhancement of female-like sexual behavior. Ward attributed these results to the production, under stress, of the weak adrenal androgen, androstenedione, by the maternal or fetal adrenal cortices, or both. This hormone would compete for the same receptor sites, possibly in the central nervous system, with a reduced quantity of the more potent androgen, testosterone; "the resulting behavior potentials would be expected to resemble those obtained by other experimental manipulations which decrease functional testosterone titers." The author suggested that the system might provide the basis for an effective population control mechanism under a stressful environment.

Psychoanalytic Theories

More complex in their analysis of psychological and social correlates of homosexuality than the conditioning or imprinting theories are the various psychoanalytic formulations. While such theories are not discussed here at length, some of the insights gained from them may help toward understanding the mechanisms of development and interaction in a genetically oriented frame.

Psychoanalytic theories are characterized by their description of an ontogenetic scheme of human psychological development, usually in a family setting, with a line of continuity between the early, largely unconscious or unformulated needs and wishes of the infant and child, and the mental, emotional, and behavioral life of the adult.

According to some theoretical concepts, there is initial freedom of sexual choice, a prerequisite for the ability to develop any sexual feelings. In the male such feelings become allied to his love for his mother who is the first to feed and care for him. In the female, the course of events is more complex. According to Freud, there may arise disappointment in the mother who is held responsible for having deprived the daughter of the male organ. When the female's primitive erotic feelings are then transferred to the father, the groundwork is laid for heterosexual love.

Concurrent with or following this development, the child passes through the stormy Oedipal period in which its presumably eroticized feelings toward the opposite-sex parent must weather the real or fantasied threat of punishment, specifically genital damage or castration, from the parent with whom these developments have placed it in competition. According to these theories, homosexuality is largely the result of excess fear generated during the given childhood period. The association of fear with sexual desires tends to forestall advances towards members of the opposite sex.

While such conceptual schemes may throw some light on the motivation for homosexuality, they do not explain why this solution presents itself only to certain persons and only in them leads to overt manifestations. Many students of psychodynamics admit to insufficient knowledge regarding this age-old problem of symptom choice.

Psychoanalytic theories differ as to the role ascribed to constitutional factors in homosexuality. In some schools, the bisexuality concept is elaborated, describing both masculine and feminine components in the instinctive sexuality of all infants. Their relative innate strength may either be exclusively responsible for a homosexual fixation, or contribute to the regression following intimidation as previously described. In other theories these factors are minimized while the role of defective parent-child relationships is stressed. Here, there is interference with the ability to identify with the parent of the same sex if this parent is hated, feared, ineffectual, or absent or if the other parent is overprotective, dominant, or seductive. Of course, it should be observed that many of the descriptions of their parents have come from homosexual patients themselves. Just as Freud heard from his early hysterical patients that they all had suffered sexual trauma at home in infancy, only to find out later that these stories were mainly fantasies created out of their fears and wishes, so it is quite possible that if there is a characterological predisposition to homosexuality, some of the attributes of the vulnerable individual may cause him to see his parents in a typical configuration. Alternatively, it must be considered that a particular child may elicit from his parents certain behavior on their part rather than simply be the victim of it.

It is possible to consider also that life experience may interact with intrinsic predispositions for the development of homosexuality which are not specifically sexual in nature, such as early defects in the ability to form identifications or relationships or specific lags in personality integration or maturation.

Predominantly Genetic Theories

Nuclear Group

A different trend in the search for the etiology of homosexuality is based on the hypothesis that at least some homosexuals are anatomically or hormonally unique, the deviation in their sexual choice being associated with an inborn intersexual constitution. In the opinion of such outstanding psychiatrists as Krafft-Ebing, Kraepelin, and Bleuler, this nuclear (central) group of homosexuals differed from the average representative of their respective sex in physical as well as mental makeup, resembling in these respects an individual of the opposite sex.

For example, a distinct tendency to engage in occupational pursuits more typical of the opposite sex has been noted in various surveys of homosexuals. Some homosexuals may show a scanty growth of body hair with a feminine pubic hair pattern, a feminine distribution of body fat, and high voices. Measurements of the pelvic bones have been shown to deviate from the norm in some homosexuals of either sex. On the whole, it has not been established that such grossly deviant individuals coincide with the small group (4%) found by Kinsey to be exclusively homosexual throughout life. Nevertheless, it is the belief in the existence of such a nuclear group that has led geneticists to search for a basic constitutional mechanism somehow related to chromosomes and genes.

Early genetic theories were derived from the finding of an apparent excess of homosexuals among the relatives of known cases of overt homosexuals. For example, Hirschfeld found fraternal evidence of the trait (brothers) in 35 percent of a group of homosexual males.

Chromosomal Basis of Sex Determination

The old hypothesis that at least some homosexuals represent a specific genetic abnormality has been investigated in connection with the chromosomal basis of sex determination. Soon after the rediscovery of Mendel's work in 1900, it was observed by geneticists working with *Drosophila* that both male and female flies possessed three like-membered pairs of chromosomes (autosomes), with the fourth pair being like-membered only in the female. In the male, this pair consisted of one rod-shaped chromosome, like the two that characteristically occur in the female, and one J-shaped chromosome. The former came to be known as the X chromosome and the latter, first called the accessory chromosome, was soon labeled the Y chromosome. Hence, ova had to be X-bearing, while spermatozoa could either be X-bearing or Y-bearing with equal probability.

The sex of a new individual, according to this simple scheme, is determined at the moment of fertilization and depends upon whether the egg is fertilized by an X-bearing or Y-bearing spermatozoon to produce a female or male, respectively. In man, the mechanism is much the same, although there are twenty-two pairs of autosomes, and the Y chromosome differs in appearance from the X by its smaller size and more distal position of the centromere.

Embryonic Development

In normal human sex development, the zygote possessing either the XX or XY pair develops in an orderly fashion into a female or a male. For the first six or seven weeks, the sex of the growing embryo cannot be distinguished morphologically. At this stage all embryos possess gonadal tissue consisting of a testis-like medulla and an ovary-like cortex, ducts of both the Müllerian (female) and Wolffian (male) type, and an undifferentiated urogenital region.

The first noticeable effect of the basic chromosomal sex pattern is the transformation of the neutral gonad into either a testis or an ovary as a result of enlargement of medulla or cortex, respectively, and atrophy of the other part. This step is followed by differentiation of the internal sexual ducts. In the female, the Müllerian ducts become the oviducts, uterus, and the upper part of the vagina; in the male, the Wolffian ducts become the sperm ducts. The opposite-sex system of ducts degenerates, leaving small remnants.

Similarly, the embryonic urogenital sinus and external genitalia differentiate into vagina, clitoris, and labia in the females; urethra, penis, and scrotum in the male. So-called secondary sex characteristics—voice, hair, body build, breasts—differentiate only at puberty under the net effect of interacting hormones of the gonadal and other endocrine glands. While the direct influence of the sex chromosomes in normal sex development apparently ends with the early differentiation of the gonad, the entire course of subsequent anatomic development is hormonally controlled.

Further insight into the process of sex development has been obtained from the study of intersexuality in animals. In cattle, for example, hormonal control is evident when fusion takes place between the fetal membranes of unlike-sex twins and hormones are carried across by the cross-circulation of the blood. The female member of such a pair regularly develops into a sterile and masculinized freemartin. Similarly, sex reversal in the chicken may follow destruction of the ovary by disease and subsequent development of a rudimentary testis.

Goldschmidt's Balance Theory

On the chromosomal level, by a classic series of experiments on *Lymantria*, the gypsy moth, Goldschmidt formulated the *balance theory* of sex determination in 1912.[9,10] This theory represented a refinement of the knowledge about chromosomal sex determination already current. *Lymantria* was found to differ from *Drosophila* in the fact that the possession of two X chromosomes characterized the male, one X characterized the female. The female-determining genes (F) in the Y chromosome were assumed to predominate over the male-determining genes (M) in one X chromosome, while two X chromosomes were thought to be sufficient to tip the balance in favor of the male-determiners.

By crossbreeding various strains of moths, however, Goldschmidt produced a set of intersexes which possessed two X chromosomes as well as female characteristics in a graded series from complete sex reversal to almost complete malehood. In these intersexes, the male determiners were found to be weak. Combination of a strong F with two weak M's made the genetic males intersexual, despite a normal male chromosome complement. Similarly, intersexual females were produced who combined one weak F with one strong M.

In *Drosophila*, a comparable situation was described by Bridges. Here, it appeared that the male determiners were in the autosomes, the female determiners in the X chromosomes. Hence, a fly with two X chromosomes and three, instead of two, sets of autosomes, developed into an intersex instead of a normal female.

In line with observations of this kind, Goldschmidt proposed his balance theory. The gist of the theory was that "the sex determiners of one sex are located in the X-chromosomes, those of the other sex outside, and that the quantities of two in the X-chromosomes win out over the action of the ever constant determiners of the other sex, while one quantity does not succeed in doing so."[10] The balance can be disturbed by increasing the relative number of determiners of one sex or the other, as in the triploid *Drosophila* with an extra set of male determiners. It can also be disturbed if the determiners of either sex are present in a weak form as in *Lymantria*. In either case an intersex develops.

The development of sexuality was attributed by Goldschmidt to the quantitative effect of the dosage of M and F factors which control the relative velocity of two simultaneous and competing chains of reactions. In the normal case, balance between the M and F dosage controls the process by which one or the other product of these reaction chains reaches a threshold of action and determines the sex alternative. With an imbalance, development proceeds first according to the chromosomal sex. Subsequently, however, the relative velocities of the two reaction chains are changed and "the wrong one (in regard to chromosomal constitution) overtakes the right one" at a particular time called the "turning point" (Drehpunkt). The time when the turning point is reached is assumed to be responsible for the degree of intersexuality, ranging from low-grade intersexuality with a late turning point to complete sex reversal if the turn occurs very early.

Sex Chromatin

The balance theory has been revised and extended to sex-determination in the human and may help to explain the development of some variations in sexual development since in recent years our understanding of the chromosomal basis of sex-determination in man has been considerably advanced. Thanks to the development of techniques for staining and counting human chromosomes, it is now possible to observe quantitative and qualitative variations in these structures. Even before this advance, interest in the problem was kindled by the discovery of Barr that the nuclei of somatic cells obtained from normal females present a darkly-stained chromatin body characteristically adjoining the nuclear membrane.[11] This body is absent or rare in cells obtained from males.

Together with the observation of sex chromatin in nondividing cells from many tissues came the finding of knob-shaped protrusions, or drumsticks, connected to the lobes of polymorphonuclear leukocytes in the female.[12] For a time, sexually underdeveloped males of the Klinefelter type, who for the most part are chromatin positive as they show the sex chromatin in their cells, were considered to be chromosomal females. In contrast, sexually underdeveloped females of the Turner type, who do not possess the chromatin (being chromatin negative) were termed chromosomal males.

Human Karyotype and Sex Determination

The incompleteness of this formulation became clear in 1959 when various groups cultured cells from these developmental sex deviants and counted the stained chromosomes in dividing cells.[13,14] In addition to two X chromosomes, phenotypically male Klinefelter patients were found to have a Y chromosome. Turner patients have only one X chromosome, but no Y chromosome; phenotypically, they are generally females. It appeared, therefore, that in man the possession of two X chromosomes results in the appearance of the chromatin body in interphase nuclei while the possession of a Y chromosome is generally necessary and sufficient to produce a male phenotype.

The balance theory as applied to human beings thus assumes the following form: the phenotypic sex (type of gonad) is the result of the balance of sex-determining genes on all the chromosomes. Male-determining genes are strongly present on the Y chromosome and probably on other chromosomes as well. Female-determining genes are probably on the X chromosome and may also exist on the autosomes. Normally, the strong male determiners on the Y chromosome shift the balance in the direction of maleness. Without this chromosome, female determiners tend to exert the stronger influence so that the individual develops as a female.

The development in man of the analogue to the intersex may take place in two ways. In the first, the imbalance may be due to the absence of a chromosome or the presence of one or more extra chromosomes. In Turner's syndrome, apparent females lack ovarian tissue and are sexually underdeveloped. Possessing only one X chromosome (XO), they are assumed to have an insufficient number of female-determining genes. On the other hand, males with underdeveloped testes and other eunuchoid features (Klinefelter's syndrome) often have an additional X chromosome (XXY). Although there are sufficient male-determining genes on the Y chromosome to produce both male gonads and male external genitalia, the extra X chromosome exerts an influence in the direction of femaleness and produces an intersex. Some true hermaphrodites have been described as being distinguished by mosaicism (some cells XO, some XY).[15]

On the other hand, some intersexes with a normal chromosome complement arise from hormonal influences. One example is the XX female with ovaries but hypertrophied clitoris and urogenital fusion seen in the adrenogenital syndrome. Another is the XY male with abnormal testes but female sex organ morphology and habitus seen in the testicular feminization (or androgen-insensitivity) syndrome, transmitted by a gene which is probably sex-linked. (For a possible regulatory system, see Ohno.[16])

Concluding this excursion into the present state of the theory of sex differentiation, it must be noted that attempts to explain homosexuality as related to intersexuality, either of the chromosomal or hormonal type, have not been generally successful. Neither the girl with Turner's syndrome nor the boy with Klinefelter's is typically homosexual; the former is usually stable and maternal, the latter usually has a lack of sexual drive with only occasionally a homosexual direction.

Sex Ratio in Sibships of Homosexuals

In 1916, on the basis of the balance theory, Goldschmidt proposed that homosexuality, at least the congenital (angeborene) type, represented a form of intersexuality.[9] As will be seen later, he changed his opinion by 1931. However, several other investigators set out in an ingenious way to test his original theory by statistical means. They hypothesized that some male homosexuals, particularly the effeminate ones, may actually be females with two X chromosomes and no Y, but with a set of weak female-determining genes. In that case, a greater proportion of males would be expected among the siblings of male homosexuals than among those of normal group of males.

While the normal sex ratio is about 106 males to 100 females, in the siblings of 1777 male homosexuals, ascertained with the help of the police of Munich and Hamburg, it was reported by Lang to be 125.8:100.[17] In Breslau and Leipzig, Jensch found the sex ratio for the siblings of 2072 male homosexuals was 113.9:100.[18] It was concluded, therefore, that in the observed groups of homosexuals there were *some* who actually were chromosomal females.

The given inferences were criticized on statistical grounds.[19, 20] Other objections were based on the absence of homosexual trends in various forms of hermaphroditism,[21, 22] as well as on the consistent failure of male homosexuals to show a positive nuclear chromatin pattern.[23, 24, 25]

In Kallmann's sample of homosexual twin subjects of the male sex, the siblings yielded a deviant sex ratio (similar to the surveys of Lang and Jensch) only in those subgroups where the twin index cases were over twenty-five years old (130.8:100) or fell into Kinsey's groups 3 and 4 (169.6:100).[26] However, the observed deviations did not reach the level of statistical significance.

It would seem, therefore, that while overt homosexuals differ from other members of their sex in their sexual object choice and behavior, chromosomally and physically they show no anomalies.

It will be necessary then to consider homosexuality in the framework of the mental apparatus, the character structure, the developmental aspects of personality, realizing that in modern psychiatric genetics attributes in these areas are equal candidates for genetic investigation as the more directly anatomical or clinical ones. Before the earlier discussion is ended, however, two sets of recent observations may be noted which raise some questions indicating that the relation between genes, chromosomes, and sexual identification is not yet a closed subject.

Birth Order

The first of these include Slater's finding that male homosexuals tend to be born late in sibship order.[27] He found an increase in mean birth order, an increase in maternal age at birth, and a variance of maternal age as great as that of patients with Down's syndrome (mongolism). This increase approached the figure obtained in the small series of Turner's and Klinefelter's cases in the literature and differed widely from that of the general population. Slater regarded these findings as supporting a hypothesis of heterogeneity in the etiology of homosexuality in the male and as suggesting that a chromosomal anomaly such as might be associated with late maternal age may play a part in causation in some instances.

To be sure, these data were reinvestigated by Abe and Moran who provided statistical evidence that a shift in *paternal* age was primary; this finding would rule out a chromosomal theory and suggest an intrafamilial environmental factor or, alternately, a genetic predisposition to sexual deviance manifesting in the fathers as a tendency to marriage at a later age than the norm.[28]

Hormonal Differences

In another area of research, there is now suggestive evidence of actual hormonal differences in adult homosexuals. Low urinary testosterone levels have been reported in male homosexuals, with female homosexuals excreting high levels of testosterone and luteinizing hormone and low levels of estrone.[29, 30] The ratio of excreted androsterone (A) to etiocholanolone (E) has been reported to differ in male homosexuals from that in heterosexual males (A greater than E in heterosexuals, E greater than A in homosexuals).[31] Finally, in a group of homosexual college students, plasma testosterone levels and sperm counts were found to be significantly reduced in subjects in Kinsey groups 5 and 6.[32] These results need further evaluation to determine the interaction of various hormones, to throw more light on whether the noted differences in steroid hormones are primary or secondary phenomena in the etiological pattern, and to relate hormonal findings to cycles of behavior.

Psychiatric Genetic Theories

At the twenty-third International Psychoanalytic Congress, Gillespie referred to two studies regarding the genetic factor which he said "should make us wary of accepting uncritically the proposition that homosexuality is simply a perversion like any other."[33] The first was the above study by Slater on birth order and maternal age,[27] the second was Kallmann's twin study showing genetic influences "rendering the individual particularly prone to a homosexual outcome."[26] This reference will serve to introduce that twin study and what may be called psychiatric-genetic theories.

Kallmann's Twin Studies

It will be remembered that of particular interest in Kallmann's study was the observation that the sharpest deviation from the expected sex ratio occurred in the sibships of those homosexual men showing an "indiscriminately promiscuous or polymorphic type of sex behavior (Kinsey groups 3 and 4)." This finding anticipated a psychiatric approach to the biology of homosexuality and retains its value even though Goldschmidt's earlier hypothesis and its subsequent applications by Lang and Jensch seem to be generally ruled out by chromosome and sex-chromatin studies.

In Kallmann's explanation, sexual behavior was considered to be part of the personality structure rather than of the gonadal or hormonal apparatus. It was in this vein that he interpreted the results of his twin study on male homosexuality which was based on a series of eighty-five predominantly or exclusively homosexual male twin index cases ascertained from a diversity of psychiatric and correctional institutions or through clandestine channels. Concordance as to the overt practice and quantitative rating of homosexual behavior after adolescence was observed in all of the monozygotic index pairs, forty in number, with no history of mutuality. In the dizygotic pairs, over one half of the cotwins of distinctly homosexual subjects yielded no evidence of overt homosexuality, and the concordance rates were only slightly in excess of Kinsey's rating for the total male population.

Kallmann did not rule out the possibility that some male homosexuals, particularly the infertile ones, might have been intersexes despite the statistically insufficient and cytologically unconfirmed evidence for a genetically oriented imbalance theory. However, in his concluding formulation, he eschewed this special genetic explanation, along with other single-factor causations, in favor of a range of genetic mechanisms capable of disturbing the labile adaptational equilibrium between the potentialities of organic sex distribution and patterns of psychosexual behavior. He wrote:

Apparently, only two males who are similar in both the genotypical and the developmental aspects of sexual maturation and personality integration are also apt to be alike in these specific vulnerabilities favoring a trend toward fixation or regression to immature levels of sexuality. The most plausible explanation of this finding is that the axis, around which the organization of personality and sex function takes place, is so easily dislocated that the attainment of a maturational balance may be disarranged at different developmental states and by a variety of disturbing mechanisms, the range of which may extend from an unbalanced effect of opposing sex genes to the equivalent of compulsive rigidity in a schizoid personality structure.

Inborn Factors and Psychosexual Development

An attempt to synthesize genetic and developmental influences may be introduced by the opinion expressed by Goldschmidt in 1931 on human homosexuality:

> As far as human homosexuality is concerned, the biologist must be extremely cautious in commenting on this much disputed field. I concede that during an earlier period (1916) I was less cautious and believed, on the basis of extensive studies of the literature, that it was justifiable to classify the clearly congenital form of homosexuality as an incipient form of intersexuality. At present I can no longer hold to this theory. What was discussed in the previous sections makes it very difficult to assign homosexuals of either sex a place in the series of intersexuals. Without pretending to be an expert in this field, I should like to point out that what has been previously said about gynecomnastia (in inherited change in the reactivity of the breast tissue to hormones) would also seem to be the most likely explanation for homosexuality, except that instead of the mammary gland the brain would be the end organ.[34]

Thus the brain, the biological mediator of personality, is dramatically brought into the foreground while the problem of homosexuality is returned to the field of psychiatric genetics and developmental psychiatry in the fullest sense.

The true function of genetics in psychiatry is to trace the process of interaction of the organism from its zygotic origin and with the potentialities that lie in the genes received from the parents. This interaction involves nuclear, cytoplasmic, enzymatic, and anatomic forces, and neural, hormonal, and total behavioral responses. All approaches to investigation, physiodynamic and psychodynamic, converge when they are focused on the biological organism with its evolutionary potentialities and its adaptive process traceable throughout its career.

Within this framework, regarding homosexuality, it may be assumed that the site of the divergent development is in the personality, and ultimately, therefore, in the mind or in the brain or the nervous system. It would be going too far afield to discuss the concept of mind. Suffice it to say that the end-organ in homosexuality may be considered to be of this nature, rather than simply gonadal or genital.

At this point we recall the previously mentioned facilitating factors which Kinsey and his associates, even in the framework of their conditioning theory, conceded as the basis for the usual heterosexual choice. To repeat, these factors were as follows: (1) greater aggressiveness in the male with a tendency to avoid sexual behavior with another individual of the same level of aggressiveness, (2) greater ease of intromission in heterosexual contact with more of the "satisfactions which intromission may bring," (3) olfactory and other anatomic and physiologic characteristics differentiating the sexes, and (4) the conditioning effect of the "more frequently successful" heterosexual contacts.

In the heterosexual individual, one may note that these factors imply a normal rate of maturation of personality development distinguished by the abilities: (1) to perceive and respond to biological sexual stimuli of a pleasurable nature, (2) to feel and recognize satisfaction and success, and (3) to utilize these experiences as integrating forces and guides to future action. Any basic defect in perception or self-perception or response to pleasure may render an individual vulnerable to accidental or socially encouraged deviant behavior.

If there are any vulnerability factors in homosexuality, they may well be of this nature, rather than "counter-fragments" (Rado) of the opposite sex.[35] A well-rounded etiological study will search for the factors which predispose as well as those which precipitate and perpetuate. The interplay of these factors may be studied on the genic or enzymatic levels or described as precisely as possible in psychological or demographic terms.

Obviously, this task has yet to be accomplished. Speculations based on evolutionary considerations were advanced by Hutchinson and Comfort.[36, 37] Hutchinson suggested that the genotype responsible for homosexuality may operate on the rates and extent of development of the neurophysiological mechanisms underlying the identification processes and other aspects of object relationship in infancy. Comfort pointed to the possible significance of the time when castration anxiety begins. In animals, this timing might be considered an adaptation phenomenon, protecting immature males from competitive harm during the period between the achievement of sexual maturity and the attainment of adequate size and strength.[38] Hence, homosexuality would be one reaction to early or excessive castration anxiety.

Co-twin Control Studies

More empirical in nature were recent observations on some individual pairs of identical twins with one clearly homosexual and one predominantly heterosexual partner. In a study of two such twin pairs by Rainer, Mesnikoff, Kolb, and Carr, psychoanalytic data on both twins were furnished for a pair of thirty-year-old male twins classified as clinically dissimilar with respect to homosexuality.[39] Designed to explore postnatal experiences associated with divergent paths of behavior in the twins, this study provided valuable insight into their similarities as well. Similarities appeared principally in the psychological test findings, where both partners were strikingly alike in underlying conflicts and unconscious motivations.[40] Marked sexual confusion was apparent in each record, and both twins showed body image distortion. Steroid excretion was normal and similar in both; sex-chromatin tests were negative.

Taking the psychological tests as a first approximation to similar underlying personality characteristics expected in identical twins, the authors then turned to interpretations of the developmental material obtained. Their working hypothesis was that divergent patterns of experience might lead to the expression of homosexual behavior in one and not the other. A determining factor in this case, as well as in some others studied subsequently, seemed to be a difference in the twins' relationships with the mother. Apparently, the twin who was preferred by the mother because of birth order or distinguishing bodily markings turned to homosexual behavior when the overprotective mother frustrated his attempts to contact girls. While derogating his father to him, she acted permissively toward the son's homosexual life. The other twin, rejected by his mother, could identify to some extent with his father, although most of his early heterosexual attachments were immature and compulsive. Genetically, it may be assumed that the structural personality potentials were relatively vulnerable in both twins but led to homosexual symptom formation in only one of them.

Other reports in the literature on twin pairs include the family described by Heston and Shields, in which three pairs of male monozyotic twins occurred in a sibship of fourteen; two of these pairs were concordant for homosexuality, one for heterosexual behavior.[41] The authors could not find any environmental factors to account for the homosexual adaptation, though the two homosexual pairs were indeed late in the birth order. There were differences, however, within the homosexual pairs, in size and weight for one pair and in history of seduction for the other, which may have interacted with presumed genetic disposition in determining the actual life patterns. Discordant monozygotic pairs have also been described by Klintworth, by Parker, and by Davison, et al.;[42, 43, 44] there is suggestion, in these various pairs, that the later homosexual twins were weaker, sicker, and/or favored by their mother although the circumstances and pathology otherwise differ greatly.

This type of investigation strives to pinpoint the most vulnerable components of the developing character structure and the most sensitive periods at which they can be affected. Of course, many more twin pairs, similar and dissimilar as to homosexuality, and of both types of zygosity, should be studied before any detailed interactional synthesis can be achieved.

CONCLUDING REMARKS

Here, then, we have the present status of research and thought in connection with the genetic theory of human homosexuality. If we consider this alternative type of behavior as a process which develops across time in a person's biological life from the moment of conception, there is definite hope that the diverse pieces of information will gradually fall into place and that synthesis of these findings will eventually lead to a better understanding of the etiologic aspects of homosexuality. Only then will the psychiatric, legal, and biosocial problems posed by homosexuality approach solution.

REFERENCES

1. Kinsey, A.C., Pomeroy, W.B., and Martin, C.E.: *Sexual Behavior in the Human Male*. Philadelphia, Saunders, 1948.
2. Kinsey, A.C., Pomeroy, W.B., Martin, C.E., and Gebhard, P.H.: *Sexual Behavior in the Human Female*. Philadelphia, Saunders, 1953.
3. Ford, C.S., and Beach, F.A.: *Patterns of Sexual Behavior*, New York, Har-Row, 1951.
4. Money, J.: Sexual dimorphism and homosexual gender identity. *Psychol Bull, 74*:425, 1970.
5. Harris, G.W., and Levine, S.: Sexual differentiation of the brain and its experimental control. *J Physiol, 181*:379, 1965.
6. Dörner, G.: Hormonal induction and prevention of female homosexuality. *J Endocrinol, 42*:163, 1968.

7. Dörner, G., and Hinz, G.: Induction and prevention of male homosexuality by androgen. *J Endocrinol*, 40:387, 1968.
8. Ward, I.L.: Prenatal stress feminizes and demasculinizes the behavior of males. *Science*, 175:82, 1972.
9. Goldschmidt, R.B.: *Physiological Genetics*. New York, McGraw, 1938.
10. Goldschmidt, R.B.: *Theoretical Genetics*. Berkeley, U of Cal Pr, 1955.
11. Barr, M.L., and Bertram, E.C.: A morphological distinction between neurones of the male and female, and the behavior of the nucleolar satellite during accelerated nucleoprotein synthesis. *Nature (Lond)*, 163:676, 1949.
12. Davidson, W.M., and Smith, D.R.: A morphological sex difference in polymorphonuclear neutrophil leukocytes. *Br Med J*, 2:6, 1954.
13. Jacobs, P.A., and Strong, J.A.: A case of human intersexuality having a possible XXY sex-determining mechanism. *Nature (Lond)*, 183:302, 1959.
14. Ford, C.E., Jones, K.W., Polani, P.E., de Almeida, J.C., and Briggs, J.H.: A sex chromosome anomaly in a case of gonadal dysgenesis (Turner's syndrome). *Lancet*, 1:711, 1959.
15. Hirschhorn, K., Decker, W.H., and Cooper, H.L.: Human intersex with chromosome mosaicism of type XY/XO. *N Engl J Med*, 263:1044, 1960.
16. Ohno, S.: Simplicity of mammalian regulatory systems inferred by single gene determination of sex phenotypes. *Nature (Lond)*, 234:134, 1971.
17. Lang, T.: Die Homosexualität als genetisches Problem. *Acta Genet Med Gemellol (Roma)*, 9:370, 1960.
18. Jensch, F.: Zur Genealogie der Homosexualität. In Bumke, O., and Spatz, H. (Eds.): *Archiv für Psychiatrie*. Berlin, Springer, 1941, pp. 527, 679, vol. CXII.
19. Koller, S.: Über die Anwendbarkeit und Verbesserung der Probandenmethode. Schlusswort zu den Bemerkungen von Th. Lang. *Z Menschl Vererb*, 26:444, 1942.
20. Darke, R.: Heredity as an etiological factor in homosexuality. *J Nerv Ment Dis*, 107:251 and 108:217, 1948.
21. Money, J., Hampson, J.G., and Hampson, J.L.: Hermaphroditism: Recommendations concerning assignment of sex, change of sex, and psychologic management. *Johns Hopkins Med J*, 97:284, 1955.
22. Money, J., Hampson, J.G., and Hampson, J.L.: An examination of some basic sexual concepts: the evidence of human hermaphroditism. *Johns Hopkins Med J*, 97:301, 1955.
23. Pare, C.M.B.: Homosexuality and chromosomal sex. *J Psychosom Res*, 1:247, 1956.
24. Raboch, J., and Nedoma, K.: Sex chromatin and sexual behavior. *Psychosom Med*, 20:55, 1958.
25. Parr, D., and Swyer, G.I.M.: Seminal analysis in 22 homosexuals. *Br Med J*, 2:1939, 1960.
26. Kallmann, F.J.: Twin and sibship study of overt male homosexuality. *Am J Hum Genet*, 4:136, 1952.
27. Slater, E.: Birth order and maternal age of homosexuals. *Lancet*, 1:69, 1962.
28. Abe, K., and Moran, P.A.P.: Parental age in homosexuals. *Br J Psychiatry*, 115:313, 1969.
29. Loraine, J.A., Ismail, A.A.A., Adamopoulos, D.A., and Dove, G.A.: Endocrine function in male and female homosexuals. *Br Med J*, 4:406, 1970.
30. Loraine, J.A., Adamopoulos, D.A., Kirkham, K.E., Ismail, A.A.A., and Dove, G.A.: Patterns of hormone excretion in male and female homosexuals. *Nature (Lond)*, 234:552, 1971.
31. Margolese, M.S.: Homosexuality: A new endocrine correlate. *Horm Behav*, 1:151, 1970.
32. Kolodny, R.C., Masters, W.H., Hendryx, J., and Toro, G.: Plasma testosterone and semen analysis in male homosexuals. *N Engl J Med*, 285:1170, 1971.
33. Gillespie, W.: Symposium on homosexuality. *Int J Psychoanal*, 45:203, 1964.
34. Goldschmidt, R. B.: *Die Sexuallen Zwischenstufen*. Berlin, Springer, 1931.

35. Rado, S.: An adaptational view of sexual behavior. In Hoch, P., And Zubin, J. (Eds.): *Psychosexual Development*. New York, Grune, 1949.
36. Hutchinson, G.E.: A speculative consideration of certain possible forms of sexual selection in man. *Am Naturalist, 93*:81, 1959.
37. Comfort, A.: Sexual selection in man—a comment. *Am Naturalist, 93*:389, 1959.
38. Szekely, L.: On the origin of man and the latency period. *Int J Psychoanal, 38*:98, 1957.
39. Rainer, J.D., Mesnikoff, A., Kolb, L.C., and Carr, A.: Homosexuality and heterosexuality in identical twins. *Psychosom Med, 22*:251, 1960.
40. Carr, A.C., Forer, B.R., Henry, W.D., Hooker, E., Hutt, M.L., and Piotrowski, Z.A.: *The Prediction of Overt Behavior Through the Use of Projective Techniques*. Springfield, Thomas, 1960.
41. Heston, L.L., and Shields, J.: Homosexuality in twins: a family study and a registry study. *Arch Gen Psychiatry, 18*:149, 1968.
42. Klintworth, G.K.: A pair of male monozygotic twins discordant for homosexuality. *J Nerv Ment Dis, 135*:113, 1962.
43. Parker, N.: Homosexuality in twins: A report on three discordant pairs. *Br J Psychiatry, 110*:489, 1964.
44. Davison, K., Brierley, H., and Smith, C.: Male monozygotic twinship discordant for homosexuality. *Br J Psychiatry, 118*:675, 1971.

Chapter 13

GENETIC AND ENVIRONMENTAL VARIABLES IN SCHIZOPHRENIA*

ROBERT CANCRO

THERE ARE CHAPTERS in this volume which review the role of genetic and environmental variables in personality, intelligence, and neuroses. While there are problems in arriving at satisfactory operational definitions of these different concepts, the problems are of a different order of magnitude. The limitations of the IQ tests can be debated in scholarly detail, but their validity, reliability, and predictive power are superb.[1] It is even possible to define intelligence as that which is measured by the tests, and there are experts who choose to do so. The topics of personality and neurosis begin to show mild signs of the malady that so obviously afflicts schizophrenia, namely confusion. There are no instruments to define or measure schizophrenia. There is profound disagreement among psychiatrists as to what schizophrenia is, and, further, as to which individuals should in any given population be so labeled. For these reasons, studies on the validity and reliability of the diagnosis are uniformly disheartening.[2, 3, 4, 5, 6]

It may be helpful to review briefly this clinical concept and some of its vicissitudes. Morel introduced the term in 1852 to describe a single case of dementia in which the onset occurred during adolescence.[7] He did not presume that this label described a disease entity but rather that it identified the remarkable features of a process in his young patient.

Despite Morel's caution, the psychiatry of that period was under the influence of men like Griesinger[8] and dominated by etiologic concepts such as brain disease, moral degeneracy, and constitutional weakness. In 1899 Kraepelin used the term as one of the two major divisions of the endogenous psychoses.[9] The basis of the division was the presence or absence of deterioration. Despite his later admission that the patients with dementia praecox need not go on to deteriorate, the division stood firm. Kraepelin's conceptualization of dementia praecox included the cases described almost a century earlier by Pinel and Haslam.[10, 11] He also included the cases identified by Hecker and Kahlbaum as hebephrenia and catatonia, respectively.[12, 13] These terms were no longer adjectives but had become nouns, i.e. the names of discrete and real entities.

Bleuler's major work on dementia praecox appeared in 1911, only twelve years after Kraepelin's classic division.[14] Bleuler argued that the term was misleading since the onset of the illness did not

---

*This chapter is a modified version of a paper read before the American College of Psychiatrists in February, 1975.

have to occur during puberty and the course did not lead inevitably to deterioration. He did more than change the name. He moved towards a syndrome concept by arguing that dementia praecox consisted of a group of disorders with similar symptomatology. This was an important step away from the disease entity model of Kraepelin.

Meyer took a more interactional view of this disorder and saw it as a reaction to the environment.[15] His stress on the life chart of the patient explicitly contained the concept of a person's reacting to environmental situations with psychological symptoms. This approach to mental illness remained strong in the United States until the 1968 revision of *The Diagnostic and Statistical Manual*.[16] It defined schizophrenia in the following fashion:

This large category includes a group of disorders manifested by characteristic disturbances of thinking, mood and behavior. Disturbances in thinking are marked by alterations of concept formation which may lead to misinterpretation of reality and sometimes to delusions and hallucinations, which frequently appear psychologically self-protective. Corollary mood changes include ambivalent, constricted and inappropriate emotional responsiveness and loss of empathy with others. Behavior may be withdrawn, regressive and bizarre.

As we review the history of the concept we are struck by the dramatic fluctuations. It has been considered everything from a single disease to a way of life. The prognosis has been described as everything from universally poor to better than most neuroses. Every effort to improve the nosological confusion has led further into the morass. It is no surprise that the student is puzzled. In 1968 we agreed to an internationally accepted classification. It specifies that schizophrenia is a group of disorders, i.e. illnesses. Not only is schizophrenia defined as an illness, but the signs and symptoms by which it can be diagnosed are listed. The student can learn to identify the particular signs and symptoms and then label the individuals who show them as schizophrenic. While it is easy to understand the desire for nosological clarity, it is doubtful that we have gained much with this new/old emphasis. It is a frighteningly static and cross-sectional view of illness as an entity which ignores its adaptive function and its process. The static clinical picture described in *The Diagnostic and Statistical Manual* is the end state of an adaptive process which has not been wholly successful. As we know, from the principle of equifinality, this end state can be arrived at from a number of different initial conditions and through a variety of pathways. The important diagnostic versus classificatory goal is increasing the reliability of the identification of a particular group of end states as schizophrenia. Yet, it is vital to understand that this increased reliability does not improve either the validity of the diagnostic category or the usefulness of the classification in comprehending the process. Is schizophrenia an end state whose clinical picture we can describe, or is it the process of reaching that end state?

It is helpful to remember the distinction between diagnosis and classification. A diagnosis is a statement of what is troubling a person at a given time, in a specified fashion, and in a particular context. It is an effort to merge general knowledge about illness with specific knowledge about the individual as a means of helping that person. A classification is quite different. It is an attempt to organize the data into a comprehensible system which matches the mode of our cognitive functioning. The ordering of data, for example, in classes and subclasses is not a function of nature but of our cognitive apparatus. It is the way we think and, therefore, the way we organize data. Classification is an essential method of science as performed by man and intends to increase our understanding of the data by ordering it in this particular fashion. An increase in reliability is a legitimate diagnostic goal since diagnosis emphasizes the end state. This task, if accomplished, would yield little value to classification, since classification emphasizes the process. This distinction between diagnosis and classification helps us to phrase the two questions mentioned earlier more clearly. Does the concept of schizophrenia as it is officially defined have validity as a diagnostic category? Does it help to improve our understanding of the process of this disorder?

In addressing the first question it is necessary to realize that the signs and symptoms found in most patients labeled schizophrenic are not myths. They are all too real. Those who work closely with psychotic individuals know they are not dealing with mythology. Yet, we can wonder about the homogeneity, if not the existence, of the grouping. Certainly, there are enormous clinical variations among people so labeled. It is helpful to understand that the current concept of schizophrenia is not useful if you conceive of the disorder as a single disease. The variation is too great for this category to include only one class of patients.[17] When we think of the category as a group or class of related end states that have been arrived at through different pathways from different initial conditions, we see that the category has some limited diagnostic usefulness.

While the category is of some value for diagnostic purposes, it is of virtually no aid in increasing our understanding of the process. We cannot infer the pathway nor the initial condition from the end state. We can draw some conclusions, however, concerning the patient's need to find a new dynamic steady state to his environment at the time of his psychotic decompensation.

His previous adaptation could no longer serve. The precise nature of the changes which exceeded the reserves of the earlier nonpsychotic adaptation is not known. They could have been chemical, intrapsychic, interpersonal, social, etc. We do know that there was a need for a new organism-environment dynamic steady state, and the process of achieving the new steady state can be called schizophrenia. Clearly, the task is to understand the process better and not just to describe the end state more lucidly. Yet, there is value in knowing that a particular end state was achieved as opposed to some other. For this reason alone it is useful to improve diagnostic reliability.

The well-known problems of diagnostic reliability are in part a function of differences in recognizing and labeling a particular end state as schizophrenic. For the remainder of the chapter the term schizophrenia will be used only in the sense of an end state. There has been a considerable lack of agreement on those criteria which are essential. Many clinicians still make the diagnosis on the basis of an intuitive feeling that they have about the patient. This cannot be an adequate basis for increasing the homogeneity of the population labeled schizophrenic. Since 1911 we have considered the fundamental signs in schizophrenia to be autism, ambivalence, the affective disturbance, and the associational disturbance. However, autism, ambivalence, and many of the affective disturbances seen in schizophrenia are not exclusive to the disorder. They can be found to a greater or lesser degree in a variety of individuals, including so-called normals. It is the particular cognitive disturbance that is probably unique to schizophrenia. For this reason it would appear wise to restrict the category to patients who show a characteristic thought disorder.[18] In this sense *The Diagnostic and*

Statistical Manual makes a real contribution since it does restrict the diagnosis to the thought-disordered patients.

EVIDENCE FOR A GENETIC CONTRIBUTION TO THE ETIOLOGY OF SCHIZOPHRENIA

The evidence for the existence of a genetic contribution to the etiology of schizophrenia is so strong as to no longer be in reasonable doubt. This does not mean that there are not individuals who deny it. It simply means there is no reasonable basis for their conclusion. There are three main lines of evidence for a genetic factor which will be summarized in this section: consanguinity, twin, and adoptive studies. The consanguinity studies consistently show that the prevalence of schizophrenia is significantly higher in the genetic relatives of schizophrenic patients than it is in the general population. This has been found in every such study performed, including those done by investigators with a markedly environmental bias.[19] (The very stability of the prevalence rate in the general population both within and between cultures argues for a genetic factor.) More important than the simple finding is the significant positive relationship between the frequency of the disorder and the degree of kinship. The closer the genetic relationship to the patient, the more likely is it that the relative will show the disorder. For the parents, siblings, and children of schizophrenics, the prevalence of schizophrenia is approximately 10 to 15 percent, while it is less than 1 percent in the general population. Even second-order relatives show a significantly higher risk than the general population, although not as high as closer relatives. Clearly, the consanguinity studies are a weak form of evidence. They can just as validly be interpreted as supporting an environmental explanation. The environmentalist can argue that the closer the degree of kinship, the more intense the psychological relationship between the individuals and, thereby, the more likely that they have shared similar predisposing environmental experiences. Yet, there is a little-known study which offers evidence for a genetic factor in the course and not just the prevalence of the disorder. Bleuler reported that relatives who had schizophrenia tended to have the same type of onset, symptomatology, duration of illness, and outcome.[20] While this finding can be interpreted in either an environmental or genetic fashion, the most parsimonious explanation is genetic, since many of the relatives were geographically well-separated and in some cases did not even know each other.

The twin studies have been helpful in adding a second and stronger body of evidence. The concordance rate for schizophrenia in monozygotic twins has consistently been found to be significantly higher than the rate in dizygotic twins. The earlier studies found concordance rates in monozygotic twins of 60 to 86 percent.[19] Unfortunately, they had certain methodological failings, not the least of which was that the index case was found in a hospital setting. More recent studies have been extended to larger populations of dizygotic twins, thereby including index cases who were less severely ill.[21, 22, 23, 24] The concordance rate in these later efforts varies between 6 to 40 percent. The exact percentage is primarily a function of the diagnostic criteria used. The more recent investigations include the total population of twins and are, therefore, methodologically superior. The concordance rate in all the studies of dizygotic twins has been no higher than for siblings. Rosenthal reviewed eleven twin studies and found that in ten the concordance rates were consistent with the predictions that would be made from genetic theory.[25] Interestingly, in the study done by Pollin on 16,000 twin pairs collected by the National Academy of Sciences from the Veterans Administration, there was a systematic bias in the opposite direction.[26] Pollin's sample included only those twins who had both served in the armed forces. If one or both members of the twin pair had been rejected, they were not included in the sample. This resulted in a population in which psychotic decompensations of late adolescence and early adulthood were excluded from the sample. This naturally produced a lower concordance rate for the monozygotic twins. Nevertheless, the concordance rate in the monozygotic twins in this sample was three times higher than in the dizygotic twins. Pollin's study also showed no significant difference in concordance between monozygotic and dizygotic pairs for psychoneuroses.

A psychological theory can still explain the twin data without having to be stretched too far. Monozygotic twins share more than their genetic makeup. There is a very special relationship between them which contributes to the formation of a strong mutual identification. They are frequently mistaken for each other, dressed alike, called by each other's names, etc. All of these experiences can contribute to a very special environment in which what one does is likely to be repeated by the other. The occurrence of identical twins who have been reared separately helps to disentangle the contributions of genes and the environment. There are at least sixteen reported cases of monozygotic twins reared apart in which one of the twins was diagnosed as schizophrenic.[19] The concordance rate for this group was 62.5 percent and is comparable to that found in monozygotic twins reared together. This finding is very supportive for the existence of a genetic factor.

The most important and compelling line of evidence comes from studies in which the parent and child generations have been separated so that the child is not reared by the biologic parent. The first such study published was done by Heston.[27] In a carefully controlled investigation, he examined the prevalence of schizophrenia in the offspring of schizophrenic women. These children had been adopted early in life by nonschizophrenic parents. As a control group he used the children of nonschizophrenic women who had also been adopted. It was the offspring of schizophrenic women who showed the higher-than-normal rate of schizophrenia and, more interestingly, showed exactly the rate that would have been predicted if they had been raised by their biologic mothers. Kety, Rosenthal, Wender, and Schulsinger approached the problem from a related, but somewhat different, direction.[28] They took a large population of adopted children and separated them into those who became or did not become schizophrenic. They then ascertained the prevalence of schizophrenia in the biologic versus the adoptive relatives of the schizophrenic children. They studied all children adopted in Copenhagen in a twenty-four-year period. They also excluded all adoptions by individuals who were biologically related to the true parents. They, too, found a significantly higher prevalence of schizophrenia and schizophrenia-like disorders in the biologic relatives. One very important finding in their study was that approximately half of the biologic relatives of index cases who showed schizophrenia spectrum disorders were paternal half-siblings and, therefore, did not share with the index case a common uterus. This finding was very strong evidence for a relatively pure genetic factor which was independent of the early intrauterine environment. It must be remembered that in the behavioral sciences we are dealing with the concept of evidence and not proof. Our hypotheses cannot be proven in a rigorous mathematical sense, but the weight of the evidence can be sufficiently strong that we can, for practical purposes, consider it proven. When one reflects on the evidence from the consanguinity, twin, and adoptive studies, it is obvious that a genetic factor has been demonstrated beyond reasonable doubt.

EVIDENCE FOR AN ENVIRONMENTAL CONTRIBUTION TO THE ETIOLOGY OF SCHIZOPHRENIA

In all the furor that has gone on in recent years concerning the existence and/or relative importance of a genetic factor in the etiology of schizophrenia, there has been a remarkable lack of discussion concerning the evidence for the existence of an environmental factor. The evidence for an environmental contribution will be briefly reviewed. It has long been known that there is a relationship between social class and the prevalence of schizophrenia. The disorder is found more frequently among the poor. While this can be interpreted as evidence for the role of impoverished social conditions in the etiology of schizophrenia, it can be explained equally well by schizophrenics tending to drift down the socioeconomic ladder because of their inability to compete effectively.

The marked increase in interest in the study of families has resulted in the careful examination of families of schizophrenics. The effects of parental psychopathology on the children and the disordered communication patterns within families has been studied in depth. The not-surprising finding has been that schizophrenic children are frequently the product of seriously and characteristically disturbed families. These studies have suffered from several methodologic limitations. Usually the observers were aware of the diagnosis of the index case, and this knowledge could easily cause them to overemphasize family interactions which might otherwise be considered insignificant. Another methodologic problem was that the families were observed after the child had become sick so that it was not certain that the disordered patterns preceded rather than followed the schizophrenic illness. A clever device for avoiding certain of the problems of halo effect was to score projective tests blindly. Wynne and co-workers have demonstrated significant differences between the Rorschachs of the parents of schizophrenic and normal children.[29] There are still methodologic problems with this approach since the psychological tester was not blind to the diagnosis. This knowledge in turn could have influenced the way the projective test data was gathered.

The most important issue is that all of the environmental data so far reviewed can support a genetic theory equally well. This is precisely the situation we saw with much of the genetic data. It is relatively nonspecific. Kety argued that: "The most incontrovertible evidence that non-genetic factors operate in the etiology of schizophrenia is the fact that the concordance rate for schizophrenia in monozygotic twins is considerably less than 100 percent."[30] He went on to add that the concordance rate of 40 to 50 percent in the newer genetic studies leaves "a large area for the operation of factors other than genetic. . . ." The moderate concordance rate tells us very little about the size of either the environmental or genetic factor. It is very difficult to divide genetic and environmental factors into separate parcels. The biologic reality is that it is essential to have both a genotype and an evoking environment. Since both are necessary, it is difficult to speak of relative importance except in the very limited and specific sense expressed by a heritability index. The traditional distinction between genetic factor and determination is essential. Having a particular genetic makeup does not determine the outcome inevitably. It takes more than a genotype to produce schizophrenia. The concordance rate confirms this fact very clearly. The concordance rate of 40 to 50 percent is also reassuring in the sense that without any planned environmental manipulation a monozygotic twin has about an even chance of not coming down with the disorder even if his cotwin has been so diagnosed. When we realize that there must be a number of monozygotic twin pairs who have the genotype but neither of whom develops the disorder, we can see that the risk is probably even less. The best theoretical argument for the existence of environmental factors is that they are necessary for the genes to express themselves. There is also some interesting empirical data which shows that in twins discordant for schizophrenia it is the one with a lower birth weight who is much more likely to develop the disorder. This parallels the IQ findings in which the lighter twin has the lower IQ. It is obvious, then, that an environmental factor, i.e. the quality of the intrauterine nutritional state, contributes to the future development of the phenotype.

When we examine a patient who shows the clinical syndrome called schizophrenia we must make a distinction between the form and content of his symptoms. The psychosocial environment contributes heavily to the determination of the content. What the voices tell him, why it is the neighbors and not someone else who persecutes him, etc., reflect the developmental realities of the patient's life. It is difficult to explain satisfactorily the form of the symptom on a psychosocial basis. It is also important to distinguish between the etiology and maintenance of the syndrome. We know from clinical experience that the psychosocial environment plays an important role in whether or not the patient improves. Suppressing the disorder through environmental manipulation, broadly defined, is the only treatment modality presently available. These environmental manipulations may be chemical, for example, through the use of drugs; or psychosocial, for example, through the use of psychotherapy and milieu treatment.

GENETIC-ENVIRONMENT INTERACTION IN SCHIZOPHRENIA

As indicated above there is no better argument for the presence of an environmental factor than our knowledge that it must exist for the genotype to act. Part of the confusion results from the word *environment* being used in multiple senses. It means everything from the concentration of critical nutrients in the maternal bloodsteam to familial rearing patterns. It would be best if we expressed the role of environmental variables in terms that allowed us to understand their interaction with the genotype more adequately. For the time being, this remains more of an aspiration than a reality in schizophrenia. A great deal of research must be done which specifies the particular evocative environmental variables and the physiologic pathways through which they act on the genotype. There have been promising leads ranging from the already mentioned family interaction patterns to nutrition. In this latter category there is evidence concerning the possible role of cereal substances in the exacerbation of schizophrenic symptoms which should not be dismissed.[31] While it is true that every other such finding has been demonstrated to be an artifact, it is inevitable that some will eventually be valid. The environment is critical in the etiology of schizophrenia for exactly the same reason that the genetic component is—they are both necessary. The question of relative importance as mentioned earlier may presently be more academic than practical. It has often been said that the genotype is necessary but not sufficient. It is better to conceive of the genotype as one member of an interacting series which produces the disorder. We may by choice divide the members of the series into two categories called genetic and environmental. This division is helpful pedagogically. It is an error to reify this division and then treat the categories as separate sources of independent variance. The interaction between gene and environment necessary to produce any phenotype is a reality of biology, and this reality is not captured in the mathematical model of an analysis of variance. Mathematical models of gene-environment interaction are possible, but good ones are not presently available. Clinical experience may offer some assistance in assessing the part played by environment in determining schizophrenia. We have clinical reason to suspect that phenocopies are possible. We know that the genotype alone is not enough to produce the disorder. When we take these facts into account, we need not be apologetic for the contribution of environment. It is more a function of scientists than of science that we argue about the relative contributions of our respective disciplines to important problems. The geneticist and the clinician both have something of value to offer. It is not essential for either to be more important.

The concepts of norm of reaction and the highly related one of phenoptions addressed themselves to the degrees of freedom existing in the genotype. The genotype can produce a relatively broad range of phenotypes as a function of differences in the environment, although there are inherent limitations as well. In practice the variation in environments to which a genotype is exposed tends to be small. It is helpful to utilize the concept of the most probable environment to which the genotype will be exposed.[32] Given the most probable environment, we can predict with relative certainty the phenotype that a particular genotype will produce. If the genotype is exposed to highly improbable environments, it becomes increasingly likely that the phenotype will differ from that produced by the most probable environment. Much research is needed to identify the environments which will prevent the potential schizophrenic genotype from unfolding in the direction of an illness. Even if this research were to be successful it still might be of limited therapeutic value since there is no reason to assume that the environments which prevent are those which suppress schizophrenia. The feasibility and legitimacy of preventing schizophrenia are complex questions involving both technical and moral issues. The social and individual benefits are obvious, the costs less so. Both the costs and benefits must be clearly identified and weighed prior to our taking any steps, no matter how well-intentioned. The following two sections will consider both the feasibility of preventing schizophrenia and some of the moral questions.

GENETIC MANIPULATION

There are two promising methods for reducing the occurrence of schizophrenia in the population through genetic manipulation. The first of these is genetic engineering. Presently this is a topic that is more suitable for the Sunday Supplement than for science. While the ability to alter the molecular structure of genes is not imminent, it is highly probable that it will be in the future. There can be little doubt that when it does we shall do as we have always done in the past, which is to use promptly any scientific development that comes along. A presently available alternative is genetic counseling or its more extreme version, eugenics, which can alter the gene pool through discouragement of reproduction in schizophrenics and thereby reduce the frequency of schizophrenia in the population. The increasing reproductive rate in schizophrenics brought about by shortened hospital stays and the effort to encourage patients to lead more normal social lives will result in a higher prevalance rate of schizophrenia in future generations.[33] This is a predictable negative consequence of our current therapeutic activities in this disorder. There is nothing being done to counter this effect on the future generations. Genetic counseling is not routinely practiced with schizophrenic patients. Even if it were, it is far from clear how effective it would be with, for example, a chronically psychotic individual. Genetic counseling may be sufficient with potential parents who are able to comprehend the issues involved and do not wish to take unnecessary risks. There is every reason to believe that the cognitive disturbance and impaired reality testing in schizophrenics would render genetic counseling ineffective.

If we wish to lower the prevalence rate of schizophrenia through genetic manipulation we have to face the very real likelihood that counseling will not be enough. It is difficult to discuss eugenics rationally because of the strong affectively-charged images the very word arouses. Eugenics is frequently denounced as undemocratic, although it has been practiced for a number of years in Denmark which is in many ways a model democracy. Another democracy which has had until recently laws restricting the rights of persons diagnosed as psychotic to marry is Sweden. Clearly, the practices of a free society do not necessarily exclude eugenics. We do not have to be Maoists to recognize that society has rights as well as the individual. We must realize eugenics and genetic counseling can be explored rationally like any other issue. The decisions can be based on the rationally arrived-at alternatives, so long as we recognize the necessity of including moral considerations. On an individual basis, does the risk of an occurrence which is less than 15 percent justify the denial of parenthood? On a broader basis, does society have the right to control parenthood? These are the critical moral questions for which scientists are no better prepared than nonscientists to answer. We can contribute some knowledge which, while not directly answering the moral questions, may be helpful in addressing them.

The major risk in manipulating the gene pool is the danger of its permanent alteration. If a gene is bred out of a population it may be impossible to reintroduce it. The negative consequences of breeding out an undesirable phenotype may not be apparent until it is too late. There are a number of studies which suggest that creativity and schizophrenia are linked. If so-called eugenic practices were to reduce the prevalence of schizophrenia at the cost of reducing the level of creativity in the population, we would indeed be facing a dilemma. It is difficult, if not impossible, to predict accurately the cost of any given genetic manipulation. The inescapable conclusion is that it is dangerous to meddle with the gene pool. This simple rule of nonaction extends beyond the small group labeled schizophrenic. Yet, we must face the equally real fact that, in the name of our humane concern for people, we have in the last thirty years influenced the human gene pool in very profound ways. Social legislation has made it possible for a number of people, including schizophrenics, to reproduce and for their offspring to survive who otherwise would not have done so. This enlightened social legislation has been primarily restricted to the technologically advanced countries, but the benefits of modern medicine have spread even to the underdeveloped nations. The introduction of antibiotics and the other medical advances since World War II have kept large numbers of individuals alive who in the past would not have survived long enough to contribute offspring. In a very real sense the development of penicillin and DDT with the resulting control of many lethal diseases have had more impact on the gene pool than any other development in the recent history of the species. The rapidly increasing world population is clear evidence of their impact. It is not enough to argue that our previous and present manipulations are in a different class since they have been to save lives. Obviously, good intentions are not an adequate justification. We are altering the genetic makeup of *Homo sapiens* as a population and doing it without plan. Perhaps this nonplan is to be preferred over a rational approach, but at least we should recognize what is being done and consider some alternatives.

The issues raised go far beyond the limited question of schizophrenia. Yet the need for rationally determined analyses of costs and benefits is the same. It is here that science can contribute without violating the democratic process. The decisions rest with the body politic, and science must not only identify the costs and benefits but must communicate them effectively and fairly to the populace.

ENVIRONMENTAL MANIPULATION

The environment, broadly defined to include the sociocultural milieu as well as the more traditional variables, interacts with the genotype to influence its norm of reaction. The precise physiologic pathways through which social variables such as rearing patterns influence the genotype are not known. There is considerable evidence in the case of IQ to suggest that early infant handling and patterns of sensory stimulation play a role in the later development of the phenotype. The family studies done with schizophrenic patients suggest that at least in some cases aberrant patterns of interaction may play a role in the development of the disorder. The evidence is not particularly compelling for the evocative role of any specific sociocultural pattern, but it is certain that the environment broadly defined plays an essential role in the evolution of schizophrenia. There is a need for further research to demonstrate the interaction of specific psychologic factors with the genotype and the particular physiological pathways through which these interactions take place.

One way of influencing the phenotype without the dangers inherent in altering the gene pool is through environmental manipulation. A disadvantage is that the environmental manipulation would have to be repeated with every generation. Yet most observers find this approach more emotionally palatable since it can be rationalized as equalizing rather than manipulating the environment. This represents more of a semantic deception than a humanistic advance. The verb "to equalize" is democratic, while "to manipulate" is fascistic. Unfortunately, the process remains the same although the verb changes. Before developing this point further it is well to reemphasize another limitation in this approach, since the environment that may evoke schizophrenia in certain individuals may evoke socially valuable traits in others with similar genetic endowments. The critical deficiency in our desire to equalize the environment in the sense of making it the same is that it does not result in the identical phenotype. Individual differences in the phenotype will still occur but will be solely on the basis of genetic differences. If we want to equalize the environment in the sense of influencing the phenotype we would have to identify environments that are equally evocative or equally suppressive in *different* individuals. We cannot assume that the environment that suppresses a schizophrenic illness in one person is necessarily going to suppress it in another. Schizophrenia is not a homogeneous entity but rather an end state (or a group of clinically similar end states) that can be arrived at either from a number of different genotypes or else through a variety of different gene interactions. We may be fortunate and find that there are a relatively small or at the very least finite number of environmental manipulations which would prevent the disorder. While we believe certain environments to be noxious, the data from the adopted child studies show that being raised in a normal home does not lower the prevalence of schizophrenia in the offspring of schizophrenic parents. This does not mean that the environment is unimportant. It does mean, however, that an environment which will evoke schizophrenia is statistically as likely in a normal as in a schizophrenic home. This suggests that the environmental experiences are quite nonspecific and likely to be encountered by most individuals in the culture. The equalization of the environment then involves considerable alteration if we are to succeed in our effort to prevent the disorder. We must create highly improbable environments to prevent schizophrenia. Even if we were to identify such environments we would have to deal with the question of the cost involved in imposing them. For example, if a child had to be taken from his natural or adoptive parents and raised by trained caretakers in special settings for high-risk children, he would have paid a high price for the vaccination. The actual damage done to the children by the techniques of prevention must be weighed against the potential benefits. This is a particularly difficult choice when we realize that the beneficiaries of the prevention number at most 15 percent of the population so treated.

DENOUEMENT

When invited to write a chapter for a book, the author usually takes pride in imparting to the reader a reasonable quantity of hard information and avoids opening Pandora's box filled as it is with ignorance and uncertainty. This particular chapter grappled with difficult moral dilemmas without offering solutions, simple or otherwise. It is clear that to speak of a single solution for any of the problems raised is misleading, since a number of viable alternatives are possible. The decisions affect society in real ways, and it is our scientific responsibility to involve the people in the decision-making process. We must avoid the danger of believing our technical expertise effectively prepares us to deal with the value and moral issues. If the questions raised stimulate the reader to think deeply about them, the chapter has been an unalloyed success. If they only irritate and disturb the reader's scientific complacency, the chapter has still been of value.

REFERENCES

1. Cancro, R. (Ed.): *Intelligence: Genetic and Environmental Influences.* New York, Grune, 1971.
2. Hoch, P.: The etiology and epidemiology of schizophrenia. *Am J Public Health,* 47:1071, 1957.
3. Kreitman, N.: The reliability of psychiatric diagnosis. *J Ment Sci, 107*:876, 1961.
4. Mehlman, B.: The reliability of psychiatric diagnosis. *J Abnorm Soc Psychol, 47*:577, 1952.
5. Pasamanick, B., Dinitz, S., and Lefton, M.: Psychiatric orientation and its relation to diagnosis and treatment in a mental hospital. *Am J Psychiatry, 116*:127, 1959.
6. Schorer, C.E.: Mistakes in the diagnosis of schizophrenia. *Am J Psychiatry, 124*:1057, 1968.
7. Morel, B.A.: *Études Cliniques: Traité Théorique et Pratique des Maladies Mentales.* Paris, Masson, 1852-1853.
8. Griesinger, W.: *Die Pathologie und Therapie der Psychischen Krankheiten.* Braunschweig, Wreden, 1871.
9. Kraepelin, E.: *Psychiatrie. Ein Lehrbuch für Studierende und Ärzte,* 6th ed. Leipzig, Barth, 1899.
10. Pinel, P.: *Traité Médico-Philosophique sur l'Aliénation Mentale, ou la Manie.* Paris, Richard, Caille et Ravier, 1801.
11. Haslam, J.: *Observations on Madness and Melancholy,* 2nd ed. London, Hayden, 1809.
12. Hecker, E.: Die Hebephrenie. *Arch für pathol Anat u Physiol u klinische Med, 52*:394, 1871.
13. Kahlbaum, K.L.: *Die Katatonie oder das Spannungsirresein.* Berlin, Hirschwald, 1874.
14. Bleuler, E.: *Dementia Praecox or the Group of the Schizophrenias.* New York, Intl Univs Pr, 1950.
15. Meyer, A.: The life chart and the obligation of specifying positive data in psychopathological diagnosis. In Winters, E.E. (Ed.): *The Collected Papers of Adolf Meyer.* Baltimore, Johns Hopkins, 1951, vol. III.
16. American Psychiatric Association: *Diagnostic and Statistic Manual of Mental Disorders,* 2nd ed. Washington, D. C., American Psychiatric Association, 1968.
17. Cancro, R. (Ed.): *The Schizophrenic Reactions: A Critique of the Concept, Hospital Treatment, and Current Research.* New York, Brunner-Mazel, 1970.
18. Cancro, R.: Thought disorder and schizophrenia. *Dis Nerv Syst, 29*:846, 1968.
19. Slater, E.: A review of earlier evidence on genetic factors. In Rosenthal, D., and Kety, S.S. (Eds.): *The Transmission of Schizophrenia.* Oxford, Pergamon, 1968.
20. Bleuler, M.: *Krankheitsverlauf, Persönlichkeit und Verwandtschaft Schizophrener und ihre gegenseitige Beziehungen.* Leipzig, Thieme, 1941.
21. Fischer, M., Harvald, B., and Hauge, M.: A Danish twin study of schizophrenia. *Br J Psychiatry, 115*:981, 1969.
22. Gottesman, I.I., and Shields, J.: Schizophrenia in twins: 16 years' consecutive admissions to a psychiatric clinic. *Br J Psychiatry, 112*:809, 1966.
23. Kringlen, E.: Schizophrenia in twins. An epidemiological-clinical study. *Psychiatry, 29*:172, 1966.
24. Tienari, P.: Psychiatric illnesses in identical twins. *Acta Psychiatr Scand, Suppl 171*:1963.
25. Rosenthal, D.: Genetic research in the

schizophrenic syndrome. In Cancro, R. (Ed.): *The Schizophrenic Reactions: A Critique of the Concept, Hospital Treatment, and Current Research*. New York, Brunner-Mazel, 1970.

26. Pollin, W., Allen, M., Hoffer, A., Stabenau, J., and Hrubec, Z.: Psychopathology in 15,909 pairs of veteran twins. *Am J Psychiatry, 126*:597, 1969.

27. Heston, L.L.: Psychiatric disorders in foster home reared children of schizophrenic mothers. *Br J Psychiatry, 112*:819, 1966.

28. Kety, S.S., Rosenthal, D., Wender, P., and Schulsinger, F.: The types and prevalence of mental illness in the biological and adoptive families of adopted schizophrenics. In Rosenthal, D., and Kety, S.S. (Eds.): *The Transmission of Schizophrenia*. Oxford, Pergamon, 1968.

29. Wynne, L.C.: Methodologic and conceptual issues in the study of schizophrenics and their families. In Rosenthal, D., and Kety, S.S. (Eds.): *The Transmission of Schizophrenia*. Oxford, Pergamon, 1968.

30. Kety, S.S.: Genetic: Environmental interactions in schizophrenia. *Trans Stud Coll Physicians Phila, 38*:124, 1970.

31. Dohan, F.C., Grasberger, J.C., Lowell, F.M., Johnston, H., Jr., and Arbegast, A.W.: Relapsed schizophrenics: More rapid improvement on a milk and cereal free diet. *Br J Psychiatry, 115*:595, 1969.

32. Cancro, R.: Genetic contributions to individual differences in intelligence: An introduction. In Cancro, R. (Ed.): *Intelligence: Genetic and Environmental Influences*. New York, Grune, 1971.

33. Erlenmeyer-Kimling, L., Rainer, J.D., and Kallmann, F.J.: Current reproductive trends in schizophrenia. In Hoch, P.H., and Zubin, J. (Eds.): *Psychopathology of Schizophrenia*. New York, Grune, 1966.

Chapter 14

GENETICS, CYTOGENETICS, DERMATOGLYPHICS, CLINICAL HISTORIES, AND SCHIZOPHRENIA ETIOLOGY

Arnold R. Kaplan, Diane Sank, Richard Allon,

Henry T. Lynch, Edward N. Hinko, Wilma E. Powell,

and Austin E. Moorhouse

CONCEPTS OF GENE-TRANSMITTED DIATHESES FOR SCHIZOPHRENIA

Basically, an individual's genotype only determines the norm or range of reactions to environmental influences. That is, a particular genotype determines an indefinite but limited assortment of potential phenotypes, different phenotypes being associated with differences in interacting environmental variables. Thus, the pathogenic relevance of any particular variable may differ in different contexts. Schizophrenia, like many other clinical entities, may involve multiple etiologies. This possibility complicates attempts to evaluate the general etiological significance of any particular variable. Environmental variables can influence correlations between primary genes and their final actuated manifestations, affecting an observed correlation between genotype and phenotypic effects. The different genetic and experiential contributing factors have different relative etiological values, dependent in each case upon the total constellation of various contributing factors. When one principal contributing factor grows stronger in its effect, the relative significance of the other factors diminishes.

It is not possible to prove that a trait is genetically caused except by showing that it is transmitted in accordance with a particular mode of biological inheritance. The genetic method consists in attempting to extract, from a possibly heterogeneous trait, one or more genetic entities associated, in increasing order of refinement, with a single mode of genetic transmission, a single locus, and a single allele. No specific mode of genetic transmission has been proved to be associated with *the* necessary and sufficient etiological basis

---

The studies described in this chapter were supported by the state of Ohio Department of Mental Health and Mental Retardation and by research grants from the Scottish Rite Committee on Research in Schizophrenia, the National Association for Mental Health, and the National Institutes of Health (MH 17302-01A1).

for schizophrenia. Nevertheless, extensive data have established the relevancies of both genetic and environmental variables for individual differences in their predispositions to schizophrenic disorders.

The major genetic theories may be divided according to the number of genetic factors hypothesized as having primary relevance to etiology of schizophrenia. Kallmann described an autosomal recessive mode of genetic transmission for the schizophrenia diathesis, supported by his data on schizophrenic twin index families.[1] Later, Kallmann elaborated his hypothesis to include a second genetic locus with modifier genes.[2] That is, the penetrance and expressivity of the primary pair of genes was hypothesized to depend largely upon the presence of one or both of the modifier genes. Böök and Slater described extensive pedigree data that supported an association between the schizophrenia diathesis and a single autosomal dominant gene.[3,4] More recently, Karlsson hypothesized that different combinations of two pairs of genes, involving two autosomal loci, determine a series of different predispositions to schizophrenias.[5]

Kety and associates emphasized the occurrence of complex systemic disruptions associated with schizophrenia which they believed was indicative of the involvement of a great many different genetic factors.[6] Accordingly, their concept of schizophrenia etiologies involves multiple heteromeric genetic factors interacting with environmental stresses. A similar concept was previously described by Ödegärd, who believed that unspecific psychotic conditions with a high degree of individual variability were associated with combinations of genes on many different loci (i.e. multifactorial inheritance).[7] There are infinite varieties of potential multiple-gene theories for genetically-transmitted schizophrenia diatheses which vary according to the numbers of hypothesized genetic loci and the numbers of alleles associated with each relevant locus. According to the different hypotheses, relevance of any specific gene is more trivial and less finite as the models involve larger numbers of genes. The ultimate in such triviality for relevance of specific gene differences to etiology of schizophrenia is encompassed by Manfred Bleuler's hypothesis that different schizophrenic disorders may result from an infinite variety of combinations of different inherited predispositions and different life experiences.[8]

The two extremes of genetic concepts regarding the etiology of schizophrenia involve: (1) the theories of simple Mendelian modes of transmission for genetic diatheses, as described by Kallmann,[1,2] Böök,[3] Slater,[4] and Karlsson;[5] and (2) the concept that there is no finite number of genotypes that are particularly associated with schizophrenia, as indicated by Bleuler's concept of an infinite variety of relevant genotype-environment combinations.[8] Between these two extremes, the multifactor concepts, previously presented by Ödegärd[7] and more recently elaborated by Gottesman and Shields,[9] may be regarded as intermediate models. Mitsuda has considered each of the different models as appropriate for some cases, not for others.[10,11] He has hypothesized multiple etiologies for schizophrenia and has accumulated data which indicate associations of schizophrenic diatheses with Mendelian modes of genetic transmission in some cases, with multiple genetic factors (i.e. involving multiple chromosomal loci) in others, and with no clearly significant genetic factors in still others.

The overall and currently available data may be fit into each of the several hypothesized modes of genetic transmission. The currently available data have *not* been demonstrated to be characteristically more consistent with any one of the genetic models than with all the others. The extensive twin and family data described in the literature have established the etiological relevance of familial factors.[1-5, 7, 10-14] In that context, the most convincing demonstrations of vertical (i.e. parent-to-offspring) transmissions of schizophrenia diatheses have come from studies of children separated from their mothers shortly after birth.[15, 16] These studies have shown that offspring separated early from their schizophrenic mothers tended to approach the schizophrenia incidence of such offspring who were not separated from their schizophrenic mothers, compared to no significantly increased incidence in the offspring who were separated early from their non-schizophrenic mothers. Kety, et al.[6] studied the occurrence of recorded mental disorder among the biological and adoptive families of adopted schizophrenics. Their findings of very significantly increased schizophrenia incidences in both maternal and paternal genetic relatives of the adopted schizophrenics also support the concepts that involve primary relevance of genetic factors for schizophrenia diatheses. One possible alternative to a genetic interpretation of these observations might be a familially transmitted and slow-acting virus or other infectious agent. Schizophrenia in some cases might conceivably be produced by interactions of gene-transmitted characteristics and a specific agent such as a virus.[17] There is no published evidence, however, that is sufficient to establish an infectious etiology for schizophrenia. A plurality of the constitutionally-oriented theoreticians seem to prefer explanations associating the schizophrenic diatheses with genetically transmitted constitutional factors. Nevertheless, no specific mode of genetic transmission for a schizophrenic diathesis has yet been established, the question of multiple etiologies for schizophrenia has not been settled, and investigators of the biology of schizophrenia have hardly begun relevant research on the cellular level.

X-CHROMOSOME-ANEUPLOIDY MOSAICISMS IN FEMALE SCHIZOPHRENICS

The following report is based upon data accumulated during more than six years of studies at the Cleveland Psychiatric Institute on female schizophrenics in several Ohio state psychiatric hospitals. The data include earlier findings in smaller samples that were previously published.[18, 19, 20] Part of this material was presented at the annual meeting of the Society for Biological Psychiatry, Montreal, in 1973.

Materials and Methods

Bilateral buccal smears were screened from 2,125 female patients who had been diagnosed as schizophrenics and were confined in state psychiatric hospitals in Ohio. The smears were fixed on coded slides and then stained with 1.0% orcein solution. Specimens were examined "blind," and patients whose buccal smears indicated possibly-abnormal sex-chromatin complements were subjected to further studies whenever possible.

Karyotype analyses were based on peripheral-blood leukocytes that were successfully cultured by the following micro method. Approximately 0.25 ml of blood was collected, in each instance, in a heparinized capillary tube from a finger stab. Each blood sample was transferred to a culture flask containing 1.5 ml of fetal calf serum, 3.5 ml of TC-199 solution containing 3.3 mM penicillin and 3.4 mM streptomycin, and five drops of phytohemogglutinin. Incubation at 36 degrees Centigrade was maintained for each culture for seventy-two hours. Colchicine was added as a mitotic-arresting agent two hours prior to harvesting the cultured cells. Afterwards, the cells were placed in a hypotonic solution for swelling; then they were fixed, air-dried on slides, stained with carbol fuchsin, and examined with a light microscope. Well-spread metaphase plates were photographed and then analyzed by cut-out karyotyping. The patients with suspicious-looking buccal smears included 216 whose blood samples were originally planned for chromosomal analyses. Of these, however, only 154 were ultimately followed-up with analyses of sufficient cultured cells.

Results

Table 14-I shows the number of patients whose buccal smears were originally screened, the number whose cultured cells were successfully analyzed, and the karyotype findings in the latter group of 154 patients. A total of seventeen patients were observed to show X-chromosomal aneuploidies. This number includes two XXX females, no one-X females, and fifteen X-chomosome-aneuploidy mosaics. Table 14-II shows the sex-chromatin evaluations that were based on the patients' buccal smears, their karyotype decisions, and which of the patients had acentric fragments in any cultured cells. Seven of the seventeen had significant numbers of cells with multiple sex-chromatin bodies, and this number included two who were found to have XXX karyotypes; the remaining five consists of four X/XX/XXX mosaics and one XX/XXX/XXXX mosaic. Ten patients who showed remarkably low counts of sex-chromatin positive cells were found to be X/XX mosaics. None of the patients studied showed stigmata consistent with Turner's syndrome or gonadal dysgenesis.

The observed incidence of females with X-chromosome aneuploidies is at least seventeen in 2,125, or 0.8 percent. This includes two with XXX karyotypes, 0.094 percent, and fifteen mosaics, 0.706 percent.

Discussion

Peripheral blood samples for karyotype analyses were solicited only from the female schizophrenic patients whose buccal smears indicated sex-chromatin abnormalities, thus screening out only those patients with X-chromosome aneuploidies manifested in the buccal cells. Subsequent karyotype analyses were based only on peripheral-blood leukocytes. Thus, only

TABLE 14-I
THE NUMBER OF HOSPITALIZED FEMALE SCHIZOPHRENIC PATIENTS SCREENED WITH BUCCAL-SMEAR EXAMINATIONS, AND NUMBERS AND PERCENTAGES OF THE SCREENED PATIENTS WHOSE CULTURED PERIPHERAL-BLOOD LEUKOCYTES SHOWED X-CHROMOSOME ANEUPLOIDIES.

	Number	Percent of Screened
Screened	2,125	100.000
Karyotyped	154	7.247
Total with X-chromosome aneuploidies	17	0.800
Triple-X	2	0.094
One-X	0	0
X-chromosome-aneuploidy mosaic	15	0.706

TABLE 14-II
CYTOGENETIC OBSERVATIONS ON 17 HOSPITALIZED FEMALE SCHIZOPHRENIC PATIENTS OBSERVED TO BE AFFECTED WITH X-CHROMOSOME ANEUPLOIDIES, TWO WITH CONSISTENT XXX COMPLEMENTS AND 15 MOSAICS.

Patient	Buccal Smear Sex Chromatin	Karyotype	Presence of Acentric Fragments
M.S.	Some Double	XXX	√
E.F.	Some Double	XXX	
B.M.	Some Double	X/XX/XXX	
E.M.	Some Double	X/XX/XXX	√
S.D.	Some Double	X/XX/XXX	
M.F.	Some Double	X/XX/XXX	√
C.V.	Some Double	XX/XXX/XXXX	√
B.H.	Low + Count	X/XX	
V.R.	Low + Count	X/XX	
A.H.	Low + Count	X/XX	
L.W.	Low + Count	X/XX	
W.M.	Low + Count	X/XX	√
F.L.	Low + Count	X/XX	√
M.S.	Low + Count	X/XX	√
I.C.	Low + Count	X/XX	√
D.S.	Low + Count	X/XX	√
A.S.	Low + Count	X/XX	

X-chromosome aneuploidies manifested in these two tissues were screened out. Many of the patients with suspicious-looking buccal smears were not available for chromosome studies. Of the 216 who were originally available, successful cultures were grown and analyzed for only 154, and the other sixty-two were not available for adequate repeated culture attempts. Thus, the positive findings described in the present report may affect a significantly larger proportion of the studied population than was observed.

The karyotype analyses were based only on peripheral-blood leukocytes only in patients who previously indicated possible X-chromosome aneuploidies in their buccal cells. The occurrence of aneuploidies other than those involving X chromosomes was not investigated, and the occurrence of X-chromosome aneuploidies involving other tissues was not investigated. Thus, our observed incidence of X-chromosome abnormalities represents a very low estimate of those who may be affected. The occurrences of cytologically less discernible mutations, such as those involving duplications and/or deletions of only small segments of X chromosomes, or gene mutations, were not investigated. The study was confined to X-chromosome aneuploidies in females because the buccal-smear studies facilitated a simple method for mass screening.

The data indicate that buccal-smear studies alone, designed to determine X-chromosome abnormalities, may be misleading because they do not discriminate between general aneuploidies and mosaicisms. Similar findings were recently described by Vartanyan and Gindilis, who screened Russian female schizophrenic patients with buccal smears and then did follow-up karyotype studies.[21] In the Ohio studies, most of the female schizophrenics whose sex-chromatin screening indicated possible X-chromosome aneuploidies were found to actually be mosaics with both normal and aneuploid cells. Our karyotype analyses have shown that the incidence of nonmosaic X-chromosome aneuploidies observed in the hospitalized schizophrenic females apparently does not differ from the expected incidence in the general population, i.e. 1.2 per thousand for X-trisomy and 0.4 per thousand for

X-monosomy females, compared to our findings of two X-trisomy and no X-monosomy females in 2,125 females.[22]

The actual incidence of X-chromosome-aneuploidy mosaics in the general population has not been determined. The general-population expectancy has been estimated at 0.2 per thousand.[20] Our female schizophrenic population, with at least fifteen such mosaics in 2,125, shows a frequency more than thirty-five times such an estimated general-population frequency. This suggests that there may be a biological association between the occurrence of such chromosomal mosaicisms and the schizophrenic disorders.

Vartanyan and Gindilis used buccal smears to screen 2,431 female schizophrenic patients only for cells indicating multiple sex-chromatin bodies and thus indicating presence of more than two X chromosomes.[21] They utilized follow-up karyotype analyses and found five cases of X-chromosome-aneuploidy mosaicism involving normal (XX) cells and cells with more than two X chromosomes. That is the same number and virtually the same frequency which we found in 2,125 Ohio female schizophrenic patients (re: Table 14-II: patients B.M., E.M., S.D., M.F., and C.V.). Unfortunately, Vartanyan and Gindilis did not include studies of their patients with abnormally low counts of normal sex-chromatin cells.[21] Such a group of patients, in our study, contained ten X/XX mosaics in the same institutionalized females' schizophrenic population that included five mosiacs with two and more-than-two X chromosomes.

Dasgupta et al. screened five hundred male schizophrenic patients for female sex-chromatin patterns.[23] They observed two XXY males and one XY/XO mosaic in that group.

Only 0.7 percent of our studied female schizophrenics were found to be affected with X-chromosome-aneuploidy mosaicisms. If the mosaicisms are interpreted as either primary or secondary contributing factors to schizophrenia etiology, then they are relatively unimportant because only a tiny proportion of schizophrenic patients are involved. Alternatively, however, it has been suggested that the mosaicisms may represent effects of, rather than causative factors for, constitutional peculiarities which also affect predisposition for schizophrenic disorders.[19, 20] Thus, we suggest that an intracellular factor, present in the early embryo and occasionally affecting mitotic errors in embryogenesis, may also later affect the etiology of schizophrenia.

ACENTRIC CHROMOSOMAL FRAGMENTS IN CULTURED CELLS OF SCHIZOPHRENIC PATIENTS

The occurrence of occasional pairs of acentric chromosomal fragments in cultured peripheral-blood leukocytes has previously been reported as being associated with schizophrenia.[18, 19, 20]

Materials and Methods

Successful cultures of peripheral-blood leukocytes, according to the method described (above), were obtained from specimens of 154 female schizophrenic patients (Table 14-I). These included two XXX females, fifteen X-chromosome-aneuploidy mosaics, and 137 with characteristically normal chromosome complements. Detailed karyotype analyses were carried out for all well-spread metaphase plates to determine the incidence of the previously-described acentric chromosomal fragments in the 154 patients whose cultured cells were available.

Results

Table 14-III lists the twenty-one patients whose cultured peripheral blood leukocytes were observed to contain acentric chromosomal fragments. This comprises 13.6 percent of the Ohio patients whose cells were cultured and investigated. The group of 154 included 133 whose cells did not show such fragments. The twenty-one patients whose cells showed acentric fragments included one of the two XXX females, eight of fifteen mosaics, and twelve schizophrenics with normal chromosome complements (i.e. twelve of the 137 patients with generally normal chromosome complements). The affected cells have been tabulated in Table 14-III to show the number occurring in each patient involved, the number of chromosomes in each cell involved, and the patients' karyotype diagnoses.

Discussion

The significance of the acentric fragments is not definitely known. Similar fragments have previously been described in one of the cultured cells from a normal male who had fathered a child affected with the G_1-trisomy syndrome involving a 22/21 translocation;[24] in one of the cultured cells of a mentally retarded mother and one of the cultured cells of her daughter;[25] and in one cultured cell from a schizophrenic male affected with Klinefelter's syndrome.[23]

During the same period in which our samples of schizophrenic patients' cells were cultured, we had also cultured similar numbers of specimens from institutionalized mental retardates and smaller numbers from various other individuals. Intensive efforts were made to observe any similar acentric fragments in cultures from nonpsychotic subjects, but none were observed. Such chromosomal abnormalities in cultured cells evidently occur only with extreme rarity in most populations. Therefore, our observation of such fragments in cultured cells from twenty-one of the 154 (i.e. nearly 14%) studied schizophrenic patients indicates an association between this normally very rare event and schizophrenia. Anders, et al. observed similar fragments in cultured cells from eight out of a group of thirty-five hospitalized psychotic females in England.[26] It is noteworthy that, in our study, eight of the fifteen X-chromosome-aneuploidy mosaics (i.e. more than 50%) showed such fragments in cultured cells. Evidently, in our sample of female schizophrenics, there is a remarkable association between occurrence of mosaicisms that involve X-chromosome aneuploidies, which are the results of abnormal mitotic events that have occurred during their embryogenesis, and occurrence of the acentric chromosomal fragments in the schizophrenics' cultured cells.

The etiology of the observed acentric chromosomal fragments is currently unknown, but their general occurrence is evidently extremely rare. Therefore, their observed occurrence in a significant proportion of schizophrenic patients indicates an association between occurrence of the fragments in cultured cells and the diagnosis of schizophrenia. The particularly striking association observed with the schizophrenics who were found to be X-chromosome-aneuploidy mosaics indicates the occurrence of a particularly strong relationship between: (1) the etiology of the chromosomal abnormality in the patients' mitosing cells in tissue culture and (2) the etiology of the abnormal mitoses which previously occurred in those individuals during embryogenesis.

TABLE 14-III
PATIENTS WHOSE CULTURED PERIPHERAL-BLOOD LEUKOCYTES WERE OBSERVED TO CONTAIN ACENTRIC CHROMOSOMAL FRAGMENTS IN ONE OR MORE CELLS, AS OBSERVED IN CELLS FROM 21 OF THE 154 PATIENTS STUDIED. EIGHT OF THE 21 PATIENTS WITH SUCH FRAGMENTS WERE X-CHROMOSOME-ANEUPLOIDY MOSAICS.*

Patient	Karyotype Diagnosis	Number of cells with acentric fragments (Chromosome number in parentheses)		
		One Pair	Two Pairs	Four Pairs
C.C.	XX	1 (46)		
H.G.	XX	1 (46)		
A.F.	XX	1 (46)		
R.W.	XX	3 (46)		
R.S.	XX	1 (46)		
G.S.	XX	1 (46)		
A.T.	XX	1 (46)		
S.S.	XX	1 (46)		
J.B.	XX	1 (45), 1 (46)		
A.P.	XX		1 (46)	
L.K.	XX		1 (47)	
D.R.	XX	1 (46)	1 (46)	
M.S.	XXX		1 (47)	
*D.S.	X/XX	2 (46)	1 (45)	
*F.L.	X/XX	1 (46)		
*M.Sc.	X/XX	1 (46)		
*W.M.	X/XX	1 (46)		
*I.C.	X/XX		1 (45)	1 (45)
*E.M.	X/XX/XXX			1 (46)
*M.F.	X/XX/XXX		1 (46)	
*C.V.	XX/XXX/XXXX		1 (46)	

DERMATOGLYPHIC PATTERNS IN THE SCHIZOPHRENIC FEMALES SHOWING THE CHROMOSOMAL PECULIARITIES IN CULTURED CELLS

The dermal ridges in the human embryo are determined before the eighth week. Dermatoglyphic peculiarities, therefore, are regarded as effects and indicators of impaired embryonic or fetal growth. The occurrence of dermatoglyphic peculiarities is not associated with any specific etiological factor but does indicate that some impairment occurred early in fetal growth or embryonic life. Numerous pathological conditions have been shown to be associated with significant deviations in finger, palm, and sole print patterns and ridge counts. The literature currently includes such reports for diseases of viral origin,[27] single-factor gene-determined traits,[28] sex-chromosome aberrations,[29, 30] autosomal abnormalities,[31, 32, 33, 34] and numerous traits for which the etiologies are multiple, disputed, or unknown.[35, 36, 37, 38]

Previously published studies of schizophrenic patients have described associations between schizophrenia and abnormal finger and palm print patterns and ridge count frequencies.[39-52] In addition to the fourteen cited studies of dermatoglyphic abnormalities in adult schizophrenics, there have also been a published study based on schizophrenic children[53] and one described for twenty "psychotic" children.[54] The results of the various studies are not in complete agreement with each other, and not all of them have shown significant dermatoglyphic differences between their schizophrenic and control populations.

Materials and Methods

We decided to study dermatoglyphics of the patients who were previously observed to be affected with one or more of the studied cytogenetic peculiarities (Tables 14-IV, 14-V, and 14-VI). Only fourteen of the adult female schizophrenic patients were still available to us for dermatoglyphic and other studies. These included the following schizophrenic patients: seven with characteristically normal karyotypes, but whose cultured cells included one or more showing, in each case, one or more pairs of acentric chromosomal fragments (R.W., R.S., G.S., A.T., J.B., A.P., and L.K. in Table 14-III); one XXX female who showed a pair of acentric fragments in one cultured cell (M.S. in Tables 14-II and 14-III); three X/XX mosaics with acentric fragments (F.L., W.M., and I.C. in Tables 14-II and 14-III); two X/XX/XXX mosiacs with acentric fragments in cultured cells (E.M. and M.F. in Tables 14-II and 14-III); and one XX/XXX/XXXX mosiac with a pair of acentric fragments in a cultured cell (C.V. in Tables 14-II and 14-III). These fourteen patients, selected for their peculiar cytogenetic findings, were compared to fourteen female schizophrenic patients who were included in our original studies and did not indicate any chromosomal abnormalities. The only other criteria for selecting "control" patients were their age, length of current institutional confinement, and current confinement in the same wards of the same hospitals as the matched patients with observed cytological peculiarities.

Dermatoglyphics were obtained with the Faurot inkless method, and the basic fingerprint patterns (arches, loops, and whorls) were identified according to the method of Cummins and Midlo.[36] The

TABLE 14-IV
DERMATOGLYPHIC COMPARISONS IN PRESENT AND PREVIOUSLY PUBLISHED DATA ON ADULT AND CHILDHOOD SCHIZOPHRENICS.

- A. Present study: 14 adult female schizophrenic patients with cytogenetic abnormalities (X-chromosome-aneuploidy mosaicism, occasional occurrence of acentric fragments).
- B. Present study: 14 adult female schizophrenic patients without observed cytogenetic abnormalities (control).
- C. Moller (1935) and Poll (1935) adult female schizophrenics (pooled data), not cytologically examined.
- D. Sank (1968) childhood female schizophrenies, not cytologically examined.

Schizophrenic Group	Arches %	Loops %	Whorls %	Fingers N
A. With cytogenetic abnormalities	2.2	68.9	28.9	135
B. Without cytogenetic abnormalities (control)	10.0	59.0	30.7	140
C. Moller (1935) and Poll (1935) adult female schizophrenics (not cytologically examined)	7.4	65.4	27.2	11280
D. Sank (1968) childhood female schizophrenics (not cytologically examined)	6.5	74.6	18.9	370

Results of Analyses

A x B: $X^2 = 7.793$, d.f. = 2, $p < 0.025$*
A x C: $X^2 = 5.254$, d.f. = 2, $0.05 > p < 0.1$
B x C: $X^2 = 2.714$, d.f. = 2, $0.05 > p < 0.1$
A x D: $X^2 = 8.360$, d.f. = 2, $p < 0.02$*
B x D: $X^2 = 11.442$, d.f. = 2, $0.001 > p < 0.005$*

*Asterisk indicates occurrence of statistically significant difference between the compared groups.

TABLE 14-V
DERMATOGLYPHIC COMPARISONS

A₁. Present study: 5 adult female schizophrenic patients with X-chromosome-aneuploidy mosaicism and acentric fragments.
A₂. Present study: 7 adult female schizophrenic patients with normal chromosome complements but with acentric fragments.
B. Present study: 14 adult female schizophrenic patients without cytogenetic abnormalities.
C. Moller (1935) and Poll (1935): 1128 adult female schizophrenics.
D. Sank (1968): 24 childhood female schizophrenics.
E. Moller (1935): 1486 normal adult females.
F. Poll (1935): 78 normal adult females.

Schizophrenics	Arches %	Loops %	Whorls %	Fingers N
A_1	2.1	83.3	14.6	48
A_2	1.4	59.4	39.2	69
B	10.0	59.3	30.7	140
C	7.4	65.4	27.2	11280
D	6.5	74.6	18.9	370
Nonschizophrenics				
E	7.5	67.2	25.3	14857
F	7.6	65.6	26.8	776

Results of Analyses
A_1 x A_2: $X^2 = 8.274$, d.f. = 2. $0.01 > p < 0.02$*
A_1 x B: $X^2 = 9.464$, d.f. = 2. $0.005 > p < 0.01$*
A_1 x C: $X^2 = 6.948$, d.f. = 2. $0.025 > p < 0.05$*
A_1 x D: $X^2 = 2.240$, d.f. = 2. $0.25 > p < 0.50$
A_1 x E: $X^2 = 84$, d.f. = 2. $p < 0.001$*
A_1 x F: $X^2 = 69$, d.f. = 2. $p < 0.001$*
A_2 x B: $X^2 = 5.686$, d.f. = 2. $0.05 > p < 0.10$
A_2 x C: $X^2 = 7.288$, d.f. = 2. $0.025 > p < 0.05$*
A_2 x D: $X^2 = 15.197$, d.f. = 2. $p < 0.001$*
A_2 x E: $X^2 = 43$, d.f. = 2. $p < 0.001$*
A_2 x F: $X^2 = 33$, d.f. = 2. $p < 0.001$*
C x D: $X^2 = 13.990$, d.f. = 2. $p < 0.001$*

*Asterisk indicates occurrence of statistically significant difference between the compared groups.

data were analyzed with a 2 X K contingency method of a Monroe 1766 electronic programmable calculator. The fourteen patients with cytogenetic abnormalities provided a total of 140 fingerprints (i.e. ten per patient), but five of the prints were not usable. Thus, only 135 fingers were analyzed from this group, compared to 140 fingers from the fourteen patient controls.

Two groups of the chronic female schizophrenic patients with cytogenetic abnormalities were compared to each other: the five patients who showed both X-chromosome-aneuploidy mosaicism and occasional acentric fragments (FL., W.M., I.C., E.M., and M.F. in Tables 14-II and 14-III) versus the seven patients with characteristically normal chromosomal complements who showed one or more cells with one or more pairs of acentric fragments (R.W., R.S., G.S., A.T., J.B., A.P., and L.K. in Table 14-III).

The dermatoglyphic findings in each of our schizophrenic groups were compared to the findings in each of several previously published studies of dermatoglyphics in schizophrenic patients.

Results

Table 14-IV shows the findings of comparisons between fourteen female schizo-

TABLE 14-VI

LENGTHS OF FIRST HOSPITALIZATIONS IN 16 CHRONIC FEMALE SCHIZOPHRENIC PATIENTS AFFECTED WITH X-CHROMOSOME ANEUPLOIDIES (i.e. 15 WITH MOSAICISMS AND ONE WITH GENERAL X-TRISOMY) AND 18 CHRONIC FEMALE SCHIZOPHRENIC PATIENTS WITH NORMAL FEMALE CHROMOSOME COMPLEMENTS. ORIGINALLY, THERE WERE 17 PATIENTS IN EACH GROUP, BUT ONE OF THOSE SUSPECTED OF BEING AFFECTED WITH X-CHROMOSOME ANEUPLOIDY SHOWED ONLY NORMAL CHROMOSOME COMPLEMENTS AND WAS THEREFORE TABULATED WITH THE 'CONTROL' GROUP. THE SECOND OF THE TWO DETERMINED X-TRISOMY-AFFECTED PATIENTS WAS NOT CONFIRMED UNTIL AFTER THIS EVALUATION WAS COMPLETED. THE PATIENTS IN THE 'CONTROL' GROUP WERE NOT MATCHED TO THE ANEUPLOIDY-AFFECTED PATIENTS FOR ANY CRITERIA OTHER THAN THE DIAGNOSIS OF CHRONIC SCHIZOPHRENIA AND CONFIRMED KARYOTYPE DETERMINATIONS.

Lengths of First Hospitalizations in Years

X-Chromosome-Aneuploidy-Affected Chronic Schizophrenic Patients	Chronic Schizophrenic Patients With Confirmed Normal Chromosome Complements
0.23	0.25
0.67	26.00
0.08	0.13
54.00	0.25
50.00	0.06
29.00	0.17
39.00	0.39
46.00	5.00
40.00	0.42
1.17	0.06
36.00	0.25
40.00	9.00
0.02	0.25
0.25	0.03
38.00	0.42
0.25	24.00
	0.25
	9.00
N: 16	18.0
x̄: 23.4	6.4
s: 21.7	11.5

Computed one-tail t value (t test for unequal variances) = 2.81601 at 23 degrees of freedom. For $p = 0.005$, t at 23 d.f. = 2.8073. Therefore, the difference is statistically significant at $p < 0.005$.

phrenics with a cytogenetic abnormality (i.e. the combined group with XXX, X-chromosome-aneuploidy mosaicism, and/or acentric fragments in one or more cultured cells) and fourteen female schizophrenic patients for whom we observed no evidence of cytogenetic abnormalities. There was a statistically significant difference between the two groups in their distributions of arches, loops, and whorls at the level of $p < .025$, indicating that these were two distinct populations dermatoglyphically as well as cytologically.

Comparisons of our adult female "cytogenetic abnormality" schizophrenics with the thirty-seven childhood schizophrenics previously studied by Sank,[53] as well as between our adult female "control" schizophrenics (i.e. with no observed chromosomal abnormalities) and the childhood schizophrenics in the Sank[53] study, revealed highly significant differences between the populations: $p < .02$ and $p < .005$, respectively. Dermatoglyphic laterality comparisons (i.e. comparisons of the characteristics in the right versus left hands), separately evaluated in our "cytogenetic abnormality" schizophrenics and "control" schizophrenics, did not reveal any significant differences within each group or between the two groups.

Table 14-IV shows the results of separate comparisons between the five adult female chronic schizophrenics whose cytogenetic studies revealed both X-chromosome-aneuploidy mosaicisms and pairs of acentric fragments (subgroup A_1), the seven patients with normal female chromosome complements but with occasional acentric fragments (subgroup A_2), and the fourteen control schizophrenics with no cytogenetic abnormalities (group B). There was a significant difference (p < .02) between subgroups A_1 and A_2. Comparison of the data in subgroup A_1 with our schizophrenic controls revealed a highly significant difference. Correlation between subgroups A_2 and the control schizophrenics (B) revealed no significant differences. There was a significant difference between our subgroup A_1 and the adult schizophrenic series of Moller[51] and Poll,[52] as well as between our subgroup A_2 and the combined series of Moller[51] and Poll.[52] Similarly, significant differences were observed between our subgroups A_1 and A_2, and both Moller's[51] and Poll's[52] normal female adults.

Discussion

The data in the present study indicate that the fourteen adult chronic female schizophrenics possessing cytogenetic abnormalities are also dermatoglyphically distinct from the fourteen apparently cytogenetically normal control schizophrenics. The statistically significant difference between the childhood schizophrenics of Sank[53] and each of the present study's two groups of adult schizophrenics (i.e with and without cytogenetic aberrations) indicates the possible existence of at least three dermatoglyphically different schizophrenic populations. The control schizophrenics (i.e. without observed cytogenetic abnormalities) are dermatoglyphically similar to the previously published data of Moller[51] and Poll[52] on schizophrenic populations.

The data indicate that there is a highly significant difference between our two subgroups of schizophrenics with various manifestations of cytogenetic abnormalities, i.e. between the schizophrenics with both X-chromosome-aneuploidy mosaicisms and occasional cells with acentric fragments versus the schizophrenics who revealed no karyotype abnormalities other than occasional acentric fragments.

The various previously published studies of dermatoglyphics and schizophrenia have not been in complete agreement with each other. Nevertheless, the overall picture reveals dermatoglyphic differences between nonschizophrenics, adult schizophrenics, and childhood schizophrenics. The significant dermatoglyphic differences observed in our studies strongly suggest that the schizophrenic patients who were cytologically differentiated from most schizophrenic patients and from nonschizophrenics and who may be subdivided into at least two different cytologically-characterized groups also represent dermatoglyphically different groups. Thus, on the basic of two different variables, cytological and dermatoglyphic, at least three different categories of schizophrenics may be separated from each other and from other schizophrenics, as well as from nonschizophrenic populations.

AN INDEX OF CHRONICITY ASSOCIATED WITH THE CYTOLOGICAL VARIABLES

The clinical histories of the chronic schizophrenic patients who showed X-chromosome aneuploidy in cultured cells were investigated for possible objectively-identifiable differences in their indices of chronicity. Their histories were compared to those of chronic schizophrenic patients who were originally suspected of being affected with cytogenetic abnormalities but whose karyotypes showed normal chromosome complements. Originally, this variable was examined for thirty-four female chronic schizophrenic patients, seventeen suspected for X-chromosome aneuploidies and seventeen with established normal chromosome complements. One of the patients in the former group showed only normal chromosome complements and was therefore tabulated with the latter group (i.e. 17 plus 1 = 18). The second of the two ultimately determined X-trisomy-affected patients was not confirmed until after this evaluation was completed and was therefore not included. One striking difference was observed, as shown in Table 14-VI; the combined group of one XXX patient and fifteen X-chromosome-aneuploidy mosaic patients had significantly longer *first* psychiatric hospitalizations compared to members of the group of schizophrenics with normal karyotypes (both with and without acentric fragments). The observation that the group of fifteen patients with X-chromosome-aneuploidy mosaicisms plus one with general X-trisomy had significantly longer *first* hospitalizations than the other chronic schizophrenic patients suggests that differences in psychometric variables may also define the cytologically categorized and dermatoglyphically different groups.

CONCLUSIONS

The present studies have shown that the cultured peripheral blood leukocytes from a significant minority of schizophrenics, nearly 14 percent, show an occasional cell containing one, two, or four pairs of acentric chromosomal fragments. This kind of chromosomal anomaly has been observed only with extreme rarity except in psychotic populations. Anders, *et al.* observed such fragments in eight out of thirty-five patients at a British psychiatric hospital.[26]

Our studies have found that a very small minority of female schizophrenics, about 0.7 percent, are X-chromosome-aneuploidy mosaics. More than one half of these particular patients also showed acentric fragments in cultured cells. A part of these data has also been replicated in another study: Vartanyan and Gindilis found the same number and incidence of female patients who were normal/supernumerary-X mosaics in a Russian population.[21] The Russian study unfortunately did not include karyotype investigations of patients whose buccal smears showed abnormally low proportions of sex-chromatin-positive cells, a group which we have found to include X/XX mosaics. The latter mosaicism category in our study occurred twice as frequently as the former (i.e. with supernumerary-X cell lines) mosaicism category.

Significant dermatoglyphic differences observed in the present studies strongly suggest that the schizophrenic patients who were cytologically differentiated from most schizophrenics and from nonschizophrenics and who may be subdivided into at least two cytologically characterized groups also represent dermatoglyphically different groups. Thus, on the basis of two different types of variables, cytological and dermatoglyphic, at least three categories of schizophrenics can be differentiated from each other and from the general population.

The findings of cytogenetic abberations and dermatoglyphic differences in selected schizophrenic patients suggest speculations about their clinical significance. These traits might, in some way, serve to differentiate schizophrenic patients into one or more subcategories. For example, associations of the cytogenetic abnormalities and the quantitative dermatoglyphic deviations may differ between patients affected with familial, as compared to so-called sporadic, schizophrenias. There may be other associations in schizophrenic patients with the types of cytogenetic aberrations and/or quantitative dermatoglyphic characterizations reported by us but with the presence of currently unidentified additional markers and/or clinical associations, e.g. cancer, diabetes mellitus, and other diseases with familial occurrence among close relatives.

These findings, in addition to potential pertinence in the differential diagnosis and etiologic classification of schizophrenia, might also have potential importance in the prognosis and in responses to certain therapeutic variables. Analogously, for example, patients with the Philadelphia chromosome and chronic myelogenous leukemia have a significantly improved response to chemotherapy and in turn a better overall prognosis than do patients with chronic myelogenous leukemia with the absence of the Philadelphia chromosome. On the other hand, patients with the familial variety of cutaneous malignant melanoma are characterized by an earlier age of onset, an increased frequency of multiple primary malignant melanomas, and a significantly better overall prognosis than do patients with the so-called sporadic form of cutaneous malignant melanoma (which accounts for about 97% of occurrences). These two analogies are suggested because, in one example, it pertains to a cytogenetic factor which appears to influence prognosis (i.e. the Philadelphia chromosome in chronic myelogenous leukemia); and, in the other, it pertains to an apparent point mutation in that familial cutaneous malignant melanoma appears to behave as an autosomal dominant, and the familial and evidently gene-transmitted variety has a different prognosis from the sporadic variety.

If additional research on schizophrenia should confirm a relationship between prognosis and differences in natural history as being associated with occurrence of the described cytogenetic abnormalities and/or dermatoglyphic characteristics, a subsequent step might then involve determination of what primary or associated factors either enhance or diminish the prognostic outlooks for these *different groups* of schizophrenic patients.

The observation that patients with X-chromosome-aneuploidy mosaicisms had significantly longer *first* hospitalizations than the other ("control") chronic patients suggests that psychometric variables, possibly including some process-reactive dimensions, might be found to differ between the various cytologically characterized groups of schizophrenics. Subsequent investigations might then be designed to determine what primary or associated factors either enhance or diminish the prognostic outlooks for members of these different categories of schizophrenic patients.

Our findings evidently support the hypothesis that the schizophrenia syndrome represents a psychobiological complex involving multiple etiologies which include genetic, developmental, psychosocial, and possibly even infectious, factors. Separation of the syndrome complex into several biologically defined subgroups may enable investigators to more successfully explore and define etiologies of, and therapies for, different schizophrenia subgroups.

REFERENCES

1. Kallmann, F.J.: The genetic theory of schizophrenia: An analysis of 691 schizophrenic twin index families. *Am J Psychiatry, 103*:309, 1946.
2. Kallmann, F.J.: *Heredity in Health and Mental Disorder.* New York, Norton, 1953.
3. Böök, J.: A genetic and neuropsychiatric investigation of a North-Swedish population, with special regard to schizophrenia and mental deficiency. *Acta Genet Statist Med, 4*:1, 1953.
4. Slater, E.: The monogenic theory of schizophrenia. *Acta Genet Statist Med, 8*:50, 1958.
5. Karlsson, J.: *The Biologic Basis of Schizophrenia.* Springfield, Thomas, 1966.
6. Kety, S.S., Rosenthal, D., Wender, P.H., and Schulsinger, F.: The types and prevalence of mental illness in the biological and adoptive families of adopted schizophrenics. In Rosenthal, D., and Kety, S.S. (Eds.): *The Transmission of Schizophrenia.* Oxford, Pergamon, 1968, pp. 345-362.
7. Ödegård, Ö.: The psychiatric disease entities in the light of a genetic investigation. *Acta Psychiatr Scand, 39*: Suppl 94, 1963.
8. Bleuler, M.: A 23-year longitudinal study of 208 schizophrenics and impressions in regard to the nature of schizophrenia. In Rosenthal, D., and Kety, S.S. (Eds.): *The Transmission of Schizophrenia.* Oxford, Pergamon, 1968, pp. 3-12.
9. Gottesman, I.I., and Shields, V.: A polygenic theory of schizophrenia. *Proc Natl Acad Sci, 58*:199, 1967.
10. Mitsuda, H.: *Clinical Genetics in Psychiatry.* Tokyo, Igaku Shoin, 1967.
11. Mitsuda, H.: Heterogeneity of schizophrenia. In Kaplan, A.R. (Ed.): *Genetic Factors in "Schizophrenia."* Springfield, Thomas, 1972, pp. 276-293.
12. Karlsson, J.: A two-locus hypothesis for inheritance of schizophrenia. In Kaplan, A.R. (Ed.): *Genetic Factors in "Schizophrenia."* Springfield, Thomas, 1972, pp. 246-255.
13. Ödegård, Ö.: The multifactorial theory of inheritance in predisposition to schizophrenia. In Kaplan, A.R. (Ed.): *Genetic Factors in "Schizophrenia."* Springfield, Thomas, 1972, pp. 256-275.
14. Slater, E.: The case for a major partially dominant gene. In Kaplan, A.R. (Ed.): *Genetic Factors in "Schizophrenia."* Springfield, Thomas, 1972, pp. 173-180.
15. Heston, L.L.: Psychiatric disorders in foster home reared children of schizophrenic mothers. *Br J Psychiatry, 112*:819, 1966.
16. Rosenthal, D., Wender, P.H., Kety, S.S., Schulsinger, F., Welner, J., and Östergaard, L.: Schizophrenics' offspring reared in adoptive homes. In Rosenthal, D., and Kety, S.S. (Eds.): *The Transmission of Schizophrenia.* Oxford, Pergamon, 1968, pp. 377-391.
17. Torrey, E.F., and Peterson, M.R.: Slow and latent viruses in schizophrenia. *Lancet, 2*:22, 1973.
18. Kaplan, A.R., and Cotton, J.E.: Chromosomal abnormalities in female schizophrenics. *J Nerv Ment Dis, 147*:402, 1968.
19. Kaplan, A.R.: Chromosomal mosaicisms and occasional acentric chromosomal fragments in schizophrenic patients. *Bio Psychiatry, 2*:89, 1970.
20. Kaplan, A.R.: Chromosomal aneuploidy, genetic mosaicism, occasional acentric fragments, and schizophrenia: Association of schizophrenia with rare cytogenetic anomalies. In Kaplan, A.R. (Ed.): *Genetic Factors in "Schizophrenia."* Springfield, Thomas, 1972.
21. Vartanyan, M.E., and Gindilis, V.M.: The role of chromosomal aberrations in the clinical polymorphism of schizophrenia. *Int J Ment Health, 1*:93, 1972.
22. MacLean, N., Harden, D.G., Court-Brown, W.M., Bond, J., and Mantle, D.J.: Sex-chromosome abnormalities in newborn babies. *Lancet, 1*:286, 1964.

23. Dasgupta, J., Dasgupta, D., and Balasubrahmanyan, M.: XXY syndrome XY/XO mosaicism and acentric chromosomal fragments in male schizophrenics. *Indian J Med Res, 61*:62, 1973.
24. Cohen, M.M., and Davidson, R.G.: Down's syndrome associated with a familial (21q−; 22q+) translocation. *Cytogenetics, 6*:321, 1967.
25. Lejeune, J., Dutrillaux, B., Lafourcale, J., Berger, R., Abongi, D., and Rethioré, M.O.: Endoréduplication sélective du bras long du chromosome 2 chez une femme et sa fille. *C R Acad Sci (Paris), 266*:24, 1968.
26. Anders, J.M., Jagiello, G., Polani, P.E., Gianelli, F., Hamerton, J.L., and Lieberman, D.M.: Chromosomal findings in chronic psychotic patients. *Br J Psychiatry, 14*:1167, 1968.
27. Achs, R., Harper, R.G., and Siegel, M.: Unusual dermatoglyphic findings associated with rubella embryopathy. *N Engl J Med, 274*:148, 1966.
28. Hodges, R.E., and Simon, J.R.: Relationship between finger print patterns and Wilson's disease. *J Lab Clin Med, 60*:629, 1962.
29. Forbes, A.P.: Finger prints and palm prints (dermatoglyphics) and palmar-flexion creases in gonadal dysgenesis, pseudohypoparathyroidism and Klinefelter's syndrome. *N Engl J Med, 270*:1268, 1964.
30. Holt, S.B., and Lindsten, J.: Dermatoglyphic anomalies in Turner's syndrome. *Ann Hum Genet, 28*:87, 1964.
31. Cummins, H.: Dermatoglyphic stigmata in mongoloid imbeciles. *Anat Rec, 73*:407, 1939.
32. Gibson, D.A., Uchida, I.A., and Lewis, A.J.: A review of the 18 trisomy syndrome. *Med Biol Illus, 13*:80, 1963.
33. Holt, S.B.: Finger-print patterns in mongolism. *Ann Hum Genet, 27*:279, 1964.
34. Uchida, I.A., Patau, K., and Smith, D.W.: Dermal patterns of 18 and D1 trisomics. *Am J Hum Genet, 14*:345, 1962.
35. Barbeau, A., Trudeau, J.G., and Coiteux, C.: Fingerprint patterns in Huntington's chorea and Parkinson's disease. *Can Med Assoc J, 92*:514, 1965.
36. Cummins, H., and Midlo, C.: *Finger Prints, Palms, and Soles.* New York, Dover, 1966.
37. Miller, J.R., and Giroux, J.: Dermatoglyphics in pediatric practice. *J Pediatr, 69*:302, 1966.
38. Sanchez, C.A.: Fingerprint patterns in congenital heart disease. *Br Heart J, 26*:524, 1964.
39. Dasgupta, J., Dasgupta, D., and Balasubrahmanyan, M.: Dermatoglyphics in the diagnosis of schizophrenia. *Indian J Psychiatry, 15*:104, 1973.
40. Polednak, A.: Dermatoglyphics of Negro schizophrenic males. *Br J Psychiatry, 120*:397, 1972.
41. Rothhammer, F., Pereira, G., Camousseight, A., and Benado, M.: Dermatoglyphics in schizophrenic patients. *Hum Hered, 21*:198, 1971.
42. Stowens, D., Sammon, J.W., and Proctor, A.: Dermatoglyphics in female schizophrenia. *Psychiatr Q 44*:516, 1970.
43. Rosner, F., and Steinberg, F.S.: Dermatoglyphic patterns of Negro men with schizophrenia. *Dis Nerv Syst, 29*:739, 1968.
44. Mellor, C.S.: Fingerprints in schizophrenia. *Nature (Lond), 213*:939, 1967.
45. Singh, S.: Dermatoglyphics in schizophrenia. *Acta Genet, 17*:348, 1967.
46. Beckman, L., and Norring, A.: Finger and palm prints in schizophrenia. *Acta Genet, 13*:170, 1963.
47. Raphael, T., and Raphael, L.G.: Fingerprints in schizophrenia. *JAMA, 180*:215, 1962.
48. Pons, J.: Relaciones entre esquizofrenia y lineas dermopapilares. *Genetica Iberica, 11*:1, 1959.
49. Wendt, G.G., and Zell, W.: Schizophrenie and Fingerleistenmuster. *Arch Psychiat Z Neurol, 186*:456, 1951.
50. Duis, B.T.: Fingerleisten bei Schizophrenen. *Z Morphol Anthropol, 36*:391, 1937.
51. Moller, N.B.: Undersøgelser over

fingerautrycket som konstitutionelt. *Kendetegn ved Sindssygdomme Hospitalstidende*: 1085-1111, 1935.
52. Poll, H.: Dactylographische Geschlëchtsonterschiede der Schizophrenen. *Monatsschr Psychiatr Neurol, 91*:65, 1935.
53. Sank, D.: Dermatoglyphics of childhood schizophrenia. *Acta Genet, 18*:300, 1968.
54. Hillbun, W.B.: Dermatoglyphic findings on a group of psychotic children. *J Nerv Ment Dis, 151*:352, 1970.

Chapter 15

PSYCHOPHARMACOGENETICS, PHYSIOLOGICAL GENETICS, AND SLEEP BEHAVIORS

Kazuhiko Abe

Although the term "human behavior genetics" has only recently been introduced, various human behavior patterns have been investigated from genetic aspects.

The oldest and perhaps the best explored field is psychiatric genetics, which is dealt with elsewhere in this book. Another well explored field is the one in which genetic factors function more or less as limiting factors, i.e. in genetics of abilities or development, for example IQ, some psychological variables,[1] and local motilities such as flexibility of the distal joints and overextension of the middle joints of fingers and tongue gymnastics such as rolling, twisting of the tongue, and formation of clover-leaf tongue (for review and an extensive twin study on local motilities see Inouye[2]), speech development,[3] and acquisition of bladder control.[4]

Still another field in which rapid progress has recently been made involves biochemical genetics and investigations of human enzymatic polymorphism, which may in the future prove much more useful in understanding human behavior than they are today. Besides the aforementioned topics there is a vast but poorly explored field of human behavior genetics.

The present writer has selected three topics which may be of interest for psychiatrists and psychologists. Firstly, the controversial field of psychopharmacogenetics will be briefly reviewed. Secondly, genetic aspects of some physiological functions will be reviewed. Since genetic aspects of EEG are reviewed in another chapter of this book, and elsewhere by Vogel,[5] they are not included in the present discussion. Thirdly, studies of sleep behaviors will be presented together with the present writer's own data recently obtained.

PSYCHOPHARMACOGENETICS

Since genetic aspects of changes in behavior due to drugs in general are dealt with elsewhere in this book, we confine ourselves here to the response to antidepressants and chlorpromazine. There are two groups of antidepressants now in use: monoamine oxidase inhibitors, such as phenelzine and isocarboxazid; and tricyclic antidepressants, such as imipramine, desmethyl imipramine, and amitriptyline. Response to the former of the depressive symptoms varies from patient to patient: in some patients it alleviates the depressive symptoms of long duration dramatically within several days while in others it exerts no beneficial effect after continuous administration for more than two weeks.

Pare et al. tested, by a double blind study, the hypothesis that response of a patient to a particular group of antidepressants might depend on the genetic type of his depression and that the patients' first degree relatives should respond in a similar manner.[6,7] First degree relatives responded to antidepressants in a similar way to the probands. He suggested that there might be two genetically specific types of depression and that one type would respond to the imipramine group and the other to the monoamine oxidase inhibitors.

Independently, based on results of a family study, Angst had suggested in 1961 that if the proband responded favorably to a certain antidepressant, the drug also tended to be effective in the probands' relatives who were affected with endogenous depression.[8] This observation was confirmed by his later investigation on forty-seven pairs of relatives with endogenous depression, of whom forty pairs were concordant and four were discordant with respect to the effect of imipramine, a statistically significant result at the chance probability level of less than 0.001.[9]

Along with the advancement of pharmacogenetics, further research was made in this field. It is well established that there are interindividual differences in isoniazid (an antituberculous agent) metabolism which are largely determined genetically.[10] Those who inactivate isoniazid rapidly are called "rapid inactivators," and the others are "slow inactivators."[11] Monoamine oxidase inhibitors which are closely related structurally to isoniazid are considered to be matabolized by the same enzyme, and as might have been expected, Price-Evans found that one of them, phenelzine, produced severe side effects only in slow inactivators.[12] However, there were no significant differences in the improvement of depressive symptoms. Sjöqvist, et al. investigated the plasma level of tricyclic antidepressants, namely desmethyl imipramine and nortriptyline, in relation to side effects in man and found that the side effects were marked in those individuals with high plasma levels of these drugs, that is, in those who metabolized the drugs slowly.[13]

Hammer, Mårtens, and Sjöqvist administered, to five hospitalized women with various types of depression, desmethylimipramine 25 mg three times a day at eight-hour intervals for seventeen days, nortriptyline 25 mg three times a day for the next eighteen days, and then again desmethylimipramine for twenty days and oxyphenylbutazone 100 mg three times a day for two days; and they compared the steady-state plasma levels of desmethylimipramine and nortriptyline with the half-life of oxyphenylbutazone in each patient.[14] The patients who achieved a relatively high steady-state plasma level with one of these drugs also achieved a high steady-plasma level with the other drug tested and showed a long half-life of oxyphenylbutazone. The findings suggested nonspecificity of the drug-hydroxylating enzymes in man and also the importance of genetic factors' influencing the steady-state plasma level of these antidepressants since it had previously been shown that an individual's half-life of oxyphenylbutazone was largely determined genetically.[15,16] However, the relationship between the antidepressant effect and plasma imipramine level is not a straightforward matter. Haidu, et al. showed that patients who responded to imipramine therapy showed low plasma levels and that refractory cases showed high levels.[17] Recent investigation by

Zeidenberg, et al. showed that although clinical improvement by imipramine correlated with (intraindividual variation of) drug levels in patients in whom the drug was effective, the same blood level produced a therapeutic response in one but not in another.[18] Their findings suggest that many factors other than blood levels governed the patients' ultimate responses.

The preceding investigations on antidepressants may be summarized as follows. The severity of side effects depends on plasma levels of the drugs and, hence, on genetically-determined types of inactivating enzymes. On the other hand, the antidepressant effect is not simply predictable from the enzyme activities. Nevertheless, there is evidence suggesting that the antidepressant effect also depends on some genetic makeup of the individual.

With regard to chlorpromazine, one of the oldest phenothiazines for the treatment of schizophrenics, the relationship between its metabolism and clinical effect is even less well-known. In man, chlorpromazine is almost entirely dependent on its rate of metabolism for its removal from the body. The rate of decline of the plasma level is a reflection of the rate of metabolism, and the half-life varies between patients over the range of two to thirty-one hours.[19] Genetic investigations were not carried out on its psychiatric effects, but on one of its side effects, namely drug-induced Parkinsonism.

Myrianthopoulos, et al. made a survey of 728 relatives of fifty-nine propositi who developed symptoms of Parkinsonism when receiving phenothiazine drugs and 777 relatives of sixty-seven controls who proved resistant to the Parkinsonian effects on high dosage (e.g. chlorpromazine 600 mg daily).[20] Among the relatives of the propositi, thirteen cases of Parkinson's disease were found and, among the relatives of the controls, three cases of Parkinson's disease were found. The frequency in the former group was much higher than among the relatives of controls or in the general population, a result which suggests that there may be a hereditary susceptibility to Parkinsonism produced by phenothiazines.

Now we change our topic from psychotropic drugs to definitely older and more popular substances, namely coffee and alcohol. Studies on the individual variations in response to them are surprisingly scanty. Goldstein, et al. found that decaffeinated coffee containing 300 mg of caffein increased the time to get to sleep in some subjects and tended to increase subjects' alertness and physical activity.[21, 22] In others, there was an increase in nervousness, and in still others, the drug was without any effect at all. The variation in these effects was not due to variation in plasma levels or rate of metabolism or excretion.

As to the response to alcohol, there are considerable variations in changes in skin color, sleepiness, mood, aggression, and, above all, in sociability. The response may vary with the circumstance in which the beverage is taken and also with the amount and alcohol concentration of the beverage, but Takala, et al. are of the opinion that there is a characteristic behavioral response for each individual.[23] In an exploratory investigation of eleven pairs of monozygotic twins, Abe asked each twin about his responses to alcohol and coffee, with particular emphasis on changes in activity, mood, sleep, and autonomic reactions.[24] Additional information was obtained from husbands, wives, and cotwins. There was a significant within-pair concordance for insomnia after drinking coffee and for overtalkativeness after drinking alcohol. However, since the concordance rate in dizygotic twins was not studied, the result cannot yet be concluded with certainty as being due to genetic factors.

SOME PHYSIOLOGICAL GENETICS

Physiological genetics contributes an important approach to the understanding of human behavior characteristics. Physiological activity involves behavior of parts or systems of an organism, regulates unconscious or involuntary activities, and also limits or modifies conscious human life. In view of the recent advances in psychophysiology, it should be useful for psychiatrists and physiologists to review how the physiological variables which are commonly employed, especially their variations within normal limits, have been studied from genetic aspects. The following is a review of some functions regulated by the autonomic nervous system, such as cardiac, vasomotor, and glandular activities and metabolism.

Research in this field started in the early 1920's. However, in the early studies present methods of complex blood typing were not available, and zygosity determinations were not as reliable as they are today. Neither were the sizes of samples of twins always satisfactory for statistical analysis. Therefore, conclusions must be drawn from them with due reservations.

With regard to cardiovascular functions, Weitz published his extensive study on forty-five pairs of twins and found that systolic blood pressure and the degree of respiratory arrhythmia are nearly equal between the twins.[25] The pulse rate at rest was nearly equal in twins except a few pairs in whom the one who had been engaged in sports or hard physical work showed significantly lower pulse rate. EKG's of the monozygotic twins were so similar that in some cases tracings from co-twins could be distinguished from one another only with difficulty. He also found that twins were always concordant with respect to cyanosis and swelling of fingers. Curtius and Korkhaus investigated pulse rate, blood pressure, and blood sugar curves after subcutaneous injection of 1 mg adrenalin in three pairs of monozygotic twins.[26] A marked similarity between the cotwins in all three pairs was demonstrated with regard to their blood pressures and sugar curves.

Werner injected 1 mg atropin, 10 mg pilocarpin, and 1 mg adrenalin, each on different days, into fifteen pairs of monozygotic and fifteen pairs of dizygotic twins.[27] Then he observed their reactions, such as dryness of the mouth, salivation, sweating or palpitation, respectively, and recorded curves for pulse and respiratory rate and blood pressure for two to three hours. He found that the maximum reactions for each of the aforementioned variables was significantly more similar within the monozygotic than within the dizygotic pairs of twins. He used the ratio, (MZ difference) ÷ (DZ difference) as an index and emphasized importance of the genetic factors in these autonomic reactions.

Ikai and Nakanishi examined the heartbeat intervals measured by cardiotachography at rest and during and after twenty seconds' breath holding in twenty pairs of monozygotic and thirteen pairs of dizygotic twins, all of whom were candidates for the Twins' Class of the Middle School attached to Tokyo University.[28] The intrapair correlation was significantly higher in the monozygotic twins compared with the dizygotic pairs with respect to the variance of heartbeat interval at rest and after the breath holding and the amount of change due to breath holding. The heartbeat interval changes observed to occur during the breath holding were divided into six types according to the shapes of fluctuations observed in the cardiotachograms. In 82 percent of the monozygotic pairs, both co-twins belonged to the same type, while only 23 percent of the dizygotic pairs were concordant in the same sense, a difference which is statistically significant at the 5.0 percent probability level.

Vandenberg, et al. found a significant hereditary component in reactions of the heartbeat frequency and breathing rate to a startling event such as the unexpected flash of light.[29]

Werner investigated basal metabolism at every thirty-minute interval for three hours in ten pairs of monozygotic and ten pairs of like-sexed dizygotic twins after fasting for ten to twenty hours and after consuming breakfasts consisting of 100 g meat, 30 g butter, 100 g bread, and 200 cc soup.[30] He found that the mean difference between values in the same individuals observed on two successive investigations is much smaller than the mean intrapair cotwin difference in dizygotic twins.

It is well-known that blood sugar levels influence psychological responses. Werner has investigated effects of carbohydrate metabolism.[31] He measured fasting blood sugar levels and the levels at thirty-minute intervals for four hours after ingestion of 50 g glucose in sixteen pairs of monozygotic and seventeen pairs of like-sex dizygotic twins. He divided the pairs into groups defined as similar, intermediate, and dissimilar, according to the shapes of the blood sugar curves. Among monozygotic pairs, he found ten similar, five intermediate, and one dissimilar pair; and, among dizygotic pairs, there were two similar, six intermediate, and nine dissimilar pairs. He found that the fasting blood sugar levels and maximum blood sugar levels were especially similar in pairs of monozygotic twins.

Kamitake studied blinking frequencies in twins.[32] He counted the numbers of blinks per minute in forty pairs of monozygotic and twenty-four pairs of dizygotic twins during psychological testing, i.e. an IQ test and the Kraepelin test, and found that monozygotic twins showed striking co-twin similarities with an intrapair coefficient of correlation of 0.935. At the same time, he found that the intrapair coefficient of correlation for dizygotic twins was 0.043. The difference between the two categories of twins was statistically significant at the 1.0 percent probability level. However, no such difference was observed according to zygosity when the subjects were asked not to blink or asked to gaze at a point. His finding suggests that blinking frequency is largely influenced by genotype when attention is not concentrated on the eye or on blinking.

As to peripheral vasomotor activities, Curtius investigated ninety-seven pairs of monozygotic and thirty-eight pairs of dizygotic twins, one or both of whom manifested some kind of so-called lability of the autonomic nervous system, and he obtained the following concordance rates.[33]

Acrocyanosis: monozygotic pairs, 92.3; dizygotic pairs, 30.8.
Frostbite: monozygotic pairs, 75.3; dizygotic pairs, 23.8.
Acroparesthesia: monozygotic pairs, 55.6; dizygotic pairs, 0.0.

There are also large interindividual differences in susceptibility to frostbite; the same individuals tend to suffer from it repeatedly. Frostbite is considered to occur as a result of an exaggeration of the normal arteriolar contraction and capillary dilatation in response to cold. Siemens noted that frostbite occurred frequently in certain families, and he regarded it as a genetic-constitutional characteristic.[34] The first systematic family study on frostbite was done by Harris.[35] He investigated Navy personnel who suffered from frostbite frequently and obtained a result which is suggestive of irregular Mendelian dominant inheritance for high susceptibility to frostbite.

Abe, Amatomi, and Kajiyama investigated frostbite in 796 three-year-old children and in childhood histories of their parents in a housing area comprising a fairly homogeneous middle-class population.[36] The data on childhood frostbite of the parents were obtained by mailed questionnaires to their respective mothers, that is, from the paternal and maternal grandmothers of the three-year-old children. The investigators found that frostbite is most frequent in the children of pairs of parents who had both suffered from frostbite during childhood; and that it is least frequent in the children of pairs of parents who were both characterized by not having suffered from frostbite. Although the effect of diet, etc. should be also taken into consideration, common environment does not play an important role here since susceptibility to frostbite in children was compared with that of the parents in a past environment about a generation ago.

Abe, Amatomi, and Kajiyama reported similar findings with regard to susceptibility to motion sickness.[36] Motion sickness occurred in 36 percent of children whose parents had both suffered from motion sickness in childhood, in 14 percent of children only one of whose parents had suffered, and 7 percent of children whose parents had both not suffered from motion sickness. There was a clear sex difference in the prevalence of motion sickness already evident at the age of three: 6 percent in boys and 12 percent in girls.

As to sweat gland activities, Siemens examined palmar sweating in thirty-five pairs of monozygotic and in twenty-three pairs of dizygotic twins; and he divided each individual into three categories, pathologically wet (hyperidrosis palmoplantaris), wet, and dry.[34] With respect to hyperidrosis, all three pairs of monozygotic twins were concordant, while only one pair out of three pairs of dizygotic twins was concordant. The rest of the twin pairs, thirty-one out of thirty-two monozygotic pairs and eleven out of twenty dizygotic pairs of twins, were concordant with respect to the normal wet-dry dichotomy. Cloward found hyperidrosis palmaris especially common in Japanese of Okinawan ancestry.[37]

A psychophysiological parameter, now widely used and which is related to sweating, is the galvanic skin reflex (also called psychogalvanic reflex). As to this reflex, Carmena found the magnitude and duration of the response to be much more similar in pairs of monozygotic twins than in pairs of dizygotic twins.[38] Jost and Sontag compared galvanic skin response records in pairs of siblings and in a few pairs of twins and concluded that there was evidence for a hereditary factor.[39] Conversely, Vandenberg, Clark, and Samuels examined galvanic skin responses to startle stimuli (flash, bell, and hammer) in thirty-four pairs of monozygotic and in twenty-six pairs of dizygotic twins and observed that the F ratio of the within-pair variances of the former and the latter were nonsignificant, that is, the monozygotic twin pairs are no more similar than those of dizygotic pairs.[29] Lader and Wing investigated various physiological measures in eleven pairs of monozygotic twins and in eleven pairs of like-sex dizygotic twins and concluded that at least three variables are subject to genetic influences: the rate of habituation of the galvanic skin response, the number of spontaneous skin conductance fluctuations, and the pulse rate.[40]

Goodman, *et al.* carried out twin studies on the rate of salivary secretion.[41] They measured parotid and sublingual-submaxillary rates of flow in sixteen pairs of monozygotic and eighteen pairs of dizygotic twins and found that the variance in monozygotic twin pairs was significantly smaller; namely the probability level was less than 0.005 for parotid and less than 0.025 for sublingual-submaxillary rates of flow, respectively (F test).

Lastly, the writer would like to present his recent findings on excessive salivary drooling of infants. It is related to the child's inability to swallow saliva. Table 15-I shows the incidence of excessive drooling in the past history of 381 three-year-old children examined in 1970, regarding childhood data on both of the parents which are available. The method of obtaining childhood data of the parents from their respective mothers, i.e. from the paternal and maternal grandmothers, will be dealt with later in the section on sleep behaviors. Contrary to motion sickness, excessive drooling occurs about twice as frequently in boys. In this study, the children were divided according to the childhood data of the parents as shown in Table 15-II. The sex difference is also apparent

TABLE 15-I
INCIDENCE OF EXCESSIVE DROOLING IN BOYS AND GIRLS

	Excessive Drooling		Total
	+	−	
Boys	69 (35%)	129	198
Girls	26 (14%)	157	183
Total	95	286	381
	$X^2 = 19.54$	$P < 0.001$	

TABLE 15-II
INCIDENCE OF EXCESSIVE DROOLING IN CHILDREN AND CHILDHOOD DATA OF THEIR PARENTS. EXCESSIVE DROOLING IS INDICATED BY A PLUS SIGN, +

Excessive Drooling in Parents		Excessive Drooling in Children		Total
Father	Mother	+	−	
+	+	3 (60%)	2	5
−	+	10 (46%)	12	22
+	−	13 (32%)	28	41
−	−	69 (22%)	244	313
Total		95 (25%)	286	381

TABLE 15-III
EXCESSIVE DROOLING IN 1017 CHILDREN EXAMINED IN 1965 AND 1969 TO 1970, AND IN THEIR PARENTS DURING CHILDHOOD

Excessive Drooling in Childhood of:	Excessive Drooling in Children		Total
	+	−	
Both of the parents	10 (59%)	7	17
One of the parents	61 (31%)	133	194
Neither of the parents	174 (22%)	632	806
Total	245 (24%)	772	1017
	$X^2 = 19.72$	$P < 0.001$	

in the childhood data of the parents where fathers had manifested excessive drooling about twice as frequently as the mothers. A marked sex difference in the incidence, already apparent in infancy, appears to be a strong suggestion favoring a genetic influence in production of the characteristic, as was also evident in susceptibility to motion sickness.

In order to see the incidence of excessive drooling in each type of parental mating, the data which had been prepared from 636 children examined in 1965 and 1969 were added, and the results are shown in Table 15-III. As is apparent from the percentage shown in the table, the incidence is highest in children where both of the parents had manifested the same phenomenon, suggesting that genetic factors, among others, are important factors in bringing about excessive drooling in children.

SLEEP BEHAVIOR GENETICS

Since humans spend about one third of their lives asleep, human behavior genetics should also include sleep behavior as one of its important topics. It was established by recent studies that there is a phase of sleep accompanied by rapid eye movement (REM sleep), which is especially associated with dreaming. Contrary to the popular view that sleepwalking and enuresis are the acting out of dreams, they are found to occur in nonrapid-eye-movement sleep in phases III and IV where dreaming is less common.[42,43] Sleep-talking occurs both in REM and non-REM sleep.[44] Teeth-grinding occurs mostly in REM sleep.[45]

Zung and Wilson made a Markov analysis of sleep and dream patterns in three monozygotic and three dizygotic pairs of twins and found no significant difference between the monozygotic and dizygotic twins with regard to the time required for onset of the sleep state, duration of sleep, distribution of the continuous sleep EEG as percent time spent in the various stages of sleep, and percent "rapid eye movement" time.[46] However, there was complete concordance in the monozygotic pairs of twins, with respect to the major shifts in sleep stages and the periodic fluctuations. The periodicity of "rapid eye movement" pattern was concordant in the monozygotic co-twins and discordant in dizygotic co-twins.

With regard to sleep behavior, Geyer made a detailed observation on thirteen pairs of monozygotic and thirteen pairs of dizygotic female twins during their sleep for three weeks at the "twin camp" (Zwillingslager).[47] In monozygotic twin pairs, there were marked co-twin concordance with regard to the tendencies to dreaming, sleep-talking, sleepwalking, teeth-grinding, posture during sleep, and rhythmic automatic movement during hypnagogic stage compared with the dizygotic twin pairs in whom cotwin discordance was the rule. There was also a marked similarity in behavior during awakening, such as slow and gradual awakening with a long confused period, or immediate awakening, or early awakening.

Pierce and Lipcon examined thirty-four naval recruits who while in training walked in their sleep and found family histories of sleep-walking in nineteen of them.[48] In a control group of sixty nonsomnambulistic electronic school students, no family history of sleepwalking was found.

Bakwin compared the degree of intrapair concordance in monozygotic and dizygotic twins with respect to sleepwalking.[49] Nine of nineteen monozygotic twin pairs were concordant while only one of the fourteen dizygotic twin pairs were concordant, statistically significant at a probability level of less than 0.04. He further found that in twenty-four families with eighty-seven children, where one of the parents gave a history of sleepwalking, twenty-four of the children were sleep-walkers, compared to 4.3 percent of 1177 children in 292 families where neither parent walked in his or her sleep. These findings suggest that sleep-walking has a genetic basis.

The present writer also has reported family studies on childhood sleep characteristics in which childhood data of the parents and those of their offspring were compared.[50,51,52,53] The findings obtained from a larger sample are as follows. In Abeno Ward (population about 163,000), which is a district in the southern part of Osaka City, a clinic called "Checkup for Three-Year-Old Children" was established in 1959, the main objective of which is early detection of physical and behavioral deviations. All children born in this area are invited to come to the clinic within ten days of their third birthday. One of the peculiarities of this clinic is that in addition to obtaining the family history from a parent of the child, there is a great emphasis placed on obtaining the childhood developmental histories of both parents from their respective mothers, that is, from the paternal and maternal grandmothers of the child. Two months prior to the child's third birthday a questionnaire is sent to his or her paternal and maternal grandmothers for the purpose of obtaining childhood data on the child's parents. Among the items explored were: developmental data, including the age of acquisition of bladder control; childhood diseases and susceptibility to motion sickness, convulsion, frost-bite, phobias, and other nervous symptoms; and sleep characteristics, such as difficulty in falling asleep, light sleep (many awakenings during the night), sleepwalking, sleep-talking, teeth-grinding, and night terror. While the physical and psychological examinations of the children are carried out, the mothers are requested to complete a questionnaire on the development and behavior of the child at home, including sleep habits. About seven hundred children are examined every year, and satisfactory childhood information on both of the parents are available for about three hundred of the children. However, since the questionnaire has been revised from time to time, the number of total records available for the statistical analyses varies from item to item.

Table 15-IV shows the incidence of sleepwalking in children, according to the childhood data of the parents. Similarly, Table 15-V shows the incidence of the difficulty in falling asleep; Table 15-VI, that of light sleep (many awakenings during the night); and Table 15-VII, for sleep-talking, respectively, in children from parental matings where both, one, or neither of the parents manifested the same problem in childhood. Each of these problems appears most frequently in children characterized by having parents who both manifested the same problem during childhood, and each appears least frequently in children characterized by parents who both did not manifest the same problem during childhood.

Since the number of children in the first row of each of the above tables is small, except in Table 15-V, the first row and the second row were combined and compared with the third row with chi-square tests involving one degree of freedom; that is, the children with one or both parents having manifested the problem and those where neither had manifested the problem were compared to each other. All of the calculated values show statistical significance at least at the probability level of 5.0 percent.

Before the results are interpreted, the

TABLE 15-IV
SLEEPWALKING IN CHILDREN AND IN THEIR PARENTS DURING CHILDHOOD
(779 CHILDREN EXAMINED IN 1965 TO 1966)

Sleepwalking in:	Sleepwalking in Children		Total
	+	−	
Both of the parents	3 (43%)	4	7
One of the parents	7 (6%)	117	124
Neither of the parents	22 (3%)	626	648
Total	32 (4%)	747	779
	$x^2 = 4.97$	$df = 1$	$P < 0.05$

TABLE 15-V
DIFFICULTY IN FALLING ASLEEP IN CHILDREN AND IN THE PARENTS DURING CHILDHOOD (661 CHILDREN EXAMINED IN 1965 AND 1968)

Childhood of the Parents Difficulty in Falling Asleep in:	Children Difficulty in Falling Asleep in:		Total
	+	−	
Both of the parents	10 (48%)	11	21
One of the parents	48 (29%)	118	166
Neither of the parents	87 (18%)	387	474
Total	145 (22%)	516	661
$X^2 = 16.35$	$df = 2$	$P < 0.001$	

TABLE 15-VI
LIGHT SLEEP (MANY AWAKENINGS DURING THE NIGHT) IN CHILDREN AND IN CHILDHOOD OF THE PARENTS (774 CHILDREN EXAMINED IN 1965 AND 1969 TO 1970)

Childhood of the Parents Light Sleep in:	Children Light Sleep in:		Total
	+	−	
Both of the parents	4 (30%)	9	13
One of the parents	22 (12%)	157	179
Neither of the parents	27 (5%)	555	582
Total	53 (7%)	721	774
$X^2 = 17.94$	$df = 1$	$P < 0.001$	

TABLE 15-VII
SLEEP-TALKING IN CHILDREN AND IN CHILDHOOD OF THE PARENTS (1965 and 1970)

Childhood of the Parents Sleep-talking in:	Children Sleep-talking in:		Total
	+	−	
Both of the parents	3 (23%)	10	13
One of the parents	30 (21%)	112	142
Neither of the parents	74 (14%)	462	536
Total	107 (15%)	584	691
$X^2 = 5.15$	$df = 1$	$P < 0.025$	

reliability and possible systematic distortion of memory of the mothers and the grandmothers of the children and their influence on the answers in the questionnaires should be considered. Since less than 5 percent of the mothers have ever worked after the childrens' birth, we can assume that most of them have been taking care of their children day and night at home and know their children well with regard to the sleep behaviors. On the other hand, the grandmothers have to answer questions on the items regarding childhood of their offspring, which involves recollections going back about ten to thirty years. Although care was taken to exclude more unreliable data, such as those coming from grandmothers more than sixty-five years of age, a significant proportion of the answers may still be incorrect. However, inclusion of random errors in the data tends to obscure the regularity and lower the chi-square values. Therefore, if the resulting data still show a significant finding after inclusion of random errors, we should naturally expect a more significant original finding. The problem here is whether or not there is the possibility of a systematic distortion of the grandmother's memory in such a way as to produce a spurious parent-child concordance with regard to the items investigated in this paper.

The data on the children and those on childhood of their parents were obtained from different observers, that is, from the mothers and the grandmothers, respectively. However, among the families of 381 children examined in 1970 and selected for the present study, the fathers' mothers were living with the families in 32 percent and the mothers' mothers were living with the families in 5 percent of them. That is, the data on childhood of the parents were obtained in 18.5 percent of the cases from grandmothers who were living together with the children in question and who knew the children's daily behaviors well. One may suspect that a grandmother's memory regarding childhood of the grandchild's parent may be affected by the grandchild's daily behavior so that it seems to approach the grandchild's behavior, especially if the infant involved is an only grandchild. More generally, the problem concerns whether or not the grandmother's answers are influenced significantly by knowing her grandchild and whether or not it can be a significant source of parent-child similarity as indicated in the tables. To examine this point, fifty father-son pairs were selected from 581 children (for 1969 to 1970) where in each case the father's mother was living together with the family; and the same number of father-son pairs were selected where the childhood information about the father was supplied from the father's mother who was living in an area which takes more than ten hours by public surface transport from Osaka, such as Southern Kyushu or Hokkaido. We can safely assume that the grandmothers in the latter group in general knew much less about their grandchildren (in fact, some of them had never seen the grandchild in question). If the aforementioned influence on the memory or on the answer is significant, one can expect much higher father-son concordance in the former compared to the latter group. The results, with regard to excessive drooling, light sleep (many awakenings during sleep), and sleep-talking, is shown in Table 15-VIII. There is no significant difference between the groups with respect to light sleep and sleep-talking; and, as to excessive drooling, unexpectedly, concordance is higher with the group characterized by the grandmother's living far apart. Therefore, there is no significant tendency for the grandmother's memory to be influenced by behavior characteristics

TABLE 15-VIII
FATHER-SON CONCORDANCE (TRAIT ABSENT IN BOTH OR PRESENT IN BOTH THE FATHER AND THE SON) IN THE FAMILIES CHARACTERIZED BY THE FATHER'S MOTHER LIVING TOGETHER WITH THE FAMILY, AND IN THOSE WHERE SHE IS LIVING FAR APART.

		Father-Son Concordance With Regard to:		
Grandmother:	Number of Families	Excessive Drooling	Light Sleep	Sleep-talking
Living far apart	50	80%	66%	90%
Living together	50	62%	68%	88%
Significance of difference, P		< 0.05	N.S.	N.S.

of her grandchild, resulting in a uniform distortion and a spurious parent-child similarity. However, errors in the grandmother's memory, assumed to be random, may be significant. Therefore, the present writer does not proceed further to speculate on the mode of inheritance from the data tabulated.

If the distortion of the memory of the grandmother cannot be responsible for the parent-child similarity indicated by the data, the next possibility to be considered is a common environment. Twenty-two percent of the fathers and 13 percent of the mothers attended an elementary school within the district, Abeno Ward. Since girls usually leave their homes when married, and since there are high numbers of transfers within the district, and since there is considerable new construction in progress, about 10 to 15 percent of the children are estimated as being reared in the house where one of the parents was reared. However, this district was a suburban area about twenty years ago, and it is now an urban area comprising the southern part of Osaka City. Although the inside of the house may be similar in some families, the traffic and noise have increased greatly. Also, the personal element in the intrafamilial environment has changed. Even in the families where the grandmothers are living, they are usually of secondary influence behind the mother unless the latter has her own job, which involves less than 5 percent of the mothers in this study. Hence, the child is usually not reared by the same person who took care of either of his or her parents. Also, the father's contact with the child is, if measured in length of time, generally much less than that of the mother: most of the fathers in this district work from 9 AM to 4 PM six days a week. It may be of interest to compare the incidence of a certain sleep characteristic in children whose fathers had manifested it in childhood with that in children whose mothers had it. The incidence of sleepwalking is 5 percent in children whose fathers had sleepwalked (and the mother had not) and 6 percent in those whose mother had. Similarly, the corresponding figures for sleep-talking are 18 percent and 22 percent, respectively; and, for light sleep, they are 14 percent and 11 percent, respectively. The apparent equality of the paternal and maternal influence (that is, with no significant difference in the above percentages) can be regarded as another finding in favor of biological rather than environmental etiology.

Recently, it has been shown that the amino acid, tryptophan, has an influence on sleep.[54] With regard to the diet, most of the parents spent their childhoods in the age of rations when the variety of food was severely restricted and uniform, which is not the case with their children. Therefore, with regard to childhood living conditions and diet, there are great parent-child differences. In addition, the traditional method of child rearing had been fading out due to "westernization" of the society. In view of the aforementioned findings, it is difficult to explain the results shown in Tables 15-IV through 15-VII, that is, with regard to difficulty in falling asleep, depth of sleep, sleepwalking, and sleep-talking in childhood, without assuming that genetic factors play important roles.

Regarding teeth-grinding, the incidence in the children of parents with childhood teeth-grinding histories is not significantly higher than in the children of parents without it (Table 15-IXA). However, there is a clear difference in the incidence in children categorized according to whether or not the parents have manifested teeth-grinding as adults (Table 15-IXB). Is it that teeth-grinding has a genetic basis, but Table 15-IXA does not reach statistical significance owing to inexact memories of the grandmothers and inclusions of random errors therefrom? Or, is current parent-child concordance due to some common environmental factor? The data available at present do not permit any definite conclusion on teeth-grinding.

TABLE 15-IXA
TEETH-GRINDING IN CHILDREN AND IN THEIR PARENTS DURING CHILDHOOD

Childhood of the Parents Teeth-grinding in:	Children Teeth-grinding in:		Total
	+	−	
Both of the parents	1 (7%)	14	15
One of the parents	28 (13%)	183	211
Neither of the parents	73 (11%)	618	691
Total	102 (11%)	815	917
$X^2 = 0.89$	df = 1	$0.3 < P < 0.5$	

TABLE 15-IXB
TEETH-GRINDING IN CHILDREN AND IN THEIR PARENTS (AFTER MARRIAGE)

Teeth-grinding (After Marriage) in:	Teeth-grinding in Children		Total
	+	−	
Both of the parents	1	0	1
One of the parents	38 (22%)	135	173
Neither of the parents	63 (8%)	680	743
Total	102 (11%)	815	917
$X^2 = 29.82$	df = 1	$P < 0.001$	

REFERENCES

1. Vandenberg, S.G.: Hereditary factor in psychological variables in man, with a special emphasis on cognition. In Spuhler, J.S. (Ed.): *Genetic Diversity and Human Behavior.* Chicago, Aldine, 1967.
2. Inouye, E.: Inheritance of localized motility traits in man. *Jap J Hum Genet, 5*:1, 1960.
3. Luchsinger, R.: Die Sprachentwicklung von ein- und zweieiigen Zwillingen und die Vererbung von Sprachstörungen in den ersten drei Lebensjahren. *Folia Phoniatr (Basel), 13*:66, 1961.
4. Hallgren, B.: Enuresis, a clinical and genetic study. *Acta Psychiatr Scand [Suppl] 114,* 1957.
5. Vogel, F.: The genetic basis of the normal human electroencephalogram. *Humangenetik, 10*:91, 1970.
6. Pare, C.M.B., Rees, L., and Sainsbury, M.J.: Differentiation of two genetically specific types of depression by response to antidepressants. *Lancet, 2*:1340, 1962.
7. Pare, C.M.B.: Differentiation of two genetically specific types of depression by the response to antidepressant drugs. *Humangenetik, 9*:199, 1970.
8. Angst, J.: A clinical analysis of the effect of Tofranil in depression. Longitudinal and follow-up studies. *Psychopharmacologia, 2*:381, 1961.
9. Angst, J.: Antidepressiver Effekt und genetische Faktoren. *Arzneimittelforschung, 14*:496, 1964.
10. Bönicke, R., and Lisboa, B.P.: Über die Erbbedingtheit der intraindividuellen Konstanz der Isoniazidausscheidung beim Menschen. *Naturwissenschaften, 44*:341, 1957.
11. Price-Evans, D.A.: Individual variations of drug metabolism as a factor in drug toxicity. *Ann NY Acad Sci, 123*:176, 1965.
12. Price-Evans, D.A., Davidson, K., and Pratt, R.T.C.: The influence of acetylator phenotype on the effects of treating depression with phenelzine. *Clin Pharmacol Ther, 6*:430, 1965.
13. Sjöqvist, F., Hammer, W., Ideström, C.-M., Lind, M., Tuck, D., and Åsberg, M.: Plasma level of monomethylated tricyclic antidepressants and side effects in man. *Proc European Soc for the Study of Drug Toxicity, 9*:246, 1967.
14. Hammer, W., Mårtens, S., and Sjöqvist, F.: A comparative study of the metabolism of desmethylimipramine, nortriptyline and oxyphenylbutazone. *Clin Pharmacol Ther, 10*:44, 1969.
15. Vesell, E.S., and Page, J.G.: Genetic control of drug levels in man: Phenylbutazone. *Science, 159*:1479, 1968.
16. Whittaker, J.A., and Price-Evans, D.A.: Genetic control of phenylbutazone in man. *Br Med J, 4*:323, 1970.
17. Haidu, G.G., Dhrymiotis, A., and Quin, G.P.: Plasma imipramine level in syndromes of depression. *Am J Psychiatry, 119*:574, 1962.
18. Zeidenberg, P., Perel, J.M., Kanzler, M., Wharton, R.N., and Malitz, S.: Clinical and metabolic studies with imipramine in man. *Am J Psychiatry, 127*:10, 1971.
19. Curry, S.H., Davis, J.M., and Janovsky, D.S.: Factors affecting chlorpromazine plasma levels in psychiatric patients. *Arch Gen Psychiatry, 22*:209, 1970.
20. Myrianthopoulos, N.C., Kurland, A.A., and Kurland, L.T.: Hereditary predisposition in drug-induced Parkinsonism. *Arch Neurol, 6*:5, 1962.
21. Goldstein, A., Warren, R., and Kaizer, S.: Psychotropic effects of caffein in man. Part I. Individual differences in sensitivity to caffein-induced wakefulness. *J Pharmacol Exp Ther, 149*:156, 1965.
22. Goldstein, A., Warren, R., and Kaizer, S.: Psychotropic effects of caffein in man. Part II. Alertness, psychomotor coordination and mood. *J Pharmacol Exp Ther, 150*:146, 1965.
23. Takala, M., Pihkanen, T.A., and Markkanen, T.: *The Effect of Distilled and Brewed*

Beverages. Helsinki, Finnish Foundation for Alcohol Studies, 1957.
24. Abe, K.: Reaction to coffee and alcohol in monozygotic twins. *J Psychosom Res, 12*:199, 1968.
25. Weitz, W.: Studien an eineiige Zwillingen. *Z Klin Med, 101*:115, 1924.
26. Curtius, F., and Korkhaus, G.: Klinische Zwillingsstudien. *Z KonstLehre, 15*:229, 1936.
27. Werner, M.: Erbunterschiede bei einigen Funktionen des vegetativen Systems nach experimentellen Untersuchungen an 30 Zwillingspaaren. *Verh Dtsch Ges Inn Med, 47*:444, 1935.
28. Ikai, M., and Nakanishi, M.: Similarity of heart beat interval fluctuations in twins. In Fujita, T. (Ed.): *Studies on Twins.* Tokyo, Japanese Assoc for the Advancement of Science, 1961, vol. III (In Japanese)
29. Vandenberg, S.G., Clark, P.J., and Samuels, I.: Psychophysiological reactions of twins: Hereditary factor in galvanic skin resistance, heartbeat, and breathing rates. *Eugen Q, 12*:7, 1965.
30. Werner, M.: Zwillingsphysiologische Untersuchungen über den Grundumsatz und die spezifisch-dynamische Eiweisswirkung. *Z Abstamm, 70*:467, 1935.
31. Werner, M.: Uber den Anteil von Erbanlagen und Umwelt beim Kohlenhydratstoffwechsel auf Grund von Zwillingsuntersuchungen. *Z Abstamm, 67*:306, 1934.
32. Kamitake, S.: A study of genetic control in blinking. In Fujita, T. (Ed.): *Studies on Twins.* Tokyo, Japanese Assoc for the Advancement of Science, 1961, vol. III (In Japanese.)
33. Curtius, F., and Feiereis, H.: Zwillingsuntersuchungen uber die Erbveranlagerung zum vegetative-endokrinen Syndrom der Frau. *Z Kreislaufforsch, 49*:44, 1960.
34. Siemens, H.: *Die Zwillingspathologie.* Berlin, Springer, 1924.
35. Harris, H.: Genetical factors in perniosis. *Ann Eugen, 14*:32, 1947.
36. Abe, K., Amatomi, M., and Kajiyama, S.: Genetical and developmental aspects of susceptibility to motion sickness and frost-bite. *Hum Hered, 20*:507, 1970.
37. Cloward, R.B.: Treatment of hyperidrosis palmaris, a familial disease in Japanese. *Hawaii Med J, 16*:381, 1957.
38. Carmena, M.: Ist die persönliche Affektlage oder "Nervosität" eine ererbte Eigenschaft? *Z Ges Neurol Psychiatr, 150*:434, 1934.
39. Jost, H., and Sontag, R.: The genetic factor in autonomic nervous system function. *Psychosom Med, 6*:308, 1944.
40. Lader, M.H., and Wing, L.: *Physiological Measures, Sedative Drugs, and Morbid Anxiety.* London, Oxford U Pr, 1966.
41. Goodman, H.O., Luke, J.E., Rosen, S., and Hachel, E.: Heritability in dental caries, certain oral microflora and salivary components. *Am J Hum Genet, 11*:263, 1959.
42. Gastaut, H., and Broughton, R.: A clinical and polygraphic study of episodic phenomena during sleep. In Wortis, J. (Ed.): *Recent Advances in Biological Psychiatry.* New York, Plenum Pr Plenum Pub, 1965, vol. VII, p. 197.
43. Jacobson, A., Lehmann, D., Kales, A., and Wenner, W.H.: Somnambule Handlungen im Schlaf mit langsamen EEG-Wellen. *Arch Psychiatr Nervenkr, 207*:141, 1965.
44. Rechtschaffen, A., Goodenough, D.R., and Shapiro, A.: Patterns of sleep talking. *Arch Gen Psychiatry, 7*:418, 1962.
45. Reding, G.R., Rubricht, W.C., Rechtschaffen, A., and Daniels, R.S.: Sleep pattern of tooth-grinding: Its relationship to dreaming. *Science, 145*:725, 1964.
46. Zung, W.W.K., and Wilson, W.P.: Sleep and dream patterns in twins. In Wortis, J. (Ed.): *Recent Advances in Biological Psychiatry.* New York, Plenum Pr Plenum Pub, 1967, vol. IX, p. 119.
47. Geyer, H.: Über den Schlaf von Zwillingen. *Z Abstamm, 73*:524, 1937.
48. Pierce, C.M., and Lipcon, H.H.: Somnambulism. Electroencephalographic studies and related findings. *US Armed*

Forces Med J, 7:1419, 1956.
49. Bakwin, H.: Sleepwalking in twins. *Lancet*, 2:446, 1970.
50. Naka, S., Abe, K., and Suzuki, H.: Childhood behaviour characteristics of the parents and certain behaviour problems of children. *Acta Paedopsychiatr (Basel)*, 32:11, 1965.
51. Abe, K., and Shimakawa, M.: Genetic and developmental aspects of sleeptalking and teeth-grinding. *Acta Paedopsychiatr (Basel)*, 33:339, 1966.
52. Abe, K., and Shimakawa, M.: Predisposition to sleepwalking. *Psychiatr Neurol*, 152:306, 1966.
53. Abe, K., and Shimakawa, M.: Genetic-constitutional factor and childhood insomnia. *Psychiatr Neurol*, 152:363, 1966.
54. Hartmann, E., Chung, R., and Chien, C.: L-tryptophan and sleep. *Psychopharmacologia*, 19:114, 1971.

Chapter 16

PSYCHOPHARMACOGENETICS

GILBERT S. OMENN AND ARNO G. MOTULSKY

THE TERM *pharmacogenetics* refers to studies of genetically determined individual differences in therapeutic and adverse responses to drugs.[1,2,3] These differences may be due to different rates of metabolism or elimination of the pharmacologically active species or may be due to different susceptibility to the drug action on specific enzymes or cell receptors. Recent reviews of such host-determined drug effects have emphasized their clinical and investigational importance as examples of gene-environment interaction.[4,5,6]

In this chapter we shall describe several approaches to genetically-determined differences in the response of individuals to behavior-modifying drugs and indicate how pharmacologic responses might be utilized to investigate certain psychiatric syndromes.

METHODS OF ANALYSIS IN PHARMACOGENETICS

Population Surveys

When a drug is tested in a general population sample or is used therapeutically in patients with any given diagnosis, considerable variability in effectiveness and in side effects is commonly noted. When drug potency, mode of administration, and pertinent dietary factors are carefully standardized, three sources of variation still must be expected. First, some individuals will need a larger or a smaller dose than the average in order to attain the same effect or same plasma concentration of drug; some individuals may fail to show the desired effect at any reasonable dose. Second, some patients may fail to respond because the diagnosis is incorrect or because the diagnostic category comprises two or more distinct subpopulations, only one of which is responsive to the drug therapy provided. Third, especially when response is measured in subjective behavioral symptoms, attitude of the patient and expectations of the volunteer subject may be highly variable and influential; however, this source of variability often can be controlled with careful double-blind and crossover protocols for administration of the drug and an appropriate placebo. Figure 16-1 presents in schematic form the patterns of variation observed when human populations are tested.

Family Studies

If a bimodal distribution of response or side effects is suggested by the population survey, families of individual probands

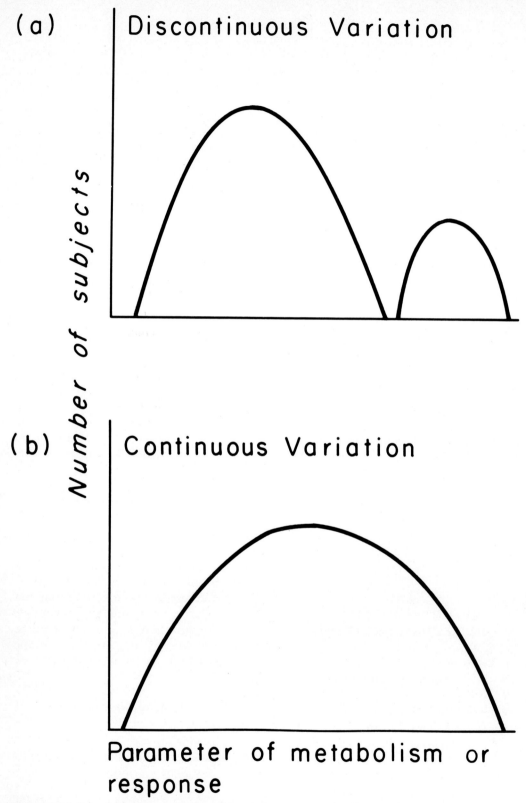

Figure 16-1. Schematic diagrams of (a) discontinuous (bimodal) and (b) continuous distributions of drug effects in population samples.

from each of the modal subpopulations should be tested. As in the case of acetylation of isoniazid (below), a simple inherited pattern may emerge from such family studies. At the least, family studies have the major advantage that the same mechanism for an unusual response to a drug or for a particular behavioral syndrome is more likely to be responsible in all affected members of one family than in a random group of individual patients. For example, investigations of the causes of mental retardation were hopelessly frustrating until subgroups of affected youngsters could be defined by associated clinical or laboratory findings and until family studies revealed specific inborn errors of metabolism in some cases. If the pattern of variation is more nearly continuously distributed (Fig. 16-1a), searching for clues to subpopulations may be especially fruitful among individuals at the extremes. Studies of the families of individuals with similarly abnormal drug reactions may reveal a bimodal distribution for specific genetic mechanisms of too low frequency in the general population to produce a discernible "hump" in the population distribution curve. Another clue that genetic factors may be involved is the finding that the distribution differs for different ethnic or racial groups.

Twin Studies

Comparison of monozygotic (identical) and dizygotic (fraternal) twins for rates of concordance of a trait or for intrapair similarity in a quantifiable measure, such as rate of elimination of a drug, permits estimation of the extent to which variation is due to inherited factors.[7] The twin method, however, does not provide evidence on mode of inheritance (i.e. recessive, dominant, or X-linked single gene effects or polygenic interaction). The contribution of environmental variables also can be assessed by changing the environment, as in chronic versus acute administration of a drug, retesting the same twin pairs, or studying monozygotic twin pairs reared apart and reared together.[8] In the last case, the extent to which the environments differ for the monozygotic twins reared apart must be estimated independently.

Biochemical Studies

The likelihood of defining specific genetic mechanisms increases as investigations reach closer to the enzyme or other protein products of genes. Thus, a mutation in the gene for plasma pseudocholinesterase or for red blood cell glucose-6-phosphate dehydrogenase is expressed directly by the altered properties of these enzymes, but only indirectly by the adverse response to certain drugs of the individual carrying such a mutation (see below). When individuals differ in plasma concentration of a drug on a given dose, the rate of absorption can be tested by comparing oral and intravenous routes. If the difference in steady-state concentration or rate of elimination persists upon intravenous administration, renal clearance and plasma protein binding must be considered. The qualitative pattern of metabolites may suggest the enzymatic conversion (hydroxylation, glucuronidation, acetylation, hydrolysis, etc.) responsible for inactivating or transforming the drug. If rates of conversion are altered, assay of the appropriate enzyme activity in blood cells or fibroblasts or liver biopsy may be possible. Electrophoretic and kinetic properties of the enzyme may be altered if a mutation in the structural gene for that enzyme is the basis for the differential drug response. Finally, difference in tissue responsiveness may be gene determined by alteration in receptor molecules. Little has been learned yet of specific drug receptor molecules, but there is likely to be variation among individuals at this level of drug action as well.

PHARMACOGENETICS OF SPECIFIC DRUGS

Succinyl Choline (Suxamethonium)

Because of its rapid onset and short duration of action, this depolarizing muscle relaxant is used widely in premedication for anesthesia and for shock treatment (electroconvulsive therapy). However, suxamethonium will cause apnea for up to several hours in about one in 2500 Caucasians genetically at risk for this possible catastrophe. These individuals have an abnormal plasma enzyme, pseudocholinesterase (PsChE), which fails to carry out the usually rapid inactivation of the drug. Measurement of plasma enzyme activity in the patients and their relatives indicated possible autosomal recessive basis.[9] However, clear differentiation of homozygotic normal (usual form $E^u E^u$), of heterozygotic carriers ($E^u E^a$), and of homozygotes for the abnormal or atypical gene ($E^a E^a$) became feasible with Kalow's method of testing percent inhibition of PsChE activity with an enzyme inhibitor called dibucaine.[10] The normal enzyme is strikingly inhibited, while the atypical enzyme is only slightly inhibited. Screening individuals with dibucaine and fluoride as inhibitors and also with electrophoresis has revealed several additional rare variant forms of PsChE. Some of these variant PsChE also predispose to suxamethonium sensitivity. On the other hand, another variant, PsCh Cynthiana causes resistance to suxamethonium; this enzyme variant (on a per molecule basis) is three times as active as the usual PsChE.[11]

In the clinical use of suxamethonium, the psychiatrist or anesthetist should inquire about personal or family history of sensitivity and should have equipment available for sustained artificial respiration. A simple screening test for PsChE sensitivity is available.[12] There is almost no need to test Negro or Oriental patients since the frequency of the E^a gene is much lower in those populations.[6] Enzyme replacement therapy for this genetic defect has been accomplished by injection of purified pseudocholinesterase into patients with prolonged apnea.[13]

Acetylation in the Liver

When a standard dose of the antituberculosis drug, isoniazid (INH), was administered to a group of individuals, the distribution histograms of blood levels of INH six hours later and of percentage of the administered INH excreted free and unacetylated in the urine were bimodal (see Price-Evans[5]). Twin studies showed great similarity between MZ twins and considerable differences between DZ twins.[14] Family studies showed that slow inactivators of isoniazid develop higher blood levels, excrete a higher proportion of free drug in the urine, and have a trait recessive to rapid inactivation.[15] The responsible enzyme is a liver acetylase which inactivates INH and phenelzine (Nardil®), as well as hydralazine (Apresoline®), dapsone, and other sulfa drugs. Rapid inactivators have no demonstrable impairment of antituberculous effect on standard dose of INH. However, slow inactivators have a greatly enhanced risk of undesirable side effects, primarily peripheral neuropathy, which can be avoided by concomitant administration of the B vitamin pyridoxine. When patients with both epilepsy and tuberculosis are given INH plus diphenylhydantoin (Dilantin®), an important drug-drug interaction may occur. Among the slow inactivators, INH concentrations reach levels which inhibit metabolism of Dilantin by hepatic microsomal oxidases, thus leading to accumulation of Dilantin to toxic concentrations with ataxia, nystagmus, and drowsiness.[16] Similarly, toxic side effects of the antidepressant phenelzine occurred only in subjects with the slow acetylator phenotype.[17] It should be noted that about 50 percent of Caucasians, 50 percent of Negroes, and only 10 to 15 percent of Orientals are slow acetylators.[4] Certain other drugs, such as para-amino-salicylic acid, are acetylated, but not by this particular polymorphic N-acetyltransferase. For other compounds, including serotonin, that can be acetylated, it is not yet clear whether the polymorphic acetylating system is responsible.

Oxidant Drugs and G6PD Deficiency

Glucose-6-phosphate dehydrogenase (G6PD) is the first enzyme of the energy-generating pentose-phosphate shunt pathway which is essential to maintaining the integrity of the red blood cell. G6PD deficiency is sex-linked, affecting up to 35 percent of males in certain Negro and Mediterranean population groups. A long list of drugs with oxidizing properties can precipitate acute hemolytic anemia in these otherwise healthy but genetically-predisposed individuals by overwhelming the capacity of a deficient G6PD. In this case, the drug does not interact directly with the abnormal enzyme; the affected tissue is more susceptible to drug injury than normal.[18] The Mediterranean-type G6PD deficiency is more severe than the Negro type, hence additional drugs with less strong oxidant properties are a threat. These drugs include primaquine and other 8-aminoquinoline antimalarials, sulfas, nitrofuran derivatives, phenacetin, acetanilide, antipyrine, probenecid, para-aminosalicylic acid, and aspirin.[5] Some of these individuals also cannot tolerate eating the broad bean *Vicia fava* or even inhaling its pollen. Interestingly enough, long before G6PD deficiency was known, the Pythagoreans were said to have surrendered to their enemy rather than flee through a field of fava beans. Also, in Greek mythology, a particular sect allowed women, but not men, to eat the fava beans,[19] consistent with the X-linked recessive inheritance of G6PD deficiency! Fava beans contain high concentrations of L-dopa; formation of a derivative of L-dopa (but not L-dopa itself) may trigger attacks of favism in the susceptible person.[20] As new drugs with oxidant properties are introduced, certain individuals can be expected to be at greater risk for such undesirable side effects.

Hydrocortisone and ACTH

These agents are used for patients with multiple sclerosis, polymyalgia rheumatica, and many other conditions. Either topical or systemic corticosteroids may elevate the intraocular pressure sufficiently to induce open-angle glaucoma. Both the basal level pressure and the extent of elevation caused by the steroids are genetically determined.[21] Family studies suggest that a single gene locus may be responsible, though the biochemical basis for this susceptibility is unknown. Especially in older patients, it may be important to test routinely for glaucoma before starting corticosteroids.

Another relevant side effect among the many associated with corticosteroid therapy is the precipitation of ACTH or steroid psychosis. Since only a small proportion of patients have this complication, it is likely that they are in some way more susceptible than others.

Tricyclic Antidepressants: Nortriptyline

Sjöqvist and his colleagues have taken the clinical observations by Angst of marked individual differences in the therapeutic effect of the tricyclic antidepressants as the basis for detailed pharmacokinetic studies of nortriptyline.[22, 23] Side effects of nortriptyline, corrected for those associated with placebo, were correlated with the steady-state plasma concentration of the drug and not with the dose administered.[24] Plasma concentrations in a group of thirty-nine twin pairs varied over a ten-fold range, and the much smaller intrapair differences for MZ as opposed to DZ twins indicated a major role for genetic factors. Three patients with very high plasma levels served as probands for family studies. No bimodality of plasma level could be found in any of these studies, and analysis of the variances in relatives of the extreme proposita versus random subjects suggests that polygenic inheritance is involved.[25] This finding appears to rule out control of the rate of elimination of nortriptyline by a single enzymatic reaction. Probably several biochemical reactions with genetic variability are involved; however, the exact number of genes and their biochemical roles are unknown. Nortriptyline is metabolized primarily by hydroxylation.

Animal Studies

It is often dangerous to assume that other species metabolize drugs by the same pathways as does man. The patterns of metabolites of amphetamines, for example, are quite different when man is compared with the dog or with rabbit and guinea pig.[26] Nevertheless, strain differences in the metabolism of specific drugs can provide useful models for study of the enzymatic steps involved and for correlation of metabolic degradation with therapeutic and adverse physiological or behavioral effects.[27]

In general, strain differences in behavioral effects of drugs have been measured at the behavioral level without determination of blood levels, patterns of excretion, or activities of enzymes involved in biotransformation. Fuller tested four inbred strains of mice for the effects of chlorpromazine and chlordiazepoxide on active and passive avoidance.[28] The complicated genotype-drug interactions observed were interpreted to reflect differences in both kinetic drive and fear-motivated responses in the different strains. The tricyclic antidepressant, imipramine, can reverse reserpine-induced hypothermia in rats. In comparison with three other strains, the Long-Evans rats were less reactive to the antireserpine activity of imipramine and had lower desipramine/imipramine ratios in the brains; their hepaticmicrosomal oxidases were less active against imipramine and other substrates *in vitro*.[29] In addition, however, the Long-Evans rats seem to be less sensitive to the desipramine formed. No breeding experiments were reported. The effects on exploratory behavior of the anticholinergic drug, scopolamine, and the anticholinesterase agent, eserine, were strikingly different in two mouse strains and in their F_1 hybrid.[30] The interpretation of these results in terms of cholinergic facilitation of exploratory tendencies is clearly speculative, though other studies do suggest differences in acetyl choline content, acetylcholinesterase activity, and biogenic amine metabolism among mouse strains.[31] The most extensive animal study of drug effects, neurotransmitter metabolism, and a specific inherited behavioral phenotype is that of audiogenic seizure susceptibility which will be discussed in the section on epilepsy.[32, 33] The fact that strains and genera of animals may differ genetically in their responsiveness to certain drugs must be recognized as a serious problem in interpreting mechanisms of drug-modified behaviors and, of course, in the practical testing of effects and toxicity of new drugs.

SIMPLY-INHERITED BEHAVIORAL DISORDERS WITH SPECIAL VULNERABILITY TO DRUGS

The Porphyrias

These genetically heterogeneous metabolic disorders of hepatic heme biosynthesis are vertically transmitted through families as autosomal dominant traits. The clinical syndromes occur in episodes of colicky abdominal pain with constipation (due to autonomic neuropathy) and variable central nervous system involvement, including flaccid paralysis, agitated and paranoid depression, and schizophrenic behavior.[34, 35, 36] Porphyria may have been the cause of the intermittent "madness" of King George III of England.[37] In the Swedish type, or intermittent acute porphyria, biochemical diagnosis during the acute attack is highly reliable. However, the increased urinary excretion of porphyrin precursors may not be present before puberty or between attacks. Increased production of delta-aminolevulinic acid (ALA) and porphobilinogen is due to higher than normal activity of the rate-limiting enzyme, ALA-synthetase, in the liver. The mechanism of the increased activity is not yet clarified, though heme feedback and possibly steroid reductase activity appear to be involved.[38, 39] In the South African type, called porphyria variegata, there is continuous fecal hyperexcretion of protoporphyrins and coproporphyrins and photosensitive dermatitis.[35]

Several common drugs induce higher activity of the ALA synthetase and may precipitate attacks in any of the predisposed individuals. These drugs include barbiturates; certain sulfonamides; the antifungal agent, griseofulvin; and possibly general anesthetics, ethanol, and chloroquine.[36] Often the diagnosis has not been suspected at the time of a drug-precipitated attack. Thus, a patient after abdominal surgery may complain of recurrent abdominal pain and develop bizarre behavior with hallucinations, paranoid delusions, and limb weakness. The significance of preoperative or postoperative medications, especially routine sleeping pills, may be overlooked unless this specific diagnosis is considered. The porphyrias are most common in individuals of European ancestry and are extremely rare among Negroes.[36] It is noteworthy that the gene for the South African type of porphyria has been traced back to a Dutch immigrant in the seventeenth century, yet the disease has been recognized only in the past thirty to forty years after introduction of phenobarital and sulfonamides.[35]

Familial Dysautonomia (Riley-Day Syndrome)

This autosomal recessive inherited disorder causes protean manifestations of neurogenic origin. Infants have difficulty swallowing, and lack of overflow tears, followed by slow achievement of developmental milestones, and recurrent pneumonia. Periodic crises of vomiting, abdominal pain, fever, flushing, sweating, and evidence of emotional lability dominate childhood. Sensitivity to pain is diminished, taste discrimination is defective (fungiform papillae are absent), and mortality risks of surgery are increased.[40, 41] Recent studies indicate a deficient release of the enzyme dopamine-beta-hydroxylase from norepinephrine-containing nerve terminals into the plasma.[42] This enzyme converts dopamine to norepinephrine (NE) and is released together with the NE upon stimulation.

Infusion of NE into dysautonomic patients produced a very exaggerated hypertensive response without bradycardia. Conversely, the parasympathomimetic drug, methacholine, gave an excessive hypotensive response without increase in heart rate, plus abdominal cramps, sweating, and overflow tears.[40] These youngsters lack the radiating pain and flare response to intradermal histamine unless pretreated with methacholine. Also, they compensate poorly for conditions of decreased oxygen saturation. The relationship of such autonomic nervous system imbalance to emotional lability is, of course, unclear. This disorder is found only among Ashkenazi Jews.[41]

Lesch-Nyhan Syndrome

This X-linked recessive condition of young boys is characterized clinically by choreoathetosis, spasticity, mental retardation, and a bizarre, compulsive behavior with self-mutilation of lips, fingers, and eyes.[43, 44] Biochemically, there is hyperuricemia and excessive production of uric acid due to complete deficiency of an enzyme in purine metabolism, hypoxanthine-guanine phosphoribosyl transferase (HGPRT). Uric acid nephropathy leading to uremia and tophaceous gout may result. The HGPRT activity is normally highest in the brain, particularly in the basal ganglia, providing a correlation with the choreoathetosis. However, the basis for the behavioral disorder is altogether unknown. Since certain drugs are transformed by the same phosphoribosyl transferase, HGPRT deficiency has pharmacogenetic consequences. The antineoplastic agent 6-mercaptopurine must be converted to its ribonucleotide by HGPRT in order to be active *in vivo*. The immunosuppressive agent azathioprine (Imuran®) is first converted to 6-mercaptopurine and then activated in the same manner. HGPRT-deficient fibroblasts in culture are resistant to inhibition by these drugs of *de novo* purine synthesis. Allopurinol is a valuable drug in the treatment of hyperuricemia and gout since it blocks the conversion of hypoxanthine and xanthine to uric acid, which is less soluble and forms uric acid stones. Allopurinol normally also decreases purine synthesis so that total xanthine excretion is less than that of uric acid. In HGPRT-deficient individuals, allopurinol ribonucleotide is not formed, so purine synthesis is not inhibited, and xanthine oxidase is subjected to a higher concentration of active free base. As a result, these patients are liable to formation of xanthine stones. Partial deficiency of HGPRT has been recognized as a rare cause of ordinary gout, without the striking childhood neurologic and behavioral syndrome.

Malignant Hyperthermia

This condition has been recognized recently as an inherited predisposition to anesthetic catastrophes.[45] Potent inhalational anesthetics (such as halothane, methoxyflurane, and ether) and muscle relaxants (such as succinylcholine) may trigger a rapid rise in body temperature and progressive muscular rigidity. Temperatures have reached 112 degrees fahrenheit with tachycardia, tachypnea, hypoxia, respiratory and metabolic acidosis, hyperkalemia, hypocalcemia, and death from cardiac arrest in about two thirds of reported cases. The first of at least fourteen familial cases was reported in 1962.[46] Apparently, individual fatal cases previously were just attributed to anoxic brain damage. The predisposition appears to be inherited as an autosomal dominant trait with variable expressivity and incomplete penetrance, depending on exposure to anesthesia.[47] The underlying basis is unknown, though patients who develop muscular rigidity often have had one of a variety of common musculoskeletal disorders.[37]

Kalow, et al. have studied muscle biopsies from survivors of malignant hyperpyrexia.[48] Such muscle preparations were more sensitive to caffeine-induced rigor than muscle of normal controls. This effect was enhanced by exposure to halothane; halothane depressed calcium uptake by the sarcoplasmic reticulum, whereas it had no such effect on that of normal controls. An important animal model for this syndrome has been discovered by halothane screening in Landrace pigs.[47] Intravenous procaine treatment, based upon the knowledge that procaine can block the effects on calcium binding by sarcoplasmic reticulum and the induction of muscle rigor by caffeine, was effective in preventing and treating this syndrome in a group of susceptible pigs.[50] No report of procaine therapy in humans has yet appeared. Others have suggested that oxidative phosphorylation or cyclic AMP metabolism may be involved.[45, 51] There appears to be no relationship between suxamethonium-induced malignant hyperthermia and plasma pseudocholinesterase activity.

Huntington's Chorea

This autosomal dominant neurologic and psychiatric disorder is one of the major problems in counseling in medical genetics. The age of onset of involuntary movements is usually in the thirties or forties, but may be delayed even longer. Thus, individuals at risk (50% risk if a parent is affected) have the dual misery of not knowing whether they will be transmitting the disease to their children and of worrying that any normal twitches or behavioral problems may be the early signs of the disease. Over a period of ten to twenty years, the affected person undergoes progressive deterioration of personality and mental function, usually requiring institutional care because of psychotic behavior or dementia or both. The pathophysiology of the disease is unknown, and no specific diagnostic test is yet available. Two indirect diagnostic approaches have been considered. First, the genetic locus for Huntington's chorea might be closely linked to some other gene whose product is easily tested, like a blood group. It is now possible in suitable pedigrees to use linkage to the secretor locus to make an early diagnosis of myotonic dystrophy, another autosomal dominant disorder with late age of onset.[52] However, no such linkage relationship is known for Huntington's chorea. The second approach is a pharmacological challenge. Since L-dopa administration to patients with Parkinsonism may induce involuntary, choreiform movements, it was speculated that carriers of the gene for Huntington's chorea might manifest such movements at a lower dose of L-dopa than do normal people or Parkinsonism patients.[53, 112] There is a reasonable fear that the symptoms induced in the preclinical stage might not be reversible.[54] Certainly, individuals at risk will differ in their desire to know or to not know whether they will be affected later.

PHARMACOGENETICS IN SPECIFIC BEHAVIORAL DISORDERS

Depression (Affective Disorders)

There is considerable evidence from twin and family studies that depression, especially manic-depressive psychosis, is conditioned by genetic factors.[55, 56] A comprehensive pathophysiologic hypothesis involving biogenic amine metabolism has been formulated on the basis of multiple, but indirect, pharmacologic effects in patients with depression and mania. In brief, depression appears to be associated with decreased action or turnover of norepinephrine (and serotonin), while manic states are associated with increased biogenic amine turnover.[57] Thus, pharmacologic agents which deplete norepinephrine from nerve terminals (reserpine) or interfere with its biosynthesis (alpha-methyl tyrosine, alpha-methyl dopa) may precipitate depression. Drugs which enhance biosynthesis of norepinephrine (L-dopa) may induce hypomanic states, and agents which prolong the action of NE by inhibiting intraneuronal monoamine oxidase (MAO) inhibitors or the neuronal reuptake of NE released into the synapse (tricyclics) are effective antidepressants. Electroconvulsive shock also acts to increase tyrosine hydroxylase activity and NE turnover.[58] Two sets of clinical observations suggest likely pharmacogenetic relationships among patients with depression.

The first involves the differential effectiveness of antidepressant drugs in different patients. Pare *et al.* reported that two groups of patients could be differentiated by their response to either MAO inhibitors (MAOIs) or tricyclic compounds.[59] In their hands, patients who responded to one class of antidepressant tended not to respond to the other. They asked two further questions as follows. (1) Does the patient have the same pharmacologic responsiveness during a subsequent episode of depression which might be precipitated by quite different life stresses? The answer was yes in twenty-seven of the twenty-eight cases. (2) Do relatives of the patient who have affective disorder share the same pattern of pharmacologic responsiveness and unresponsiveness? Again, the findings were positive, with twelve of the twelve and ten of the twelve relatives concordant with the pattern of response in the proband patients in two separate studies.[59, 60] Angst used only imipramine and found thirty-four relatives concordant with proband for a positive antidepressant response, four concordant for lack of response, and only five of the forty-one first-degree relatives discordant.[23] There is some skepticism about the distinctiveness of the treatment responsiveness in these studies. Pairs concordant for positive response might reflect, at least in part, the frequent favorable outcome with placebo administration in depressed patients.[61] Rarely were the relatives and probands both tested with both classes of drugs. The drug was not tested in unselected relatives, but only in those who were already identified as being affected with depression. Thus, the pattern of drug response and the predisposition to depression might be due to some common genetic mechanism, and the drug would serve to delineate two classes of depression rather than just a difference in the metabolism of the particular drug. No biochemical studies to test this hypothesis have been carried out. It is feasible to test the susceptibility of brain monoamine oxidase to inhibition by typical MAO inhibitors since patients who fail to respond might have a variant form of the enzyme which is not inhibited by usual concentrations of the drug. Such a situation would be analogous to the test of dibucaine inhibition of pseudocholinesterase in suxamethonium sensitivity (above).

The second clinical observation of pharmacogenetic interest is the risk that about 10 percent of patients treated for high blood pressure with reserpine will develop depression.[62, 63] These patients tend to have a personal history of affective disorder more often than do the 90 percent of reserpine-treated patients who do not develop depression.[63] Reserpine in animals is known to deplete neuronal stores of NE and serotonin and to induce an increase in activity of tyrosine hydroxylase, the rate-limiting step for biosynthesis of NE. It is conceivable that individuals differ in their capacity to step up NE biosynthesis or that individuals with low-normal stores of NE might be more severely depleted at similar doses of reserpine. Whatever the mechanism, it is likely that reserpine unmasks a predisposition to depression.

We have already noted that at least two types of antidepressant agents are subjected to genetic variation in their metabolism. Since phenelzine, an MAO inhibitor, is acetylated by the polymorphic hepatic acetylase system, individuals with the slow acetylator phenotype are more likely to experience side effects.[17] Likewise, standard doses of the tricyclic agent nortriptyline lead to variability in blood level and risk of side effects because of genetically determined differences in rate of metabolism.[24]

Finally, it should be emphasized that so general a phenotype as depression is likely to be predisposed to or mediated by multiple mechanisms, even if various alterations in biogenic amine metabolism serve as a common pathogenetic pathway. We should be alert to the possibility that differential responses to therapy and differential susceptibility to precipitation of attacks with reserpine or alpha methyl dopa or ACTH may provide insights and investigational "handles" into the heterogeneous causes of the syndrome.

Schizophrenia

Despite claims to the contrary, certain phenothiazines do not appear to be relatively better than others for different forms of schizophrenic illness.[64] Longitudinal follow-up of individual patients and family studies also fail to support clinical subtypes as a basis for sorting presumed heterogeneity in schizophrenia.[65]

Phenothiazines induce extrapyramidal side effects in an increasing proportion of patients as dosage is raised. The risk of Parkinsonian symptoms from these agents is significantly higher in those patients with a positive family history of spontaneous Parkinson's disease.[66, 67] Presumably, the drugs unmask an inherited predisposition. Phenothiazines may also induce dramatic dystonic syndromes, including lockjaw, torticollis, and oculogyric crisis. It is not known whether the predisposition is the same as for the Parkinson syndrome. Furthermore, it is not known whether those patients most susceptible to extrapyramidal side effects are also at greatest risk for cholestatic liver damage. If so, individual differences in blood levels might be the crucial variable.

Given the evidence for genetic predisposition to schizophrenia,[65, 68] it is reasonable to wonder whether individuals who have a schizophrenia-like reaction to amphetamines are genetically predisposed to such psychotic reactions and would have been at relatively high-risk for development of spontaneous schizophrenia.[69] Pharmacologic deductions may provide insight into possible neurotransmitter mediation of at least some types of schizophrenia. The active groups of potent antipsychotic phenothiazines, when viewed in a three-dimensional molecular model, appear to resemble the molecular conformation of dopamine.[70] Amphetamines also act on biogenic amines, primarily through inhibition of the neuronal reuptake mechanism. *In vitro* studies of isolated synaptosomes from dopamine-rich and from norepinephrine-rich regions have shown that D-amphetamine and L-amphetamine are equipotent in the action on dopamine reuptake (dopamine lacks an optically active carbon), whereas D-amphetamine is ten times more potent than its L-steroisomer on NE reuptake (both amphetamine and norepinephrine have an asymmetric, optically active carbon). A locomotor activity measure thought to be mediated by NE in rats enhanced *in vivo* at a ratio of 10:1 by D and L amphetamine, whereas a stereotyped gnawing behavior thought to be mediated by dopamine is elicited at a 1:1 ratio.[71] Finally, D and L amphetamine appear to be equipotent in inducing amphetamine psychosis, many features of which resemble schizophrenic syndromes.[72] Extension of such studies of genotype-drug interactions may reveal clinical heterogeneity and biochemical mechanisms for schizophrenia and for other behavioral disorders.

Experimental psychologists have noted that magnitude of LSD and psilocybin effects on perception is related to the variability that the particular subject showed on the perceptual or behavioral test before ingesting the drug.[73] Genetic factors in the biological substrate may be involved in the lability or variability of such specific behaviors. It is not known, of course, whether such genetic factors might be the same as those predisposing to schizophrenia.

Seizure Disorders

The normal pattern of electrical activity in the brain, as measured by the electroencephalogram (EEG), is determined almost entirely by genetic factors in a polygenic system, according to Vogel.[74] In addition, several single-gene mediated variants of the normal EEG have been described, affecting altogether about 15 percent of the general population (Table 16-I). The clinical significance of these variants is unknown. No studies have yet been carried out with groups of individuals having different baseline EEG patterns to see whether there are different responses to various psychopharmacologic agents. The EEG variation provides an intermediate level of evidence between the observable variation in human behavior and the biochemical variation at the level of individual enzymes in brain tissue of different people. One older study of phenobarbital effects found induction of beta waves in thirteen of the eighteen similarly treated subjects.[75] All thirteen became tolerant by clinical criteria, but six of the thirteen failed to show reversal to normal EEG activity (EEG tolerance). Withdrawal caused EEG abnormalities in only five of the eighteen subjects; none had preexisting or permanent EEG changes. It is likely that similar differences in EEG response occur with alcohol and meprobamate withdrawal. Population studies indicate polygenic predisposition to epilepsy in man and probable autosomal dominant predisposi-

TABLE 16-I
VARIANTS OF THE NORMAL HUMAN EEG

Rhythm	Genetic Basis	Population Frequency	Comment
Normal alpha (8-13 cps)	Polygenic		
Low voltage alpha	Auto Dom	7.0%	
Quick alpha (16-19 cps)	Auto Dom	0.5%	
Occipital slow (4-5 cps)	?	0.1%	?Psychopathy
Monotonous Tall Alpha	Auto Dom	4.0%	?Assortative Mating
Beta waves	Multifactorial	5.0-10.0%	Sex, Age; ?Assortative Mating
Frontal beta groups (25-30 cps)	Auto Dom	0.4%	
Fronto-precentral beta (20-25 cps)	Auto Dom	1.4%	

tion to the 3cps spike-and-wave centrencephalic pattern of childhood seizure disorder.[76,77]

A variety of anticonvulsive agents has been employed in clinical seizure disorders without biochemical elucidation of mechanisms of therapeutic effectiveness. In mice, however, an excellent model of seizure susceptibility has been analyzed in great detail. Mice of certain genotypes, containing usually the autosomal recessive allele d for dilute pigmentation, are particularly susceptible to audiogenic seizures. The generality of the disorder extends to caffeine, strychnine, nicotine, pentylenetetrazol, and electrical induction of seizures. The phenomenon is strikingly age-dependent, and the age of greatest susceptibility is different in different strains. Pharmacologic manipulation can greatly alter the seizure susceptibility; age-correlated reductions in NE, 5-hydroxytryptamine (5HT) and gamma-aminobutyric acid (GABA), and abnormalities in oxidative phosphorylation have been reported.[32] The primary defect is still unknown, however.

The anticonvulsant most commonly used in man is diphenylhydantoin (Dilantin), which is eliminated slowly by biotransformation.[16] Since this drug is lipid-soluble, only negligible amounts are excreted unchanged by the kidney. Two thirds of administered Dilantin is hydroxylated by the hepatic microsomal oxidase system, the rate of which is influenced by genetic factors and by other drugs which induce activity (especially phenobarbital) or inhibit hydroxylation (isoniazid). Both of these combinations are encountered commonly in practice. Most patients are capable of metabolizing up to 10 mg/kg/day of Dilantin. However, Kutt has observed members of several families who, when given the usual adult dose of 4 to 5 mg/kg/day, continuously accumulated the unmetabolized drug without stabilization of the blood level and developed progressive signs of toxicity.[16] At the other extreme, unusually rapid metabolism of Dilantin has been found in several patients, but no family studies were carried out.[16] Such abnormal patterns of drug metabolism may be enlightening about the specific enzymes involved in drug metabolism.

Mental Retardation

Several of the inborn errors of metabolism that lead to mental retardation can be ameliorated by treatment with special dietary regimens which might be considered a particular form of interaction between genotype and exogenous agent. Thus, diets restricted in phenylalanine and in branched-chain amino acids are effective in phenylketonuria and in maple-syrup urine disease, respectively. When an inherited enzyme abnormality causes decreased affinity for an essential coenzyme, exogenous administration of the coenzyme in massive doses may be effective. Thus, some patients with homocystinuria or with cystathioninuria have a favorable metabolic response to pyridoxine; there is no proof that the mental status has been improved.[78] Likewise, high doses of vitamin B12 may overcome the enzyme abnormality in methylmalonic aciduria.[79] Administration of concentrates of protein factors from normal plasma or leukocytes or urine may correct the accumulation of mucopolysaccharides in such disorders as Hurler's syndrome.[80] These rare autosomal recessive diseases are mentioned primarily to point out that for each disease that occurs at a frequency of only one in 40,000, the frequency of heterozygotic carriers is 1 percent. Such carriers may be predisposed to mental illness or to adverse psychological effects from drugs.[81]

The finding of rare enzyme abnormalities in cofactor binding that is responsive to administration of massive doses of vitamins has refueled the suggestion that individual variability in requirements for such dietary factors might lead to dietary deficiencies on the "usual recommended daily dose" of these vitamins. For example, Pauling has proposed that nicotinic acid or vitamin C in massive doses be used to ameliorate schizophrenia or enhance intelligence.[82] No biochemical evidence of such mechanisms and no definite clinical responses have been reported. However, no systematic study of enzymes with specific cofactors has been carried out in search of enzyme variants, possibly occurring at low frequency, that might require abnormally high concentrations of vitamin cofactors.

Drugs of Abuse and Addiction Syndromes

The most commonly abused agent, of course, is ethanol. There is a common lay impression that individuals differ markedly in tolerance for alcohol and in susceptibility to its effects. Twin studies in Scandinavia and in a sample of American high school students indicate considerable heritability for at least some aspects of drinking behavior.[83, 84, 85] Family studies demonstrate high frequencies of alcoholism among close relatives of alcoholics but do not necessarily distinguish genetic and environmental influences. Such distinction could be obtained with the comparison of biological and adoptive relatives of children adopted early in life, as in the extensive studies of schizophrenia.[86] An alternative approach is the study of half-sibs. Schuckit has reported that alcoholism in biological relatives, but not alcoholism in adoptive relatives, predisposes to alcoholism in their patients.[87] In fact, the half-sib analysis allows comparison of maternally-related and paternally-related half-sibs, so that the possible effects of the intrauterine environment can be assessed, too.

Table 16-II lists several levels at which genetic factors could play a role in the total picture of alcohol use.[88] Acute intoxicating

TABLE 16-II
LEVELS OF GENETIC INFLUENCE IN ALCOHOLISM
1. Susceptibility to acute intoxicating effects
2. Metabolism of ethanol
3. Central nervous system cellular adaptation to chronic intake-addictability
4. Predisposing personality factors
5. Susceptibility to medical and behavioral complications

effects of alcohol are related to the blood level. Studies of rate of elimination of ethanol from the blood in MZ and DZ twins indicate that genetic factors are almost entirely responsible for individual differences (heritability estimated at 0.98).[89] Similar studies of antipyrine, phenylbutazone, bishydroxycoumarin, and, to a lesser extent, halothane, show that environmental contributions may be negligible compared to genetic factors in nonmedicated, nonhospitalized healthy twins. This extent of genetic control over the variation in drug metabolism has come as a surprise to many experimental pharmacologists familiar with the striking effects of dosage schedules, cage conditions, diet, etc.

As noted above, gene frequencies are often different in different ethnic or racial groups. Casual police reports and lay impressions indicated that Eskimos and Indians in Canada take longer to sober up after an alcoholic binge than do Caucasians in the same area.[90] Differences in rate of metabolism of ethanol were shown to underlie these observations. The amount of alcohol required per unit body weight to induce or maintain an intoxicated state was similar in all three groups, but the rate of decline in blood alcohol levels was significantly lower in both the Eskimo and Indian samples than in the whites.[90] Wolff found ethnic differences in vascular reactivity to alcohol ingestion in humans.[91] Japanese, Taiwanese, and Korean (Mongoloid) adults and infants responded with marked facial flushing, mild symptoms of intoxication, and increased pulse pressure to doses of alcohol that had little or no effect on Caucasian adults and infants. Presumably, these differences in reactivity of the autonomic nervous system are genetically determined.

Extensive studies in mice and rats have demonstrated strain differences in drinking behavior in ethanol/water choice situations. Furthermore, differences in brain sensitivity to alcohol have been found in inbred mouse strains, using "sleep time" as a measure.[92] These differences are not accounted for by strain differences in alcohol or aldehyde dehydrogenase activities.

The relationship between acute intoxication and development of tolerance and physical dependence (addicted state) is still unclear, especially for alcohol. It is likely that differences in brain metabolism cause differences in susceptibility to addiction. Also, it is not known whether there is some common metabolic process that underlies addictability to such different agents as alcohol, opiates, and even barbiturates. *In vitro,* acetaldehyde can condense with norepinephrine or serotonin to form Schiff bases that undergo spontaneous rearrangment to tetrahydroisoquinolines or tetrahydropapaveraline, which are structures similar to plant alkaloids with high addictive potency.[93] Also, there are claims that strains of rats obtained by selective breeding for susceptibility to opiate addiction manifest similar relapsing behavior when offered alcohol after periods of drinking and abstinence.[94] Each agent might interact with different cell receptors to initiate a common process. As an analogy, it has recently been shown that many polypeptide hormones interact with specific cell surface receptors, then activate the enzyme adenyl cyclase and release cyclic AMP, a mediator for intracellular effects of all these hormones.[95] Inhibitors of protein and RNA synthesis inhibit the development of tolerance to morphine without blocking the acute reaction in nontolerant mice.[96] The same inhibitors have been used to explore the range of macromolecules that might be involved in information processing for general memory. Drug tolerance thus may be a special case of memory functions.

Minimal Brain Dysfunction: Hyperkinetic Youngsters

This clinical syndrome consists of motor hyperactivity, distractibility, impulsivity, and learning performance below objective expectations in grade school children.[97, 98] Some 3 to 15 percent of school children are thought to merit this diagnostic categorization. Boys are affected ten to twelve times more frequently than girls. Some children have histories of brain damage, but many have no history or clinical or EEG signs of brain damage and seem to merge with the normal range of childhood behavior. Two family studies have strongly suggested predisposing genetic factors.[98a, 98b] Adoptive relatives of MBD children who had been adopted early in life did not have any excess of behavioral abnormalities.[98c] One twin study has been reported.[99] Four pairs of monozygotic male twins were concordant for MBD, while only one of six dizygotic pairs was concordant. Unfortunately, four of the discordant dizygotic pairs were of opposite sex, and there is a marked preponderance of expression of this disorder in males.

It seems likely that this syndrome is as heterogeneous in causes and mechanisms as mental retardation. Laboratory and clinical criteria are needed desperately to sort the presumed heterogeneity. The most promising approach appears to be pharmacogenetic. Bradley noted in 1937 that amphetamine had a paradoxical quieting effect on such children and phenobarbital was observed to excite, rather than sedate, hyperkinetic children.[100, 101] Both amphetamines and methylphenidate (Ritalin®) tend to subdue the obnoxious behaviors in these youngsters, while allowing desirable behaviors to be expressed.[102] However, there are wide differences among patients in the clinical response to these central nervous system stimulants; dramatic improvement is the exception, and the usual improvement may not be significantly greater than that obtained with careful behavioral and environmental modification programs. No systematic comparison of the potency of D-amphetamine and methylphenidate or of D-amphetamine and L-amphetamine in the same patients has been attempted; nor have the paradoxical responses to amphetamine and phenobarbital been examined in the same patients. Neurophysiological studies of responses to these drugs may be feasible. For example, intravenous amphetamine or methylphenidate was reported to suppress photic driving responses in six of seven hyperkinetic youngsters who had such abnormal EEG responses before administration of the drug.[103] Studies of drug metabolism and responses to the drugs in family members may distinguish whether the differences in clinical response are due to differential handling of the drug or to different mechanisms of the underlying disorder.

DRUG-DRUG INTERACTIONS

When two drugs are administered simultaneously, the effects of one or both drugs may be potentiated, or inhibited, or made toxic. Often the drugs are given for different indications and started at different times; the physician ordering the second agent may be unaware that his patient already is taking the first drug. If one of these agents has an important genotype-drug interaction as well, then the drug-drug interaction may vary indirectly with the genotype of the patients.

In this chapter, we will limit discussion of this large and clinically significant topic to examples of three types of such drug-drug-genotype interactions.[104]

Competition for Plasma Protein Binding Sites

Trichloracetic acid, the major metabolite of the commonly used sedative chloral hydrate, displaces the anticoagulant drug warfarin (Coumadin®) from its binding sites on plasma albumin.[105] As a result, warfarin has higher unbound plasma levels, shorter half-life, lower doses required to induce the anticoagulated state, and greater risk of adverse effects due to overanticoagulation.[106] A number of other acidic drugs may displace warfarin similarly.[105] The coumarin anticoagulants alone are metabolized at different rates in different individuals on a genetic basis, according to twin studies.[107]

Alteration of Hepatic Drug-metabolizing Activity

Phenobarbital in clinically used doses markedly enhances hepatic microsomal hydroxylase, dehydrogenase, and glucuronidating activities. The extent of this inducibility may be controlled genetically, at least in strains of rats.[108] Phenobarbital depresses plasma levels of Dilantin by increasing the rate of hydroxylation of Dilantin.[16] Conversely, isoniazid, which decreases microsomal oxidase activity, increases Dilantin levels on a standard dose and predisposes in slow acetylators of INH to Dilantin toxicity.[16] High blood levels of L-dopa and nortriptyline also inhibit the microsomal oxidase system.[109] Methylphenidate appears to inhibit hepatic demethylase and hydroxylase activities necessary for inactivation of potent narcotics.[110] Strain differences in domestic and wild rats in aniline hydroxylase and ethylmorphine N-demethylase activities suggest that genetic differences, as well as drug-induction, may be involved.[111]

Interaction of Drugs at Site of Action

The hazardous potentiation of action of tricyclic antidepressants and of sympathomimetic amines like tyramine by monoamine oxidase inhibitors is so well recognized that MAO inhibitors are little used clinically as a result.[62] We have noted above the genetically determined variability in plasma concentration of the tricyclic compound nortriptyline and the possibility that individuals may differ in susceptibility of brain monoamine oxidase to inhibition. In addition to a marked hypertensive reaction, the combination of MAO inhibitor and tricyclic antidepressant may induce a hyperpyrexic state clinically similar to malignant hyperthermia.[43]

SUMMARY

Genetically-determined differences between individuals in metabolism of or response to psychoactive agents may be termed psychopharmacogenetics. A variety of biochemical and population approaches is available to establish the pattern of metabolism of a particular drug and the basis of observed individual differences.

Suxamethonium, Dilantin, oxidant drugs, and corticosteroids are examples of commonly used drugs which carry defined risks of marked adverse effects in genetically susceptible individuals. Patients with such simply inherited rare diseases as porphyria, dysautonomia, Lesch-Nyhan syndrome, malignant hyperthermia, and Huntington's chorea are particularly vulnerable to certain drugs. For each of these conditions, certain ethnic or racial groups are at high or low risk, an important consideration in identifying patients to be screened. Pharmacologic responses also offer a potentially powerful tool to dissect heterogeneity of mechanisms for the common psychiatric syndromes, including depression, schizophrenia, epilepsy, mental retardation, drug addiction, and minimal brain dysfunction.

Investigation of the clinical observation of individual variability in response to psychotherapeutic agents may provide guidelines for rational therapy and insight into disease mechanisms.

REFERENCES

1. Motulsky, A.G.: Drug reactions, enzymes, and biochemical genetics. *JAMA*, 165:835, 1957.
2. Kalow, W.: *Pharmacogenetics. Heredity and the Response to Drugs.* Philadelphia, Saunders, 1962.
3. Vogel, F.: Moderne Probleme der Humangenetik. *Ergeb Inn Med Kinderheilkd*, 12:52, 1959.
4. Motulsky, A.G.: Pharmacogenetics. In Steinberg, A.G., and Bearn, A.G. (Eds.): *Progress in Medical Genetics.* New York, Grune, 1964, vol. 3:49.
5. Price-Evans, D.A.: Pharmacogenetics. In Clark, C.A. (Ed.): *Selected Topics in Medical Genetics.* Oxford, Oxford U Pr, 1969, pp 69.
6. Motulsky, A.G.: History and current status of pharmacogenetics. *Proc 4th Intl Congr Human Genetics, Paris, 1971.* Excerpta Medica, 1972, p. 381.
7. Vandenberg, S.G.: Contributions of twin research to psychology. *Psychol Bull*, 66:327, 1966.
8. Shields, J.: *Monozygotic Twins Brought Up Apart and Brought Up Together.* Oxford, Oxford U Pr, 1962.
9. Lehmann, H., and Ryan, E.: The familial incidence of low pseudocholinesterase level. *Lancet*, 2:124, 1956.
10. Kalow, W., and Genest, K.: A method for the detection of atypical forms of human serum cholinesterase. Determination of dibucaine numbers. *Can J Biochem*, 35:339, 1957.
11. Yoshida, A., and Motulsky, A.G.: A pseudocholinesterase variant (E$_{Cynthiana}$) associated with elevated plasma enzyme activity. *Am J Hum Genet*, 21:486, 1969.
12. Morrow, A.C., and Motulsky, A.G.: Rapid screening method for the common atypical pseudocholinesterase variant. *J Lab Clin Res*, 71:350, 1968.
13. Goedde, H.W., and Altland, K.: Suxamethonium sensitivity. *Ann NY Acad Sci*, 179:695, 1971.
14. Bönicke, R., and Lisboa, B.P.: Über der Erbbedingtheir der intraindividuellen Konstanz der Isoniazidausscheidung beim Mensch. *Nauturwissenschaften*, 44:314, 1957.
15. Knight, R.Z., Selin, M.J., and Harris, H.W.: Genetic factors influencing isoniazid blood levels in humans. *Proc Trans Conf Chemother Tuberculosis, 18th Conf*, 52, 1959.
16. Kutt, H.: Biochemical and genetic factors regulating Dilantin metabolism in man. *Ann NY Acad Sci*, 179:704, 1971.
17. Price-Evans, D.A., Davison, K., and Pratt, R.T.C.: The influence of acetylator phenotype on the effects of treating depression with phenelzine. *Clin Pharmacol Ther*, 6:430, 1965.
18. Beutler, E.: Glucose-6-phosphate dehydrogenase deficiency. In Stanbury, J.B., Wyngaarden, J.B., and Fredrickson, D.S. (Eds.): *The Metabolic Basis of Inherited Disease*, 3rd ed. New York, McGraw, 1972, p. 1358.
19. Graves, R.: *The Greek Myths.* Baltimore, Penguin, 1955, vol. I, p. 348.
20. Beutler, E.: L-dopa and favism. *Blood*, 36:523, 1970.
21. Armaly, M.G.: Genetic factors related to glaucoma. *Ann NY Acad Sci*, 151:861, 1968.
22. Alexanderson, B., and Sjöqvist, F.: Individual differences in the phar-

macokinetics of monomethylated tricyclic anti-depressants: Role of genetic and environmental factors and clinical importance. *Ann NY Acad Sci, 179*:739, 1971.
23. Angst, J.: Antidepressiver Effekt und genetische Faktoren. *Arzneimittelforschung, 14*:496, 1964.
24. Åsberg, M., Cronholm, B., Sjöqvist, F., and Tuck, D.: The correlation of subjective side-effects with plasma concentrations of nortriptyline. *Br Med J, 4*:18, 1970.
25. Åsberg, M., Price-Evans, D.A., and Sjöqvist, F.: Genetic control of nortripyline kinetics in man: A study of relatives of propositi with high plasma concentrations. *J Med Genet, 8*:129, 1971.
26. Davis, J.M., Kopin, I.J., Lemberger, L., and Axelrod, J.: Effects of urinary pH on amphetamine metabolism. *Ann NY Acad Sci, 179*:493, 1971.
27. Meier, H.: *Experimental Pharmacogenetics.* New York, Acad Pr, 1963.
28. Fuller, J.L.: Strain differences in the effects of chlorpromazine and chlordiazepoxide upon active and passive avoidance in mice. *Psychopharmacologia, 16*:261, 1970.
29. Jori, A., Bernardi, D., Pugliath, C., and Garattini, S.: Strain differences in the metabolism of imipramine by rat. *Biochem Pharmacol, 19*:1315, 1970.
30. Van Abeelen, J.H.F., and Strijbosch, H.: Genotype-dependent effects of scopolamine and eserine on exploratory behavior in mice. *Psychopharmacologia, 16*:81, 1969.
31. Karczmar, A.G., and Scudder, C.L.: Behavioral responses to drugs and brain catecholamine levels in mice of different strains and genera. *Fed Proc, 26*:1186, 1967.
32. Schlesinger, K., and Griek, B.J.: The genetics and biochemistry of audiogenic seizures. In Lindzey, G., and Thiessen, D.D. (Eds.): *Contributions to Behavior-Genetic Analysis: The Mouse as a Prototype.* New York, Appleton, 1970.
33. Fuller, J.L., and Collins, R.L.: Genetics of audiogenic seizures in mice: A parable for psychiatrists. *Semin Psychiatry, 2*:75, 1970.
34. Waldenström, J.: Studien ueber Porphyrie. *Acta Med Scand, [Suppl], 82*:1, 1937.
35. Dean, G.: *The Porphyrias. A Story of Inheritance and Environment.* Philadelphia, Lippincott, 1963.
36. Marver, H.S., and Schmid, R.: The porphyrias. In Stanbury, J.B., Wyngaarden, J.B., and Fredrickson, D.S. (Eds.): *The Metabolic Basis of Inherited Disease,* 3rd ed. New York, McGraw, 1972.
37. MacAlpine, I., and Hunter, R.: Porphyria and King George III. *Sci Am, 221*:38, 1969.
38. Strand, L.J., Felsher, R.F., Radeker, A.G., and Marver, H.S.: Heme biosynthesis in intermittent acute porphyria: Decreased hepatic conversion of porphobilinogen to porphyrin and increased delta aminolevulinic acid synthetase activity. *Proc Natl Acad Sci USA, 67*:1315, 1970.
39. Kappas, A., Bradlow, H.L., Gillette, P.N., and Gallagher, T.F.: Abnormal steriod hormone metabolism in the genetic liver disease acute intermittent porphyria. *Ann NY Acad Sci, 179*:611, 1971.
40. Dancis, J.: Altered drug response in familial dysautonomia. *Ann NY Acad Sci, 151*:876, 1968.
41. Brunt, P.W., and McKusick, V.A.: Familial dysautonomia: A report of genetic and clinical studies with a review of the literature. *Medicine (Baltimore), 49*:343, 1970.
42. Weinshilboum, R.M., and Axelrod, J.: Reduced plasma dopamine beta-hydroxylase activity in familial dysautonomia. *N Engl J Med, 285*:938, 1971.
43. Lesch, M., and Nyhan, W.L.: A familial disorder of uric acid metabolism and central nervous system function. *Am J Med, 36*:561, 1964.
44. Kelley, W.N., and Wyngaarden, J.B.: The

Lesch-Nyhan syndrome. In Stanbury, J.B., Wyngaarden, J.B., and Fredrickson, D.S. (Eds.): *The Metabolic Basis of Inherited Disease*, 3rd ed. New York, McGraw, 1972.
45. Britt, B.A., and Kalow, W.: Malignant hyperthermia. A statistical review. Aetiology unknown. *Can Anaesth Soc J, 17*:293, 316, 1970.
46. Denborough, M.A., Forster, J.F.A., Lovell, R.R.H., Maplestone, P.A., and Villers, J.D.: Anaesthetic deaths in a family. *Br J Anaesth, 34*:395, 1962.
47. Britt, B.A., Locher, W.G., and Kalow, W.: Hereditary aspects of malignant hyperthermia. *Can Anaesth Soc J, 16*:89, 1969.
48. Kalow, W., Britt, B.A., Terreau, M.E., and Haist, C.: Metabolic error of muscle metabolism after recovery from malignant hyperthermia. *Lancet, 2*:895, 1970.
49. Harrison, G.G., Biebuyck, J.F., Terblanche, J., Dent, D.M., Hickman, R., and Saunders, S.J.: Hyperpyrexia during anesthesia. *Br Med J, 3*:594, 1968.
50. Harrison, G.G.: Anaesthetic-induced malignant hyperpyrexia: A suggested method of treatment. *Br Med J, 3*:454, 1971.
51. Pollock, R.A., and Watson, R.L.: Malignant hyperthermia associated with hypocalcemia. *Anesthesiology, 34*:188, 1971.
52. Harper, P.S., Rivas, M.L., Bias, W.B., Hutchinson, J.R., Dyken, P.R., and McKusick, V.A.: Genetic linkage confirmed between the loci for myotonic dystrophy, ABH secretion, and Lutheran blood group. *Am J Hum Genet, 24*:310, 1972.
53. Klawans, H.L., Paulsen, G.W., and Barbeau, A.: A predictive test for Huntington's chorea. *Lancet, 2*:1185, 1970.
54. Cederbaum, S.: Tests for Huntington's chorea. *N Engl J Med, 284*:1045, 1971.
55. Winokur, G., Clayton, P.J., and Reich, T.: *Manic-Depressive Illness*. St. Louis, Mosby, 1969.
56. Gershon, E.S., Dunner, D.L., and Goodwin, F.K.: Toward a biology of affective disorders. Genetic considerations. *Arch Gen Psychiatry, 25*:1, 1971.
57. Schildkraut, J.J.: Neuropsychopharmacology and the affective disorders. *N Engl J Med, 281*:197, 248, 302, 1969.
58. Musacchio, J.M., Julou, L., Kety, S.S., and Glowinski, J.: Increase in rat brain tyrosine hydroxylase activity produced by electroconvulsive shock. *Proc Natl Acad Sci USA, 63*:1117, 1969.
59. Pare, C.M.B., Rees, L., and Sainsbury, M.J.: Differentiation of two genetically specific types of depression by the response to anti-depressants. *Lancet, 2*:1340, 1962.
60. Pare, C.M.B., and Mack, J.W.: Differentiation of two genetically specific types of depression by the response to antidepressant drugs. *J Med Genet, 8*:306, 1971.
61. King, L.J.: Chemotherapy of mental depression. *Ann Rev Med, 21*:367, 1970.
62. Harris, T.H.: Depression induced by Rauwolfia compounds. *Am J Psychiatry, 113*:950, 1957.
63. Muller, J.C., Pryor, W.W., Gibbons, J.E., and Orgain, E.S.: Depression and anxiety occurring during Rauwolfia therapy. *JAMA, 159*:836, 1955.
64. Hollister, L.E.: Choice of antipsychotic drugs. *Am J Psychiatry, 127*:104, 1970.
65. Rosenthal, D.: *Genetic Theory and Abnormal Behavior*. New York, McGraw, 1970.
66. Myrianthopoulos, N.C., Kurland, A.A., and Kurland, L.T.: Hereditary predisposition in drug-induced Parkinsonism. *Arch Neurol, 6*:5, 1962.
67. Myrianthopoulos, N.C., Waldrop, F.N., and Vincent, B.L.: A repeat study of hereditary predisposition in drug-induced Parkinsonism. *Excerpta Medica Intl Cong, Series 175*:486, 1967.
68. Heston, L.L.: The genetics of schizophrenia and schizoid disease. *Science,*

167:249, 1970.
69. Ellinwood, E.H., Jr.: Amphetamine psychosis: A multi-dimensional process. *Semin Psychiatry, 1*:208, 1969.
70. Horn, A.S., and Snyder, S.H.: Chlorpromazine and dopamine: Conformational similarities that correlate with the anti-schizophrenic activity of phenothiazine drugs. *Proc Natl Acad Sci USA, 68*:2325, 1971.
71. Snyder, S.H., Taylor, K.M., Coyle, J.T., and Meyerhoff, J.L.: The role of brain dopamine in behavioral regulation and the actions of psychotropic drugs. *Am J Psychiatry, 127*:117, 1970.
72. Angrist, B.M., Shopsin, B., and Gershon, S.: The comparative psychotomimetic effects of stereoisomers of amphetamine. *Nature (Lord), 234*:152, 1971.
73. Fischer, R.: A cartography of the ecstatic and meditative states. *Science, 174*:897, 1971.
74. Vogel, F.: The genetic basis of the normal human electroencephalogram (EEG). *Humangenetik, 10*:91, 1970.
75. Essig, C.F., and Fraser, H.F.: Electroencephalographic changes in man during use and withdrawal of barbiturates in moderate doses. *Electroencephalogr Clin Neurol, 10*:649, 1958.
76. Pratt, R.T.C.: *The Genetics of Neurological Disorders.* Oxford, Oxford U Pr, 1967.
77. Metrakos, J.D., and Metrakos, K.: Genetic studies in clinical epilepsy. In Jasper, H.H., Ward, A.A., and Pope, A. (Eds.): *Basic Mechanisms of the Epilepsies.* Boston, Little, 1969, p. 700.
78. Mudd, S.H., Edwards, W.A., Loeb, P.M., Brown, M.S., and Laster, L.: Homocystinuria due to cystathionine synthase deficiency: The effect of pyridoxine. *J Clin Invest, 49*:1762, 1970.
79. Rosenberg, L.E.: Disorders of propionate, methylmalonate, and vitamin B_{12} metabolism. In Stanbury, J.B., Wyngaarden, J.B., and Fredrickson, D.S. (Eds.): *The Metabolic Basis of Inherited Disease.* New York, McGraw, 1972, p. 440.
80. Knudson, A.G., Jr., DiFerrante, N., and Curtis, J.E.: Effect of leukocyte transfusion in a child with type II mucopolysaccharidosis. *Proc Natl Acad Sci USA, 68*:1738, 1971.
81. Williams, R.J.: *Biochemical Individuality.* New York, Wiley, 1956.
82. Pauling, L.: *Vitamin C and the Common Cold.* San Francisco, Freeman, 1970.
83. Kaij, L.: *Alcoholism in Twins.* Stockholm, Almqvist & Wiksell, 1960.
84. Partanen, J., Bruun, K., and Markkanen, T.: *Inheritance of Drinking Behavior—A Study on Intelligence, Personality, and Use of Alcohol of Adult Twins.* Helsinki, Finnish Foundation for Alcohol Studies, 1966.
85. Loehlin, J.C.: An analysis of alcohol-related questionnaire items from the National Merit twin study. In Seixas, F.A., Omenn, G.S., Burk, E.D., and Eggleston, S.A. (Eds.): *Nature and Nurture in Alcoholism.* New York, NY Acad Sci, 197:117, 1972.
86. Rosenthal, D., and Kety, S.S. (Eds.): *The Transmission of Schizophrenia.* Oxford, Pergamon, 1968.
87. Schuckit, M.A.: Family history and half-sibling research in alcoholism. In Seixas, F.A., Omenn, G.S., Burk, E.D., and Eggleston, S.A. (Eds.): *Nature and Nurture in Alcoholism.* New York, NY Acad Sci, 197:121, 1972.
88. Omenn, G.S., and Motulsky, A.G.: A biochemical and genetic approach to alcoholism. In Seixas, F.A., Omenn, G.S., Burk, E.D., and Eggleston, S.A. (Eds.): *Nature and Nurture in Alcoholism.* New York, NY Acad Sci, 197:16, 1972.
89. Vesell, E.S., Page, J.G., and Passananti, G.T.: Genetic and environmental factors affecting ethanol metabolism in man. *Clin Pharmacol Ther, 12*:192, 1971.
90. Fenna, D., Mix, L., Schaefer, O., and Gilbert, J.A.L.: Ethanol metabolism in

various racial groups. *Can Med Assoc J, 105*:472, 1971.
91. Wolff, P.H.: Ethnic differences in alcohol sensitivity. *Science, 175*:449, 1972.
92. Kakihana, R., Brown, D.R., McClearn, G.E., and Tabershaw, I.R.: Brain sensitivity to alcohol in inbred mouse strains. *Science, 154*:1574, 1966.
93. Davis, V.E., and Walsh, M.J.: Alcohol, amines, alkaloids: A possible biochemical basis for alcohol addiction. *Science, 167*:1005, 1970.
94. Nichols, J.R., and Hsaio, S.: Addiction liability of albino rats: Breeding for quantitative differences in morphine drinking. *Science, 157*:561, 1967.
95. Sutherland, E.W.: On the biological role of cyclic AMP. *JAMA, 214*:1281, 1970.
96. Way, E.L., Loh, H.H., and Shen, F.H.: Morphine tolerance, physical dependence and synthesis of brain 5-hydroxytryptamine. *Science, 162*:1290, 1968.
97. Conners, C.K.: The syndrome of minimal brain dysfunction: psychological aspects. *Pediatr Clin N Am, 14*:749, 1967.
98. Paine, R.S.: Syndromes of "minimal cerebral damage." *Pediatr Clin N Am, 15*:779, 1968.
98a. Morrison, J.R., and Stewart, M.A.: A family study of the hyperactive child syndrome. *Biol Psychiatry, 3*: 189, 1971.
98b. Cantwell, D.P.: Psychiatric illness in families of hyperactive children. *Arch Gen Psychiatry, 27*:414, 1972.
98c. Morrison, J.R., and Stewart, M.A.: The psychiatry status of the legal families of adopted hyperactive children. *Arch Gen Psych, 28*:888, 1973.
99. Lopez, R.E.: Hyperactivity in twins. *Can J Psychol, 10*:421, 1965.
100. Bradley, C.: The behavior of children receiving benzedrine. *Am J Psychiatry, 94*:577, 1937.
101. Lindsley, D.B., and Henry, C.F.: The effect of drugs on behavior and the electroencephalogram of children with behavior disorders. *Psychosom Med, 4*:140, 1942.
102. Sulzbacher, S.I.: *Diagnosis and Treatment with Medication of Learning and Behavior Problems in the School-Age Child. Working paper #5, Child Development and Mental Retardation Center.* Seattle, U of Wash Pr, 1971.
103. Shetty, T.: Photic responses in hyperkinesis of childhood. *Science, 174*:1356, 1971.
104. Vesell, E.S.: Drug metabolism in man. *Ann NY Acad Sci, 179*:1, 1971.
105. Sellers, E.M., and Koch-Weser, J.: Potentiation of warfarin-induced hypoprothrombinemia by chloral hydrate. *N Engl J Med, 283*:827, 1970.
106. Boston Collaborative Drug Surveillance Program: Interaction between chloral hydrate and warfarin. *N Engl J Med, 286*:53, 1972.
107. Vesell, E.S., and Page, J.G.: Genetic control of dicoumarol levels in man. *J Clin Invest, 47*:2657, 1968.
108. Deitrich, R.A.: Genetic basis of phenobarbital induced increase in aldehyde dehydrogenase: Implication for alcohol research. In Seixas, F.A., Omenn, G.S., Burk, E.D., and Eggleston, S.A. (Eds.): *Nature and Nurture in Alcoholism.* New York, NY Acad Sci, 197:73, 1972.
109. Vesell, E.S., Ng, L., Passananti, G.T., and Chase, T.N.: Inhibition of drug metabolism by levodopa in combination with a dopa-decarboxylase inhibitor. *Lancet, 2*:370, 1971.
110. Dayton, P.G., and Perel, J.M.: Physiological and physiochemical bases of drug interactions in man. *Ann NY Acad Sci, 179*:67, 1971.
111. Page, J.G., and Vesell, E.S.: Sex and strain differences in basal and induced aniline hydroxylase and ethylmorphine N-demethylase activities from domestic and wild rats. *Pharmacology, 2*:321, 1969.
112. Klawans, H.L., Jr., Paulson, G.W., Ringel, S.A., and Barbeau, A.: Use of L-dopa in the detection of presymptomatic Huntington's chorea. *N Engl J Med, 286*:1332, 1972.

Chapter 17

HUMAN BIOCHEMICAL VARIATION

Charles R. Shaw

Genes make polypeptides [or RNA (tRNA, rRNA)]. So far as is known, this is their only function. Some genes determine the structures of polypeptides (structural genes); others control the rates of activities of the structural genes (regulator genes), and thus determine the amounts of polypeptides produced.

Gene variation, therefore, is expressed primarily through variation in structure or in amount of a protein. The proteins which genes produce have various functions. Some comprise the structural materials of cells and organs, such as components of cell membranes. Many of the proteins are enzymes, regulating metabolism, digestion, physiological activities, etc. Some are hormones, such as the luteinizing hormone. A few proteins have no known function, although it seems unlikely that any protein is produced for no good reason, and probably the function simply remains to be discovered. Some of the polypeptides appear to have regulatory actions themselves in conjunction with the regulatory genes; at least this is so in bacteria, and this group may eventually be found to comprise a very large category of polypeptides.

ENZYME DEFICIENCY DISEASES

The first mutations to be described were necessarily those which had some visible effects on the organism. The very first, of course, were those studied by the father of modern genetics, Gregor Mendel, in his researches on the garden pea. These included such traits as color (green vs yellow) and skin surface (wrinkled vs smooth). The same point applied to studies of human genetic variation where easily recognized traits, such as eye color, were shown to be genetically determined. Somewhat later, hereditary diseases in man began to be recognized, and analyzed. But this aspect of human genetic variation was remarkably slow in developing. Garrod, in a classical and foresighted study, described the first "inborn error in metabolism" (his own term) in 1908.[1] This was the metabolic disorder alkaptonuria. He showed that affected persons secreted an abnormal product, homogentisic acid, in the urine and reasoned that there was a metabolic block in the pathway of phenylalanine and tryosine metabolism, of which homogentisic acid was deficient. He further demonstrated from family studies that afflicted individuals were homozygous for the disorder, and that both parents were heterozygous carriers. This was, of course, long before it was known that genes determine the structure of enzymes. Garrod's description of this disease constituted the prototype for the many metabolic errors subsequently described. It was not until fifty years after Garrod's publication that the enzyme deficiency itself was demonstrated.[2]

There has now been a large number of enzyme deficiency diseases described in man. These are listed in Table 17-I (adapted from Harris[3]). Most have been discovered within the past decade and a half. Most are of rare occurrence. They range in their phenotypic effects from no demonstrable effect to early mortality, and many produce severe mental retardation. It is likely there are others not discovered because they result in early embryonic death.

Most of these enzyme deficiency diseases are similar in principle to the prototypic alkaptonuria. That is, the mutation produces a polypeptide with very low or absent enzyme activity. The heterozygous carrier is phenotypically normal or almost normal since the normal allele produces active enzyme to carry out the metabolic steps. When two carriers reproduce, each offspring has one chance in four of being homozygous for the mutant form. He would thus produce only the inactive enzyme.

In many cases, discovery of the enzyme defect was based on demonstration of an abnormal metabolite in the urine, as in the case of tyrosinuria. A continuing search for such abnormal metabolites represents an ongoing and very active human genetics research, and we may expect the discovery of many additional such diseases during the next several years.

Being as they are almost universally deleterious, at least in the homozygous state, these diseases would be expected to occur in the population in relatively low frequency, as the occurrence of such morbidity would be tolerated evolutionarily only if there were a selective advantage in the heterozygous state. The latter does not appear to be the case in most of these deficiency diseases and, as expected, they are of low frequency. Exceptions are sickle cell anemia and G6PD deficiency, where the heterozygous states do appear to provide protection against malarial infection. Cystic fibrosis is a common recessively inherited disease in which there has been speculation but no facts concerning the advantage of the heterozygote.

An interesting variation of the enzyme deficiency type disease is that in which an enzyme occurs in two or more molecular forms (isozymes). This is illustrated in the case of pyruvate kinase deficiency.[4] This enzyme converts phosphoenolpyruvate to pyruvate with the generation of ATP from ADP. At least two distinct forms of pyruvate kinase have been demonstrated in mammals and are apparently controlled by separate gene loci. One form is found only in red cells and liver; the other is more widely distributed throughout a number of tissues, including liver but not red cells. A deficiency has been found in the first form of the enzyme so that in the homozygote the red cells have no pyruvate kinase, and this results in a severe anemia due to disturbances in red cell glycolysis. This form of the enzyme is also absent in the liver, but the second form remains normal in liver tissue.

Enzyme deficiency may be due to either an alternation in the enzyme so that its activity is diminished or to a decreased amount of enzyme. In most cases of human deficiency disease, it has not been determined which form of the alteration has occurred. In one instance, citrullinemia, it has been shown that the responsible enzyme is produced in normal amounts but has diminished activity, the enzyme being argininosuccinate synthetase.[5]

While the enzyme deficiency may be known for a particular genetic disease, the mechanisms leading to the various phenotypic effects usually remain obscure.

In some cases, energy-producing metabolism is diminished and this results in a variety of effects. In some cases, there is an accumulation of metabolites which has a number of adverse effects on the cells and tissues. Sometimes both effects occur, as in glucose-6-phosphatase deficiency, producing a characteristic form of glycogen storage disease known as von Gierke's. Here, glycogen accumulates in the liver and kidney which becomes very enlarged, and there is also hypoglycemia with retardation of growth.

Heterozygotes

With the development of sensitive enzyme assays of various body tissues and fluids, it has become possible to detect the heterozygous, or carrier, state for certain of the enzyme deficiency diseases. This is due to the fact that less than the normal amount of enzyme activity occurs in many of these cases, and the ranges of the normal and heterozygous state are sufficiently narrow that there is little or no overlap between the two groups. An example is galactosemia in which the enzyme galactose-1-phosphate uridyl transferase is diminished in the red cells and in cultured skin cells of heterozygous carriers.[6, 7] Heterozygotes can also be detected in some cases by intermediate levels of the pertinent metabolite. For example, in phenylketonuria, a phenylalanie tolerance curve is elevated in the heterozygote.[8]

Asymptomatic Enzyme Deficiencies

While most of the hereditary enzyme deficiencies have gross phenotypic effects, a few do not produce any demonstrable morbidity. An example is acatalasia, in which many individuals have less than 1 percent of normal catalase activity, yet suffer no apparent ill effect. A few such cases, on the other hand, develop ulceration of the oral mucosa. As screening procedures develop for detection of more abnormal or unusual metabolites, we may expect to find an increasing number of enzyme deficiencies in this category.

PROTEIN POLYMORPHISM

The hemoglobin of persons with sickle cell anemia was found by Pauling and his associates to differ from normal hemoglobin in a number of physical properties, including rate of migration in an electrophoretic field.[9] The hemoglobin of heterozygous carriers of sickle cell anemia, when electrophoresed on starch block, showed two different zones, one corresponding to the normal hemoglobin A, the other to the abnormal hemoglobin S (Fig. 17-1). Subsequently a number of other abnormal hemoglobins have been found which have usually been detected by zone electrophoresis. Starch gel electrophoresis has for the most part now replaced starch block, as it gives higher resolution.

The starch gel technique was developed

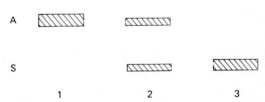

Figure 17-1. Starch gel electrophoresis patterns of hemoglobin in three individuals.
1. Normal A homozygote
2. AS heterozygote
3. S homozygote (sickle-cell anemia)

by Smithies, who first applied the method to the separation of human serum haptoglobins.[10] Here the proteins were demonstrated in the gel after electrophoresis by the use of a general protein stain such as amido black.

Shortly thereafter, Hunter and Markert adapted the starch gel technique to the electrophoretic analysis of specific enzymes by applying specific histochemical stains to the gel following electrophoresis.[11] This led to the finding that many enzymes occur in multiple forms within the same tissue, the multiple forms being termed isozymes, or isoenzymes.[12]

The electrophoretic analysis of proteins and enzymes has effected a minor revolution in genetics research. The reasons for this are two-fold. First, the zone electrophoresis technique provides a simple way of analyzing a large number of genetic loci in a large number of individuals with a relatively small expenditure of time and money. Second, since the genetic variants thus demonstrated retain their catalytic activity (otherwise they would not be detectable), the method allows the study of mutants, which, for the most part, have no obvious or significant phenotypic effects. They are therefore applicable for a wide variety of genetic studies since they do not produce deleterious or negatively selective influences. This point may be argued, and there are certainly exceptions; but for a majority of the electrophoretic variants, it is probably true. These methods thus provide a powerful new tool, and they have only begun to be exploited. Nonetheless, they have already provided important new insights into genetics and evolution.

One of the exciting findings to come from the application of the electrophoretic technique has occurred in the studies of natural populations. Such studies have demonstrated a much higher frequency of genetic polymorphism than had hitherto been thought to occur. This is true for man as well as for other animal species. The first clear evidences of the high frequency of polymorphisms came from the studies of Hubby and Lewontin in *Drosophila*.[13] Soon afterwards, data began to accumulate on human populations, with a large portion of the information coming from the laboratory of Harris and his co-workers.[3] The results of all surveys to date show, with a fair degree of regularity, that about one third of the genes in a population are polymorphic.[3,13,14,15,16] It has been further shown that in each individual within the population, some 17 percent of his gene pairs are in the heterozygous state, a figure surprisingly consistent among several different species.

Polymorphism has an arbitrary definition and is usually taken to mean inherited variation in a population of 1 percent or higher at a particular locus. When one considers less common variants, such as the rare form of human erythrocyte esterase found in only one family out of more than 6000, then the number of genes which show variation become much greater.[17] In fact, it has been conjectured that if one looks at enough individuals in a population, a variant form can be found for nearly every gene examined.[18]

A word of reservation about frequencies based on the electrophoretic method. The method probably underestimates frequencies, mainly for the reason that not all mutations result in amino acid changes (due to the degeneracy in the genetic code), nor do all amino acid substitutions result in changes in the net charge of the polypeptide molecule (which alters electrophoretic migration). Accurate figures are not obtainable, but probably less than one third of all mutations are detectable electrophoretically.

Table 17-II (adapted from Harris[3]) lists the human enzymes and proteins which have been shown to be polymorphic. Gene frequencies are not listed, although these have been determined for various populations for most of the enzymes listed. Most of the variations have been detected by electrophoretic methods, but several of the serum proteins were demonstrated by immunological techniques. Appropriate references in the table are necessarily few and do not begin to cover the field for such proteins as hemoglobin, for which thousands of references are available.

IMPLICATIONS FOR HUMAN BEHAVIOR

There are two aspects of behavior genetics to be considered here. One is normal range of behavior, the other is behavioral pathology.

A number of the genetic enzyme deficiencies listed in Table 17-I produce serious behavioral aberrations such as mental retardation and hyperkinesis. The mechanism by which the altered metabolism exerts its effect on the nerve cells in these disorders is little understood, but for the most part the effects are of a massive, nonspecific type, with major disruptions in cell function. Elucidation of these mechanisms are unlikely to contribute importantly to our understanding of behavior genetics.

Of more pertinence to this discussion is normal variation in behavior. A table of behavioral functions affected by heredity has been compiled by Shaw and Lucas, and includes such parameters as visual acuity, musical memory, motor skill, hoarding, aggressiveness, and a wide variety of other functions.[19] Little is known about the genetic control of these functions beyond the demonstration that they are affected, either in part or in whole, by heredity. Most of these character traits are probably polygenically determined. A recent, exquisitely performed study on variation in cricket songs has shown that these sounds, which are quite discrete and which vary among species, are controlled by a number of genes, some of which have been located on certain chromosomes.[20] Certain of the genes exert very discrete effects such as determining the rate of discharge of the impulse from the neurons controlling the song. And variation has been found in many of these song-controlling genes.

The point is, normal variation in behavior, a subject which is only recently coming under careful analysis and which is the basic substance of behavior genetics, is controlled by a very large number of gene loci. The high frequency of genetic variation described in this chapter provides a basis, at the molecular level, for the requirement of a range of variation in a very large number of genes within a population.[21] There had been no previous data available on a sufficient number of gene loci, and estimates of variability based on the genetic load provided a figure too small by several orders of magnitude. The recent findings that over one third of the genes vary within a population, with some 17 percent polymorphic in each individual, presents us with a very large number of varying genes. The total number of functional genes is not estimable with any accuracy, but it is perhaps on the order of 100,000, 17 percent of which makes 17,000 variable loci per individual. While some of these, perhaps a majority, are neutral, producing no abnormal changes, they may in fact produce normal variations in their host. Considering the fact that many genes have pleiotrophic functions, the combinational range of phenotypic effects in all parameters of function, including behavior, becomes astronomical. Is it any wonder, then, that every human being is unique in looks and actions?

TABLE 17-I
HEREDITARY ENZYME DEFICIENCIES

Enzyme Deficiency	Clinical Disorder	Remarks	Reference
Hexokinase (red cell isozyme)	Hexokinase deficiency haemolytic anemia		(22)
Phosphohexose isomerase (glucose phosphate isomerase)	Phosphohexoseisomerase (glucose phosphate isomerase deficiency)	Severe hemolytic anemia	(23, 24)
Phosphofructokinase (muscle type isozyme)	Phosphofructokinase deficiency	Muscle weakness	(25, 26)
Triosephosphate isomerase	Triosephosphate isomerase deficiency	Anemia, progressive neurological damage	(27, 28)
Adolase B (liver isozyme)	Hereditary fructose intolerance		(29, 30)
Phosphoglycerate kinase	Phosphoglycerate kinase deficiency	Hemolytic anemia	(31, 32)
Pyruvate kinase (red cell isozyme)	Pyruvate kinase deficiency	Hemolytic anemia	(33, 34, 35)
Diphosphoglycerate mutase	Diphosphoglycerate mutase deficiency	Severe hemolytic disease	(36)
Glucose-6-phosphate dehydrogenase	Glucose-6-phosphate dehydrogenase deficiency, primaquine sensitivity, favism	Primaquine sensitivity, favism, *etc.*	(37, 38, 3)
Glutathione reductase	Glutathione reductase deficiency	Chronic hemolytic anemia	(39, 40)
Galactose-1-phosphate uridyl transferase	Galactosemia	Mental retardation, failure to thrive, cirrhosis, cataracts	(41, 42, 43)
Galactokinase	Galactokinase deficiency	Cataracts	(44)
Fructokinase	Essential fructosuria	No clinical symptoms	(45, 46)
Glucose-6-phosphatase	von Gierke's disease (glycogen storage disease Type I)	Enlargement of heart and liver, hypoglycemia	(47)
α-1, 4 glucosidase (lysosomal)	Pompe's disease (glycogen storage disease Type II)	Enlargement of heart	(48)
Amylo-1, 6-glucosidase	Forbes' disease (glycogen storage disease Type III)	Enlargement of liver acidosis	(49)
Amylo-(1, 4→1, 6)-transglucosidase	Andersen's disease (glycogen storage disease Type IV)	Cirrhosis of liver	(50)
Phosphorylase (muscle type)	McArdle's disease (glycogen storage disease Type V)	Muscle cramps	(51, 52)
Phosphorylase kinase	Glycogen storage disease (one type)	Enlargement of liver	(53)
Glycogen synthetase	Glycogen synthetase deficiency	Hypoglycemia	(54)
L-xylulose reductase	Congenital pentosuria	No clinical symptoms	(55, 56)
Isomaltase (maltase la) and sucrase (malt)	Sucrose and isomaltose intolerance	Chronic diarrhea frothy stools	(57)
Lactase	Congenital lactose intolerance	Intolerance of milk, chronic diarrhea	(58)
Phenylalanine 4-hydroxylase	Phenylketonuria	Mental retardation, often severe	(59, 60, 61)
p-Hydroxyphenylpyruvic acid oxidase (hydroxylase)	Tyrosinaemia	Severe liver damage, kidney tubule damage, nickets	(62, 63)
Homogentisic acid oxidase (homogentisate oxygenase)	Alkaptonuria	No clinical symptoms	(2)
Histidase (histidine α-deaminase)	Histidinaemia	Mental retardation, speech defects	(64)
Cystathionine synthetase (serine dehydratase)	Homocystinuria	Mental retardation, skeletal abnormalities	(65)
Cystathionase (homoserine dehydratase)	Cystathioninuria	Mental retardation	(66)
Iodotyrosine deiodinase	Goitrous cretinism (one type)	Severe hypothyroidism	(67)
'Branched chain ketoacid decarboxylase(s)'	Maple syrup urine disease	Progressive neurological degeneration, early death	(68, 69)
Valine transaminase	Hypervalinaemia	Mental retardation	(70)

Enzyme Deficiency	Clinical Disorder	Remarks	Reference
		and failure to thrive	
Isovaleryl-coenzyme A dehydrogenase	Isovaleric acidaemia	Intermittent acidosis	(71)
Proline oxidase	Hyperprolinaemia	Sometimes mental retardation and kidney abnormalities	(72)
Hydroxyproline oxidase	Hydroxyprolinaemia	Severe mental retardation	(73)
Argininosuccinase (argininosuccinate lyase)	Argininosuccinic aciduria	Mental retardation and liver dysfunction	(74)
Argininosuccinate synthetase [L-citruline: L-asparate ligase (AMP)]	Citrullinaemia	Mental retardation and liver dysfunction	(75)
Ornithine carbamoyl transferase [Carbamoyl phosphate: carbamoyl transferase]	Hyperammonaemia	Mental retardation and liver dysfunction	(76)
Xanthine oxidase [Xanthine: O_2 oxidoreductase]	Xanthinuria	Renal calculi	(77, 78)
Reduced nicotinamide-adenine dinucleotide (NADH) oxidase	Chronic granulomatous disease	Suppurative lymphadenitis, lung infiltration, hepatosplenomegaly	(79)
Sulphite oxidase [Sulphite: O_2 oxidoreductase]	Sulphite oxidase deficiency		(80)
Catalase [H_2O_2: H_2O_2 oxidoreductasel]	Acatalasia	Often asymptomatic; may be ulceration of buccal mucosa	(81, 82)
Glutathione peroxidase (Glutathione:H_2O_2 oxidoreductase)	Red cell glutathione peroxidase deficiency	Mild hemolytic disease	(83)
Methaemoglobin reductase (DPNH: methaemoglobin oxidoreductase)	Congenital methaemoglobinaemia	Cyanosis	(84, 85)
Orotidine-5′-phosphate pyrophosphorylase [Orotidine-5′-phosphate: pyrophosphate phosphoribosyltransferase] and Orotidine-5′-phosphate decarboxylase [Orotidine-5′-phosphate carboxylyase]	Orotic Aciduria	Severe retardation of growth and development	(86, 87)
Glucocerebrosidase	Gaucher's disease	Hepatosplenomegaly, pathological fractures, neurological damage in "infantile" form	(88, 89)
Hexosaminidase A	Tay-Sachs	Blindness, severe retardation	(90)
Hexosaminidase A and B	Sandhoff's	Blindness, retardation, renal disease	(91, 92)
Arylsulphatase-A [Arylsulphate sulphodyrolase]	Methachromatic leucodystrophy	Progressive brain degeneration	(93, 94)
Arylsulfatase A, B, and C	"Degenerative disorder"		(95)
Ceramide trihexosidase	Fabry's disease (angiokeratoma corporis diffusum)	Skin lesions, cataracts, neurological disorders	(96)
Sphingomyelinase	Niemann-Pick disease	Progressive brain degeneration	(97)
'Acid' β-galactosidase	Generalized gangliosidosis	Brain and liver degeneration	(98)
Galactocerebroside β-galactosidase	Krabbe's	Severe retardation	(99)
α-fucosidase	Fucosidosis		(100)
Serum cholinesterase (pseudocholinesterase) [Acylcholine acyl-hydrolase]	Suxamethonium sensitivity	Prolonged response to succinyl choline	(101, 102, 103)
Alkaline phosphatase	Hypophosphatasia	Skeletal	(104, 105)

Enzyme Deficiency	Clinical Disorder	Remarks	Reference
Acid phosphatase (lysosomal)		abnormalities Vomiting, hypotonia, lethargy, early death	(106)
Adenosine triphosphatase	Red cell ATPase deficiency	Hemolytic anemia	(107)
Pancreatic lipase [Glycerol ester hydrolase]	Congenital pancreatic lipase deficiency	Indigestion of fats	(108)
Acid lipase	Wolman's disease	"Foamy" cells in visceral organs, xanthomatosis	(109)
Lecithin: cholesterol acyltransferase	Familial serum cholesterol ester deficiency	Anemia, proteinuria, corneal opacity	(110)
Hypoxanthine-guanine phosphoribosyl transferase	Lesch-Nyhan syndrome	Severe mental retardation, spastic palsy, choreoathetosis	(111, 112)
2-oxo-glutarate: glyoxalate carboligase	Primary hyperoxaluria (one type)	Urinary calculi	(113, 114)
D-glyceric dehydrogenase	Primary hyperoxaluria (one type)	Urinary calculi	(115)
Trypsinogen	Trypsinogen deficiency disease	Protein indigestion	(116)
Enterokinase (enteropeptidase)	Intestinal enterokinase deficiency	Protein indigestion	(117)
Methylmalonyl-Co A carbonyl mutase [Methylamalonyl-Co-A Co A—Carbonyl mutase]	Methylmalonic acidemia	Severe acidosis, failure to thrive, mental retardation	(118, 119)
Lysyl hydroxylase		Scoliosis, hyperextensible skin and joints	(120)

TABLE 17-II
ENZYME AND PROTEIN POLYMORPHISMS

Enzyme or Protein	Reference
1. Hemoglobin	(121, 3)
2. Haptoglobin	(122, 3)
3. Transferrin	(123)
4. Serum α-globulin (Gc)	(124, 125)
5. Immunoglobulins Heavy chains (Gm system)	(126, 127)
6. Immunoglobulins Light chains (Inv system)	(128, 127)
7. Serum β-lipoproteins (Ag system)	(129, 130)
8. Serum β-lipoproteins (Lp system)	(131, 132)
9. Ceruloplasmin	(133)
10. Serum α_1 trypsin inhibitor (Pi system)	(134)
11. Third component of complement (C'3 system)	(135)
12. Serum α_2-macroglobulin (Xm system)	(136)
13. Serum α_1-acid glycoprotein (oroseromucoid)	(137)
14. Glucose-6-phosphate dehydrogenase [D-glucose-6-phosphate: NADP oxidoreductase]	(3)
15. Phosphoglucomutase (locus PGM$_1$) [α-D-glucose-1, 6-diphosphate: α-D-glucose-1-phosphate phosphotransferase]	(138, 139)
16. Phosphoglucomutase (Locus PGM$_3$)	(140)
17. Placental alkaline phosphatase	(141, 142)
18. Serum cholinesterase (Locus E$_1$) [acylcholine acyl-hydrolase]	(143, 144)
19. Serum cholinesterase (Locus E$_2$)	(145, 146)
20. Peptidase A (dipeptidase)	(147, 148)
21. Red cell acid phosphatase	(149, 150, 151)
22. Adenylate kinase	(152)
23. Adenosine deaminase	(153)
24. Phosphogluconate dehydrogenase [6-phospho-D-gluconate:NADP oxidoreductase (decarboxylating)]	(154, 155)
25. Erythrocyte nicotinamide adenine dinucleotide nucleosidase (NADase) [NAD glycohydrolase]	(156)
26. Galactose-1-phosphate uridyl transferase (Duarte variant [UDP glucose: α-D-galactose-1-phosphate uridylyl transferase]	(157, 158)
27. Glutathione reductase	(159)
28. Pancreatic amylase	(160)
29. Peptidase D (prolidase)	(161)

SUMMARY

Our knowledge of the pathway from gene to behavior, certainly a long and tortuous one, is as yet little understood. A few glimmerings are being obtained. What has become quite clear from recent molecular genetics is that whatever genes do, and however many of them operate in the designation of behavior, a large proportion of the total genes, those affecting behavior as well as those controlling all other parameters of life and health, clearly vary within any natural population. Their variation provides the genetic basis for the wide and complex differences in behavior seen in man as well as among lower organisms.

REFERENCES

1. Garrod, A.E.: *Inborn Errors of Metabolism.* Oxford, Oxford U Pr, 1909.
2. La Du, B.N., Zannoni, V.G., Lasten, L., and Seegmiller, J.E.: The nature of the defect in tyrosine metabolism in alkaptonuria. *J Biol Chem, 230*:251, 1958.
3. Harris, H.: *The Principles of Human Biochemical Genetics.* New York, Elsevier, 1970.
4. Bigley, R.H., and Koler, R.D.: Liver pyruvate kinase (PK) isozymes in a PK-deficient patient. *Ann Hum Genet, 31*:383, 1968.
5. Tedesco, T.A., and Mellman, W.J.: Argininosuccinate synthetase activity and citrulline metabolism in cells cultured from a citrullinemic subject. *Proc Natl Acad Sci USA, 56*:829, 1967.
6. Kirkman, H.N., and Bynum, E.: Enzymic evidence of a galactosaemic trait in parents of galactosaemic children. *Ann Hum Genet, 23*:117, 1959.
7. Russell, J.D., and DeMars, R.: UDP-glucose: α-D-galactose-1-phosphate uridylyltransferase activity in cultured human fibroblasts. *Biochem Genet, 1*:11, 1967.
8. Hsia, D.Y.Y., Driscoll, K., Troll, W., and Knox, W.E.: Detection by phenylalanine tolerance tests of heterozygous carriers of phenylketonuria. *Nature (Lond), 178*:1239, 1956.
9. Pauling, L, Itano, H.A., Singer, J., and Wells, I.C.: Sickle cell anemia, a molecular disease. *Science, 110*:543, 1949.
10. Smithies, O.: Zone electrophoresis in starch gels: Group variations in the serum proteins of normal human adults. *Biochem J, 61*:629, 1955.
11. Hunter, R.L., and Market, C.L.: Histochemical demonstration of enzymes separated by zone electrophoresis in starch gels. *Science, 125*:1294, 1957.
12. Market, C.L., and Moller, F.: Multiple forms of enzymes: tissue, ontogenetic, and species specific patterns. *Proc Natl Acad Sci USA, 45*:753, 1959.
13. Hubby, J.L., and Lewontin, R.C.: A molecular approach to the study of genic heterozygosity in natural populations. I. The number of alleles at different loci in *Drosophila pseudoobscura. Genetics, 54*:577, 1966.
14. Selander, R., Hunt, W.G., and Yang, S.Y.: Protein polymorphism and genic heterozygosity in two European subspecies of the house mouse. *Evolution, 23*:379, 1969.
15. Selander, R., Yang, S.Y., Lewontin, R.C., and Johnson, W.E.: Genetic variation in the horseshoe crab *(Limulus polyphemys),* a phylogenetic "relic." *Evolution, 24*:402, 1970.
16. Shaw, C.R.: How many genes evolve?

Biochem Genet, 4:275, 1970.
17. Shaw, C.R., Syner, F.N., and Tashian, R.E.: New genetically-determined molecular form of erythrocyte esterase in man. *Science, 138*:31, 1962.
18. Shaw, C.R.: Electrophoretic variation in enzymes. *Science, 149*:936, 1965.
19. Shaw, C.R., and Lucas, A.R.: *The Psychiatric Disorders of Childhood,* 2nd ed. New York, Appleton, 1970.
20. Bentley, D.R.: Genetic control of an insect neuronal network. *Science, 174*:1139, 1971.
21. Shaw, C.R.: Frequency of genetic polymorphism: Implications for mental disease. In Beng, T.H., and McIsaac, W.M. (Eds.): *Brain Chemistry and Mental Disease.* New York, Plenum Pr, Plenum Pub, 1971, pp. 275-280.
22. Valentine, W.N., Oski, F.A., Paglia, D.E., Baughan, M.A., Schneider, A.S., and Naiman, J.L.: Hereditary hemolytic anemia with hexokinase deficiency. *N Engl J Med, 276*:1, 1967.
23. Baughan, M.A., Valentine, W.N., Paglia, M.D., Ways, P.O., Simon, E.R., and DeMarsh, Q.B.: Hereditary hemolyptic anemia associated with blucosephosphate isomerase (GPI) deficiency—a new enzyme defect of human erythrocytes. *Blood, 32*: 236, 1968.
24. Detter, J.C., Ways, P.O., Giblett, E.R., Baughan, M.A., Hopkinson, D.A., Povey, S., and Harris, H.: Inherited variations in human phosphohexose isomerase. *Ann Hum Genet, 31*:329, 1968.
25. Tarui, S., Okuno, G., Ikura, Y., Tanaka, T., Suda, M., and Nishikawa, M.: Phosphofructokinase deficiency in skeletal muscle. A new type of glycogenosis. *Biochem Biophys Res Commun, 19*:517, 1965.
26. Layzer, R.B., Rowland, L.P., and Ranney, H.M.: Muscle phosphofructokinase deficiency. *Arch Neurol, 17*:512, 1967.
27. Schneider, A.S., Valentine, W.N., Hattori, M, and Heins, H.L.: Hereditary hemolytic anemia with triosephosphate isomerase deficiency. *N Engl J Med, 272*:229, 1965.
28. Valentine, W.N., Schneider, A.S., Baughan, M.A., Paglia, D.E., and Heins, H.L., Jr.: Hereditary hemolytic anemia with trisoe-phosphate isomerase deficiency. *Am J Med, 41*:27, 1966.
29. Froesch, E.R., Wolf, H.P., Baitsch, H., Prader, A., and Labhart, A.: Hereditary fructose intolerance. An inborn defect of hepatic fructose-1-phosphate splitting aldolase. *Am J Med, 34*:151, 1963.
30. Chambers, R.A., and Pratt, R.T.C.: Idiosyncrasy to fructose. *Lancet, 2*:340, 1956.
31. Kraus, A.P., Langston, M.F., and Lynch, B.L.: Red cell phosphoglycerate kinase deficiency. *Biochem Biophys Res Commun, 30*:173, 1968.
32. Valentine, W.N.: Hereditary hemolytic anemias associated with specific erythrocyte enzymopathies. *Cal Med, 108*:280, 1968.
33. Tanaka, K.R., and Valentine, W.N.: Pyruvate kinase deficiency. In Beutler, E. (Ed.): *Hereditary Disorders of Erythrocyte Metabolism.* New York, Grune, 1968.
34. Grimes, A.J., Meisler, A., and Dacie, J.V.: Hereditary non-spherocytic haemolytic anemia. A study of red-cell carbohydrate metabolism in twelve cases of pyruvate-kinase deficiency. *Br J Haematol, 10*:403, 1964.
35. Keitt, A.S.: Pyruvate kinase deficiency and related disorders of red cell glycolysis. *Am J Med, 41*: 742, 1966.
36. Schroter, W.: Kongenitale Nichtspäracytäre Hämolytische Anämic bei 2, 3-Diphosphoglyceratmutase Mangel der Erythrocyten im frühen Säuglingsalter. *Klin Wochenschr 43*:1147, 1965.
37. Kirkman, H.N., McCurdy, P.R., and Naiman, J.L.: Functionally abnormal G6PD. *Cold Spring Harbor Symp Quant Biol, 29*:391, 1964.

38. Boyer, S.H., Porter, I.H., and Weibacher, R.G.: Electrophoretic heterogeneity of glucose-6-phosphate dehydrogenase and its relationship to enzyme deficiency in man. *Proc Natl Acad Sci USA, 48*:1868, 1962.
39. Löhr, G.W., and Waller, H.D.: Eine neue Enzymopenische Hämolytische Anämie mit Glutathionreductase-Mangel. *Med Klin, 57*:1521, 1962.
40. Waller, H.D.: Glutathione reductase deficiency. In Beutler, E. (Ed.) : *Hereditary Disorders of Erythrocyte Metabolism.* New York, Grune, 1968.
41. Townsend, E.H., Mason, H.H., and Strong, P.S.: Galactonsmia and its relation to Laennec's cirrhosis: review of literature and presentation of six additional cases. *Pediatrics, 7*:760, 1951.
42. Kalckar, H.M., Anderson, E.P., and Isselbacher, K.J.: Galactosemia, a congenital defect in a nucleotide transferase. *Biochem Biophys Acta, 20*:262, 1956.
43. Schwarz, V., Goldberg, L., Komrower, G.M., and Holzel, A.: Some disturbance of erythrocyte metabolism in galactosaemia. *Biochem J, 62*:34, 1956.
44. Gitzelmann, R.: Hereditary galactokinase deficiency, a newly recognized cause of juvenile cataracts. *Pediatr Res, 1*:14, 1967.
45. Lasker, M.: Essential fructosuria. *Hum Biol, 13*:51, 1941.
46. Schapira, F., Schapira, G., and Dreyfus, J.C.: La lésion enzymatique de la fructosurie benigne. *Enzyme 1*:170, 1961.
47. Cori, G.T., and Cori, C.F.: Glucose-6-phosphatase of the liver in glycogen storage disease. *J Biol Chem, 199*:661, 1952.
48. Hers, H.G.: α-glucosidase deficiency in generalized glycogen-storage disease (Pompe's disease). *Biochem J, 86*:11, 1962.
49. Illingworth, B., Cori, G.T., and Cori, C.F.: Amylo 1,6 glucosidase activity in muscle tissue in generalized glycogen storage disease. *J Biol Chem, 218*:123, 1956.
50. Brown, B.I, and Brown, D.H.: Lack of an α-1, 4-glucan: α-1, 4-glucan 6-glycosyl transferase in a case of type IV glycogenosis. *Proc Natl Acad Sci USA, 56*:725, 1966.
51. Mcardle, R.B.: Myopathy due to a defect in muscle glycogen breakdown. *Clin Sci, 10*:13, 1951.
52. Schmid, R., Robbins, P.W., and Taut, R.R.: Glycogen synthesis in muscle lacking phosphorylase. *Proc Natl Acad Sci USA, 45*:1236, 1959.
53. Hug, G., Schubert, W.K., and Chuck, G.: Phosphorylase kinase of the liver: Deficiency in a girl with increased hepatic glycogen. *Science, 163*:1534, 1966.
54. Lewis, G.M., Spencer-Peet, J., and Stewart, K.M.: Infantile hypoglycaemia due to inherited deficiency of glycogen synthetase in liver. *Arch Dis Child, 38*:40, 1963.
55. Lasker, M., Enklewitz, M., and Lasker, G.W.: The inheritance of L-xyloketosuria (essential pentosuria). *Hum Biol, 8*:243, 1936.
56. Hiatt, H.H.: Pentosuria. In Stanbury, J.S., Wyngaarden, J.B., and Fredrickson, D.S. (Eds.): *The Metabolic Basis of Inherited Disease,* 2nd ed. New York, McGraw, 1966.
57. Weijers, H.A., Van de Kamer, J.H., Mossell, D.A.A., and Dicke, W.K.: Diarrhoea caused by deficiency of sugar splitting enzymes. *Lancet, 2*:296, 1960.
58. Lifschitz, F.: Congenital lactase deficiency. *J Pediatr, 69*:229, 1966.
59. Jervis, G.A.: Phenylpyruvic oligophrenia: deficiency of phenylalanine oxidizing system. *Proc Soc Exp Biol Med, 82*:514, 1953.
60. Mitoma, C., Auld, R.M., and Undenfriend, S.: On the nature of enzymatic defect in phenylpyruvic oligophrenia. *Proc Soc Exp Biol Med, 94*:634, 1957.
61. Berman, J.L., Cunningham, G.C., Day, R.W., Ford, R., and Hsia, D.Y.Y.: Causes for high phenylalanine with normal tyrosine. *Am J Dis Child, 117*:54, 1969.
62. La Du, B.N.: Histidinemia. *Am J Dis Child, 113*:88, 1967.

63. Scriver, C.R., LaRochelle, J., and Silverberg, M.: Hereditary tyrosinemia and tyrosyluria in a French Canadian geographic isolate. *Am J Dis Child, 113*:41, 1967.
64. Ghadmimi, H., and Partington, M.W.: Salient features of histidinemia. *Am J Dis Child, 113*:83, 1967.
65. Gerritsen, T., and Waisman, H.A.: Homocystinuria: An error in the metabolism of methionine. *Pediatrics, 33*:413, 1964.
66. Harris, H., Penrose, L.S., and Thomas, D.H.H.: Cystathioninuria. *Ann Hum Genet, 23*:442, 1959.
67. Murray, P., Thomson, J.A., McGirr, E.M., Wallace, T.J., MacDonald, E.M., and MacCabae, H.J.: Absent and defective iodotyrosine deiodination. *Lancet, 1*:183, 1965.
68. Dancis, J., Hutzler, J., and Rokkones, T.: Intermittent branched-chain ketonuria. *N Engl J Med, 276*:84, 1967.
69. Dancis, J., Hutzler, J., and Levitz, M.: The diagnosis of maple syrup urine disease (branch chain ketoaciduria) by the *in vitro* study of the peripheral leucocyte. *Pediatrics, 32*:234, 1963.
70. Dancis, J., Hutzler, J., Tada, K., Wada, Y., Morikawa, T., and Arakawa, T.: Hypervalinemia. A defect in valine transamination. *Pediatrics, 39*:813, 1967.
71. Tanaka, K.R., Budd, M.A., Efron, M.L., and Isselbacher, K.J.: Isovalerid acidemia: a new genetic defect of leucine metabolism. *Proc Natl Acad Sci USA, 56*:236, 1966.
72. Shafer, I.A., Scriver, C.R., and Efron, M.L.: Familial hyperprolinaemia, cerebral dysfunction, and renal anomalies occurring in a family with hereditary nephropathy and deafness. *N Engl J Med, 267*:51, 1962.
73. Efron, M.L.: Hydroxyprolinemia. II. A rare metabolic disease due to a deficiency of the enzyme "hydroxyproline oxidase." *N Engl J Med, 272*:1299, 1965.
74. Miller, A.L., and McLean, P.: Urea cycle enzymes in the life of a patient with argininosuccinid aciduria. *Clin Sci, 32*:385, 1967.
75. McMurray, W.C., Rathbun, J.C., Mohyuddin, F., and Koegler, S.J.: Citrullinuria. *Pediatrics, 32*:347, 1963.
76. Russell, A., Levin, B., Oberholzer, V.G., and Sinclair, L.: Hyperammonemia. A new instance of an inborn enzymatic defect of the biosynthesis of urea. *Lancet, 2*:699, 1962.
77. Dent, C.E., and Philpot, G.R.: Xanthinuria, an inborn error (or deviation) of metabolism. *Lancet, 1*:182, 1954.
78. Watts, R.W.E, Engleman, K., Klinenberg, J.R., Seegmiller, J.E., and Sjoerdsma, A.: Enzyme defect in a case of xanthinuria. *Nature (Lond), 201*:395, 1963.
79. Baehner, R.L., and Karnovsky, M.L.: Deficiency of reduced nicotinamide-adenine dinucleotide oxidase in chronic granulomatous disease. *Science, 162*:1277, 1968.
80. Mudd, S.H., Irreverre, F., and Laster, L.: Sulfite oxidase deficiency in man: demonstration of the enzymatic defect. *Science, 156*:1599, 1967.
81. Takahara, S.: Progressive oral gangrene, probably due to a lack of catalase in the blood (acatalasaemia). *Lancet, 2*:1101, 1952.
82. Aebi, H., Bossi, E., Cantz, M., Matsubara, S., and Suter, H.: Acatalas (em) ia in Switzerland. In Beutler, E. (Ed.): *Hereditary Disorders of Erythrocyte Metabolism.* New York, Grune, 1968.
83. Necheles, T.F., Maldonado, N., Barquet-Chediak, A., and Allen, D.M.: Homozygous erythrocyte glutathione-peroxidase deficiency. Clinical and biochemical studies. *Blood, 33*:164, 1969.
84. Barcroft, H., Gibson, Q., Harrison, D., and McMurray, J.: Familial idiopathic methaemoglobinaemia and its treatment with ascorbic acid. *Clin Sci, 5*:145, 1967.
85. Gibson, Q.: The reduction of methaemoglobin in red blood cells and studies on the cause of idiopathic methaemoglobinaemia. *Biochem J, 42*:13, 1948.

86. Huguley, C.M., Jr., Bain, J.A., Rivers, S., and Scoggins, R.: Refractory megaloblastic anemia associated with excretion of orotic acid. *Blood, 14*:615, 1959.
87. Smith, L.H., Jr., Sullivan, M., and Huguley, C.M., Jr.: Pyrimidine metabolism in man. IV. The enzymatic defect of oroticaciduria. *J Clin Invest, 40*:656, 1961.
88. Patrick, A.D.: A deficiency of glucocerebrosidase in Gaucher's disease. *Biochem J, 97*:17, 1965.
89. Brady, R.O., Kanfer, J.N., Shapiro, D., and Bradley, R.M.: Demonstration of a deficiency of glucocerebroside-cleaving enzyme in Gaucher's disease. *J Clin Invest, 45*:1112, 1966.
90. O'Brien, J.S.: Five gangliosidoses. *Lancet, 2*:805, 1969.
91. Sandhoff, K., Andreae, U., and Jatzkewitz, H.: Deficient hexosaminidase activity in an exceptional case of Tay-Sachs disease with additional storage of kidney globoside in visceral organs. *Life Sci, 7*:283, 1968.
92. Suzuki, K.: G$_{M2}$-gangliosidosis with total hexosaminidase deficiency. *Neurology, 21*:313, 1971.
93. Austin, J., Balasubramanian, A., Pattabiraman, T., Sarswathi, S., Basu, D., and Bachhawat, B.: A controlled study of enzymic activities in three human disorders of glycolipid metabolism. *J Neurochem, 10*:805, 1963.
94. Jatzkewitz, H., and Mehl, E.: Cerebroside-sulphatase and arylsulphatase: a deficiency in metachromatic leukodystrophy (ML). *J Neurochem, 16*:19, 1969.
95. Murphy, J.V., Williams, M., and Moser, H.W.: Multiple sulfatase deficiencies, the enzymatic basis of a new disorder. Paper presented at meeting of the American pediatrics Society and the Society for Pediatric Research, Atlantic City, New Jersey, April-May, 1971 (Program and Abstracts, p. 16).
96. Brady, R.O., Gal, A.E., Bradley, R.M., Martensson, E., Warshaw, A.L., and Laster, L.: Enzymatic defect in Febry's disease. Ceramidetrihexosidase deficiency. *N Engl J Med, 276*:1163, 1967.
97. Brady, R.O., Kanfer, J.N., Mock, M.B., and Fredrickson, D.S.: The metabolism of sphingomyelin. II. Evidence of an enzymatic deficiency in Niemann-Pick disease. *Proc Natl Acad Sci USA, 55*:366, 1966.
98. Okada, S., and O'Brien, J.S.: Generalized gangliosidosis: β-galactosidase deficiency. *Science, 160*:1002, 1968.
99. Suzuki, K., and Suzuki, Y.: Globoid cell leukodystrophy (Krabbe's disease): deficiency of galactocerebroside β-galactosidase. *Proc Natl Acad Sci, 66*:302, 1970.
100. Van Hoff, F., and Hers, H.G.: The abnormalities of lysosomal enzymes in mucoplysaccharidoses. *Eur J Biochem, 7*:34, 1968.
101. Evans, F.T., Gray, P.W.S., Lehmann, H., and Silk, E.: Sensitivity of succinylcholine in relation to serum cholinesterase. *Lancet, 1*:1229, 1952.
102. Kalow, W., and Genest, K.: A method for the detection of atypical forms of human serum cholinesterases. Determination of dibucaine numbers. *Can J Biochem Physiol, 35*:339, 1957.
103. Kalow, W.: Cholinesterase types. In Wolstenholme, E.W., and O'Connor, C.M. (Eds.): *Biochemistry of Human Genetics. Ciba Foundation Symp.* London, Churchill, 1959.
104. Rathbun, J.C.: Hypophosphatasia, a new development anomaly. *Am J Dis Child, 75*:822, 1948.
105. Rathbun, J.C., MacDonald, J.W., Robinson, H.M., and Wanklin, J.M.: Hypophosphatasia: A genetic study. *Arch Dis Child, 36*:540, 1961.
106. Nadler, H.L., and Egan, T.J.: Deficiency of lysosomal acid phosphatase. A new familial metabolic disorder. *N Engl J Med, 282*:302, 1970.
107. Harvald, B., Hanel K.H., Squires, R., and

Trap-Jensen, J.: Adenosine-triphosphatase deficiency in patients with non-spherocytic haemolytic anaemia. *Lancet, 1*:18, 1964.

108. Seldon, W.: Congenital pancreatic lipase deficiency. *Arch Dis Child, 39*:268, 1964.

109. Patrick, A.D., and Lake, B.D.: Deficiency of an acid lipase in Wolman's disease. *Nature (Lond), 222*:1067, 1969.

110. Norum, K.R., and Gjone, E.: Familial serum-cholesterol esterification failure. A new inborn error of metabolism. *Biochem Biophys Acta, 144*:698, 1967.

111. Lesch, M., and Nyhan, W.L.: A familial disorder of uric acid metabolism and central nervous system function. *Am J Med, 36*:561, 1964.

112. Seegmiller, J.E., Rosenbloom, F.M., and Kelley, W.N.: Enzyme defect associated with a sex-linked human neurological disorder and excessive purine synthesis. *Science, 155*:1682, 1967.

113. Archer, H.E., Dormer, A.E., Scowen, E.F., and Watts, R.W.E.: Primary hyperoxaluria. *Lancet, 2*:320, 1967.

114. Hockaday, R.D.R., Clayton, J.E., Frederick, E.W., and Smith, L.H., Jr.: Primary hyperoxaluria. *Medicine (Baltimore), 43*:315, 1964.

115. Williams, H.E., and Smith, L.H., Jr.: L-glyceric aciduria. A new genetic variant of primary hyperoxaluria. *N Engl J Med, 278*:233, 1968.

116. Townes, P.L.: Trypsinogen deficiency disease. *J Pediatr, 66*:275, 1965.

117. Hadorn, B., Tarlow, M.J., Lloyd, J.K., and Wolfe, O.H.: Intestinal enterokinase deficiency. *Lancet, 1*:812, 1969.

118. Oberholzer, V.G., Levin, F., Burgess, E.A., and Young, W.F.: Methylmalonic aciduria. Inborn error of metabolism leading to metabolic acidosis. *Arch Dis Child, 42*:492, 1967.

119. Rosenberg, L.E., Lilljeqvist, A.-Ch., and Hsia, Y.E.: Methylmalonic aciduria. *N Engl J Med, 278*:1319, 1968.

120. Krane, S.M., Pinnell, S.R., and Erbe, R.W.: Decreased lysyl-protocollagen hydroxylase activity in fibroblasts from a family with a newly recognized disorder: hydroxylysine-deficient collagen. *Am Soc Clin Invest*, in press.

121. Livingstone, F.B.: *Abnormal Hemoglobins in Human Populations.* Chicago, Aldine, 1967.

122. Kirk, R.L.: *The Haptoglobin Groups in Man.* Basel, Karger, 1968.

123. Smithies, O., and Hiller, O.: The genetic control of transferrin in humans. *Biochem J, 72*:121, 1959.

124. Hirschfeld, J.: The Gc system. Immunoelectrophoretic studies of normal human sera with special reference to a new genetically determined serum system (Gc). *Prog Allergy, 6*:155, 1962.

125. Bearn, A.G., Bowman, B.H., and Kitchin, F.D.: Genetical and biochemical considerations of the serum group-specific component. *Cold Spring Harbor Symp Quant Biol, 29*:435, 1964.

126. Grubb, R., and Laurell, A.B.: Hereditary serological human serum groups. *Acta Pathol Microbiol Scand, 39*:390, 1956.

127. Steinberg, A.G.: Genetic variations in human immunoglobulins. The Gm and InV types. In Greenwalt, T.J. (Ed.): *Advances in Immunogenetics.* Philadelphia, Lippincott, 1967.

128. Ropartz, C., Lenoir, J., and Rivat, L.: A new inheritable property of human sera: the InV factor. *Nature (Lond), 189*:586, 1961.

129. Allison, A.C., and Blumberg, B.: An isoprecipitation reaction distinguishing human serum protein types. *Lancet, 1*:634, 1961.

130. Morganti, C., Beolchini, P.E., Vierucci, A., and Butler, R.: Contributions to the genetics of the serum β-lipoproteins in man. I. Frequency, transmission and penetrances of factors Ag(x) and Ag(y). *Humangenetik, 4*:262, 1967.

131. Berg, K.: A new serum type system in man—the Lp system. *Acta Pathol Microbiol Scand*, *59*:369, 1963.
132. Berg, K.: A new serum type system in man—the Ld system. *Vox Sang*, *10*:513, 1965.
133. Shreffler, D.C., Brewer, G.J., Gall, J.C., and Honeyman, M.S.: Electrophoretic variation in human serum cerulophasmin: A new genetic polymorphism. *Biochem Genet*, *1*:101, 1967.
134. Fagerhol, M.K., and Laurell, C.B.: The polymorphism of 'prealbumins' and a_1-antitrypsin in human sera. *Clin Chim Acta*, *16*:199, 1967.
135. Alper, C.A., and Propp, R.P.: Genetic polymorphism of the third component of human complement (C'3). *J Chin Invest*, *47*:2181, 1968.
136. Berg, K., and Bearn, A.G.: An inherited X-linked serum system in man. The Xm system. *J Expl Med*, *123*:379, 1966.
137. Schmid, K., Binett, J.P., Tokita, K., Moroz, L., and Yoshizaki, H.: The polymorphic forms of a_1-acid glycoprotein of normal Caucasian individuals. *J Clin Invest*, *43*:2347, 1964.
138. Spencer, N., Hopkinson, D.A., and Harris, H.: Quantitative differences and gene dosage in the human red cell acid phosphatase polymorphism. *Nature (Lond)*, *201*:299, 1964.
139. Harris, H., Hopkinson, D.A., Luffman, J.E., and Rapley, S.: Electrophoretic variation in red cell enzymes. In Beutler, E. (Ed.): *Hereditary Disorders of Erythrocyte Metabolism*. New York, Grune, 1968.
140. Hopkinson, D.A., and Harris, H.: A third phosphoglucomutase locus in man. *Ann Hum Genet*, *31*:359, 1968.
141. Boyer, S.H.: Alkaline phosphatase in human sera and placentae. *Science*, *134*:1002, 1961.
142. Robson, E.B., and Harris, H.: Further studies on the genetics of placental alkaline phosphatase. *Ann Hum Genet*, *30*:219, 1967.
143. Kalow, W., and Staron, N.: On the distribution and inheritance of atypical forms of human serum cholinesterase as indicated by dibucaine numbers. *Can J Biochem Physiol*, *35*:1305, 1957.
144. Harris, H., Whittaker, M., Lehmann, H., and Silk E.: The pseudocholinesterase variants. Esterase levels and dibucaine numbers in families selected through suxamethonium sensitive individuals. *Acta Genet Statist Med*, *10*:1, 1960.
145. Harris, H., Robson, E.B., Glen-Bott, A.M., and Thornton, J.A.: Evidence for nonallelism between genes affecting human serum cholinesterase. *Nature (Lond)*, *200*:1185, 1963.
146. Simpson, N.E.: Factors influencing cholinesterase activity in a Brazilian population. *Am J Hum Genet*, *18*:243, 1966.
147. Lewis, W.H.P., and Harris, H.: Human red cell peptidases. *Nature (Lond)*, *215*:351, 1967.
148. Lewis, W.H.P., Corney, G., and Harris, H.: Pep A 5-1 and Pep A 6-1: Two new variants of peptidase A with features of special interest. *Ann Hum Genet*, *32*:35, 1968.
149. Hopkinson, D.A., Spencer, N., and Harris, H.: Red cell acid phosphatase variants; a new human polymorphism. *Nature (Lond)*, *199*:969, 1963.
150. Hopkinson, D.A., Spencer, N., and Harris, H.: Genetical studies on human red cell acid phosphatase. *Am J Hum Genet*, *16*:141, 1964.
151. Luffman, J.E., and Harris H.: A comparison of some properties of human red cell acid phosphatase in difference phenotypes. *Ann Hum Genet*, *30*:387, 1967.
152. Fildes, R.A., and Harris, H.: Genetically determined variation of adenylate kinase in man. *Nature (Lond)*, *209*:261, 1966.
153. Spencer, N., Hopkinson, D.A., and Harris, H.: Adenosine deaminase

polymorphism in man. *Ann Hum Genet, 32*:9, 1968.
154. Fildes, R.A., and Parr, C.W.: Human red cell phosphogluconate dehydrogenase. *Nature (Lond), 200*:890, 1963.
155. Carter, N.D., Fildes, R.A., Fitch, L.I., and Parr, C.W.: Genetically determined electrophoretic variations of human phosphogluconate dehydrogenase. *Acta Genet, 18*:109, 1968.
156. Ng, W.G., Donnell, G.N., and Bergren, W.R.: Deficiency of erythrocyte nicotinamide adenine dinucleotide nucleosidase (MADase) activity in the Negro. *Nature (Lond), 217*:64, 1968.
157. Beutler, E., Baluda, M.C., Sturgeon, P., and Day, R.: The genetics of galactose-1-phosphate uridyl transferase deficiency. *J Lab Clin Med, 64*:646, 1966.
158. Mathai, C.K., and Beutler, E.: Electrophoretic variation of galactose-1-phosphate uridyltransferase. *Science, 154*:1179, 1966.
159. Long, W.K.: Glutathione reductase in red blood cells: variant associated with gout. *Science, 155*:721, 1967.
160. Kamaryt, J., and Laxova, R.: Amylase heterogeneity. Some genetic and clinical aspects. *Humangenetik, 1*:579, 1965.
161. Lewis, W.H.P., and Harris, H.: Peptidase D (Prolidase) variants in man. *Ann Hum Genet, 32*:317,1969.

Chapter 18

TASTE SENSITIVITY AND HUMAN VARIATION

Arnold R. Kaplan, Wilma E. Powell,

Austin E. Moorhouse, and Edward N. Hinko

INTRODUCTION

There have been numerous pedigree and population studies of taste thresholds for the bitter phenylthiourea ("PTC") type antithyroid compounds containing the characteristic H-N-C=S grouping. A population's taste threshold distribution for this class of compounds, which includes 6-n-propylthiouracil and 1-methyl-2-mercaptomidazole, tends to approach a bimodal curve. Taste thresholds for the majority of other compounds tend to follow monomodal and approximately "Gaussian" distributions.

Early studies indicated that taste sensitivity for phenylthiourea has a bimodal population distribution.[1, 2, 3, 4, 5, 6] The literature has largely supported the hypothesis that taste *in*sensitivity (i.e. occurrence of an individual's taste acuity level within the insensitive mode of the population's bimodal distribution) is the effect of a homozygous pair of recessive alleles. The hypothesis was independently derived by Snyder[5] and Blakeslee and Salmon.[3] Reports of discrepancies in twin and pedigree data have, however, prevented definite confirmation of the hypothesis.[7, 8, 9, 10, 11, 12, 13, 14, 15, 16]

The studies cited above have involved use of the compound, phenylthiourea, in one of three forms—in solution, as crystal, and impregnated filter paper. Hartmann demonstrated that both the crystal and paper tests were significantly less reliable than the solution test.[17] Harris and Kalmus devised a more reliable method for determining taste threshold which involves using "a few c.c." of each sample in a tumbler, double blind, and random presentation of solution and placebo samples for the subject's final sorting-out procedure.[18] Merton, Kalmus, and Leguèbe described extensive genetic studies which utilized the improved methodology.[13, 19, 20] Sinnot and Rauth used distilled water instead of tap-water, and a between-sample mouth rinse.[21] Fischer, *et al.* used 6-n-propylthiouracil instead of phenylthiourea, and they observed that a sample volume of at least 5.0 ml. was generally necessary for a reliable taste test.[22] They also reintroduced the use of distilled, instead of tap, water and the use of a

---

The studies described in this chapter were supported by the State of Ohio, Department of Mental Health and Mental Retardation and by research grants from the Council for Tobacco Research, United States of America, and from the National Institutes of Health (HD 00591-02 and GRS 05563).

between-sample mouth rinse. The principal reason for using 6-n-prophythiouracil is the odorless character of solutions that are half-saturated or less concentrated. Phenylthiourea solutions, on the other hand, manifest a distinct aromatic odor for some individuals which affects detection of the difference and determination of apparent taste threshold.[23]

METHODS

Taste thresholds were determined according to modifications of the procedure described by Harris and Kalmus.[18, 22] Their basic procedure, double-blind presentation of solution and placebo samples for the subject's final sorting-out, has been observed to be the most reliable of the published methods for determining taste thresholds.[24, 25, 26] Serial dilutions of the compounds were prepared by dissolving each substance in distilled water in concentrations ranging from 7.32×10^{-7} M to 6.00×10^{-3} M. The most concentrated solution of PROP (6-n-propylthiouracil), number 14, consists of 1.0212 grams of the compound dissolved in 1.0 liter of distilled water; that of quinine (1-quinine sulfate), number 13, consists of 1.11744 grams of the compound in 1.0 liter of distilled water; and the highest concentration of hydrochloric acid is a 0.012 M solution, number 15. Each solution number represents twice the concentration of its preceding solution number, as shown in Table 18-I. A subject's threshold, expressed as a solution number, is defined as the lowest concentration at which the samples of solution are correctly differentiated from the placebos. Each threshold is first estimated by providing the subject with solutions in progressively doubled concentrations until a definite taste is reported. The threshold is then determined by final sorting out of eight cups, four containing solution and four containing distilled water placebo (5.0 ml. per cup), presented in a random and double-blind order. A subject is required to utilize a mouth rinse with distilled water after tasting the contents of each cup. Due to the strong affinity of PROP and the other phenylthiourea-type compounds for taste receptor cells, the thresholds for subsequently-tasted compounds may be higher than when determined without prior tasting of a phenylthiourea type compound. Therefore, the compounds were tested in the following sequence: hydrochloric acid, quinine, and PROP. The testing of a subject for all three compounds required about forty-five minutes. Data regarding age and other demographic variables—food and beverage consumption, smoking habits, drug and hormone therapy, menstruation, and pregnancy—were obtained from questionnaires completed by the subjects. Some of the subjects also completed the 566-item long-form MMPI (Minnesota Multiphasic Personality Inventory) questionnaire to facilitate analysis of twenty-four personality scales.

TABLE 18-I
CONCENTRATIONS OF TASTE SOLUTIONS IN MOLARITY

Solution No.	Molarity
15	1.20×10^{-2}
14	6.00×10^{-3}
13	3.00×10^{-3}
12	1.50×10^{-3}
11	7.50×10^{-4}
10	3.75×10^{-4}
9	1.88×10^{-4}
8	9.38×10^{-5}
7	4.69×10^{-5}
6	2.34×10^{-5}
5	1.17×10^{-5}
4	5.86×10^{-6}
3	2.93×10^{-6}
2	1.46×10^{-6}
1	7.32×10^{-7}

SUBJECTS

Most of our subjects were volunteers contacted through the Cleveland Area Twin Registry and Mothers-of-Twins Clubs in the greater Cleveland area. Additional volunteers were obtained from the staffs and students at our institution and neighboring institutions and from the general public in response to publicity which solicited such volunteers. Ulcer patients were contacted through cooperation with staff members of the Case Western Reserve University School of Medicine, Cleveland Metropolitan General Hospital, and the Crile Veterans Administration Hospital. Currently active (wet) alcoholics were contacted through cooperation with staff members of Saint Vincent's Charity Hospital, and controlled (dry) alcoholics were contacted through cooperation of Alcoholics Anonymous clubs.

PHENOTYPES AND GENOTYPES

Taste thresholds for most substances have a continuous and normal or nearly "Gaussian" distribution in the population. Such patterns have been described for population distributions of taste thresholds for hydrochloric acid, quinine sulfate, and over two dozen other studied compounds.[26] Population distributions of thresholds for phenylthiourea type compounds such as PROP, however, manifest a bimodal tendency.

Figure 18-1 is a schematic for a population's quantitative distribution of a multifactorial trait. The factors are cumulative; the trait's distribution tends to be continuous and normal. The factors determining taste thresholds for quinine and those for hydrochloric acid are evidently multifactorial and cumulative.

Figure 18-2 compares schematically the

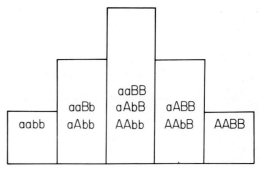

Figure 18-1. Schematic to show continuous distribution in a population of a trait based on multiple factors that are cumulative in effect.

distribution in a population of a trait based on two different monofactorial genetic mechanisms. One indicates a continuous and normal distribution of the trait, the result of cumulative gene action. The other schematic indicates the kind of distribution that occurs when the gene action is not cumulative. In this case, the dominant homozygote and the heterozygote each manifest the trait while the recessive homozygote is different, and there is no clearly observed intermediate group. In such cases, as our measuring techniques are improved, controlled, and made more specific for the genetic factor, the curve of population distribution more clearly approaches discontinuity between the modes. As other variables affect the parameter being measured, however, the antimode becomes less distinct. Taste thresholds for a phenylthiourea type compound, such as PROP, exhibit such a population distribution. This is characteristic of the class comprising dozens of antithyroid compounds, natural and synthetic, all bitter-tasting.[26] Evidently, taste sensitivity for the group of compounds involves a single pair of genetic factors,[13, 19, 20] but numerous other variables also affect the trait. Control of these other variables is essential for any genetic study to be clear. Several of these variables

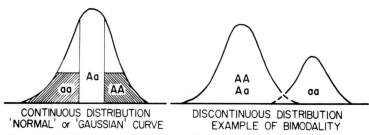

Figure 18-2. Schematic to show continuous distribution in a population of a trait based on genetic factors at a single locus which are cumulative in effect (left); and discontinuous, bimodal distribution in a population of a trait based on genetic factors at a single locus which are dominant and recessive rather than cumulative in effect (right).

have been studied and are discussed in the present paper. The incompleteness and imperfection of our controls may be reflected by the incompleteness of the antimodes in curves of population distribution for the variables. The bimodal classification according to the population distribution based on the measured phenotype or trait does not completely correspond to the genotypic distribution and there is some phenotypic overlap.

TWIN AND SIBLING STUDIES

Intrapair differences in taste thresholds have been compared for hydrochloric acid, quinine, and PROP. Comparative distributions of intrapair threshold differences in monozygotic and same-sex dizygotic twin pairs have previously been described.[27] Figure 18-3 indicates the distribution of intrapair differences in thresholds for hydrochloric acid, based on twenty-five monozygotic (sixteen female and nine male) and twenty-six dizygotic (fifteen female and eleven male) pairs of same-sex twins. Figures 18-4 and 18-5 indicate the distributions of intrapair differences in thresholds for quinine and PROP, respectively. The graphs are based on sixty-nine monozygotic (forty-eight female and twenty-one male) and forty-five dizygotic (twenty-nine female and sixteen male) pairs. The distribution of intrapair differences for quinine (Figure 18-4) indicates a greater difference between the two main categories than that observed for hydrochloric acid. The difference of PROP intrapair threshold differences between the two twin categories (Figure 18-5) was the greatest of the three compounds. The relative importance of genetic factors in taste threshold differences was also suggested by a comparison of the average intrapair differences observed in the above groups of monozygotic and dizygotic twins, respectively, for each of the three compounds: for hydrochloric acid, 1.00 and 1.04; for quinine, 1.22 and 1.64; for PROP, 0.75 and 2.87. There was very little average intrapair differences between the pairs of genetically identical and those of the same-sex fraternal twins for hydrochloric acid. A much greater difference of average intrapair differences was observed between the two twin categories for quinine, and the greatest difference was observed for PROP. Intrapair variance of taste thresholds for hydrochloric acid, for

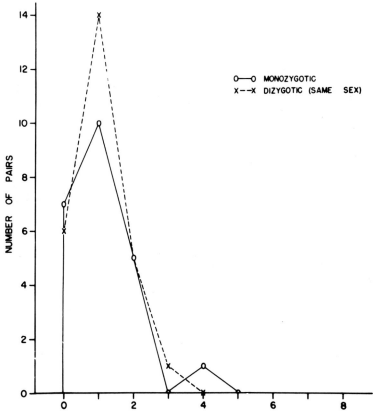

Figure 18-3. Comparative distribution of intrapair taste-threshold differences for hydrochloric acid in twenty-five monozygotic (sixteen female, nine male) and twenty-six dizygotic (fifteen female, eleven male) twin pairs.

quinine, and for PROP were later determined for larger numbers of monozygotic-twin, dizygotic-twin, and nontwin sibling pairs.[28] The numbers of respective types of pairs studied were twenty-six, forty-five, and forty-five for hydrochloric acid; seventy-five, seventy, and seventy-eight for each of the other two substances. The dizygotic-monozygotic intrapair threshold differences were not significant for hydrochloric acid ($P > .10$) or for quinine ($P > .05$) but was highly significant for PROP ($P < .001$). There was, however, no significant difference between the intrapair variance observed in dizygotic twin pairs and that in same-sex sibling pairs for PROP ($P > .10$). A principal role for genetic factors in etiology of individual taste-threshold differences was not demonstrated for sour-tasting hydrochloric acid or for bitter-tasting quinine, but was emphatically demonstrated for bitter-tasting PROP.

CORRELATIONS BETWEEN THRESHOLD FOR DIFFERENT COMPOUNDS

The PROP thresholds of monozygotic twins with differing thresholds were

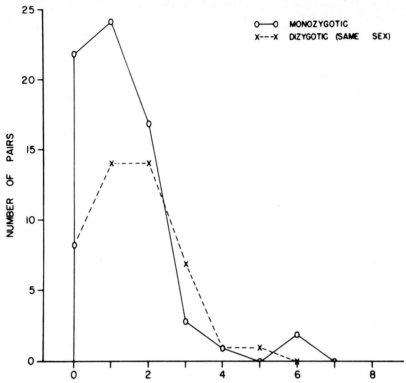

Figure 18-4. Comparative distribution of intrapair taste-threshold differences for quinine in sixty-nine monozygotic (forty-eight female, twenty-one male) and forty-five dyzygotic (twenty-nine female, sixteen male) twin pairs.

examined to determine whether the cotwins with the higher guinine thresholds also tended to have higher PROP thresholds. The quinine thresholds of fifty-three pairs of monozygotic twins in our population differed from each other by one or more thresholds. In fourteen of the pairs, the same individuals manifested the higher threshold for each of the two substances; in twenty-two pairs, both cotwins had the same PROP threshold despite their differences in quinine threshold, and in seventeen pairs, the cotwin with the lower quinine threshold had the higher PROP threshold. Apparently independent factors influence the intrapair taste threshold differences observed in monozygotic twins for the two different bitter-tasting drugs, quinine and PROP.

Taste thresholds for the three compounds investigated were significantly and positively correlated with each other. Analyses based on the threshold data for 308 individuals indicated the following correlation coefficients: between PROP and quinine, $r = 0.444 \pm .046$, significant at a chance probability level of P less than .01; between quinine and hydrochloric acid, $r = 0.35 \pm .05$, significant at a probability level of P less than .01; between PROP and hydrochloric acid, $r = 0.166 \pm 0.55$, significant at a probability level of P less than .05.

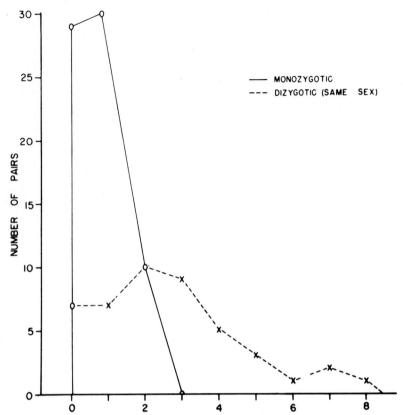

Figure 18-5. Comparative distribution of intrapair taste-threshold differences for PROP in sixty-nine monozygotic (forty-eight female, twenty-one male) and forty-five dizygotic (twenty-nine female, sixteen male) twin pairs.

DIETARY PREFERENCES RELATED TO TASTE THRESHOLDS

The flavor exhibited by a particular food is the product of the interacting effects of various tastes and odors and might not, therefore, be expected to be highly correlated with taste threshold for one particular compound. Food preferences and rejections are related to various factors, including social custom, experience, and psysiological state. Individual differences in taste acuity for particular compounds have also been correlated with food likes and dislikes. Fischer, et al. found that in a group of forty-eight college students, proportions of foods disliked from a given list of 118 specific items were apparently associated with taste thresholds for quinine and PROP; but they found no associations with taste thresholds for sucrose, sodium chloride, or hydrochloric acid.[29] A later study involved 187 adults (51 males, 136 females) ranging in age from twenty-two to sixty-six years and with an average age of thirty-eight.[30] The sample included thirty-nine husband-wife pairs, sixteen pairs of monozygotic twins, and ten pairs of dizygotic twins. Our questionnaire was designed to minimize the influence of social custom through use of a list of only a few carefully selected foods which are in wide use locally and are commonly prepared in several different forms. The subject was rated according to preference for mild, moderate, or strongly tasting preparations. Part A involved a list of widely used foods which are commonly prepared in a variety of ways, providing a graded series ranging from mild through intermediate to strongly tasting. Subjects were advised that the listed foods and drinks could each be prepared in a number of different ways and that the preferred way should be checked in each case from a group of choices. The listed alternatives were as follows, except that the choices were randomized: (1) Coffee: with more than one spoon of cream/with one spoon of cream / black, no cream. (2) Coffee: with more than one spoon of sugar / with one spoon of sugar / no sugar. (3) Cheese: American / Longhorn or Swiss / blue cheese. (4) Cheddar cheese: mild / medium / sharp. (5) Salad dressing: mild / oil and vinegar / roquefort or blue cheese. Scores were determined as follows: questions (1) and (2), both of which involve coffee, were treated as a single question. One point was given for a selection of: more than one spoon of either cream or sugar or one each of cream and sugar. One spoon of cream or one of sugar only was given two points, and black coffee without sugar was scored as three points. For questions (3), (4), and (5), one point was given for the first choice, two points for the second, and three for the third. The minimum score was therefore four points (mildest choice) and the maximum was twelve (strongest tasting choice). Part B of the questionnaire involved a list of five foods. Subjects were asked to answer "yes" or "no" to questions inquiring whether they liked grapefruit juice, lemon juice, sauerkraut, vinegar, and horseradish. Verbal explanations of the questionnaire were given if required and subjects were asked to indicate those foods not customarily consumed. The entire food-preference questionnaire was given to 181 subjects. Of this number fifteen could not complete Part B because they were unfamiliar with one or more of the foods listed. Incomplete forms were not counted, leaving a total of 166 completed scores on Part B. Six subjects could not complete Part A for the same reason, but an additional twelve subjects were tested for Part A only (since Part B was added after the investigation had begun). The total number who completed Part A was therefore 187. Table 18-II shows the mean taste thresholds with their standard

TABLE 18-II
MEAN TASTE THRESHOLDS WITH THEIR STANDARD ERRORS FOR SUBJECTS SELECTING MILD, MEDIUM AND STRONG TASTING CHOICES IN QUESTIONS 1-5 ON THE FOOD QUESTIONNAIRE. THE NUMBER OF SUBJECTS IN EACH GROUP IS SHOWN WITHIN PARENTHESES.

Threshold Number		Choice of Foods Tasting	
	Mild	Medium	Strong
Quinine 1 and 2	(100) 5.87 ± 0.19	(42) 5.65 ± 0.32	(45) 5.82 ± 0.27
3	(78) 5.21 ± 0.19	(82) 6.17 ± 0.19	(27) 6.33 ± 0.48
4	(84) 5.23 ± 0.20	(46) 5.96 ± 0.25	(57) 6.49 ± 0.24
5	(52) 5.58 ± 0.29	(84) 5.66 ± 0.20	(51) 6.24 ± 0.24
PROP 1 and 2	(100) 9.29 ± 0.25	(42) 9.28 ± 0.41	(45) 10.24 ± 0.37
3	(78) 8.68 ± 0.26	(82) 10.20 ± 0.27	(27) 10.11 ± 0.54
4	(84) 8.74 ± 0.25	(46) 9.52 ± 0.32	(57) 10.77 ± 0.26
5	(52) 8.84 ± 0.37	(84) 9.68 ± 0.25	(51) 10.06 ± 0.38

errors for quinine and PROP, for subjects who preferred the mild, medium, and strongly tasting choices in questions (1) through (5) of Part A. Taste sensitivity for quinine and PROP apparently is more important in determining the answers to questions (3), (4), and (5), than in determining preferences regarding the ways coffee may be prepared (the first two questions). The correlation coefficients observed between taste thresholds for the drugs and food-preference scores (Part A of the questionnaire) were as follows: Quinine/food score, + 0.262 ± 0.068; PROP/food score, + 0.366 ± 0.064. The Quinine/PROP threshold correlation coefficient was + 0.540 ± 0.052. Thus, there are highly significant positive correlations between high scores on the food questionnaire (that is, preference for strongly tasting foods) and high taste thresholds (that is, low taste sensitivity) for both quinine and PROP. Also, the thresholds for quinine and PROP are themselves correlated with each other.

Results from Part B of the food-preference questionnaire were analyzed by calculation of the linear regression coefficients of taste thresholds on the number of foods disliked. The coefficient with quinine is 0.319 ±0.114, and that with PROP is 0.480 ± 0.146. Both of these values are significantly different from zero and support the conclusion that there is a positive association between increasing number of foods disliked and increasing taste sensitivity (i.e. decreasing taste threshold). Both males and females showed a positive correlation between preference for mild-tasting foods and relatively sensitive perception of the taste of quinine and PROP. The males showed a somewhat higher correlation than the females, but the difference was not statistically significant.

The sixteen pairs of monozygotic twins in the sample were all married, and the co-twins had lived apart for varying lengths of time. Their scores on Part A of the food-preference form were found to be highly correlated (0.70, significant for N = 16 at the 0.001 level). The scores on Part A of the questionnaire for ten pairs of dizygotic twins, eight pairs of whom lived apart, showed an insignificant correlation of 0.18.

The sample included thirty-nine husband-wife pairs. The members of these pairs are not closely related genetically and there was no significant intrapair correlation between their scores for quinine and for PROP despite their having shared a common home environment for varying lengths of time. A significant positive correlation was observed between the scores on the food questionnaire of husbands and wives. On Part A of the questionnaire, the correlation coefficient was found to be 0.484 (significant for N = 39 at the 0.01 level). The correlation between the indicated food preferences of spouses observed in our data, despite absence of a correlation between their taste thresholds, demonstrates that social factors, as well as individual taste thresholds, are important for the determination of individual food preferences.

AGE, SEX, SMOKING, AND TASTE THRESHOLDS

Previous studies have shown that the proportion of smokers is lower in groups of sensitive quinine tasters and higher among the insensitive tasters.[24, 31] These observations appeared to be analogous to those indicating large numbers of reported food dislikes associated with low taste threshold for quinine and small numbers of food dislikes associated with high quinine taste threshold, as well as the differences in preferences for mildly or strongly tasting foods (discussed above).[24]

Harris and Kalmus observed that taste sensitivity for phenylthiourea solutions decreased at a rate of one doubling in concentration (i.e. increase in one threshold) for each twenty-year age span.[18] Kalmus and Trotter observed a decline in phenylthiourea taste sensitivity corresponding to a mean annual increase of 0.03 threshold.[32] They also noted a more rapid deterioration in men than in women. Leguèbe observed that women were approximately 0.8 and 0.5 threshold more sensitive tasters than men for phenylthiourea and quinine, respectively.[20] He observed that age was related to threshold differences only for the insensitive tasters (nontasters) of phenylthiourea, and that age was not related to differences in quinine threshold. Previously, Byrd and Gertman had reported no significant age-related difference in taste threshold for quinine.[33]

Extensive investigations of taste threshold distributions have demonstrated an age-related exponential decline in sensitivity (i.e. increase in threshold) for quinine and PROP, tabulated by age for 308 male and 368 female subjects. Table 18-III indicates the comparative rates of decrease in taste sensitivity for the male and female subjects. Figures 18-6 and 18-7, respectively for quinine and PROP, illustrate the increase in taste threshold (i.e. decrease in sensitivity) associated with increase in age and the more rapid change occurring in males.

The studies reviewed above were continued until sufficient numbers of subjects had been investigated to facilitate categorization according to smoking habits as well as age and sex.[34] The final sample included eighty-four males and 184 females who had never smoked regularly, as well as seventy-eight males and forty-nine females who had regularly smoked twenty or more cigarettes per day at the time they were tested. The mean thresholds for quinine and PROP, distributed according to age in the heavy smokers and, separately, in the nonsmokers, are shown in Figures 18-8 and 18-9 for the females, Figures 18-10 and 18-11 for the males. These four illustrations indicate the mean thresholds with their standard errors, together with the fitted regression lines. The regression coefficients calculated for taste threshold on age are shown in Table 18-IV.

Separate analyses of quinine and PROP threshold distributions for each sex indicated again that the heavy cigarette smokers were significantly less sensitive than the nonsmokers. In each case, the statistical significance occurred below the 5.0 percent level of chance probability.

Among the tested nonsmokers, the regression coefficients did *not* differ significantly from zero, and there was no significant difference between the males and females either in mean threshold or in the apparent effect of age upon taste. There was, however, a highly significant decline in taste sensitivity among the heavy smokers with increasing age. The rates of decline and the mean scores were not significantly different in the two sexes after

TABLE 18-III
COMPARATIVE RATES OF DECREASE IN TASTE SENSITIVITY IN MALES AND FEMALES, AGED 16 TO 55. COLUMN A SHOWS THE DECREASES IN SENSITIVITY IN THRESHOLDS PER YEAR, TOGETHER WITH THE STANDARD ERROR. COLUMN B SHOWS THE NUMBERS OF YEARS REQUIRED TO INCREASE MEAN SCORE BY ONE THRESHOLD, ACCORDING TO THE RATE INDICATED IN COLUMN A.

Substance	Sex	A	B
PROP	M	0.052 ± 0.011	19
	F	0.026 ± 0.013	38
Quinine	M	0.066 ± 0.013	15
	F	0.043 ± 0.009	23
Hydrochloric Acid	M	0.071 ± 0.028	14
	F	0.018 ± 0.014	55

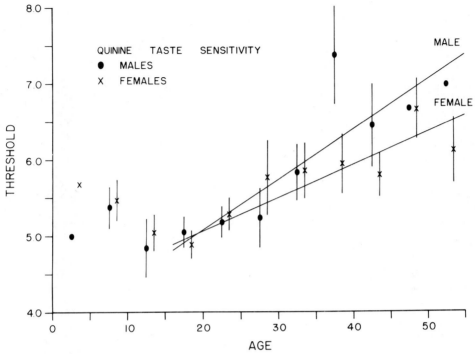

Figure 18-6. Influence of age on taste threshold for L-quinine sulfate ● Males; x females. The mean score and standard error for each age group are shown, together with the fitted regression line for the scores of subjects aged sixteen to fifty-five years. Thresholds are based on solution numbers. Vertical lines represent ± 1σMη.

the samples were controlled for smoking habits.

Previous studies of the influence of age and sex on taste sensitivity have been based on populations which were heterogeneous regarding smoking habits. The separate examination of the influence of age and sex on taste sensitivity in nonsmokers, and in heavy smokers of cigarettes, has indicated no significant difference in taste threshold for quinine and PROP related to age or sex in the nonsmokers. The regression lines for the nonsmokers and heavy smokers diverge near or within the sixteen-to twenty-year age range, and threshold differences become increasingly pronounced with advancing age. Since heavy smokers in the different age groups had been smoking for different average lengths of time, the divergence of the regression lines could be interpreted as being due to a cumulative effect of heavy smoking. The previously reported sex differences, based on population data which were analyzed without regard to smoking habits, might be ascribed to the occurrence of different proportions of smokers in women and men. Men, as a group, smoke more than women.[35, 36] In our own sample collected without regard for smoking habits, women comprised 68.7 percent of our nonsmokers, but only 38.6 percent of our heavy smokers. Aging is not associated with significant increase in taste threshold unless combined with smoking. The sex differences sometimes observed in popula-

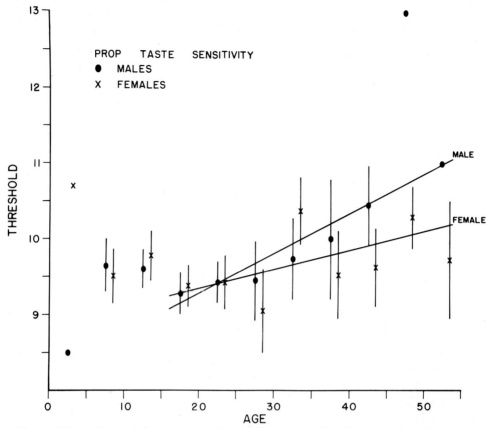

Figure 18-7. Influence of age on taste threshold for prop (6-N-Propylthiouracil). ● Males; x females. The mean score and standard error for each age group are shown, together with the fitted regression line for the scores of subjects aged sixteen to fifty-five years. Thresholds are based on solution numbers. Vertical lines represent ± 1σMη.

tion surveys apparently are the result of different smoking habits between the two sexes. Statistical control of smoking history affects the population's threshold distribution and location of the antimode; and the antimode is not the same for all populations, but tends to be floating or itinerant.

MENSTRUATION AND TASTE THRESHOLDS

Variations in numerous physiological and psychological characteristics have been correlated with phases of the menstrual cycle, but reports on associated changes in sensitivity of sensory perception are rare. The relative concentrations of certain hormones are known to be capable of affecting taste thresholds.[37] Fluctuations in taste sensitivity at the time of menstruation may be correlated with fluctuations in endocrine balance at this time. The observation that hormonal changes such as those associated with the menstrual cycle might modify taste sensitivity was suggested by data from repeated tests of the same female subjects and female monozygotic twins.[25] Beiguelman, however, observed no difference in taste threshold for

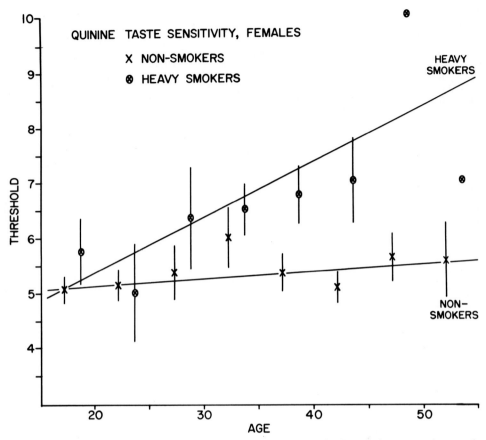

Figure 18-8. Influence of age on taste threshold for quinine in female heavy smokers and nonsmokers. The mean score and standard error for each group are shown, together with the fitted regression line for the scores of subjects aged sixteen to fifty-five years. Thresholds are based on solution numbers. Vertical lines represent ±1 standard error.

phenylthiourea associated with the menstrual cycle.[38] His study involved 100 subjects who were each tested twice, once during the menstruation phase and again between periods. The wide variation which we have observed in individual response indicates the necessity for utilizing extreme caution in generalizing from limited amounts of taste threshold data. Repeated testing under controlled conditions provided a reliable method of examining the relationship between the menstrual cycle and constancy of taste thresholds.

Taste thresholds for quinine and PROP were determined throughout one or more menstrual cycles in a group of nineteen subjects.[39, 40] Taste tests were carried out at the same time each day, three days per week, for periods ranging from four to nine weeks. The subjects were apparently healthy and normal females, and all but two were student nurses. Their ages ranged from nineteen to twenty-seven (average 20.7). Subjects were requested not to take drugs during the experiment. Four additional subjects were excluded from the

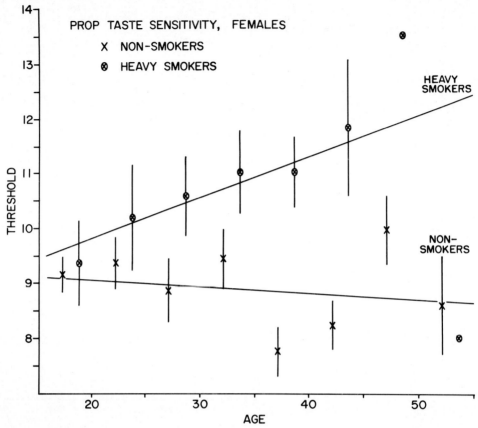

Figure 18-9. Influence of age on taste threshold for PROP (6-N-Propylthiouracil) in female heavy smokers and nonsmokers. The mean score and standard error for each group are shown, together with the fitted regression line for the scores of subjects aged sixteen to fifty-five years. Thresholds are based on solution numbers. Vertical lines represent ± 1 standard error.

analysis because of illness or medication. The days in the menstrual cycle were numbered forward and backward from the first day of menstruation. In those instances in which two cycles were recorded, the days were counted from the first day of menstruation in the second cycle. Sensitivities within three phases of the cycle were compared: the premenstrual (days −9 to −5), menstrual (days −1 to +4) and postmenstrual (days +6 to +10).

Several subjects continued to improve after the first two tests; therefore the most conservative comparison was between the menstrual and postmenstrual phases of the cycle. In the majority of individuals, thresholds tended to be lower (i.e. taste tended to be more sensitive) during the menstrual period. Analysis by the Wilcoxon test for matched pairs and signed ranks showed that these differences were statistically significant: in the two-tailed test for PROP, $P < 0.05$; for quinine, $P = 0.05$; and for both combined, $P = 0.003$. The responses of the subjects were markedly heterogeneous. The average increase

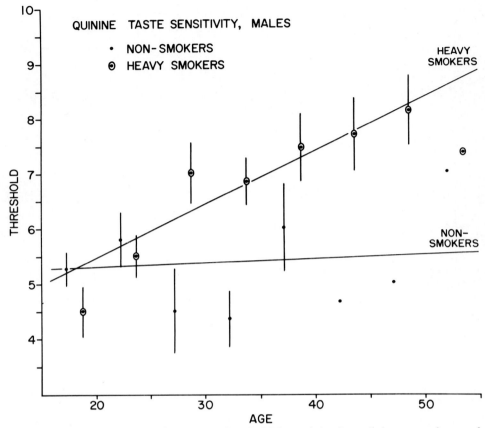

Figure 18-10. Influence of age on taste threshold for quinine in male heavy smokers and nonsmokers. The mean score and standard error for each group are shown, together with the fitted regression line for the scores of subjects aged sixteen to fifty-five years. Thresholds are based on solution numbers. Vertical lines represent ± 1 standard error.

in taste sensitivity during menstruation was 0.68 thresholds for PROP and 0.45 for quinine. When the scores for PROP and quinine were considered together, 18 percent showed a decreased sensitivity during the menstrual phase compared with the post-menstrual; 16 percent showed no change, and 66 percent increased in sensitivity. An increase of 0.5 thresholds or more was shown by 47 percent of the total sample; 18 percent had an increase of one threshold or more and 8 percent increased by two thresholds or more. The most extreme change was shown by one of the subjects:[40] six days before the onset of menstruation, her scores for both compounds changed markedly and reached maximum sensitivity on day −3, returning to their usual values by day +6. Her fluctuation in taste sensitivity, as indicated by the differences in menstrual and intermenstrual scores, represent a 1,024-fold change in PROP sensitivity and a 362-fold change in quinine sensitivity. This subject was tested over two cycles and, in both, showed the same change immediately be-

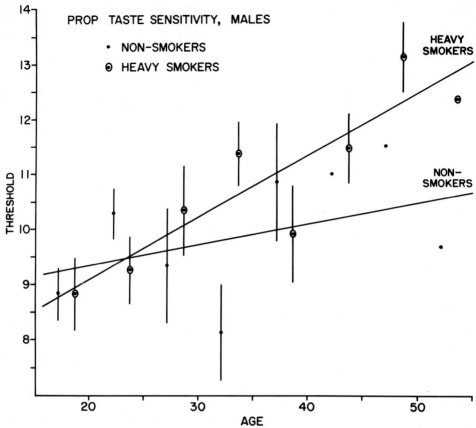

Figure 18-11. Influence of age on taste threshold for PROP (6-N-Propylthiouracil) in male heavy smokers and nonsmokers. The mean score and standard error for each group are shown, together with the fitted regression line for the scores of subjects aged sixteen to fifty-five years. Thresholds are based on solution numbers. Vertical lines represent ± 1 standard error.

TABLE 18-IV
REGRESSION COEFFICIENTS OF TASTE THRESHOLDS FOR INCREASING AGE.
NUMBERS OF SUBJECTS: MALE NONSMOKERS = 84; MALE HEAVY SMOKERS = 78;
FEMALE NONSMOKERS = 184; FEMALE HEAVY SMOKERS = 49;
HEAVY SMOKERS EACH SMOKED 20 OR MORE CIGARETTES DAILY AT TIME OF TEST.

		Regression Coefficients with Standard Errors	
		Nonsmokers	Heavy Smokers
Quinine	Males	+0.007 ± 0.023	+0.097 ± 0.020
	Females	+0.012 ± 0.011	+0.100 ± 0.029
PROP	Males	+0.038 ± 0.029	+0.114 ± 0.023
	Females	+0.014 ± 0.015	+0.075 ± 0.034

fore the onset of menstruation. The change, if any, shown by other subjects was less dramatic. The majority of the subjects reached maximum sensitivity after the onset of menstruation on days +1 to +5.

In large populations, as discussed above,

taste thresholds for quinine tend to approach a normal or Gaussian distribution, but the curve of distribution for PROP thresholds tends toward bimodality. Individuals have been classified as tasters or nontasters of PROP according to whether their thresholds occurred below or above the antimode between thresholds 9 and 10.[26] Both tasters and nontasters showed increased sensitivity at the time of menstruation. It is interesting that one subject (discussed above) would according to the antimode-determined dichotomy, have been classified usually as a nontaster, but during menstruation she would clearly qualify as a taster. She also changed from being unable to taste phenylthiourea paper (Carolina Biologicals, Inc.) to being able to taste it as bitter during menstruation. The degree of change associated with menstruation does not appear to be associated with the position of the individual's taste threshold on the population distribution curve.[39, 40]

In our studies, information was obtained from each subject with a simple questionnaire relating to menstrual cycle regularity, symptoms experienced, and the time at which the symptoms were most severe. The subjects were asked to rate symptoms on a four-point scale: none, slight, moderate, and severe. The percentages of subjects who reported symptoms as moderate or severe were as follows: pain, 47 percent; headache, 0 percent; irritability, 58 percent; depression, anxiety, nervousness, or tension, 37 percent. In addition, 47 percent reported some swelling in various parts of the body. Pain was most severe on the first day in all but one of the subjects who experienced it, but other symptoms were most commonly experienced immediately prior to the onset of menstruation. These figures are comparable to those obtained by Coppen and Kessel for a much larger and somewhat older population.[41, 42] No significant correlation was detected between increased taste sensitivity at menstruation and the occurrence of any of these symptoms.

Inquiries were also made into the length and regularity of the menstrual cycle. Subjects were asked to rate themselves as regular or irregular and to state the maximum and minimum interval between menstrual periods in recent months. Cycles varying in length by more than six days were classified as irregular. There was complete agreement between the subjects' self-rating and the latter method of evaluation for the group in the present study. No correlation was found between change in sensitivity and length of the cycle at the time of the tests. A significant positive correlation was detected, however, between increased taste sensitivity during menstruation and irregularity in length of the cycle as defined above. Eleven of the subjects were classified as regular and eight as irregular. The magnitude and direction of the changes were compared with a two-tailed Mann-Whitney U test: when the scores of the two groups were compared for PROP, $P = .090$ and for quinine, $P = .046$; when the data for PROP and quinine were combined, $P = .010$. The results indicate an association between irregularity in length of the cycle and increased taste sensitivity at or immediately prior to the time of menstruation.

TASTER GENOTYPES, PEPTIC ULCERS, AND THE INTINERANT ANTIMODE

The taste thresholds of sixty-eight duodenal ulcer patients and thirty-four gastric ulcer patients, compared to those of the general population, showed that the 6-n-propylthiouracil taste thresholds of the gastric ulcer patients tended to be higher than those of the other two groups.[43] When the taster genotypes were interpolated from the modal locations of the patients' thresholds *in the general population distribution,* it appeared that the duodenal ulcer patients included an abnormally high proportion of sensitive tasters, but that the distribution of taster genotypes of the gastric ulcer patients was similar to that of the general population.[43] A reevaluation of these data, in context with the phenomenon of the intinerant antimode (discussed above), provides a revision of the previous interpretation of the data. The sample of duodenal ulcer patients includes forty-two (61.8%) in the relatively sensitive mode, four (5.9%) at the antimode, and twenty-two (32.3%) in the relatively insensitive mode. The taster genotypes of the sample of gastric ulcer patients, determined according to *that specific group's* bimodal distribution of taste thresholds, shows the same distribution as that of the duodenal ulcer patients: twenty-one (61.8%) two (5.9%) and eleven (32.3%) in the sensitive-taster mode, at the antimode and in the insensitive-taster mode, respectively. Thus, contrary to the previously-published interpretation of the ulcer patients' taste-threshold data, the two categories of ulcer patients show identical distributions of the taster genotypes; and both categories include significantly higher proportions with the sensitive taster genotype and significantly lower proportions with the insensitive taster genotype, compared to the general population.[43] Superimposed upon the genotype-determined threshold modalities, there is a two-threshold increase associated with the gastric-ulcer pathology.

An individual's genotypically determined taste-threshold modality for 6-n-propylthiouracil and related compounds may be interpolated only with taste-threshold data from a population controlled for the various other factors which influence taste threshold. Such factors have been shown to include recent alcoholism; heavy smoking of cigarettes; gastric ulcer pathology, but not duodenal ulcer pathology. Studies of taste thresholds for phenylthiourea or for 6-n-propylthiouracil or for other compounds in their class (i.e. characterized by presence of the H-N-C=S grouping) involve nongenetic effects on taste thresholds superimposed upon the genotype-associated taste-threshold modalities. Previous failure to observe and correct for the phenomenon of the itinerant antimode has led to observed discrepancies with the classical taster-genetics hypothesis and to erroneous genetical and anthropological assessments.

TASTER GENOTYPES IN ALCOHOLICS AND HEROIN ADDICTS

Taste tests were performed with 537 male alcoholics. This group includes 100 who were followed-up for six months after their participation in an institutional sobriety program, who did not consume any alcoholic beverage during that time; and seventy-seven who were followed-up and were inebriated at least once in the six months following participation in the program. During the same period, we also taste-tested seventy-one male heroin addicts and 505 unselected male volunteers. Table 18-V shows the PROP threshold numbers (i.e. associated with the concentrations as molarities for the solutions used in determining taste thresholds as shown in Table 18-I), and the threshold distributions for each of the cited groups. Table 18-VI shows the percentage distributions of the two modes in which the subjects at the antimode in each case were included with the sensitive mode. Table 18-VII de-

TABLE 18-V

TASTE THRESHOLD DISTRIBUTIONS IN 505 CONTROL MALES, 71 MALE HEROIN ADDICTS, AND 537 MALE ALCOHOLICS. THE LATTER CATEGORY INCLUDES 100 F.N., FOLLOWED-UP MALE ALCOHOLICS WHO RETAINED SOBRIETY FOR AT LEAST SIX MONTHS AFTER COMPLETING AN INSTITUTIONAL SOBRIETY PROGRAM; AND 77 F.R., FOLLOWED-UP MALE ALCOHOLICS WHO WERE INEBRIATED AT LEAST ONCE DURING THE SIX MONTHS FOLLOWING THE INSTITUTIONAL PROGRAM.

PROP Threshold Number	Control Males (n = 505)	Male Heroin Addicts (n = 71)	Male Alcoholics, Total (n = 537)	Male Alcoholics, F.N. (n = 100)	Male Alcoholics, F.R. (n = 77)
1	1	0	3	1	1
2	1	0	0	0	0
3	4	0	0	0	0
4	6	0	0	0	0
5	11	0	2	0	1
6	37	4	6	1	2
7	62	4	34	7	4
8	94	18	56	12	8
9	36	8	46	6	3
10	55	17	79	16	21
11	58	4	62	10	9
12	67	5	61	12	5
13	40	6	65	12	7
14	23	1	64	9	9
>14	10	4	69	14	7

TABLE 18-VI

PERCENTAGES OF SENSITIVE AND INSENSITIVE TASTERS OF PROP IN MALE ALCOHOLICS, HEROIN ADDICTS, AND CONTROLS.

	Percentage of PROP Sensitive Tasters	Percentage of PROP Insensitive Tasters
Followed-up Sober Alcoholic Males (n = 100)	27.0	73.0
Followed-up Recidivist Alcoholic Males (n = 77)	24.7	75.3
Total Alcoholic Males (n = 537)	27.4	72.6
Heroin Addicts (n = 71)	47.9	52.1
Control Males (n = 505)	49.9	50.1

tails the chi-square analysis comparing the modal distributions of the alcoholic and control males.

The data from our population studies on unselected male volunteers indicate a distribution of PROP taste thresholds that is consistent with other studied populations. If the antimode population is included with those in the sensitive mode (i.e. those whose thresholds occur at concentrations at or below solution number nine), then the population is nearly equally divided between those with sensitive taster

TABLE 18-VII

CHI-SQUARE ANALYSIS OF COMPARATIVE DISTRIBUTIONS OF SENSITIVE AND INSENSITIVE TASTERS OF PROP IN 537 ALCOHOLIC AND 505 CONTROL MALES. THE VALUE OF CHI-SQUARE AT TWO DEGREES OF FREEDOM EQUALS 66.852. THEREFORE, THE DIFFERENCES BETWEEN THE TWO DISTRIBUTIONS ARE STATISTICALLY SIGNIFICANT AT A CHANCE PROBABILITY LEVEL OF LESS THAN 0.001.

	Control Males	Alcoholic Males	Total
Sensitive Tasters	Obs., 252 (Exp., 193)	Obs., 147 (Exp., 206)	399
Insensitive Tasters	Obs., 253 (Exp., 312)	Obs., 390 (Exp., 331)	643
Total	505	537	1042

genotypes and those with insensitive taster genotypes. Our data clearly indicate, however, that about three-fourths of the tested alcoholic males possess insensitive taster genotypes. The distribution difference between the 537 alcoholic males and the 505 unselected males is extremely significant statistically (p less than 0.001). Among the alcoholics who were followed-up for six months following participation in an institutional sobriety program, the one hundred who consumed no alcohol and the seventy-seven who succumbed to inebriation at least once showed similar distributions for the two taster categories. The seventy-one tested heroin addicts, however, showed a distribution similar to that in the general population.

Two conclusions have been derived from the above research. Firstly, among alcoholics there is no difference in prognosis for sobriety associated with the studied genotype difference. Secondly, there is a very significant association between alcoholism and the single gene difference which can be measured with standardized taste-threshold tests: while the general population is about evenly distributed between the two genotype categories, individuals in one of these genotype categories are three times as likely to become alcoholics as individuals in the other category. That is, the insensitive-tasting males are three times as likely as the sensitive-tasting males to become alcoholics.

TASTE THRESHOLDS AND BIOLOGICAL ACUITY

Fischer and Griffin observed that subjects who were very sensitive tasters and very insensitive tasters of quinine were also sensitive and insensitive tasters, respectively, for numerous other compounds.[26] The solutions which they tested and found to conform to this pattern included the following compounds: phenylthiourea, thioacetamide, thiourea, L-ergothioneine, Chlorpromazine® (Smith, Kline, & French), Mellaril® (Sandoz), Triflupromazine® (Squibb), methylene blue, Tofranil® (Geigy), desmethylimipramine (Geigy), acetamide, urea, L-phenylalanine, DL-phenylalanine, D-amphetamine sulfate, L-amphetamine sulfate, DL-dopa, sucrose, sodium chloride, potassium chloride, hydrochloric acid, Niamid ® (Pfizer), Antistine® (Ciba). They concluded that the subjects who were very sensitive and very insensitive tasters of quinine tended to be, respectively, sensitive and insensitive tasters of drugs in general; sensitive and insensitive tasters, respectively, of compounds which primary taste qualities are sweet, salty, and sour, as well as bitter. Thus, particularly high and low taste acuities for quinine were observed to be associated, respectively, with high and low taste acuities in general.

Previous studies indicated that there may be a relationship between a population's distributions of taste thresholds for different forms of a compound and the comparative systemic activities of those different forms. Subjects generally display lower thresholds for L-quinine, and higher ones for quinidine or D-quinine.[44] The former analog displays the higher toxicity of the two. The oral LD_{50} (i.e. lethal dose for 50 %) of the L form was found to be 214.8 ±25.1 mg/kg for the mouse, [45] whereas the corresponding value for the "D" form was 535 mg/kg.[46] Most subjects tested displayed lower thresholds for D-amphetamine, higher ones for L-amphetamine.[26] The LD_{50} rating for L-amphetamine after very slow intravenous injection (1.0 cc. in two minutes) was 79.2 ± 8.5 mg/kg in the mouse, compared to only 5.0 ± 1.3 mg/kg for

D-amphetamine.[26, 44] The latter form, for which most subjects display lower taste thresholds, was sixteen times as active as the former. In the above two cases, quinine versus quinidine and L-amphetamine versus D-amphetamine, the drug which elicited the greater taste sensitivity in man also manifested the lower or more powerful lethal dose in mice.[44] The oral dosage of psilocybin which induced measurable neurological effects, as shown by finger-tapping tests, in a very sensitive taster of quinine and PROP manifested no such changes in a very insensitive taster of quinine and PROP.[47] Lower taste threshold (i.e. higher taste sensitivity) evidently tends to be associated with a higher general systemic reactivity.

TASTE THRESHOLD AND PERSONALITY

The complete Wechsler Adult Intelligence Scale (W.A.I.S.) was administered to twenty-seven college students whose taste thresholds had previously been determined for quinine and PROP. The group included four very insensitive tasters with quinine thresholds of 7 or more and PROP thresholds of 11 or more and five very sensitive tasters with quinine thresholds of 4 or less and PROP thresholds of 9 or less. These categories are based on the observations of Fischer and Griffin that members of those two classes tend to generally be sensitive or insensitive tasters, respectively, for many other compounds as well as quinine and PROP.[44] An intermediate group with quinine thresholds of 5 or 6 and PROP thresholds equal to or more than 5 but not more than 13 showed less consistent responses for the different compounds. The investigator who administered the W.A.I.S. tests and then determined their personality profiles was not aware of his subjects' taste thresholds.[44, 47] The results indicated that the insensitive tasters of quinine and PROP (n=4) showed a "compensated pattern of scores on the W.A.I.S.," while the sensitive tasters (n=5) showed a W.A.I.S. pattern previously described as "internalized."[47] In other words, the insensitive tasters (who may be relatively low in general systemic reactivity) yielded W.A.I.S. score profiles characteristic of subjects who easily maintain contact with other individuals, whereas the very sensitive tasters (who may be relatively high in general systemic reactivity) yielded patterns characteristic of relatively introverted subjects.

REFERENCES

1. Fox, A.L.: Tasteblindness. *Science, 73* Suppl. (April 17):14, 1931.
2. Fox, A.L.: The relation between chemical constitution and taste. *Proc Natl Acad Sci USA, 18*:115, 1932.
3. Blakeslee, A.F., and Salmon, M.R.: Odor and taste-blindness. *Eugen News, 16*:105, 1931.
4. Blakeslee, A.F.: Genetics of sensory thresholds: Taste for phenylthiocarbamide. *Proc Natl Acad Sci USA, 18*:120, 1932.
5. Snyder, L.H.: Inherited taste deficiency. *Science, 74*:151, 1931.
6. Snyder, L.H.: Studies in human inheritance. IX. The inheritance of taste deficiency in man. *Ohio J Sci, 32*:436, 1932.
7. Ardashnikov, S.N., Lichtenstein, E.A., Martynova, R.P., Soboleva, G.V., and Postnikova, E.N.: The diagnosis of

zygosity in twins. (Three instances of differences in taste acuity in identical twins.) *J Hered, 27*:465, 1936.
8. Rife, D.C.: Contribution of the 1937 national twins convention to research. *J Hered, 29*:83, 1938.
9. Harris, H., and Kalmus, H.: The distribution of taste thresholds for phenylthiourea of 384 sib pairs. *Ann Eugen, 16*:226, 1951.
10. Das, S.R.: A contribution to the heredity of the P.T.C. taste character based on a study of 845 sib-pairs. *Ann Hum Genet, 20*:334, 1956.
11. Das, S.R.: Inheritance of the P.T.C. taste character in man: An analysis of 126 Rarhi Brahmin families of West Bengal. *Ann Hum Genet, 22*:200, 1957.
12. Kalmus, H.: Defective colour vision, P.T.C. tasting, and drepanocytosis in samples from fifteen Brazilian populations. *Ann Hum Genet, 21*:313, 1957.
13. Merton, B.B.: Taste sensitivity to PTC in 60 Norwegian families with 176 children. Confirmation of the hypothesis of single gene inheritance. *Acta Genet, 8*:114, 1958.
14. Dencker, S.J., Hauge, M., and Kaij, L.: An investigation of the PTC taste character in monochorionic twin pairs. *Acta Genet, 9*:236, 1959.
15. Verkade, P.E., Wepster, B.M., and Stegerhoek, L.J.: Investigations on taste blindness with thiocarbamides. Intrapair discrepancy of taste in pairs of identical twins. *Acta Genet Med Gemellol (Basel), 8*:361, 1959.
16. Sutton, H.E., de Lamadrid, E.G., and Esterer, M.B.: The hereditary abilities study: Genetic variation in human biochemical traits. *Am J Hum Genet, 14*:4, 1962.
17. Hartmann, G.: Application of individual taste difference towards phenyl-thiocarbamide in genetic investigations. *Ann Eugen, 9*:123, 1939.
18. Harris, H., and Kalmus, H.: The measurement of taste sensitivity to phenylthiourea (P.T.C.). *Ann Eugen, 15*:24 and 32, 1949.
19. Kalmus, H.: Improvements in the classification of the taster genotype. *Ann Hum Genet, 22*:222, 1958.
20. Leguèbe, A.: Génètique et anthropologie de la sensibilité à phénylthioucarbamide. *Bull Inst Roy Sci Natl Belgique, 26*:1, 1960.
21. Sinnot, J.J., and Rauth, J.E.: The effect of smoking on taste thresholds. *J Gen Psychol, 19*:151, 1937.
22. Fischer, R., Griffin, F., England, S., and Pasamanick, B.: Biochemical-genetic factors in taste polymorphism and their relation to salivary thyroid metabolism in health and mental retardation. *Med Exp, 4*:356, 1961.
23. Skude, G.: Some factors influencing taste perception for phenylthiourea (P.T.C.). *Hereditas, 50*:203, 1963.
24. Fischer, R., Griffin, F., and Kaplan, A.R.: Taste thresholds, cigarette smoking, and food dislikes. *Med Exp, 9*:151, 1963.
25. Kaplan, A.R., Powell, W., Fischer, R., and Marsters, R.: Re-examination of genetic aspects of taste thresholds for phenylthiourea type compounds. In Geerts, J. (Ed.): *Genetics Today*. London, Pergamon, 1964, p. 292, Vol. I.
26. Fischer, R., and Griffin, F.: Pharmacogenetic aspects of gustation. *Arzneim Forsch, 14*:673, 1964.
27. Kaplan, A.R., and Fischer, R.: Taste sensitivity for bitterness: some biological and clinical implications. In Wortis, J. (Ed.): *Recent Advances in Biological Psychiatry*. New York, Plenum Pr, 1965, Vol. VII, p. 183.
28. Kaplan, A.R., Fischer, R., Karras, A., Griffin, F., Powell, W., Marsters, R., and Glanville, E.V.: Taste thresholds in twins and siblings. *Acta Genet Med Gemellol (Basel), 16*:229, 1967.
29. Fischer, R., Griffin, F., England, S., and Garn, S.M.: Taste thresholds and food dislikes. *Nature (Lond), 191*:1328, 1961.
30. Glanville, E.V., and Kaplan, A.R.: Food preference and sensitivity of taste for bitter compounds. *Nature (Lond), 205*:851, 1965.
31. Krut, L.H., Perrin, M.J., and Bronte-

Steward, B.: Taste perception in smokers and non-smokers. *Br Med J, 1*:384, 1961.
32. Kalmus, H., and Trotter, W.R.: Direct assessment of the effect of age on P.T.C. sensitivity. *Ann Hum Genet, 26*:145, 1962.
33. Byrd, E., and Gertman, S.: Taste perception in the aged. *Geriatrics, 14*:381, 1959.
34. Kaplan, A.R., Glanville, E.V., and Fischer, R.: Cumulative effect of age and smoking on taste sensitivity in males and females. *J Gerontol, 20*:334, 1965.
35. Matarazzo, J.D., and Saslow, G.: Psychological and related characteristics of smokers and non-smokers. *Psychol Bull, 57*:493, 1960.
36. Horn, D.: Behavioral aspects of cigarette smoking. *J Chronic Dis, 16*:383, 1963.
37. Henkin, R.I., Gill, J.R., Jr., and Bartter, F.C.: Studies on taste thresholds in normal man and in patients with adrenal cortical insufficiency: The role of adrenal cortical steroids and of sodium concentration. *J Clin Invest, 42*:727, 1963.
38. Beiguelman, B.: Taste sensitivity to phenylthiourea and menstruation. *Acta Genet Med Gemellol (Basel), 13*:197, 1964.
39. Glanville, E.V., and Kaplan, A.R.: Taste perception and the menstrual cycle. *Nature (Lond), 205*:930, 1965.
40. Glanville, E.V., and Kaplan, A.R.: The menstrual cycle and sensitivity of taste perception. *Am J Obstet Gynecol, 92*:189, 1965.
41. Coppen, A., and Kessel, N.: Menstrual disorders and personality. *Br J Psychiatry, 109*:711,1963.
42. Coppen, A., and Kessel, N.: Menstrual disorders and personality. *Acta Psychotherapeutica, 11*:174, 1963.
43. Kaplan, A.R., Fischer, R., Glanville, E.V., Powell, W., Kamionkowski, M., and Fleshler, B.: Differential taste sensitivities in duodenal and gastric ulcer patients. *Gastroenterology, 47*:604, 1964.
44. Fischer, R., and Griffin, F.: Quinine dimorphism: A cardinal determinant of taste sensitivity. *Nature (Lond), 200*:343, 1963.
45. Pfeiffer, C.C.: Optical isomerism and pharmacological action, a generalization. *Science, 124*:29, 1956.
46. Schallek, W.: Quinidine-like activity of thephorin. *J Pharmacol Exp Ther, 105*:291, 1952.
47. Fischer, R., Griffin, F., and Pasamanick, B.: The perception of taste: some psychophysiological, pathophysiological, and clinical aspects. In Hoch, P.H., and Zubin, J. (Eds.): *Psychopathology of Perception.* New York, Grune, 1965, p. 129.

GLOSSARY

Acentric Descriptive term referring to a chromosome or chromosome segment without a centromere.

Acrocentric Descriptive term referring to a chromosome in which the centromere is located very close to one end, so that one chromosome arm is relatively very small and the other one is much larger.

Acuity Sharpness or clearness.

Additive genes Genes which interact with each other in trait manifestation (e.g. cumulative effects rather than dominance-recessiveness).

Affective disorder Personality disorder characterized by recurring and alternating periods of depression and elation not readily attributable to external circumstances.

Akinetic Diminishing the power of the muscles.

Allele Shortened term for allelomorph. One of two or more alternative forms of a gene occupying the same locus or point on a particular chromosome. The activities of alleles involve the same biochemical and/or developmental process(es). A haploid cell has only a single representative for each locus. A diploid cell has two representatives for each locus, two alleles. If the two alleles at a given locus are identical, the individual is homozygous for the locus involved; if the two alleles at a given locus are different, then the individual is heterozygous for that locus.

Amine Organic (i.e. carbon-containing) compound containing one or more amino group ($-NH_2$).

Amphetamine Synthetic racemic desoxynonephedrine or alpha-methyl-phenethylamine, $C_6H_5 \cdot CH_2CHNH_2CH_3$. It is used in inhalation and in 1% solution in liquid petrolatum as a spray in head colds, hay fever, etc. Amphetamine is a sympathomimetic amine with a marked central stimulating effect on which most of its uses depend. The amoretic (appetite suppressant) drugs are chemically related to amphetamine.

Aneuploidy Deviations from the normal diploid number of chromosomes that are not multiples of this number, for example, 45 or 47. The deviation may be of either a hypoploid or hyperploid type.

Anthropometry The study of human body measurements, especially on a comparative basis.

Antisocial Personality Personality disorder characterized by a basic lack of socialization and by behavior patterns that bring the individual repeatedly into conflict with society. People with this disorder are incapable of significant loyalty to individuals, groups, or social values and are grossly selfish, callous, irresponsible, impulsive, and unable to feel guilt or to learn from experience and punishment. Frustration tolerance is low, and such individuals tend to blame others or offer plausible rationalizations for their behavior.

Anxiety Apprehension and tension, primarily of intrapsychic origin, the source(s) of which may be largely unrecognized. Fear, on the other hand, is an emotional response to a consciously recognized and usually external threat of danger.

A priori Presumptive; being without examination or analysis; relating to or derived from reasoning from self-evident propositions; presupposed by experience (Latin).

Assortive Mating Nonrandom mating. The preferential selection of a spouse with a particular genotype.

Assortment The genetic consequence of the random distribution of nonhomologous chromosomes to daughter cells in meiosis. See *segregation*.

Assortment, independent See *independent assortment*.

Asthenic Characterized by slender build and slight muscular development (i.e. ectomorphic, compared to athletic, pyknic). See *asthenic personality*.

Asthenic personality Personality disorder characterized by easy fatigability, low energy level, lack of enthusiasm, marked incapacity for enjoyment, and oversensitivity to physical and emotional stress.

Attenuation The reduction in size of a correlation coefficient due to measurement error.

Audioradiography A technique applied in cytogenetics for the study particularly of DNA synthesis by the chromosome, using for example radioactive tritium with labeled thymidine as the marker.

Autosome Any chromosome other than a sex chromosome, i.e. any chromosome other than an X chromosome or a Y chromosome.

Backcross Term from experimental genetics to indicate mating between heterozygote and homozygote (or between F_1 hybrid and one of the two parental strains). Double backcross is the mating between a double heterozygote and a homozygote. The mating most informative for linkage analysis is the double backcross mating involving the recessive homozygote.

Balanced load That which depresses the overall fitness of a *population* owing to the segregation of inferior or defective genotypes, the component genes of which are maintained in the population because they add to fitness in different combinations, e.g. as heterozygotes. See *balanced polymorphism*.

Balanced polymorphism A condition in which different alleles of the same genetic locus occur in relatively high frequencies in a population and in which the disadvantages produced by a double dose of one allele (homozygosity) are balanced by the advantages produced by a single dose of the allele (heterozygosity). Genetic polymorphism can be maintained in a population if the heterozygotes for the allele under consideration manifest a higher adaptive value than either homozygote.

Balanced translocation Chromosomal structural change characterized by the change in position (location) of one or more chromosome(s) or chromosome segment(s) within the normal chromosome complement. That is, some of the chromosome material is abnormally located, but the chromosome complement involves no duplication or deletion of any chromosomal material.

Barr body Sex chromatin. The condensed single X chromosome seen in the nucleus of a somatic cell in a female mammal; named for Murray Barr, its discoverer.

Barr-negative See *sex-chromatin-negative*.

Barr-positive See *sex-chromatin-positive*.

Base pair The guanine-cytosine and adenine-thymine pairs of purine-pyrimidine bases which make up DNA. In RNA, uracil substitutes for thymine. One of the pair is on one chain, the other on the complementary chain.

Behaviorism A direction in psychology

where attention is focused on the overt behavior of organisms only, where the usefulness of introspective and holistic methods is questioned or denied.

Biometry Statistical study of biological observations and phenomena.

Bimodality Refers to having two modes or peaks.

Bisexuality The presence in an individual of both male and female characteristics (either psychological or physical).

Catecholamine Epinephrine and norepinephrine, elaborated by the cells of the adrenal medulla, are often referred to collectively as catecholamines. Each of these hormone molecules contains an asymmetric carbon atom and therefore can exist in two optically active forms. The hormones secreted by the gland are levorotatory (L-forms), whereas those produced by laboratory synthesis are racemic (DL-forms). The two hormones are very similar in chemical properties, but the presence of an additional methyl group (in epinephrine) changes the side chain from a primary amine (norepinephrine) to a secondary amine (epinephrine). Adrenal extracts contain both epinephrine and norepinephrine. Dopamine, a precursor of norepinephrine, is also found in medullary extracts. Both hormones increase the heart rate, although epinephrine is the more potent in this respect. Both increase systolic blood pressure, whereas epinephrine has no effect on diastolic pressure. The hypertensive effect of norepinephrine is a consequence of increased peripheral resistance; that of epinephrine results from the increased cardiac output in spite of decreased total peripheral resistance. Epinephrine increases the blood flow through skeletal muscle, liver, and brain, whereas norepinephrine has no effect or decreases it. Both hormones produce constriction of the skin capillaries and cause pallor. Renal blood flow is diminished by both hormones. The net peripheral vascular effect of epinephrine is to cause vasodilation; norepinephrine exerts limited vasodilator actions, but its overall effect is to produce vasoconstriction. Epinephrine is very potent in increasing oxygen consumption and glucose output from the liver, but norepinephrine is relatively weak in these respects. Anticipatory states tend to elevate the release of norepinephrine more than epinephrine, but in intense emotional reactions both of the amines are elevated. Epinephrine acts to prevent hypoglycemia by stimulating metabolism and mobilizing glycogen as glucose, and redistributing the blood, draining it out of the skin, and forcing it into important organs, such as skeletal muscle, liver, and brain. Norepinephrine is predominantly present at sympathetic nerve endings and seems to function largely as a pressor hormone normally required for the maintenance of blood pressure. With the exception of the coronary arteries, norepinephrine produces general vasoconstriction and stimulates the heart but is relatively impotent in its metabolic actions. The injection of epinephrine to normal human subjects produces a feeling of restlessness and anxiety and a sense of fatigue, whereas norepinephrine does not produce such symptoms. Hence, the two hormones have different effects on the central nervous system.

Causality A relation assumed between two events in order to explain one of them in terms of the other.

Centromere The constricted portion of the chromosome, separating it into a short arm and a long arm, to which the spindle fibers are attached in mitosis and meiosis.

Centromeric Index An expression of the position of the centromere of a chromosome, determined as length of the short arm divided by the total length of the entire chromosome.

Chiasma An x-shaped configuration of chromosomal material caused by the breakage and exchange and reciprocal fusion of equivalent segments of homologous chromatids during the meiosis.

Chiasma formation The cytologic basis of genetic recombination, or crossing over, occurring between homologous chromosomes at meiosis.

Chi-square The ratio of an observed sum of squares to the corresponding variance fixed by hypotheses.

Chromatid One of the two parallel components of the metaphase chromosome joined at the centromere, each destined to form one chromosome in the daughter cell. One half of a replicated chromosome.

Chromatin Nuclear substance which takes basic stain and becomes incorporated in the chromosomes; so called because of the readiness with which it becomes stained with certain dyes (i.e. chromaticity). Chromatin is that part of the nuclear material that makes up the genetic material and contains the genetic information (i.e. genes) of the cell. The chromatin of a cell is considered to organize and contract into the stainable chromosomes during cell division. Larger chromatin bodies are called chromocenters.

Chromatin-negative See *sex-chromatin-negative*.

Chromatin-positive See *sex-chromatin-positive*.

Chromomeres Areas of different optical density and/or different diameters along the length of a chromosome, especially clearly detectible during the prophase of a cell division.

Chromosomal aberration Chromosomal abnormality; a deviation from the normal morphology or number of the chromosomes.

Chromosomal deficiency See *deletion, chromosomal*.

Chromosomal mosaicism Mosaicism in which the gentically different cell lines contain observable morphological differences in the chromosome complements. See *mosaicism*.

Chromosomal mutation Any morphologically visible structural change in the chromosomal complement involving the gain, loss, or relocation of chromosomal segments. Descriptively, all such structural changes include deletions or deficiencies, duplications, inversions, and other translocations. See *mutation*.

Chromosomal translocation See *translocation*.

Chromosome Microscopically observable nucleoprotein bodies, darkly stained with basic dyes, observable in the cell during division. The chromosomes carry the genes of the cell arranged in a specific linear order. They are autoreduplicating structures whose number per cell, shape, and organization are species-specific characteristics. Chemically, a chromosome is a DNA-histone thread residing in the nucleus of a cell.

Chromosome complement Group of chromosomes derived from a particular gametic (i.e. haploid), or zygotic or somatic (i.e. diploid), nucleus.

Chromosome disjunction Assortment of chromosomes during the first meiotic division.

Chromosome fusion Union of two or more chromosomes by means of chromosome structural changes to form a single chromosome. The change in the chromosome material which occurs with chromosome fusion is one form of chromosomal mutation. Among chromosomes with localized centromeres, such unions are stable if the fusion takes place in the centromere region.

Chromosome map Graphic, linear representation of a chromosome in which the genes belonging to particular linkage groups (i.e. genetic markers) are plotted according to their relative distances. There are two kinds of chromosome maps, genetic maps, and cytological maps. See *genetic maps, cytological maps*.

Cistron The smallest functional unit of genetic material, that is, that which determines the amino acid sequence of one polypeptide chain. The gene as usually conceived is synonymous with the cistron.

Clone Cells all derived from a single cell by repeated mitosis and all having the same genetic constitution.

Cohort A group of individuals, or vital statistics about them, having a statistical factor in common in a demographic study. A group of related families.

Complete Penetrance Situation in which a dominant gene always produces a particular phenotypic effect, or a recessive gene in the homozygous state always produces a detectable effect. See *penetrance*.

Concordance Occurrence of a particular trait in both members of a pair of individuals. Agreement between the two members of a pair regarding presence or absence of a particular character. Usually used in reference to twins to indicate that both have a given trait. Discordance is the antithetical term.

Congenital Present at birth. The etiology of a congenital condition may or may not involve genetic factors as well as other variables relevant to embryonic and fetal development and the birth process.

Consanguineous union Union between related individuals, i.e. individuals having one or more ancestors in common.

Consanguinity Involving genetic relationship (i.e. at least one common ancestor in the preceding few generations).

Constant A quantity whose value is assumed fixed during an investigation.

Constitution The basic structure of an individual, embracing various aspects of the phenotype which are comparatively constant over time and, in the process of maturation, usually become increasingly more resistant to external influences. Constitutional research deals preferably with (a) traits characterized by a high degree of interindividual variability, and (b) the correlations that those traits show among themselves and with other variables (e.g. actual disturbances of somatic or mental processes). Thus, constitutional variables can serve as markers for the identification of individuals from a biological point of view and as predictors of their spontaneous development and their reactions toward environmental stimuli.

Continuous distribution Manifestation, by a collection of data, of a continuous spectrum of values. Graphic distribution of the data in a population is monomodal, i.e. includes only a single mode in the distribution curve. See *discontinuous distribution*.

Correlation coefficient One of a number of measures of correlation, usually assuming values from $+1$ to -1.

Coupling In a double heterozygote, linkage is said to be in the coupling phase when the mutant alleles of interest at the two loci are on the same chromosome. See *repulsion*, the antithetical term.

Crisscross inheritance Mode of inheritance involving transmission of a gene which occurs on the X chromosome (i.e. a sex-linked gene).

Crossing over Reciprocal exchange of segments at corresponding positions along pairs of homologous linkage units (i.e. chromatids, chromosomes) by symmetrical breakage and crosswise rejoining. Crossing over results in an exchange of genes and therefore produces combinations which are different from the combinations characteristic of the parents.

Cytogenetics Biological discipline concerned with chromosomes and their implications in genetics, i.e. the behavior of the chromosomes during mitosis and meiosis, their origin, and their relation to the transmission and recombination of genes.

Cytokinesis Cytoplasmic division and other changes exclusive of mitosis or meiosis. Cell division involves both nuclear division (i.e. mitosis or meiosis) and cytokinesis.

Cytological map Graphic, linear representation of a chromosome in which the genes belonging to particular linkage groups (i.e. genetic markers) are plotted according to their relative distances. The distances are determined on the basis of cytological observations obtained with the aid of chromosome mutations (i.e. deletions, inversions, and other translocations) experimentally induced by mutagenic agents. Another method of preparing cytological maps is based upon microscopic analysis of the giant chromosomes of *Diptera* in which the position and extent of the chromosomal structural changes may be determined morphologically.

Cytology Study of the structure and function of the cell.

Cytoplasm Protoplasm of a cell outside the nucleus, i.e. all living parts of the cell except the nucleus.

Cytoplasmic inheritance Hereditary transmission dependent on the cytoplasm or structures in the cytoplasm rather than the nuclear genes, i.e. extranuclear inheritance.

Deficiency, chromosomal See *deletion, chromosomal*.

Deletion, chromosomal Actual loss of a portion or portions of one or more chromosomes and the included genes.

Deoxyribonucleic acid See *DNA*.

Dermatoglyphics Patterns of the ridged skin of palms, fingers, soles, and toes; the study and systematic classification of such patterns.

Desoxyribonucleic acid See *DNA*.

Developmental genetics Branch of genetics primarily concerned with transmission of genetic variables which control or modulate developmental processes.

Dichotomy Sharp distinction.

Dimension Within the framework of factor-analytic models, this term refers to the geometric representation of a factor as a coordinate in a (formal) multivariate space. Factor and dimension are often used synonymously when describing fundamental tendencies to which the covariations within a set of variables are attributed. In this context, they have the meaning of complex traits. See *factor*.

Diploid Containing two homologous sets of chromosomes, one of paternal origin and the other of maternal origin, in which each type of chromosome except the sex chromosomes of the heterogametic sex (i.e. the male sex in mammals) is normally represented in duplicate. (Symbol: 2N.)

Discontinuous distribution Manifestation by a collection of data into discontinuous groups of values. Graphic distribution of the data is bimodal or multimodal.

Discordance Disagreement between the two members of a pair regarding the presence or absence of a particular character.

Disjunction See *chromosome disjunction*.

Dizygotic Refers to fraternal twins derived by fertilization of two different ova by different spermatozoa.

DNA Abbreviation for deoxyribonucleic acid (desoxyribonucleic acid), one of the key chemical compounds governing life functions. DNA is found in the cell nucleus and is an essential constituent of the genes. Structurally, DNA consists of complementary but oppositely polarized polynucleotide chains which are hydrogen-bonded together and wound helically around a common axis in a double helix. Genetic information is apparently encoded in the sequence of the nucleotides. Replication is believed to be accomplished by separation of the complementary nucleotide chains with each single chain's retaining its integrity while serving as a template against which new complementary chains are built for the four nucleotide components (i.e. adenine and thymine, guanine and cytosine). The antiparallel strands form a right-handed helix which undergoes one complete revolution with each ten nucleotide pairs. DNA molecules are the largest biologically active molecules known, having molecular weights greater than 1×10^8. In the translation of the genetic information contained in DNA, it is commonly assumed that RNA molecules are synthesized, which then serve as templates to form specific proteins.

Dominance (a) In genetics, dominance is a manifestation of a gene's ability to be expressed in the phenotype of an individual, even though that (dominant) gene is paired with a different gene. A dominant gene is manifested in the heterozygote (i.e. the individual whose genotype includes the dominant gene and another allele); a recessive gene is not manifested in the heterozygote, but only in the homozygote (i.e. the individual whose genotype includes a pair of the same recessive genes). Dominance and recessiveness ae not properties of the genes per se but result from action of the particular gene within the total reaction system of the particular genotype. See *recessive*. (b) In psychiatry, dominance refers to an individual's disposition to play a prominent or controlling role in interactions with others.

Double-blind Study Research procedure in which both the investigator and the subject are unable to discriminate between the experimental factor and the control factor at the time of administering the test to obtain data for evaluating differences in effects of two factors.

Double helix See *DNA*.

Duplication, chromosomal Chromosomal structural change which involves presence of a group of genes (i.e. of a chromosome or chromosomal segment) more than once in the haploid genotype or more than twice in the diploid genotype. A chromosomal duplication may result from a chromosomal structural change (i.e. mutation), resulting in the doubling of some chromosomal material in a cell and in cells derived from the cell.

Dyad Pair of chromatids connected at the centromere, derived from one chromosome in the first meiotic division.

Dysgenic Tending to be harmful to the hereditary qualities of a species.

DZ See *dizygotic*.

Ectomorph An individual characterized by predominance of the structures developed from the ectodermal layer of the embryo, i.e. skin, nerves, sense organs, brain; of a light or asthenic type of body build (compared to endomorph, mesomorph).

EEG Electroencephalogram. A graphic recording of minute electrical impulses arising from activity of cells in the brain.

Electroencephalogram See *EEG*.

Electrophoresis The migration of charged particles in an electric field. Sometimes used in separating a compound into a number of fractions.

Empiric risk Prediction of the probability that a particular abnormality will occur in a particular individual. The risk that an individual who is characterized by a particular genetic relationship to an affected individual will manifest the trait (i.e. will be affected). Empiric risk figures are utilized in human genetics for traits which apparently involve genetic predisposition but for which no definitive pattern of genetic transmission can be demonstrated. Empiric risk figures are based upon recurrence histories in families characterized by occurrence of affected members, and they are predictions of the probabilities that the abnormality will recur in the families in which it has already occurred.

Encephalitis General term used to designate a diffuse inflammation of the brain. The condition may be acute or chronic and may be caused by a variety of agents, such as viruses, bacteria, spirochetes, fungi, protozoa, and chemicals (e.g. lead).

Encephalopathy Broad term disignating any of the diffuse degenerative diseases of the brain (e.g. Alzheimer's disease, Pick's disease, encephalitis, Parkinsonism, Huntington's chorea, organic brain syndromes).

Endogamy Mating within the group. Synonym: *inbreeding*.

Endomorph Individual distinguished by predominance of the structures developed from the endodermal layer of the embryo (e.g. the internal organs); of a pyknic type of body build (compared to ectomorph, mesomorph).

Epinephrine Adrenalin. One of the catecholamines secreted by the adrenal gland and by fibers of the sympathetic nervous system. It is responsible for many of the physical manifestations of fear and anxiety. See *catecholamine*.

Epistasis A form of interaction between genes situated at different loci where one gene masks or prevents the expression of another gene at a different locus. (Dominance is the term used when one gene masks one of its own alleles; that is, another gene at the same locus.)

Etiology Causation, particularly with reference to disorder or disease.

Eugenic Descriptive term relevant to measures or trends which improve the genetic endowment of a human population, as opposed to dysgenic.

Exogenous Produced from without; originating from or resulting from external causes. Having a cause external to the body; not primarily resulting from structural or functional failure of the body.

Expressivity Degree of phenotypic expression or severity or manifestation of a penetrant gene or genotype. Expressivity refers to the degree and type of expression of a trait or traits controlled by a particular penetrant gene which may produce different degrees and types of expression in different individuals. Thus, expressivity of a particular penetrant gene in a particular individual may be slight or severe and may be described in qualitative and/or quantitative terms. Some genes show variable expressivities in different individuals, and some genes are relatively consistent in their manifestations.

Extrachromosomal Descriptive term to

describe structures that are not part of the chromosomes.

Extraversion State in which attention and energies are largely directed outward from the self, as opposed to inward toward the self as in introversion. See *introversion*.

F_1 First filial generation; the first generation of descent from a given mating.

Factor A hypothetical variable presumed to account for the variance which a series of observed variables have in common. Factors may be derived by formalized procedures of factor analysis from a table of intercorrelations among the variables under investigation. General factors are related to all these variables, group factors to some of them. The degree of affinity between a factor and an observed variable is expressed as a correlation coefficient. This coefficient is designated the factor loading of the variable or the weight of the factor for this variable. If loadings on one factor bear opposite signs (indicating that the respective variables seem influenced by this factor in the opposite direction), the factor is termed "bipolar." See *dimension*

Factor analysis Statistical technique for evaluations of intercorrelations between different items in the data. Factor analysis identifies the items which are correlated with each other and those which are independent of each other. The various items of the intercorrelation matrix are examined for indications of occurrence of different clusters of items. The identification of different clusters may facilitate interpretations of data as measuring several different and unrelated variables, with each of the different variables reflected by one of the clusters.

Familial Refers to the occurrence of two or more affected individuals in a family group or pedigree.

Fertility Reproductive potential, as measured by the quantity or percentage of developing eggs or of fertile matings or offspring produced.

Fixation (a) In *psychiatry,* arrest of psychosexual maturation. (b) In *biology,* procedure of killing cells and preventing subsequent decay with minimal distortion of structure, as a step in making permanent preparation for microscopic study.

Folie à deux Condition in which two closely related persons, usually in the same family, share the same delusion or delusions (French).

Forme fruste An incomplete, partial, or mild form of trait or syndrome (French).

Fraternal twins Dizygotic, or two-egg, twins.

Functional In *medicine,* changes in the way an organ system operates that are not attributed to known structural alterations. Psychogenic disorders are functional in that their symptoms are not based on any detectable alterations in the structure of the brain, but not all functional disorders of the psyche are of emotional origin.

Galvanic skin response (GSR) A measure of changes in the electrical conductance of the skin of an individual's response to emotion-arousing stimuli.

Gamete Mature reproductive cell. The gametes are produced from gametocytes during the process of gametogenesis, during which a diploid cell undergoes meiosis and produces haploid gametes. The female's gametes are ova and the male's gametes are spermatozoa.

Gaussian distribution See *normal distribution*.

Gene Ultimate unit of genetic material, not further subdivisible by either genetic recombination or chromosomal structural rearrangement. A gene is composed of DNA and may be defined as a specific sequence of nucleotides. Genes are arranged in characteristic linear order on chromosomes within cells. The total complement of genes makes up the genotype. Genes in diploid organisms occur as pairs of alleles. Gene action is manifested by control of the specificity and rate of biosynthetic processes. The gene is the fundamental unit of biological heredity.

Gene expression Phenotypic manifestation of a gene.

Gene flow Spread of genes from one breeding population to others through the dispersal of gametes. Exchange of genetic factors between populations.

Gene frequency Proportion of one particular type of genetic allele among the total of all alleles at the particular genetic locus in a breeding population or the probability of finding the particular gene when a gene is randomly chosen from the population.

Gene interaction Interaction between allelic or nonallelic genes of the same genotype in the production of particular phenotypic characters.

Gene locus The point or position occupied by a gene in the chromosome. Different genes which may occupy the same locus (on homologous chromosomes) are alleles of each other.

Gene, major See *major gene*.

Gene mutation Heritable change within the limits of a single gene; point mutation, as opposed to changes in chromosome structure or number. As a result of gene mutation, there are alternative states of a gene, i.e. alleles. Since the genetic information carried by a gene is encoded in a specific nucleotide sequence of DNA and is replicated by forming complementary nucleotide chains, a gene mutation may result from any change in the normal sequence of entire nucleotides or their component bases.

Gene pool Totality of genes of a given *population* existing at a given time.

Genetic (a) In *biology*, pertaining to genes or to traits associated with genes transmitted from one generation to the next via the chromosomal material. (b) In *psychiatry* and *psychology*, pertaining to the historical development of an individual's psychological attributes and/or disorders.

Genetic drift Change in gene frequency in a population which may or may not be directed. Irregular and random fluctuation in gene frequency in a population from one generation to another, most likely to occur in a small breeding population, may lead to random fixation of one allele and extinction of another without regard to their adaptive values.

Genetic homeostatis Tendency of a population to equilibrate its genetic composition and to resist sudden change.

Genetic isolate Breeding population which does not exchange genes (i.e. does not interbreed) with any other such group.

Genetic map Graphic, linear representation of a chromosome in which the genes belonging to particular linkage groups (i.e. genetic markers) are plotted according to their relative distances. The relative distances are based upon the frequencies of intergenic crossing-over between any two linked markers. The accuracy of genetic mapping is dependent upon the precision with which the crossing-over frequencies may be estimated from the recombination frequencies. See *chromosome map, cytological map*.

Genetic marker Gentically controlled phenotypic difference used in genetic analysis.

Genetics Scientific study of biological heredity, transmission of genetic material from one generation to the next via the gametes. See *genetic*.

Genocopy Refers to an individual who is phenotypically like another individual who is genotypically different—that is, the same phenotype is produced by a different genotype. Different combinations of different genetic and nongenetic factors may produce the same effect or character.

Genome The total genetic endowment.

Genotype Sum total of the genetic material in a particular individual, i.e. the total genetic constitution, either at one specific locus or more generally.

Germ line Pertaining to the cells from which gametes are derived. The cells of the germ line, unlike somatic cells, bridge the gaps between generations.

Germ Plasm Hereditary material transmitted to offspring through the germ cells or gametes.

GSR See *galvanic skin response*.

Gyandromorph Mosaic individual whose tissues include cells of both male genotype and female genotype.

Habitus The external aspect of constitution comprising both the somatic and mental spheres: physical habitus (physique) and habitual forms of behavior (personality).

Haploid Descriptive term referring to presence of only a single set of chromosomes, such as occurs in a gamete produced by a single diploid individual. (Symbol 1 N.)

Hardy-Weinberg Law If two alleles (A and a) occur in a randomly mating population with the frequency of p and q, respectively, where $p + q = 1$, then the expected proportions of the three genotypes $AA = p^2$, $Aa = 2pq$, and $aa = q^2$, remain constant from one generation to the next. Mutation, selection, migration, and genetic drift can disturb the Hardy-Weinberg equilibrium.

Hemizygous The genetic state of the heterogametic sex (i.e. in mammals, the male) with regard to the X chromosome.

Hereditary, Heritable, and Heredofamilial Essentially synonymous terms for genetically transmissible traits.

Heritability Proportion of the total phenotypic variance that is genetic for a particular character in a particular population at a single generation under one particular set of conditions. Heritability is commonly expressed as a percentage and decreases with an increasing environmental component of variance for the character under observation. Mathematically, it is the genetic variance divided by the total phenotypic variance. The heritability of a particular disease or disorder may be defined as the extent to which the variation in individual risk of acquiring the disease or disorder is due to genetic differences.

Heterogametic Capable of producing two kinds of gametes. In man the male is the heterogametic sex since X-bearing and Y-bearing gametes are formed.

Heterozygote Diploid individual who has inherited two different alleles at one or more loci. An individual with different genetic factors (i.e. alleles) at the homologous (i.e. corresponding) loci of a pair of homologous (i.e. which pair during meiosis) chromosomes.

Holandric Descriptive term referring to genes occurring on the Y chromosome, linked to the Y chromosome.

Homeostasis Tendency of a system to maintain a dynamic equilibrium and, in case of disturbance, to restore the equilibrium by its own regulatory mechanisms. See *genetic homeostasis*.

Homogametic Descriptive term regarding that sex which produces only one basic type of gamete in contrast to the heterogametic sex. In humans, the female is homogametic and normally produces only ova which are characterized by one set of autosomes and one X chromosome.

Homologous Corresponding in structure, position. Descriptive term regarding chromosomes or chromosome segments which are identical with respect to their constituent genetic loci (i.e. the same loci in the same sequence) and their visible structure. Homologous chromosomes are morphologically paired during the process of meiosis. The descriptive term may also be used to refer to the specificity relationship between an antigen and/or its hapten and an antibody.

Homozygote Diploid individual who has inherited identical, rather than different, alleles in the corresponding loci of a pair of homologous chromosomes.

Hybrid A heterozygote, i.e. a diploid individual who has inherited two different alleles at one or at each of more than one locus. In common usage, an individual produced by two parents who are genetically unlike.

Hypothesis The initial assumption of any investigation; something assumed to be true for the purpose of tracing the consequences of the assumption. If the deduced consequences are verified, then support for the hypothesis is strengthened. In modern scientific philosophy, all synthetic assertions of science are considered hypotheses in the broadest sense of the term, although in ordinary usage the most firmly established hypotheses are called "facts."

Idiogram Diagrammatic representation of the karyotype or chromosomal complement of an individual. Idiogram construction is based upon measurements of total chromosome length, arm-length ratio, and centromere position for each chromosome during metaphase of mitosis. The chromosomes in an idiogram are characteristically arranged in pairs in descending order of size. See *karyotype*.

Immunogenetics Genetics of antigens, antibodies, and their reactions.

Imprinting The determination of future behavior by exposure to a specific stimulus at a critical age.

Inborn error or metabolism Genetically determined biochemical disorder resulting in an enzyme defect that produces a metabolic block which may have pathological consequences.

Inbred Result of matings between genetic relatives.

Incidence Number of cases of a particular trait or disorder in a given population over a set period of time.

Independent assortment, Mendel's Principle of Members of different allele pairs assort independently of each other when germ cells are randomly formed, provided the genes in question are unlinked (i.e. not located on the same chromosome).

Indeterminancy In factor analysis, the problem arising because current factor analytic methods cannot uniquely determine the exact position of factors in a multidimensional space.

Index Case See *proband*.

Intelligence The capacity for problem solving on the basis of reasoning.

Intelligence quotient A score on a test designed to measure intelligence (IQ).

Interchange, chromosomal Exchange of segments between nonhomologous chromosomes resulting in translocations.

Interphase That part of the cell cycle during which metabolism and synthesis occur without visible evidence of division, and the nucleus of the cell is not visibly engaged in mitosis or meiosis.

Intranuclear Descriptive term to describe structures or processes which occur inside the nucleus.

Introversion Preoccupation with oneself and accompanying reduction of interest in the outside world; state in which attention and energies are largely directed inward toward the self, as opposed to outward from the self as in extraversion. See *extraversion*.

Inversion, chromosomal Chromosomal structural change characterized by reversal of a chromosome segment or chromatid segment and the gene sequence contained therein (i.e. relative to the standard arrangement of the linkage group involved). Reinsertion of a chromosomal segment into its original position but in reversed sequence.

IQ See *intelligence quotient*.

Irregular dominance A condition in which a generally dominant gene is of variable expressivity and/or variable frequency of penetrance.

Isoalleles Alleles that produce such slight phenotypic differences that special techniques are required to reveal their presence.

Isochromosome An anomalous chromosome, with median centromere and identical (duplicate) arms, which arose through transverse splitting of the centromere rather than longitudinal splitting between two chromatids. The occurrence of such anomalous division by a chromatid pair produces two anomalous daughter chromosomes, each characterized by duplication of one arm and deletion of the other arm.

Isograft Graft between genetically identical individuals; in man, monozygotic twins.

Isolate Segment of a population within which assortative mating occurs.

Karyotype The term is used in two ways: (a) the somatic chromosomal complement of an individual or species; (b) photomicrographs of the metaphase chromosomes arranged in a standard sequence—an idiogram. See *idiogram*.

Kindred Family in the larger sense. The term family is usually restricted to the nuclear family (parents and children).

Kinetochore See *centromere*.

Klinefelter's syndrome A condition in man that characteristically produces sterile males with small testes lacking spermatozoa, is sometimes associated with mental retardation, and is due to occurrence of two or more X chromosomes in addition to one or more Y chromosome(s).

Late-replicating X chromosome X chromosome which completes its replication later than the functional X chromosome and the autosomes. In the mammalian somatic cell nucleus, all X chromosomes except one apparently coil up and condense and apparently do not function in transcription. The condensed mass of a late-replicating X chromosome in a mammalian somatic cell nucleus is believed to be the source of each sex-chromatin (i.e. Barr) body.

Leptomorphic Leptosomic. Asthenic, ectomorphic.

Leptosome An ectomorphic individual.

Lethal genes Alleles which cause early death of the affected individual, at the stage of the embryo, the fetus, or the infant. Such genes can never be passed on to the offspring unless they occasionally fail to penetrate. Most cases with a lethal trait are therefore due to new mutations.

Linkage Occurrence of genetic loci on the same chromosome. Loci which occur on the X chromosome are termed X-linked, or sex-linked; and loci which occur on the Y chromosome are termed Y-linked, or holandric. Linkage refers to the association of nonallelic genes which reside on the same chromosome.

Load, genetic Deleterious recessive genes

carried in heterozygous state. Five genes which in the homozygote each reduces the fitness by 20 percent make one lethal equivalent. Genetic load is measured by observing the effects of inbreeding.

Locus The site in a chromosome occupied by a member of a specific set of alleles.

Lyon hypothesis The genetic inactivation of all X chromosomes in excess of one on a random basis in each of the cells at an early stage of embryogenesis.

Major gene Gene which is individually associated with pronounced phenotypic effects. Major genes control production of discontinuous or qualitative characters in contrast to "minor genes" or "polygenes" which manifest individually small effects. Major genes segregate clearly and are subject to Mendelian analyses. The dichotomy of genes into major and minor categories represents an arbitrary division in a continuous spectrum of gene actions and interactions. See *minor gene*.

Manic-depressive disorder Major affective disorder characterized by severe mood swings and a tendency to remission and recurrence.

MAOI Monoamine oxidase inhibitor. Refers to a group of antidepressant drugs that appear to ameliorate the emotional state by inhibiting certain brain enzymes and raising the level of serotonin.

Map, genetic See *genetic map*.

Matrix A rectangular array of quantities or other symbols convenient for representing relations between each pair of an aggregate. Strictly speaking, such an array can be treated as a generalized quantity, that is, subjected to certain rules of calculation.

MAVA Multiple Abstract Variance Analysis. This hypothetical method designates four sources of individual differences in any trait: between-family environmental differences, between-family hereditary differences, within-family environmental differences, and within-family hereditary differences. The method was developed to facilitate determination of nature-nurture ratios for primary personality factors in objective tests.

Meiosis Process involving two successive divisions of the nuclear chromosomes preceding the formation of mature gametes. Homologous chromosomes pair, replicate once only, and undergo assortment so that each of the four meiotic products resulting from one complete meiosis receives one representative of each chromosome pair. Thus, the diploid chromosome number is meiotically reduced to the haploid number characteristic for the gametes or haploid phase of the life cycle.

Mendelian character Character which, in biological inheritance or genetic transmission, follows Mendel's laws.

Mendelism Particular inheritance of chromosomal genes according to Mendel's laws.

Mendel's laws (a) *The law of segregation:* The allelic factors of a pair are segregated, i.e. the factors separate into two different gametes (and, thence, into different offspring). Segregation occurs in meiosis for the two members of each pair of alleles possessed by the diploid parental organism. (b) *The law of independent assortment:* The members of different pairs of allelic factors assort independently, i.e. the members of different pairs of alleles are assorted into gametes during gametogenesis independently of other pairs of alleles which reside on different chromosomes; the subsequent pairing of alleles in combinations of male and female gametes is at random.

Mesomorphic Characterized by predominance of the structures developed from the mesodermal layer of the embryo, i.e.

muscle, bone, connective tissue; of the muscular or athletic type of body-build.

Metacentric Refers to a chromosome in which the centromere occurs in an approximately median position which leads to a "V" or "J" shape in the metaphase appearance of the chromosome. A metacentric chromosome is one in which the centromere is located at or close to the median point or midpoint of the chromosome.

Minnesota multiphasic personality inventory See *MMPI*.

Monoamine oxidase inhibitor See *MAOI*.

Monogenic Monomeric. Refers to character differences genetically controlled by the alleles of one particular genetic locus, i.e. as opposed to polygenic control exerted by two, three, or many nonallelic genes. The genetic transmission of a factor in monogenic inheritance involves alleles which occur at only one genetic locus, and phenotypic variation is discontinuous.

Monohybrid Refers to a cross between parents differing with respect to a single specified pair of allelic genes.

Monomeric See *monogenic*.

Monosomy The abnormal condition in which one chromosome of one pair is missing, i.e. deletion of a whole chromosome from the normal chromosome number. The presence, in an otherwise normally diploid complement, of only one member of a particular chromosome pair.

Monotonic transformation A transformation of data from one scale of measurement to another without altering the rank order of the data.

Monozygotic Refers to twins derived from one egg (identical twins).

Morbidity Relative incidence of a disease or disorder.

Morbid Risk The risk that an individual, influenced by specific variables, will develop a particular disease or disorder.

Mosaic Organism consisting of two or more genetically different cell types, involving either the same tissue type (i.e. chimera) or different tissues. The condition may occur as a result of a somatic mutation in a cell during mitosis, or from fusion (i.e. grafting) of cells from one individual upon another, or from a double fertilization (i.e. fertilization of two different pronuclei within the same oocyte or within two attached oocytes or within an oocyte and a polar body which are attached, by two different spermatozoa). If the mosaicism involves the same tissue type, then the mosaic is a chimera. If the mosaicism does not involve the same tissue type, then the mosaic is not a chimera. See *chimera*.

Mosaicism The occurrence of cells of two or more genotypes in the same individual.

Multifactorial inheritance See *polygenic inheritance*.

Multigenic inheritance See *polygenic inheritance*.

Multiple allelic series A series comprising three or more possible alleles at the same genic locus.

Multiple genes Two or more nonallelic genes (i.e. involving two or more different loci) with similar or complementary cumulative effects on a single character. See *polygenic inheritance*.

Multiple-factor hypothesis Hypothesis according to which the inheritance of a particular quantitative character involves an absence of clear-cut segregation into readily recognizable classes showing typical Mendelian ratios. The multiple factor hypothesis assumes that quantitative characters result from the cumulative action of multiple sets of independently transmitted genes, each of which produces only a small effect. See *polygenic inheritance.*

Mutagen Physical or chemical agent which significantly increases mutational events and, thus, increases mutation rates above the spontaneous background level. Mutagens include ionizing radiations, ultraviolet rays, base analogues, and alkylating agents.

Mutation Any detectable and heritable change in the genetic material not caused by segregation or genetic recombination which is transmitted to daughter cells and even to succeeding generations. Mutations in the germ line of sexually reproducing organisms may be transmitted by the gametes to the next generation. A mutation may be any change affecting the chemical or physical constitution, mutability, replication, phenotypic function, or recombination of one or more deoxyribonucleotides. Nucleotides may be added, deleted, inverted, or transposed to new positions with and without inversion. A chromosomal mutation involves duplication, deletion, or alteration of chromosomal material which is visible with currently available cytological methods. A gene mutation or point mutation is a mutation which is not associated with any visible change in the chromosomal material.

MZ Monozygotic.

N Number of subjects.

Neonatal Relating to or affecting.

Nondisjunction Abnormality of nuclear division in which a pair of newly divided chromosomes fails to separate with the result that the daughter cells contain unequal numbers of chromosomes. Nondisjunction can occur during a meiotic or mitotic division; results in one daughter cell receiving both of the unseparated chromosomes, and one daughter cell receiving none of the chromosomes involved.

Nonparametric statistics Methods of statistical analysis which are free from assumption, regarding the shape of the underlying probability distribution.

Norepinephrine Norepinephrine is the chief neurohumor that mediates adrenergic impulses and acts mainly at adrenergic neuronal terminals. Although small amounts of this hormone may diffuse from the nerve endings into the general circulation and some is released from the adrenal medulla, this hormone is less well suited than epinephrine for promotiong emergency adjustments. Norepinephrine is currently regarded as functioning in the normal organism in the maintenance of vascular tone and blood pressure. It is probably secreted continuously by the sympathetic endings and acts locally to effect rapid adjustments essential for the maintenance of blood pressure. See *catecholamine.*

Norm of reaction Range of phenotypic reactions of a particular genotype, i.e. the variety of phenotypes which may be produced by a particular genotype interacting with different nongenetic (i.e. environmental) influences.

Normal distribution As commonly used, a unit normal distribution having the following characteristics: (a) symmetry about the mean; (b) a mean of zero and a standard deviation of one; (c) approximately 34 percent of the area under the distribution falling between the mean and 1 σ on each side of the mean so that $\pm 1\sigma$ involves about 68 percent, approximately 14 percent of the area falling between 1 σ and 2 σ on each side of the mean so that $\pm 2\sigma$ involves about 96 percent, and approximately 2 percent of the area falling between 2 σ and 3 σ on each side of the mean so that deviations exceeding $\pm 2\sigma$ involve about 2 percent on each side for a total of about 4 percent of the area in the distribution curve.

NREM sleep The term given to the longer period of sleep that begins as the subject passes from wakefulness into a light sleep with *no rapid eye movements* (NREM's). REM sleep interrupts NREM sleep about once in every ninety minutes and lasts for about twenty minutes. See *REM sleep*.

Nucleic acid Nucleotide polymer composed of subunits which are either deoxyribonucleotides (i.e. in deoxyribonucleic acid, DNA) or ribonucleotides (i.e. in ribonucleic acid, RNA).

Nucleoprotein Compound of nucleic acid and protein. Either one of two main classes of basic proteins are found combined with DNA: one of low molecular weight, protamine; and one of high molecular weight, histone. The basic amino acids of these proteins neutralize the phosphoric acid residues of the DNA.

Nucleoside Purine or pyrimidine base attached to ribose or deoxyribose.

Nucleotide Nucleoside—phosphoric acid complex.

Nucleus Spheroidal structure present in most cells which contains the chromosomes.

Null hypothesis The standard hypothesis used in testing the statistical significance of the differences between the means of samples drawn from two populations. The null hypothesis states that there is no difference between the populations from which the samples are drawn. One then determines the probability that one will find a difference equal to or greater than the one actually observed. If this probability is 0.05 or less, the null hypothesis may be rejected and the difference said to be statistically significant.

One-gene-one-enzyme hypothesis The hypothesis based on studies in biochemical genetics that each gene controls the synthesis or the activity of only a single protein with catalytic activity (i.e. enzyme). A large class of genes exist in which one gene controls the synthesis or activity of a single enzyme. There are, however, individual enzymes and other proteins for which the respective syntheses are controlled by more than one gene. Thus, the one-gene-one-enzyme hypothesis has been appended to include the one-gene-one-polypeptide hypothesis, according to which each gene controls the synthesis of a single polypeptide. The polypeptide may function independently or as a subunit of a more complex protein.

One-gene-one-polypeptide hypothesis The hypothesis that a large class of genes exist in which each gene controls the synthesis of a single polypeptide. See *one-gene-one-enzyme hypothesis*.

Ontogency The biological development or course of development of an individual (i.e. distinguished from phylogeny).

Oogenesis Development of the female germ cell (i.e. egg cell or ovum) of an animal which takes place in the gonad; includes meiosis of the oocyte, vitellogenesis, and formation of egg membranes. An inclusive term covering both female meiosis and ovagenesis.

Oxidation Gain of oxygen or loss of hydrogen or loss of electrons by a compound.

Oxidative phosphorylation Enzymatic phosphorylation of ADP to ATP, which is coupled to the electron transport chain, thus transforming respiratory energy into phosphate-bond energy.

Panmictic population Interbreeding population of individuals who mate at random, i.e. each individual is equally likely to mate with any individual of the opposite sex.

Panmixis Random mating. See *panmictic population*.

Paradigm Example, pattern.

Parameter Any measurable characteristic of a population.

Parameter estimation In factor analysis, determination of the values of the diagonal elements in the correlation matrix.

Parametric statistics Methods of statistical analysis which assume that the data follow a defined probability distribution and the results of the calculations are valid only if the data are so distributed.

Parkinson's syndrome Parkinsonism, Parkinsonian syndrome. Muscular rigidity, immobile facies, and tremor which tends to disappear on volitional movement; associated with lesions in the *globus pallidus*.

Parsimony A principle in the philosophy of science according to which one strives to keep the number of concepts and assumptions at a minimum. Hence, of two theories explaining a phenomenon equally well, the simpler one is to be preferred. (The simpler one does not necessarily mean the one which is easier to understand in terms of previously accepted concepts but, rather, the one which uses fewer independent concepts and assumptions.)

Partial trisomy The presence in triplicate of a chromosome segment in an otherwise diploid complement.

Pathognomonic Refers to a symptom or group of symptoms that are specifically diagnostic or typical of a particular disease entity.

Pedigree Table, chart, or diagram setting forth the ancestral or genealogical history of an individual. Diagram of a family tree to show the occurence of one or more traits in different members of a family.

Penetrance When a gene or an allelic pair shows an effect in the phenotype, it is said to *penetrate*. The penetrance of a gene is the number of individuals showing the phenotypic trait expressed as a percentage of all those possessing the gene. When a dominant gene fails to produce any effect in an individual, there is said to be a failure of penetrance. Failure of penetrance is the extreme degree of reduced expressivity of a gene. A dominant gene which fails to penetrate from time to time is an irregular dominant gene. Penetrance is the term used to indicate the frequency with which a particular gene or combination of genes is/are manifested in the phenotypes of the carriers under a set of specified environmental conditions. Expressivity of a particular gene may depend upon both the (total) genotype and the environment.

Personality The mental structure of an individual composed of affective and conative (temperament) as well as cognitive (intelligence) characteristics. Some authors restrict the term to the affective and conative structure alone, others extend it to include the individual contents of interests and attitudes and even the somatic constitution, particularly body build (physique) and physiological reactivity.

Peptide A compound formed by the union of two or more amino acids. When two amino acids unite, the result is a dipeptide; when three, a tripeptide; when more than three, a polypeptide.

Pharmacogenetics The area of biochemical genetics dealing with genetically controlled variations in response to drugs.

Phenothiazine derivatives Group of psychotropic drugs that chemically have in common the phenothiazine nucleus, but differ from one another through variations in chemical structure. The phenothiazines are known as major tranquilizers, possessing marked antipsychotic properties.

Phenotype The observable properties of an organism, structural and functional, produced by interaction of the organism's genotype and the particular environmental influences involved. The term phenotype may be applied either to the totality of expressions of the genotype in a particular environmental context or to only particular characters or traits.

Phenotypic variance Total variance observed in a particular character.

Phenylketonuria (PKU) Biochemically, the main feature of phenylketonuria is the high concentration of phenylalanine in the blood, the cerebrospinal fluid, and the urine of affected individuals. This is associated with the excretion, in the urine, of phenylpyruvic acid, phenyllactic acid, and phenylacetylglutamine in grossly abnormal amounts, and also the abnormal excretion of certain hydroxyphenolic acids and indoles. The metabolic lesion in phenylketonuria involves an enzymatic defect which affects conversion of phenylalanine to tyrosine. The block in intermediary metabolism of phenylalanine, derived from protein in the food or from breakdown of tissue proteins, causes its accumulation in the blood and cerebrospinal fluid and excretion of abnormal amounts in the urine. Phenylpyruvic acid, phenyllactic acid, and phenylacetylglutamine are formed in excess, but since renal thresholds for these substances are very low they are excreted in large amounts in the urine and occur hardly at all in the blood. The enzymatic abnormality is transmitted by a pair of autosomal recessive genetic alleles, and heterozygous carriers of the recessive gene for phenylketonuria may be distinguished biochemically.

Phylogenetic Pertaining to the evolutionary history of the species.

Phylogeny The evolutionary history of a genetically related group of organisms (i.e. as distinguished from ontogeny, the development of the individual organism).

Physique (physical habitus) In the restricted sense used in this book, the term refers mainly to body build. In a broader sense, it may include the entire physical constitution (body mass and composition, inner structures, developmental and functional characteristics of the individual organism).

Phytohemagglutinin A mucoprotein derived from the string bean, *Phaseolus vulgaris,* which stimulates mammalian cells to undergo mitosis in culture.

PKU See *phenylketonuria.*

Pleiotrophy Production, by one particular gene, of multiple phenotypic effects.

Plethysmograph An instrument for recording variations in the volume of parts.

Point mutation Gene or intragenic mutation, i.e. mutation associated with genetically transmitted change in phenotype, but not observable cytogenetically as a gross chromosomal aberration. See *gene mutation.*

Poisson distribution A function that assigns probabilities to the sequence of outcomes of observing no events of a specified type, one event, two events, and so on without limit. Events following a Poisson distribution are completely randomized and will not be found if the events are correlated positively (in the case of clumping) or negatively (in the case of mutual repulsion). The Poisson is specified by the average number of events per observation, and its mean and variance are equal.

Polygene Gene which individually exerts a slight effect on the phenotype, controlling a quantitative character in conjunction with several or many other equivalent genes which involve multiple loci.

Polygenic inheritance Mode of genetic transmission in which the character(s) is/are affected by genes which occur on several genetic loci. Polygenic or multigenic inheritance, involving two or more different genetic loci and phenotypic continuity, may be contrasted to monogenic or simple Mendelian inheritance involving a single genetic locus and phenotypic discontinuity. Threshold phenomena, however, may provide for a discontinuous (bimodal) distribution for a trait associated with polygenic inheritance. The different genes involved in such a trait may manifest a similar share in control of the trait, or they may manifest individually different shares (heteromeric genes).

Polymorphism Refers to the coexistence of two or more alleles in a population in frequencies which are too high to be explained merely by new mutations. See *balanced polymorphism*.

Population A community of potentially interbreeding individuals in a given (geographic) location which share a common gene pool.

Polypeptide See *peptide*.

Population genetics A branch of genetics dealing with the study of populations in terms of the percentages of specific genes.

Postulate An assertion whose validity is assumed throughout a discussion, especially in a mathematical system.

Prevalence Number of cases of a disorder or trait that *currently* exists in a given population.

Probability A measure of the subjective expectancy of an event expressed as a fraction which usually denotes the ratio of the number of ways an event can occur to the number of all possible outcomes of the situation in question.

Proband An affected individual through whom a family is first brought to the attention of the geneticist. Index case. Propositus.

Proband method A method of comparing the proportion of progeny in families which were selected by occurrence of a proband (index case, propositus) in each family, with the proportion expected if this character were the effect of a genotype transmitted by a particular Mendelian genetic mode.

Projective technique A personality measure consisting of vague or incomplete stimuli based on the theory that the subject will complete the meaning of the stimulus by projecting his personality onto it.

Projective test Psychological test used as a diagnostic tool in which the test material is so unstructured that any response will reflect a projection of some aspect of the subject's underlying personality and psychopathology.

Propositus See *proband*.

Protein structure The primary structure of a protein refers to the number of polypeptide chains in it, the amino acid sequence of each, and the posititions of interchain and intrachain disulfide bridges. The secondary structure refers to the type of helical configuration possessed by each polypeptide chain resulting from the formation of intramolecular hydrogen bonds along its length. The tertiary structure refers to the manner in which each chain folds upon itself. The quaternary structure refers to the way two or more of the component chains may interact.

Psychogenic Refers to production or causation of a symptom or disorder by mental or psychic factors as opposed to organic ones.

Psychopharmacogenetics Genetically determined differences between individuals in metabolism of or response to psychoactive agents.

Psychotropic Refers to drugs that have special actions upon the psyche.

Pyknic Characterized by shortness of stature, broadness of girth, and powerful muscularity; endo-mesomorphic.

Quantitative inheritance See *polygenic inheritance*.

Random Arrived at by chance without the exercise of any choice.

Random mating The situation in which an individual of one sex has an equal probability of mating with any individual of the opposite sex.

Random sample A sample selected such that every element in the population has an equal opportunity of being chosen and every combination of the desired number of elements has an equal opportunity of being chosen.

Range Distance between the lowest and highest values observed.

Recessive Nonmanifestation of a gene which is present together with an allele which is manifested. The failure of a gene to express phenotypically its presence in the heterozygous genotype is called recessivity, as opposed to dominance. (See *dominance*.) A recessive allele is one that does not produce a phenotypic effect when heterozygous with a dominant allele.

Recombination Occurrence of progeny with combinations of genes other than those that occurred in the parents because of independent assortment or crossing over.

Refute Prove to be false or erroneous; overthrow by argument, evidence, or proof.

Regulator gene A gene whose primary function is to control the rate of synthesis of the product(s) of another or other distant gene(s).

REM sleep One of two kinds of sleep. The term designates the "deep sleep" periods during which the sleeper makes coordinated rapid eye movements (REM's) resembling purposeful fixation shifts, as might be seen in the waking state. REM sleep is also called "dreaming sleep" since there appears to be an intimate relationship with dreaming activity, as if the dreamer were watching the visual imagery of his sleep. NREM sleep is the term given to the longer period of sleep that begins as the subject passes from wakefulness into a light sleep with *no rapid eye movements* (NREM's). REM sleep interrupts NREM sleep about once in every ninety minutes and lasts for about twenty minutes. REM sleep is believed to account for one fifth to one fourth of the total sleep time. Between the two forms of sleep, there are distinct differences in the EEG patterns and in the oculomotor, cardiovascular, respiratory, muscular, and other bodily activities.

Remission Abetment of an illness or disorder.

Representative sample A sample selected according to a quota system so that diverse elements of the population are included.

Repulsion In a double heterozygote, linkage is said to be in the repulsion phase when the mutant alleles of interest at the two loci are on opposite chromosomes.

Resting cell Cell not undergoing division.

Ribonucleic acid See *RNA*.

Ribonucleoprotein A complex macromolecule containing both RNA and protein, and symbolized by RNP.

Ribonucleotide An organic compound that consists of a purine or pyrimidine base bonded to ribose which in turn is esterified with a phosphate group. Such monomers are polymerized to form RNA. See *RNA*.

RNA Ribonucleic acid. Polymer of ribonucleotides which is chemically very similar to DNA. RNA is a long, unbranched molecule consisting of four types of nucleotides linked together by phosphodiester bonds. The sugar component of RNA is ribose. RNA contains no thymine, which is replaced by the closely related pyrimidine, uracil. In contrast to DNA, most RNA molecules are single-stranded, but have the potential for forming complementary helices of the DNA-type. The RNA molecules or their descendants serve as templates against which amino acids are properly ordered to form specific proteins, the amino acids being transported to their proper site on the template by specific carrier or soluble RNA molecules. See *ribonucleotide*.

RNP See *ribonucleoprotein*.

Sample In *statistics*, a finite series of observations of individuals taken from the hypothetical infinitely large population of possible observations or individuals.

Sampling error Variability resulting from the limited size of the samples.

Satellite In *cytogenetics*, a small segment of a chromosome connected to the main body of the chromosome by a constricted segment.

Schizophrenia A large group of disorders, usually of psychotic proportion, manifested by characteristic disturbances of thought, mood, and behavior. Thought disturbances are marked by alterations of concept formation that may lead to misinterpretation of reality and sometimes to delusions and hallucinations. Mood changes include ambivalence, constriction, inappropriateness, and loss of empathy with others. Behavior may be withdrawn, regressive, and bizarre.

Schizophrenia, process Schizophrenia attributed more to organic factors than to environmental ones; typically begins gradually, continues chronically, and progresses (either rapidly or slowly) to an irreversible psychosis; contrasted with reactive schizophrenia.

Schizophrenia, reactive Schizophrenia attributed primarily to strong predisposing and/or precipitating environmental factors; usually of rapid onset and brief duration, with the affected individual appearing well both before and after the schizophrenic episode; contrasted with process schizophrenia.

Scopolamine A myriadic alkaloid, the scopoline ester of tropic acid, from the root of *Scopolia atropoides, Atropa belladonna,* and other solanaceous plants. It is a poisonous nerve depressant, myriadic, and hypnotic.

S.D. See *standard deviation*.

S.E. See *standard error*.

Segregation, Mendel's principle of Members of an allele pair separate from each other when a diploid individual forms haploid gametes, with one of the two alleles normally represented in each haploid gamete produced by meiosis of a diploid primary gametocyte.

Selection The exercise of discrimination in sampling or in arrangement, opposed to randomness.

Serotonin Biogenic amine derived from tryptophan, present in the intestine and in the brain; a smooth-muscle constrictor or stimulator which may influence nervous system activity.

Sex cells Gametes. Ova and spermatozoa.

Sex-chromatin-negative Refers to a sample of cells from an individual in which the nuclei possess no sex chromatin or Barr bodies, indicating a chromosome complement characterized by the inclusion of only one X chromosome.

Sex-chromatin-positive Refers to a sample of cells in which a significant proportion of the nuclei show sex chromatin or Barr bodies. The characteristic presence of a sex-chromatin body within the nucleus indicates a chromosome complement characterized by the inclusion of two X chromosomes. Multiple sex-chromatin bodies indicate the presence of more than two X chromosomes.

Sex chromosomes The homologous chromosomes that are dissimilar in the heterogametic sex (i.e. the X and Y chromosomes).

Sex-limited A trait which expression is restricted to one sex or reduced in one sex.

Sex ratio The ratio (usually at birth) in a population of males to females.

Sib method In human genetics, a method of deriving the proportions of individuals with and without a particular character under observation from those members of the sibship other than the proband by whom the sibship was identified.

Sibling Term for a full brother or sister.

Sibs A single word for brothers and sisters in the same family.

Significance The measure of reality of an apparent discrepancy between observation and expectation. A departure is said to be *statistically significant*, i.e. to be judged real, if the probability of obtaining one as large or larger is lower than some chosen level, which is then referred to as the *level of significance*.

Significance Test A test designed to assess significance and so to distinguish between deviations resulting from sampling error and those indicating real discrepancies between observations and hypotheses.

Single-factor inheritance See *monogenic inheritance*.

Skewness Asymmetry in a frequency distribution.

Soma The body of an organism apart from the germ-line cells.

Somatic Referring to the body, or vegetative cells and tissues of an organism, as opposed to the germ-line cells which give rise to the germ cells or gametes.

Somatic cells All the body cells except the sex cells.

Somatic mutation Mutation occurring in a somatic cell, i.e. in any cell that is not destined to become a germ cell. If the mutated cell continues to divide, the individual will come to contain a patch of tissue characterized by a genotype different from the cells of the rest of the body and may manifest mosaic effects. A somatic mutation will not be transmitted to the progeny, since the germ cells are not affected.

Somatotype Body type, physique; a classification of human body-build in terms of the relative development of ectomorphic, endomorphic, and mesomorphic components.

Spermatogenesis Development of the male germ cell (i.e. spermatozoan) of an animal, which takes place in the gonad. An inclusive term covering both male meiosis and spermiogenesis.

Sphygmomanometer An instrument for measuring arterial blood pressure.

Standard deviation A measure of the variability in a population of items. The standard deviation of a sample is equal to the square root of the arithmetic average of the squares of the differences between the observations and their mean. The distance, measured along the abscissa, of the point of inflection, or maximum slope from the mean in a normal curve.

Standard error Measure of the variability that a value would show in taking repeated samples from the same universe of observations, i.e. the variation which might be expected to occur merely by chance in the various characteristics of samples drawn equally randomly from one and the same population. The standard error is a measure of the variation of a population of means. It is equal to the standard deviation divided by the square root of one less than the total number of items in the population.

Standard Score A score expressed as a deviation from the mean in terms of the standard deviation as the unit.

Standardizing a distribution Transforming the raw scores from a distribution of data into standard scores which are distributed according to the unit normal distribution.

Stanford-Binet intelligence scale An individually administered intelligence test emphasizing verbal facility. Used with individuals from age two through adulthood, the test yields a mental age and an IQ.

Statistic The estimate of a parameter arrived at from observed samples. It bears the same relation to the sample as the parameter does to the population.

Statistics The discipline concerned with the collection, analysis, and presentation of data. The analysis of data depends upon the application of probability theory. Statistical inference involves the selection of one conclusion from a number of alternatives according to the result of a calculation based on observations. See *parametric statistics, nonparametric statistics*.

Stratified sample A sample selected so that each stratum defined in the population is included.

Stratum Defined by one or more specifications which will divide a population into mutually exclusive segments.

Supernatant The fluid lying above a precipitate (e.g. following the centrifugation of a suspension).

Sympathetic nervous system The system of nerves supplying viscera and blood vessels and intimately connected with spinal and some cerebral nerves. This portion of the autonomic nervous system is adapted for general diffussion of autonomic impulses.

Synapsis The pairing of homologous chromosomes, such as occurs normally in meiosis and which permits crossing over to take place.

Temperament An individual's emotional organization (personality).

Template Refers to a macromolecular mold for the synthesis of another macromolecule. Through template processes, a limited number of building blocks are polymerized into macromolecular structures whereby the actual arrangement of the building blocks is uniquely determined by a preexisting one which is identical, complementary, or otherwise related to the newly-synthesized macromolecule.

Teratogen Any agent that produces or increases the incidence of congenital malformations in a population.

Tetrachoric correlation coefficient The degree of association calculated between two sets of variables, with each set dichotomized into two groups, each (group) of which is assumed to show a normal distribution.

Threshold character A term used for those phenotypic characters whose segregating distributions are phenotypically discontinuous but whose inheritance is multigenic like that of continuously varying (i.e. quantitative) characters. The discontinuous segregation of such characters result from threshold effects, i.e. the characters in question have an underlying continuity with a threshold which imposes a discontinuity on their apparent expression. Threshold characters may segregate into many discontinuous phenotypic classes.

Topological Concerned with relations between objects abstracted from exact qualitative measurement.

Tranquilizer A drug that decreases anxiety and agitation, usually without causing drowsiness. Tranquilizers may be divided into two groups: (a) major tranquilizers, drugs such as phenothiazines, which produce relief symptoms of psychosis; and (b) minor tranquilizers, drugs that are used predominantly to diminish neurotic anxiety.

Translocation Relocation of chromosomal segment(s) within one or between nonhomologous chromosomes. A translocation which involves nonhomologous chromosomes may or may not be reciprocal (i.e. be characterized by an exchange of segments between the two different chromosomes). A translocation may or may not be balanced (i.e. characterized by a normal complement of chromosomal material despite the abnormal location of part of that complement). See *balanced translocation*.

Transposition Transfer of a chromosomal segment to another position within the same chromosome, i.e. interchromosomal structural change.

Triploid Possession of three sets of chromosomes instead of the usual two sets or diploid number.

Trisomy Occurrence of a chromosome in triplicate in an otherwise diploid complement. See *partial trisomy*.

Turner's syndrome A condition in man resulting from monosomy for the X chromosome. An affected individual is female in phenotype, but usually sterile.

Twin method In human genetics, a method using monozygotic and dizygotic twins for evaluating the influences of heredity and environment on the phenotype.

Twins Pair of individuals who develop simultaneously within one uterus.

Typology A doctrine or theory of types; study of, or study based on, types.

Type A reference pattern representing characteristics in which a series of concrete patterns resemble each other. The term can be applied to either a group of similar objects or to a group of interrelated traits (syndromes). It may indicate the mode(s) of a multivariate distribution (modal type or types) as well as its extremes (extreme or polar types).

Uracil A pyrimidine base found only in RNA.

Validity A characteristic of a behavior measure which indicates the degree to which the measure is assessing the behavioral dimensions for which it was designed.

Variable A quantity whose value is not fixed in an investigation.

Variance Numerical calculation describing the extent of dispersion of data around a mean. When all values in a population are expressed as plus or minus deviations from the population mean, the variance is the mean of the squared deviations.

Variance analysis See *analysis of variance*.

Variate A variable quantity whose measurements or frequencies form all or part of the data for analysis.

Vector A quantity which can be represented on a many-dimensional scale and is subject to mathematical analysis.

Vicissitude The quality or state of being changeable. An accident of fortune. Alternating change, succession.

Vitalism A philosophy holding that phenomena exhibited in living organisms are the result of special forces distinct from chemical and physical ones.

WAIS Wechsler Adult Intelligence Scale. A verbal and performance test especially designed to measure intelligence in adults.

Wechsler adult intelligence scale See *WAIS*.

Withdrawal symptoms Term used to describe physical and mental effects of withdrawing drugs from patients who have become habituated or addicted to them. The physical symptoms may include nausea, vomiting, tremors, abdominal pain, and convulsions.

X chromosome The sex chromosome found in double dose in the homogametic sex (in mammals, the females), and in single dose in the heterogametic sex (in mammals, the males).

X-linkage Linkage due to the presence of a gene on the X-chromosome: the term is applied especially to genes on the nonhomologous segment of the X-chromosome.

Y chromosome The sex chromosome found only in the heterogametic sex (in mammals, the males).

Y-linkage Linkage due to the presence of a gene on the Y-chromosome: the term is applied especially to genes on the nonhomologous segment of the X-chromosome.

Zygosity Term applied to the number of zygotes from which a pair of twins or larger set of multiple births has resulted (e.g. monozygosity, dizygosity).

Zygote The diploid cell resulting from the union of the haploid male and female gametes.

REFERENCE AUTHOR INDEX

A

Abe, K.: 1965. Ch. 15, #50, p. 355.
  1966. Ch. 15, #51, #52, #53, p. 355.
  1968. Ch. 15, #24, p. 349.
  1969. Ch. 2, #32, p. 20; Ch. 12, #28, p. 311.
  1970. Ch. 15, #36, p. 352.
Abongi, D.: 1968. Ch. 14, #25, p. 336.
Achs, R.: 1966. Ch. 14, #27, p. 337.
Adamopoulos, D. A.: 1970. Ch. 12, #29, p. 311.
  1971. Ch. 12, #30, p. 311.
Adkins, M. M.: 1971. Ch. 10, #164, p. 257.
Aebi, H.: 1968. Ch. 17, #82, p. 391.
Alexanderson, B.: 1971. Ch. 16, #22, p. 368.
Allen, D. M.: 1969. Ch. 17, #83, p. 391.
Allen, G.: 1962. Ch. 6, #72, p. 123.
  1967. Ch. 6, #36, p. 100.
Allen, M.: 1969. Ch. 11, #1, p. 279, p. 284, p. 286; Ch. 13, #26, p. 321.
  In press. Ch. 11, #64, p. 297.
Allison, A. C.: 1961. Ch. 17, #129, p. 392.
Allport, G.: 1933. Ch. 9, #99, p. 217.
Alper, C. A.: 1968. Ch. 17, #135, p. 392.
Altland, K.: 1971. Ch. 16, #13, p. 366.
Amatomi, M.: 1970. Ch. 15, #36, p. 352.
American Psychiatric Association: 1968. Ch. 11, #3, p. 280; Ch. 13, #16, p. 318.
American Psychological Association: 1963. Ch. 5, #9, p. 75.
  1966. Ch. 5, #10, p. 75.
Anastasi, A.: 1961. Ch. 3, #18, p. 40.
  1965. Ch. 10, #14, p. 231, p. 248, p. 249, p. 260, p. 266
  1968. Ch. 5, #13, p. 79.
Anders, J. M.: 1968. Ch. 14, #26, p. 336.
Anderson, E. P.: 1956. Ch. 17, #42, p. 390.
Anderson, V. E.: 1968. Ch. 8, #42, p. 172.
Andreae, U.: 1968. Ch. 17, #91, p. 391.
Angrist, B. M.: 1971. Ch. 16, #72, p. 374.
Angst, J.: 1961. Ch. 15, #8, p. 348.
  1964. Ch. 15, #9, p. 348; Ch. 16, #23, p. 368, p. 372.
  1966. Ch. 10, #192, p. 262.
Anthony, A.S.: 1958. Ch. 4, #14, p. 54, p. 61.
Arakaki, D. T.: 1967. Ch. 2, #27, p. 19.
Arakawa, T.: 1967. Ch. 17, #70, p. 390.
Arbegast, A. W.: 1969. Ch. 13, #31, p. 324.
Archer, H. E.: 1967. Ch. 17, #113, p. 392.
Ardashnikov, S. N.: 1936. Ch. 18, #7, p. 401.
Arellano, M. G.: 1967. Ch. 6, #20, p. 96.
Armaly, M. G.: 1968. Ch. 16, #21, p. 367.

Arnoff, F. N.: 1954. Ch. 5, #26, p. 86.
Asakawa, T.: 1950. Ch. 10, #71, p. 234, p. 236.
Asberg, M.: 1967. Ch. 15, #13, p. 348.
  1970. Ch. 16, #24, p. 368, p. 373.
  1971. Ch. 16, #25, p. 368.
Ash, P.: 1949. Ch. 5, #27, p. 86.
Atkins, L.: 1968. Ch. 2, #31, p. 20.
Atkinson, G. F.: 1967. Ch. 6, #16, p. 95.
Auld, R. M.: 1957. Ch. 17, #60, p. 390.
Austin, J.: 1963. Ch. 17, #93, p. 391.
Axelrod, J.: 1971. Ch. 16, #26, p. 368, #42, p. 370.
Aya, T.: 1966. Ch. 2, #44, #46, p. 29.

B

Bachhawat, B.: 1963. Ch. 17, #93, p. 391.
Baehner, R. L.: 1968. Ch. 17, #79, p. 391.
Bain, J. A.: 1959. Ch. 17, #86, p. 391.
Baitsch, H.: 1963. Ch. 17, #29, p. 390.
Bajema, C.: 1963. Ch. 8, #109, p. 192.
  1968. Ch. 8, #86, p. 185.
Bakwin, H.: 1970. Ch. 15, #49, p. 354.
Balasubrahmanyan, M.: 1973. Ch. 14, #23, p. 335, p. 336; #39, p. 337.
Balasubramanian, A.: 1963. Ch. 17, #93, p. 391.
Ball, R. S.: 1938. Ch. 8, #81, p. 184.
Baluda, M. C.: 1966. Ch. 17, #157, p. 392.
Banik, N. D. D.: 1962. Ch. 6, #8, p. 93.
Barbeau, A.: 1965. Ch. 14, #35, p. 337.
  1970. Ch. 16, #53, p. 372.
  1972. Ch. 16, #112, p. 372.
Barcroft, H.: 1967. Ch. 17, #84, p. 391.
Barquet-Chediak, A.: 1969. Ch. 17, #83, p. 391.
Barr, A.: 1961. Ch. 6, #23, p. 97.
Barr, M. L.: 1949. Ch. 2, #33, p. 20; Ch. 12, #11, p. 308.
Bartsocas, C. S.: 1968. Ch. 2, #31, p. 20.
Bartter, F. C.: 1963. Ch. 18, #37, p. 412.
Basu, D.: 1963. Ch. 17, #93, p. 391.
Baughan, M. A.: 1966. Ch. 17, #28, p. 390.
  1967. Ch. 17, #22, p. 390.
  1968. Ch. 17, #23, #24, p. 390.
Baumrind, D.: 1967. Ch. 6, #110, p. 138.
Bayer, L. M.: 1946. Ch. 10, #85, p. 237.
  1959. Ch. 3, #13, p. 37.
Bayley, N.: 1946. Ch. 10, #85, p. 237.
  1950. Ch. 3, #9, p. 36.
  1959. Ch. 3, #13, p. 37.
  1963. Ch. 6, #107, p. 138.
  1964. Ch. 6, #108, p. 138.
Beach, F. A.: 1951. Ch. 12, #3, p. 308.

Beadle, G. W.: 1941. Ch. 3, #3, p. 34.
Bearn, A. G.: 1964. Ch. 17, #125, p. 392.
　1966. Ch. 17, #136, p. 392.
Beck, L. H.: 1950. Ch. 10, #141, p. 251.
Becker, J.: 1963. Ch. 10, #129, p. 249, #132, p. 249.
Becker, W. C.: 1956. Ch. 5, #25, p. 85.
　1964. Ch. 6, #112, p. 138.
Beckman, L.: 1963. Ch. 14, #46, p. 337.
Bee, L. S.: 1960. Ch. 9, #39, p. 202.
Behnke, A. R.: 1951. Ch. 10, #68, p. 234.
Beiguelman, B.: 1964. Ch. 18, #38, p. 413.
Bell, R. Q.: 1965. Ch. 7, #11, p. 153.
　1968. Ch. 8, #66, p. 179.
Bellugi, U.: 1964. Ch. 4, #5, p. 51, p. 61.
Benado, M.: 1971. Ch. 14, #41, p. 337.
Benda, C. E.: 1969. Ch. 10, #26, p. 230.
Bennett, E. L.: 1968. Ch. 3, #16, p. 38.
Bentley, D. R.: 1971. Ch. 17, #20, p. 389.
Beolchini, P. E.: 1967. Ch. 17, #130, p. 392.
Berg, K.: 1963. Ch. 17, #131, p. 392.
　1965. Ch. 17, #132, p. 392.
　1966. Ch. 17, #136, p. 392.
Berger, R.: 1968. Ch. 14, #25, p. 336.
Bergren, W. R.: 1968. Ch. 17, #156, p. 392.
Beringer, K.: 1921. Ch. 10, #181, p. 260.
Berman, J. L.: 1969. Ch. 17, #61, p. 390.
Bernardi, D.: 1970. Ch. 16, #29, p. 368.
Bertram, E. G.: 1949. Ch. 2, #33, p. 20; Ch. 12, #11, p. 308.
Beutler, E.: 1966. Ch. 17, #157, #158, p. 392.
　1970. Ch. 16, #20, p. 367.
　1972. Ch. 16, #18, p. 367.
Bias, W. B.: 1972. Ch. 16, #52, p. 371.
Biebuyck, J. F.: 1968. Ch. 16, #49, p. 371.
Bigley, R. H.: 1968. Ch. 17, #4, p. 386.
Bilbro, W. C.: 1966. Ch. 6, #15, p. 95.
Binet A.: 1905. Ch. 5, #17, p. 95; Ch. 8, #19, p. 165.
Binett, J. P.: 1964. Ch. 17, #137, p. 392.
Birch, H.: 1963. Ch. 7, #4, p. 152, #15, p. 158.
　1968. Ch. 7, #3, p. 152, p. 153, p. 161.
Black, A. E.: 1967. Ch. 6, #110, p. 138.
Blakeslee, A. F.: 1931. Ch. 18, #3, p. 401.
　1932. Ch. 18, #4, p. 401.
Blau, P.: 1967. Ch. 8, #84, p. 185.
Bleuler, E.: 1950. Ch. 13, #14, p. 317.
Bleuler, M.: 1929. Ch. 9, #83, p. 216.
　1941. Ch. 13, #20, p. 320.
　1968. Ch. 14, #8, p. 331.
Bliss, E. L.: 1953. Ch. 5, #6, p. 91.
Bloom, B. S.: 1964. Ch. 8, #48, p. 173.
Blumberg, B.: 1961. Ch. 17, #129, p. 392.
Bock, R. D.: 1968. Ch. 6, #118, p. 140.
　In press. Ch. 6, #25, p. 98.
Bodmer, W. F.: 1971. Ch. 8, #74, p. 182.
Bond, J.: 1964. Ch. 2, #21, p. 19, p. 25.
　Ch. 14, #22, p. 335.
Bönicke, R.: 1957. Ch. 15, #10, p. 348.
　Ch. 16, #14, p. 366.
Böök, J.: 1953. Ch. 14, #3, p. 331, p. 332.
Bookbinder, L. J.: 1964. Ch. 10, #170, p. 259.
Borhani, N. O.: 1967. Ch. 6, #20, p. 96.
Bossi, E.: 1968. Ch. 17, #82, p. 391.
Boston Collaborative Drug Surveillance Program: 1972. Ch. 16, #106, p. 379.
Bouchard, T. J., Jr.: 1968. Ch. 8, #8, p. 165.
Bourne, C. P.: 1965. Ch. 10, #246, p. 268.
Bowlby, J.: 1969. Ch. 4, #9, p. 52, p. 61.
Bowman, B. H.: 1964. Ch. 17, #125, p. 392.
Boyer, S. H.: 1961. Ch. 17, #141, p. 392.
　1962. Ch. 17, #38, p. 390.
Brace, G. L.: 1962. Ch. 10, #156, p. 255.
Braconi, Z.: 1961. Ch. 11, #24, p. 284.
Bradley, C.: 1937. Ch. 16, #100, p. 378.
Bradley, R. M.: 1966. Ch. 17, #89, p. 391.
　1967. Ch. 17, #96, p. 391.
Bradlow, H. L.: 1971. Ch. 16, #39, p. 369.
Brady, R. O.: 1966. Ch. 17, #89, p. 391; #97, p. 391.
　1967. Ch. 17, #96, p. 391.
Bräutigam, W.: 1964. Ch. 10, #36, p. 231.
Brenglemann, J. C.: 1960. Ch. 10, #160, p. 256.
Brenglemann, L.: 1960. Ch. 10, #160, p. 256.
Brewer, G. J.: 1967. Ch. 17, #133, p. 392.
Bridges, P. K.: 1967. Ch. 10, #93, p. 237, p. 253.
Brierley, H.: 1971. Ch. 12, #44, p. 314.
Briggs, J. H.: 1959. Ch. 12, #14, p. 308.
Britt, B. A.: 1969. Ch. 16, #47, p. 371.
　1970. Ch. 16, #45, p. 371; #48, p. 371.
Britton, R. S.: 1969. Ch. 11, #37, p. 289, p. 290.
Bronte-Stewart, B.: 1961. Ch. 18, #31, p. 409.
Brooksbank, B. W. L.: 1970. Ch. 10, #184, p. 260.
Brouha, L.: 1943. Ch. 10, #159, p. 256.
Broverman, I.: 1968. Ch. 5, #28, p. 99.
Brown, A. M.: 1967. Ch. 9, #73, p. 209.
　1968. Ch. 6, #27, p. 99.
　1971. Ch. 6, #81, p. 126.
Brown, B. I.: 1966. Ch. 17, #50, p. 390.
Brown, D. H.: 1966. Ch. 17, #50, p. 390.
Brown, D. R.: 1966. Ch. 16, #92, p. 377.
Brown, F. W.: 1942. Ch. 11, #29, p. 286, p. 287.
Brown, J. L.: 1964. Ch. 7, #9, p. 153.
Brown, M. S.: 1970. Ch. 16, #78, p. 376.
Brown, R.: 1964. Ch. 4, #5, p. 51, p. 61.
Browne, A.: 1971. Ch. 9, #6, p. 199.
Broughton, R.: 1965. Ch. 15, #42, p. 354.
Brožek, J.: 1953. Ch. 10, #67, p. 234; #100, p. 239.
　1963. Ch. 10, #105, p. 239.
Brunt, P. W.: 1970. Ch. 16, #41, p. 370.
Bruun, K.: 1966. Ch. 6, #119, p. 140; Ch. 9, #67, p. 208; Ch. 16, #84, p. 376.
Budd, M. A.: 1966. Ch. 17, #71, p. 391.

Bull, K. R.: 1958. Ch. 10, #140, p. 251.
Bulmer, M. G.: 1970. Ch. 6, #12, p. 94, p. 96.
Burdette, W. J.: 1962. Ch. 2, #10, p. 13.
Burgess, E. A.: 1967. Ch. 17, #118, p. 392.
Burgess, E. W.: 1944. Ch. 9, #23, p. 201, p. 202.
Burks, B.: 1938. Ch. 2, #17, p. 16.
Burks, B. S.: 1928. Ch. 8, #47, p. 172, p. 173, p. 174; Ch. 9, #32, p. 201.
Buros, O. K.: 1965. Ch. 5, #16, p. 95.
Burt, C.: 1946. Ch. 8, #105, p. 218.
  1949. Ch. 10, #154, p. 255.
  1955. Ch. 8, #5, p. 164, p. 165.
  1956. Ch. 8, #21, p. 166, p. 172.
  1957. Ch. 8, #36, p. 168.
  1958. Ch. 8, #41, p. 169, p. 183.
  1961. Ch. 8, #88, p. 186.
  1963. Ch. 8, #37, p. 168, p. 169.
  1966. Ch. 6, #85, p. 131; Ch. 8, #77, p. 183; Ch. 10, #220, p. 264.
  1971. Ch. 8, #22, p. 166.
  1972. Ch. 8, #23, p. 166.
Butler, R.: 1967. Ch. 17, #130, p. 392.
Bynum, E.: 1959. Ch. 17, #6, p. 387.
Byrd, E.: 1969. Ch. 18, #33, p. 410.

C

Camousseight, A.: 1971. Ch. 14, #41, p. 338.
Campbell, R. C.: 1967. Ch. 10, #123, p. 249.
Cancro, R.: 1968. Ch. 13, #18, p. 319.
  1970. Ch. 13, #17, p. 319.
  1971. Ch. 8, #1, p. 164; Ch. 13, #1, p. 317; #32, p. 324.
Cantwell, D. P.: 1972. Ch. 16, #98b, p. 378.
Cantz, M.: 1968. Ch. 17, #82, p. 391.
Carey, W. B.: 1970. Ch. 7, #5, p. 152.
Carmena, M.: 1934. Ch. 15, #38, p. 352.
Carpenter, A.: 1941. Ch. 10, #92, p. 237.
Carr, A.: 1960. Ch. 12, #39, p. 313.
Carr, A. C.: 1960. Ch. 12, #40, p. 313.
Carrigan, P. M.: 1960. Ch. 10, #150, p. 253.
Carter, C. H.: 1965. Ch. 10, #31, p. 230.
Carter, H. D.: 1933. Ch. 9, #55, p. 204; Ch. 11, #39, p. 292.
Carter, J. E. L.: 1967. Ch. 10, #69, p. 234, p. 236.
Carter, N. D.: 1968. Ch. 17, #155, p. 392.
Caspari, E.: 1967. Ch. 3, #2, p. 33; Ch. 8, #97, p. 191; Ch. 11, #54, p. 294.
Casselman, W. G. B.: 1961. Ch. 2, #36, p. 20.
Cattell, R. B.: 1937. Ch. 8, #103, p. 191.
  1950. Ch. 7, #2, p. 151.
  1951. Ch. 8, #107, p. 191.
  1960. Ch. 6, #28, p. 99; Ch. 9, #16, p. 201.
  1963. Ch. 8, #67, p. 179.
  1965. Ch. 6, #29, p. 99; Ch. 9, #17, p. 201; Ch. 10, #121, p. 249.
  1966. Ch. 5, #11, p. 78.
  1967. Ch. 9, #42, p. 202.
  1971. Ch. 8, #11, p. 165.
Cavalli-Sforza, L. L.: 1971. Ch. 8, #74, p. 182.
Cederbaum, S.: 1971. Ch. 16, #54, p. 372.
Cederlöff, R.: 1961. Ch. 6, #14, p. 94.
Chambers, R. A.: 1956. Ch. 17, #30, p. 390.
Chance, M. R. A.: 1953. Ch. 4, #13, p. 54, p. 61.
Chang, P. T.: 1972. Ch. 10, #97, p. 237, p. 255, p. 259, p. 260, p. 262, p. 263.
Chase, T. N.: 1971. Ch. 16, #109, p. 379.
Cheek, D. B.: 1968. Ch. 10, #238, p. 268.
Chess, S.: 1963. Ch. 7, #4, p. 152.
  1964. Ch. 7, #16, p. 161.
  1968. Ch. 7, #3, p. 152, p. 153, p. 161; #18, p. 162.
  1969. Ch. 7, #6, p. 152.
  1970. Ch. 7, #7, p. 152.
  1971. Ch. 7, #8, p. 152.
Chien, C.: 1971. Ch. 15, #54, p. 358.
Child, I. L.: 1941. Ch. 10, #116, p. 248, p. 251, p. 253.
  1950. Ch. 10, #141, p. 251; #144, p. 251, p. 257, p. 258.
Ching, S. C.: 1967. Ch. 9, #25, p. 201.
Chomsky, N.: 1966. Ch. 4, #17, p. 58, p. 61.
  1968. Ch. 4, #2, p. 51, p. 61.
Christie, R. G.: 1968. Ch. 10, #171, p. 259, p. 262.
Chuck, G.: 1966. Ch. 17, #53, p. 390.
Chung, R.: 1971. Ch. 15, #54, p. 358.
Churchill, J. A.: 1965. Ch. 6, #73, p. 123.
  1967. Ch. 6, #74, p. 124.
Ciocco, A.: 1936. Ch. 10, #48, p. 231.
Ciolfi, F.: 1949. Ch. 9, #89, p. 216.
Claney, H.: 1969. Ch. 3, #31, p. 43.
Clark, P. J.: 1955. Ch. 6, #11, p. 94.
  1956. Ch. 6, #37, p. 101.
  1965. Ch. 15, #29, p. 351, p. 352.
Clayton, J. E.: 1964. Ch. 17, #114, p. 392.
Clayton, P. J.: 1969. Ch. 16, #55, p. 372.
Cliquet, R. L.: 1963. Ch. 6, #70, p. 123.
Cloward, R. B.: 1957. Ch. 15, #37, p. 352.
Cobb, E. A.: 1943. Ch. 10, #164, p. 258.
Coffin, T. E.: 1944. Ch. 10, #142, p. 251.
Cohen, D.: In press. Ch. 11, #64, p. 297.
Cohen, M. M.: 1967. Ch. 14, #24, p. 336.
Coiteux, C.: 1965. Ch. 14, #35, p. 337.
Collins, R. L.: 1970. Ch. 16, #33, p. 369.
Comfort, A.: 1959. Ch. 12, #37, p. 313.
Comrey, A. L.: 1966. Ch. 6, #98, p. 136.
  1967. Ch. 9, #64, p. 208.
Conners, C. K.: 1967. Ch. 16, #97, p. 378.
Conrad, H. S.: 1940. Ch. 8, #44, p. 172.
Conrad, K.: 1963. Ch. 10, #52, p. 233, p. 250, p. 262.
  1967. Ch. 10, #1, p. 230, p. 231.
Conway, J.: 1959. Ch. 8, #87, p. 185.

Cooper, B.: 1966. Ch. 10, #214, p. 263.
Cooper, H. L.: 1960. Ch. 12, #15, p. 309.
Cooper, J. E.: 1969. Ch. 9, #77, p. 210.
Cooper, R.: 1958. Ch. 8, #59, p. 177.
Coppen, A.: 1963. Ch. 18, #41, #42, p. 417.
  1965. Ch. 9, #115, p. 220; Ch. 11, #49, p. 292.
  1966. Ch. 10, #147, p. 252.
  1967. Ch. 10, #95, p. 237, p. 262.
  1970. Ch. 10, #184, p. 260.
Coppen, A. J.: 1961. Ch. 10, #94, p. 237, p. 263.
Corbin, J. E.: 1962. Ch. 3, #5, p. 35.
Cori, C. F.: 1952. Ch. 17, #47, p. 390.
  1956. Ch. 17, #49, p. 390.
Cori, G. T.: 1952. Ch. 17, #47, p. 390.
  1956. Ch. 17, #49, p. 390.
Corney, G.: 1968. Ch. 17, #148, p. 392.
Cortés, J. B.: 1965. Ch. 10, #145, p. 251.
Cotton, J. E.: 1968. Ch. 2, #38, p. 20, p. 26; Ch. 14, #18, p. 332, p. 335.
Count, E.: 1967. Ch. 4, #8, p. 52.
Court-Brown, W. M.: 1964. Ch. 2, #21, p. 19, p. 25; Ch. 14, #22, p. 335.
Cowie, V.: 1965. Ch. 9, #115, p. 220; Ch. 11, #49, p. 292.
  1971. Ch. 10, #225, p. 264; Ch. 11, #46, p. 292, p. 293.
Coyle, J. T.: 1970. Ch. 16, #71, p. 374.
Cravioto, J.: 1968. Ch. 10, #32, p. 230.
Cresseri, A.: 1954. Ch. 2, #43, p. 29.
Cronbach, L. J.: 1951. Ch. 5, #4, p. 72.
  1969. Ch. 8, #65, p. 179.
  1970. Ch. 5, #14, p. 79.
Cronholm, B.: 1970. Ch. 16, #24, p. 368, p. 373.
Crook, M. N.: 1937. Ch. 9, #114, p. 219, p. 220.
Cullen, F.: 1781. Ch. 11, #2, p. 280.
Cummins, H.: 1939. Ch. 14, #31, p. 337.
  1961. Ch. 6, #18, p. 96.
  1966. Ch. 14, #36, p. 337, p. 338.
Cunningham, A. E.: 1970. Ch. 10, #184, p. 260.
Cunningham, G. C.: 1969. Ch. 17, #61, p. 390.
Curry, S. H.: 1970. Ch. 15, #19, p. 349.
Curtis, J. E.: 1971. Ch. 16, #80, p. 376.
Curtus, F.: 1936. Ch. 15, #26, p. 350.
  1960. Ch. 15, #33, p. 351.

D

Dacie, J. V.: 1964. Ch. 17, #34, p. 390.
Dalton, K.: 1968. Ch. 10, #233, p. 226.
Damaree, R. G.: 1968. Ch. 9, #7, p. 199.
Dancis, J.: 1963. Ch. 17, #69, p. 390.
  1967. Ch. 17, #68, #70, p. 390.
  1968. Ch. 16, #40, p. 370.
Daniels, R. S.: 1964. Ch. 15, #45, p. 354.
Darke, R.: 1948. Ch. 12, #20, p. 310.
Darlington, C.: 1954. Ch. 11, #14, p. 283.
  1963. Ch. 11, #15, p. 283.
Das, S. R.: 1956. Ch. 18, #10, p. 401.
  1957. Ch. 18, #11, p. 401.
Dasgupta, D.: 1973. Ch. 14, #23, p. 335, p. 336; #39, p. 337.
Dasgupta, J.: 1973. Ch. 14, #23, p. 335, p. 336; #39, p. 337.
Dave, R. H.: 1963. Ch. 6, #104, p. 137.
Davidson, R. G.: 1967. Ch. 14, #24, p. 336.
Davidson, M. A.: 1955. Ch. 10, #202, p. 263.
Davidson, W. M.: 1954. Ch. 12, #12, p. 308.
Davis, J. M.: 1970. Ch. 15, #19, p. 349.
  1971. Ch. 16, #26, p. 368.
Davis, K. E.: 1962. Ch. 9, #41, p. 202.
Davis, V. E.: 1970. Ch. 16, #93, p. 377.
Davison, K.: 1965. Ch. 15, #12, p. 348; Ch. 16, #17, p. 367, p. 373.
  1971. Ch. 12, #44, p. 314.
Day, R. W.: 1966. Ch. 17, #157, p. 392.
  1969. Ch. 17, #61, p. 390.
Dayton, P. G.: 1971. Ch. 16, #110, p. 379.
deAlmeida, J. C.: 1959. Ch. 12, #14, p. 308.
Dean, G.: 1963. Ch. 16, #35, p. 369.
Decker, W. H.: 1960. Ch. 12, #15, p. 309.
DeFries, J.: 1972. Ch. 6, #47, p. 109.
DeGeorge, F. V.: 1959. Ch. 6, #24, p. 97; Ch. 10, #9, p. 230, p. 264.
deLamadrid, E. G.: 1962. Ch. 18, #16, p. 401.
Delbrück, H.: 1926. Ch. 10, #126, p. 249, p. 260.
DeMars, R.: 1967. Ch. 17, #7, p. 387.
DeMarsh, Q. B.: 1968. Ch. 17, #23, p. 390.
Denborough, M. A.: 1962. Ch. 16, #46, p. 371.
Dencker, S. J.: 1959. Ch. 18, #14, p. 401.
Denenberg, V. H.: 1971. Ch. 11, #57, p. 295.
Dennmark, H. G. F.: 1940. Ch. 9, #95, p. 216.
Dennis, W.: 1935. Ch. 6, #57, p. 116.
Dent, C. E.: 1954. Ch. 17, #77, p. 391.
Dent, D. M.: 1968. Ch. 16, #49, p. 371.
Denver Report: 1960. Ch. 2, #39, p. 25.
Detrich, R. A.: 1972. Ch. 16, #108, p. 379.
Detter, J. C.: 1968. Ch. 17, #24, p. 390.
Dhrymiotis, A.: 1962. Ch. 15, #17, p. 348.
Diamond, M. C.: 1968. Ch. 3, #16, p. 38.
Diamonds, S.: 1957. Ch. 10, #135, p. 250.
Dibble, E.: In press. Ch. 11, #64, p. 297.
Dicke, W. K.: 1960. Ch. 17, #55, p. 390.
DiFerrante, N.: 1971. Ch. 16, #80, p. 376.
Dimascio, A.: 1959. Ch. 10, #241, p. 268.
Dinitz, S.: 1959. Ch. 13, #5, p. 317.
Doering, W.: 1971. Ch. 10, #37, p. 231.
Dohan, F. C.: 1969. Ch. 13, #31, p. 324.
Donnell, G. N.: 1968. Ch. 17, #156, p. 392.
Dörmer, A. E.: 1967. Ch. 17, #113, p. 392.
Dorner, G.: 1968. Ch. 12, #6, #7, p. 304.
Dove, G. A.: 1970. Ch. 12, #29, p. 311.

Draper, G.: 1941. Ch. 10, #84, p. 237.
Drash, P. W.: 1968. Ch. 10, #21, p. 230, p. 260.
Dreyfus, J. C.: 1961. Ch. 17, #46, p. 390.
Drillien, C. M.: 1964. Ch. 6, #71, p. 123.
Driscoll, K.: 1956. Ch. 17, #8, p. 387.
Duis, B. T.: 1937. Ch. 14, #50, p. 337.
Duncan, O. D.: 1968. Ch. 8, #79, p. 184, p. 185.
Dunner, D. L.: 1971. Ch. 16, #56, p. 372.
Dupertuis, C. W.: 1950. Ch. 10, #98, p. 239.
  1951. Ch. 10, #68, p. 234.
  1954. Ch. 10, #76, p. 234, p. 237, p. 242.
Düser, G.: 1921. Ch. 10, #181, p. 260.
Dutrillaux, B.: 1968. Ch. 14, #25, p. 336.
Dyken, P. R.: 1972. Ch. 16, #52, p. 371.

E

Eastham, R. D.: 1968. Ch. 10, #25, p. 230.
Eaves, L. J.: 1969. Ch. 9, #117, p. 224.
  1970. Ch. 9, #118, p. 224.
  1972. Ch. 8, #76, p. 183; Ch. 9, #119, p. 224
Eckland, B. K.: 1970. Ch. 8, #85, p. 185.
  1971. Ch. 8, #82, p. 184, p. 185.
Edgson, A. C.: 1959. Ch. 10, #229, p. 264.
Edwards, A. L.: 1959. Ch. 10, #161, p. 256.
Edwards, J. H.: 1970. Ch. 6, #80, p. 126.
Edwards, W. A.: 1970. Ch. 16, #78, p. 376.
Efron, M. L.: 1962. Ch. 17, #72, p. 391.
  1965. Ch. 17, #73, p. 391.
  1966. Ch. 17, #71, p. 391.
Egan, T. J.: 1970. Ch. 17, #106, p. 392.
Ehrhardt, A. A.: 1968. Ch. 10, #230, p. 264, p. 265, p. 266.
Ehrman, L.: 1972. Ch. 8, #2, p. 164.
Eik-Nes, K.: 1953. Ch. 5, #6, p. 73.
  1960. Ch. 5, #7, p. 73.
Ekman, G.: 1951. Ch. 10, #56, p. 233; #74, p. 234, p. 235, p. 236.
Ellinwood, E. H., Jr.; 1969. Ch. 16, #69, p. 374.
Elliot, O.: 1965. Ch. 3, #17, p. 39.
Elmgren, J.: 1951. Ch. 9, #84, p. 216.
Elässer, G.: 1951. Ch. 10, #34, p. 231, p. 232, p. 233, p. 249, p. 259, p. 260, p. 262.
  1952. Ch. 10, #189, p. 262.
Elston, R. C.: 1968. Ch. 6, #44, p. 106.
  1972. Ch. 6, #99, p. 137.
England, S.: 1961. Ch. 18, #22, p. 401, p. 402; #29, p. 407.
Engleman, K.: 1963. Ch. 17, #78, p. 391.
Enke, W.: 1936. Ch. 10, #125, p. 249.
Enklewitz, M.: 1936. Ch. 17, #55, p. 390.
Entsch, M.: 1969. Ch. 3, #31, p. 43.
Epstein, S. S.: 1971. Ch. 2, #49, p. 29.
Erbe, R. W.: In press. Ch. 17, #120, p. 392.
Erlenmeyer-Kimling, L.: 1963. Ch. 3, #19, p. 40; Ch. 8, #24, p. 166, p. 171.
  1966. Ch. 13, #33, p. 325.
  1971. Ch. 8, #101, p. 191.
  1972. Ch. 8, #69, p. 181.
Eron, L. D.: 1965. Ch. 9, #10, p. 200, p. 209.
'Espinasse, P. G.: 1942. Ch. 2, #19, p. 16.
Essig, C. F.: 1958. Ch. 16, #75, p. 374.
Esterer, M. B.: 1962. Ch. 18, #16, p. 401.
Evans, F. T.: 1952. Ch. 17, #101, p. 391.
Eysenck, H. J.: 1945. Ch. 10, #66, p. 233, p. 241, p. 247, p. 248, p. 256, p. 258, p. 259, p. 263.
  1947. Ch. 9, #4, p. 198.
  1951. Ch. 9, #12, p. 200, p. 210, p. 216; Ch. 10, #222, p. 264.
  1952. Ch. 9, #78, p. 210; Ch. 11, #45, p. 292.
  1956. Ch. 9, #79, p. 210, p. 213, p. 215.
  1958. Ch. 10, #162, p. 257, p. 263.
  1962. Ch. 9, #100, p. 217.
  1963. Ch. 10, #151, p. 253.
  1964. Ch. 10, #207, p. 263.
  1965. Ch. 10, #57, p. 233, p. 235, p. 240, p. 249, p. 250, p. 255, p. 263.
  1967. Ch. 9, #11, p. 200, p. 203, p. 204, p. 214, p. 217; Ch. 11, #11, p. 283, p. 292.
  1968. Ch. 9, #48, p. 202, p. 220.
  1969. Ch. 9, #3, p. 198, p. 199, p. 210; #49, p. 202, p. 220; Ch. 10, #153, p. 253.
  1970. Ch. 6, #114, p. 139; Ch. 9, #2, p. 198, p. 199, p. 205; Ch. 10, #221, p. 264.
  1972. Ch. 9, #9, p. 199; #43, p. 202.
  1973. Ch. 9, #76, p. 210.
Eysenck, S. B. G.: 1963. Ch. 10, #151, p. 253.
  1965. Ch. 9, #54, p. 204.
  1968. Ch. 9, #48, p. 202, p. 220.
  1969. Ch. 9, #3, p. 198, p. 199, p. 210; #49, p. 202, p. 220; Ch. 10, #153, p. 253.

F

Fagerhol, M. K.: 1967. Ch. 17, #134, p. 392.
Fahrenberg, J.: 1966. Ch. 10, #157, p. 255.
  1967. Ch. 10, #240, p. 268.
Falconer, D. S.: 1960. Ch. 6, #39, p. 103; Ch. 8, #72, p. 181.
  1965. Ch. 6, #34, p. 100.
  1966. Ch. 8, #75, p. 182.
  1967. Ch. 2, #7, p. 12.
Falek, A.: 1971. Ch. 8, #111, p. 192, p. 193.
Falkner, F. T.: 1962. Ch. 6, #8, p. 93.
Farkas, T.: 1969. Ch. 9, #77, p. 210.
Feiereis, H.: 1960. Ch. 15, #33, p. 351.
Felsher, R. F.: 1970. Ch. 16, #38, p. 369.
Fenton, G. W.: 1971. Ch. 9, #80, p. 211; #107, p. 217; Ch. 10, #223, p. 264.
Fenna, D.: 1971. Ch. 16, #90, p. 377.
Ferguson, L. R.: 1956. Ch. 6, #105, p. 138.
Fernandez, P. B.: 1971. Ch. 7, #8, p. 152.

Fields, V.: 1954. Ch. 10, #78, p. 235.
Filds, R. A.: 1963. Ch. 17, #154, p. 392.
　1966. Ch. 17, #152, p. 392.
　1968. Ch. 17, #155, p. 392.
Finegold, M.: 1968. Ch. 10, #172, p. 260.
Fischer, C. L.: 1969. Ch. 2, #30, p. 20.
Fischer, M.: 1968. Ch. 6, #17, p. 95.
　1969. Ch. 13, #21, p. 320.
Fischer, R.: 1961. Ch. 18, #22, p. 401, p. 402; #29, p. 407.
　1963. Ch. 18, #24, p. 402, p. 410; #44, p. 420, p. 421.
　1964. Ch. 18, #25, p. 402; #26, p. 402, p. 403, p. 417, p. 420, p. 421; #43, p. 418.
　1965. Ch. 18, #27, p. 404; #34, p. 410; #47, p. 421.
　1967. Ch. 18, #28, p. 405.
　1971. Ch. 16, #73, p. 374.
Fisher, R. A.: 1918. Ch. 6, #42, p. 105; Ch. 9, #18, p. 201.
　1963. Ch. 2, #4, p. 9, p. 13.
Fisk, N.: 1973. Ch. 10, #232, p. 265.
Fiske, D. W.: 1971. Ch. 10, #122, p. 264.
Fitch, L. I.: 1968. Ch. 17, #155, p. 392.
Fleshler, B.: 1964. Ch. 18, #43, p. 418.
Floud, J. E.: 1958. Ch. 8, #90, p. 187.
Forbes, A. P.: 1964. Ch. 14, #29, p. 337.
Ford, C. E.: 1959. Ch. 12, #14, p. 308.
Ford, C. S.: 1951. Ch. 12, #3, p. 303.
Ford, E. B.: 1964. Ch. 3, #34, p. 44.
Ford, R.: 1969. Ch. 17, #61, p. 390.
Forer, B. R.: 1960. Ch. 12, #40, p. 313.
Forster, J. F.: 1962. Ch. 16, #46, p. 371.
Fox, A. L.: 1931. Ch. 18, #1, p. 401.
　1932. Ch. 18, #2, p. 401.
Fox, R.: 1967. Ch. 4, #11, p. 54, p. 61.
　1971. Ch. 4, #15, p. 56, p. 61.
　1972. Ch. 4, #12, p. 54, p. 61.
Frank, J.: 1958. Ch. 11, #9, p. 282.
Fraser, H. F.: 1958. Ch. 16, #75, p. 374.
Fred, H. L.: 1960. Ch. 5, #7, p. 73.
Frederick, E. W.: 1964. Ch. 17, #114, p. 392.
Fredrickson, D. S.: 1966. Ch. 17, #97, p. 391.
Freedman, D.: 1965. Ch. 9, #74, p. 209.
　1966. Ch. 11, #5, p. 281.
Freedman, D. G.: 1963. Ch. 6, #66, p. 120.
　1971. Ch. 7, #14, p. 157.
Freeman, F. N.: 1937. Ch. 6, #83, p. 131; Ch. 8, #52, p. 174; Ch. 9, #52, p. 203, p. 204, p. 213, p. 216; Ch. 11, #40, p. 294.
Freud, S.: 1938. Ch. 10, #232, p. 265.
　1961. Ch. 11, #53, p. 293, p. 296.
Frieberg, L.: 1961. Ch. 6, #14, p. 94.
Fritsch, W.: 1972. Ch. 10, #187, p. 262.
Froesch, E. R.: 1963. Ch. 17, #29, p. 390.
Fry, D. E.: 1967. Ch. 10, #95, p. 237, p. 262.
Fujinaga, K.: 1966. Ch. 2, #47, p. 29.

Fulker, D. W.: 1970. Ch. 6, #31, p. 99, p. 217; Ch. 8, #63, p. 179, p. 181, p. 183, p. 192; Ch. 9, #20, p. 201, p. 213, p. 214, p. 215, p. 222, p. 224; Ch. 11, #12, p. 283.
Fuller, J. L.: 1956. Ch. 3, #8, p. 36.
　1960. Ch. 10, #224, p. 264; Ch. 11, #50, p. 292, p. 295.
　1965. Ch. 3, #4, p. 35, p. 36, p. 41, p. 42, p. 44.
　1966. Ch. 9, #103, p. 217.
　1967. Ch. 3, #29, p. 43.
　1970. Ch. 11, #55, p. 295; Ch. 16, #28, p. 368, #33, p. 369.
Furth, H. G.: 1966. Ch. 8, #17, p. 165.
　1971. Ch. 8, #18, p. 165.
Furusho, T.: 1968. Ch. 6, #45, p. 108.

G

Gal, A. E.: 1967. Ch. 17, #96, p. 391.
Gall, J. C.: 1967. Ch. 17, #133, p. 392.
Gallagher, T. F.: 1971. Ch. 16, #39, p. 369.
Galton, F.: 1962. Ch. 8, #102, p. 191.
　1975. Ch. 6, #5, p. 91.
Garattini, S.: 1970. Ch. 16, #29, p. 368.
Gardner, L. I.: 1972. Ch. 10, #21, p. 230; #22, p. 230.
Garn, S. M.: 1961. Ch. 18, #29, p. 407.
Garrison, R. J.: 1968. Ch. 8, #42, p. 172.
Garrod, A. E.: 1909. Ch. 17, #1, p. 385.
Gastaut, H.: 1965. Ch. 15, #42, p. 354.
Gatti, F. M.: 1965. Ch. 10, #145, p. 251.
Gautier, M.: 1959. Ch. 2, #40, p. 26.
Gebhard, P. H.: 1953. Ch. 12, #2, p. 301.
Gedda, L.: 1961. Ch. 6, #4, p. 91.
Genest, K.: 1957. Ch. 16, #10, p. 366; Ch. 17, #102, p. 391.
Gerritsen, T.: 1964. Ch. 17, #65, p. 390.
Gershon, E. S.: 1971. Ch. 16, #56, p. 372.
Gershon, S.: 1971. Ch. 16, #72, p. 374.
Gershowitz, H.: 1967. Ch. 6, #16, p. 95.
Gertman, S.: 1959. Ch. 18, #33, p. 410.
Gesell, A.: 1929. Ch. 6, #87, p. 134.
　1941. Ch. 6, #90, p. 134.
Geyer, H.: 1937. Ch. 15, #47, p. 354.
Ghadmini, H.: 1967. Ch. 17, #64, p. 390.
Ghiselli, E.: 1966. Ch. 8, #20, p. 165.
Gianelli, F.: 1968. Ch. 14, #26, p. 336.
Gibbens, T. C. N.: 1963. Ch. 10, #208, p. 263.
Gibbons, J. E.: 1955. Ch. 16, #63, p. 373.
Gibbs, E. C.: 1945. Ch. 9, #104, p. 217.
Gibbs, F. A.: 1945. Ch. 9, #104, p. 217.
Giblett, E. R.: 1968. Ch. 17, #24, p. 390.
　1969. Ch. 6, #9, p. 93.
Gibson, D. A.: 1963. Ch. 14, #32, p. 337.
Gibson, J. B.: 1970. Ch. 8, #89, p. 186, #92, p. 188, p. 190.
Gibson, Q.: 1948. Ch. 17, #85, p. 391.

1967. Ch. 17, #84, p. 391.
Giellis, S. S.: 1968. Ch. 10, #172, p. 260.
Gilbert, J. A. L.: 1971. Ch. 16, #90, p. 377.
Gill, J. R., Jr.: 1963. Ch. 18, #37, p. 412.
Gillespie, W.: 1964. Ch. 12, #33, p. 311.
Gillette, P. N.: 1971. Ch. 16, #39, p. 369.
Gindilis, V. M.: 1972. Ch. 14, #21, p. 334, p. 335, p. 342.
Ging, R.: 1964. Ch. 10, #170, p. 259.
Giroux, J.: 1966. Ch. 14, #37, p. 337.
Gitzelmann, R.: 1967. Ch. 17, #44, p. 390.
Gjone, E.: 1967. Ch. 17, #110, p. 392.
Glanville, E. V.: 1964. Ch. 18, #43, p. 418.
    1965. Ch. 18, #28, p. 405, #30, p. 407, #34, p. 410, #39, p. 413, p. 417, #40, p. 413, p. 415, p. 417.
Glen-Bott, A. M.: 1963. Ch. 17, #145, p. 392.
Glowinski, S. S.: 1969. Ch. 16, #58, p. 372.
Glucksberg, S.: 1960. Ch. 9, #40, p. 202.
Glueck, E.: 1956. Ch. 10, #212, p. 263.
    1964. Ch. 10, #210, p. 263.
Glueck, S.: 1956. Ch. 10, #212, p. 263.
    1964. Ch. 10, #210, p. 263.
Gneist, J.: 1969. Ch. 10, #130, p. 249.
Goedde, H. W.: 1971. Ch. 16, #13, p. 366.
Goffman, E.: 1961. Ch. 5, #34, p. 86.
Goldberg, L.: 1956. Ch. 17, #43, p. 390.
Goldschmidt, R. B.: 1931. Ch. 12, #34, p. 312.
    1938. Ch. 12, #9, p. 307, p. 309.
    1955. Ch. 12, #10, p. 307, p. 308.
Goldschmidt, W.: 1966. Ch. 1, #1, p. 3.
Goldstein, A.: 1965. Ch. 15, #21, #22, p. 349.
Goldstein, M. J.: 1969. Ch. 5, #30, p. 86.
Gooch, P. C.: 1969. Ch. 2, #30, p. 20.
Goodenough, D. R.: 1962. Ch. 15, #44, p. 354.
Goodman, H. O.: 1959. Ch. 15, #41, p. 352.
Goodwin, F. K.: 1971. Ch. 16, #56, p. 372.
Goodwin, M. S.: 1965. Ch. 7, #12, p. 153.
Gordon, E. M.: 1967. Ch. 7, #19, p. 162.
Gottesman, I. I.: 1963. Ch. 8, #25, p. 166, p. 168, p. 177, p. 179; Ch. 9, #56, p. 205; Ch. 11, #42, p. 292.
    1965. Ch. 9, #34, p. 202, p. 205, p. 209; Ch. 11, #43, p. 292.
    1966. Ch. 9, #62, p. 207; Ch. 13, #22, p. 320.
    1967. Ch. 14, #9, p. 331.
    1968. Ch. 6, #44, p. 106; Ch. 8, #26, p. 166, p. 184, p. 190, p. 192.
    1971. Ch. 3, #33, p. 44; Ch. 5, #29, p. 86; Ch. 8, #101, p. 191.
Gottlieb-Jensen, K.: 1968. Ch. 6, #17, p. 95.
Gottschaldt, K.: 1960. Ch. 6, #95, p. 135.
Gough, H. G.: 1965. Ch. 8, #99, p. 191.
Goy, R. W.: 1968. Ch. 10, #231, p. 264.
Grasberger, J. C.: 1969. Ch. 13, #31, p. 324.
Graves, R.: 1955. Ch. 16, #19, p. 367.
Gray, H.: 1949. Ch. 9, #35, p. 202.

Gray, P. W. S.: 1952. Ch. 17, #101, p. 391.
Green, M.: 1966. Ch. 2, #47, p. 29.
Green, R.: 1973. Ch. 10, #32, p. 230.
Greenberg, N. E.: 1968. Ch. 10, #21, p. 230, p. 260.
Greenblatt, M.: 1959. Ch. 10, #241, p. 268.
Gregor, A. J.: 1968. Ch. 6, #50, p. 109.
Greif, S.: 1970. Ch. 9, #8, p. 199.
Griek, B. J.: 1970. Ch. 16, #32, p. 369, p. 375.
Griesinger, W.: 1871. Ch. 13, #8, p. 317.
Griffin, F.: 1961. Ch. 18, #22, p. 401, p. 402, #29, p. 407.
    1963. Ch. 18, #24, p. 402, #44, p. 420, p. 421.
    1964. Ch. 18, #26, p. 402, p. 403, p. 417, p. 420, p. 421.
    1965. Ch. 18, #47, p. 421.
    1967. Ch. 18, #28, p. 405.
Grimes, A. J.: 1964. Ch. 17, #34, p. 390.
Grubb, R.: 1956. Ch. 17, #126, p. 392.
Grüsser, O. -J.: 1971. Ch. 10, #50, p. 232.
Guilford, J. P.: 1959. Ch. 7, #1, p. 151; Ch. 10, #58, p. 233, p. 249, p. 253, p. 255.
    1965. Ch. 5, #3, p. 68.
    1967. Ch. 8, #13, p. 165.
Gurland, B. J.: 1969. Ch. 9, #77, p. 210.
Guttmacher, A. F.: 1953. Ch. 6, #2, p. 90.
Guze, S. B.: 1967. Ch. 11, #33, p. 287.

H

Hachel, F.: 1959. Ch. 15, #41, p. 352.
Hadorn, B.: 1969. Ch. 17, #117, p. 392.
Häfner, H.: 1950. Ch. 10, #183, p. 260.
Hagnell, O.: 1966. Ch. 10, #215, p. 263; Ch. 11, #7, p. 282.
Haidu, G. G.: 1962. Ch. 15, #17, p. 348.
Haist, C.: 1970. Ch. 16, #48, p. 371.
Haldane, J. B. S.: 1946. Ch. 8, #70, p. 181.
    1947. Ch. 2, #18, p. 16.
Hall, C. S.: 1957. Ch. 10, #136, p. 250, p. 255; Ch. 11, #61, p. 296.
    1968. Ch. 9, #1, p. 198.
Hall, K.: 1971. Ch. 5, #32, p. 86.
Halldörsson, S.: 1972. Ch. 10, #173, p. 260.
Hallgren, B.: 1967. Ch. 15, #4, p. 347.
Halsey, A. H.: 1958. Ch. 8, #90, p. 187.
    1959. Ch. 8, #91, p. 187.
Hambert, G.: 1966. Ch. 10, #226, p. 264.
Hamburger, V.: 1970. Ch. 3, #15, p. 38.
Hamerton, J. L.: 1968. Ch. 14, #26, p. 336.
Hammer, W.: 1967. Ch. 15, #13, p. 348.
    1969. Ch. 15, #14, p. 348.
Hampson, J. G.: 1955. Ch. 12, #21, #22, p. 310.
Hampson, J. L.: 1955. Ch. 12, #21, #22, p. 310.
Hanel, K. H.: 1964. Ch. 17, #107, p. 392.
Hanley, C.: 1951. Ch. 10, #146, p. 251, p. 266.
Hardy, G. H.: 1908. Ch. 2, #12, p. 13.
Harlow, H. F.: 1965. Ch. 3, #30, p. 43.

Harlow, M. K.: 1965. Ch. 3, #30, p. 43.
Harman, H. H.: 1967. Ch. 5, #37, p. 87.
Harnden, D. G.: 1964. Ch. 2, #21, p. 19, p. 25; Ch. 14, #22, p. 335.
Harper, P. S.: 1972. Ch. 16, #52, p. 371.
Harper, R. G.: 1966. Ch. 14, #27, p. 337.
Harris, G. W.: 1965. Ch. 12, #5, p. 304.
Harris, H.: 1947. Ch. 15, #35, p. 351.
  1949. Ch. 18, #18, p. 401, p. 402, p. 403, p. 410.
  1951. Ch. 18, #9, p. 401.
  1959. Ch. 16, #15, p. 366; Ch. 17, #66, p. 390.
  1960. Ch. 17, #144, p. 392.
  1963. Ch. 17, #145, p. 392.
  1964. Ch. 17, #150, p. 392.
  1966. Ch. 17, #152, p. 392.
  1967. Ch. 17, #142, p. 392; #147, p. 392; #151, p. 392.
  1968. Ch. 17, #24, p. 390; #139, #140, p. 392; #153, p. 392.
  1969. Ch. 17, #161, p. 392.
  1970. Ch. 17, #3, p. 386, p. 388, p. 390, p. 392.
Harris, J. A.: 1912. Ch. 9, #27, p. 201.
Harris, T. H.: 1957. Ch. 16, #62, p. 373, p. 379.
Harrison, D.: 1967. Ch. 17, #84, p. 391.
Harrison, G. G.: 1968. Ch. 16, #49, p. 371.
  1971. Ch. 16, #50, p. 371.
Hartl, E. M.: 1949. Ch. 10, #82, p. 236, p. 239, p. 250, p. 263.
Hartmann, E.: 1971. Ch. 15, #54, p. 358.
Hartmann, G.: 1939. Ch. 18, #17, p. 401.
Harvald, B.: 1964. Ch. 17, #107, p. 392.
  1967. Ch. 6, #36, p. 100.
  1968. Ch. 6, #17, p. 95.
  1969. Ch. 13, #21, p. 320.
Harvard Educational Review: 1969. Ch. 8, #3, p. 164.
Haseman, J. K.: 1972. Ch. 6, #99, p. 137.
Haslam, J.: 1809. Ch. 13, #11, p. 317.
Hattori, M.: 1965. Ch. 17, #27, p. 390.
Hauge, M.: 1959. Ch. 18, #14, p. 401.
  1968. Ch. 6, #17, p. 95.
  1969. Ch. 13, #21, p. 320.
Hayes, K. J.: 1962. Ch. 8, #68, p. 179.
Healey, M. J. R.: 1959. Ch. 10, #229, p. 264.
Heath, B. H.: 1967. Ch. 10, #69, p. 234, p. 236.
Heath, H.: 1952. Ch. 10, #112, p. 241.
Hechter, H. H.: 1967. Ch. 6, #20, p. 96.
Hecker, E.: 1871. Ch. 13, #12, p. 317.
Heidbreder, E.: 1926. Ch. 10, #118, p. 248.
Heilbrun, A. B.: 1965. Ch. 6, #109, p. 138, p. 139.
Heins, H. L.: 1965. Ch. 17, #27, p. 390.
  1966. Ch. 17, #28, p. 390.
Helmstadter, G. C.: 1964. Ch. 5, #15, p. 79.
Henderson, N. D.: 1970. Ch. 8, #60, p. 177.
Hendryx, J.: 1971. Ch. 12, #32, p. 311.
Henkin, R. I.: 1963. Ch. 18, #37, p. 412.

Henry, C. F.: 1942. Ch. 16, #101, p. 378.
Henry, W. D.: 1960. Ch. 12, #40, p. 313.
Hernstein, R.: 1971. Ch. 8, #83, p. 184, p. 190.
Hers, H. G.: 1962. Ch. 17, #48, p. 390.
  1968. Ch. 17, #100, p. 391.
Hertzig, M. E.: 1963. Ch. 7, #4, p. 152.
Hesketh, F. E. A.: 1925. Ch. 10, #54, p. 233.
Heston, L. L.: 1966. Ch. 13, #27, p. 321; Ch. 14, #15, p. 332.
  1968. Ch. 12, #41, p. 314.
  1970. Ch. 16, #68, p. 373.
Hewitt, L. E.: 1946. Ch. 6, #113, p. 139.
Hiatt, H. H.: 1966. Ch. 17, #56, p. 390.
Hickman, R.: 1968. Ch. 16, #49, p. 371.
Hierneaux, J.: 1963. Ch. 10, #8, p. 230.
Higgins, J. V.: 1961. Ch. 8, #43, p. 172.
  1962. Ch. 8, #108, p. 191.
Hilgard, J. R.: 1933. Ch. 6, #89, p. 134.
  1951. Ch. 11, #38, p. 290.
Hiller, O.: 1959. Ch. 17, #123, p. 392.
Hill, M.: 1968. Ch. 9, #33, p. 202.
Hillbun, W. B.: 1970. Ch. 14, #54, p. 337.
Hinde, R. A.: 1971. Ch. 3, #27, p. 43.
Hinz, G.: 1968. Ch. 12, #7, p. 304.
Hirsch, J.: 1962. Ch. 1, #5, p. 3.
  1968. Ch. 1, #8, p. 3.
  1970. Ch. 1, #9, p. 3; Ch. 2, #5, p. 11.
Hirschfeld, J.: 1962. Ch. 17, #124, p. 392.
Hirschhorn, K.: 1960. Ch. 12, #15, p. 309.
Hoch, P.: 1957. Ch. 13, #2, p. 317.
Hockaday, R. D. R.: 1964. Ch. 17, #114, p. 392.
Hodges, R. E.: 1962. Ch. 14, #28, p. 337.
Hodgkin, W. E.: 1967. Ch. 2, #28, p. 19.
Hoffer, A.: 1969. Ch. 11, #1, p. 279, p. 284, p. 286; Ch. 13, #26, p. 321.
  1970. Ch. 11, #28, p. 286.
Hofmann, G.: 1973. Ch. 10, #195, p. 262.
Hofstätter, P. R.: 1944. Ch. 10, #106, p. 241.
Hollaender, A.: 1971. Ch. 2, #48, p. 29.
Hollingshead, A. B.: 1950. Ch. 9, #21, p. 201.
Hollister, L. E.: 1970. Ch. 16, #64, p. 373.
Holmes, L. B.: 1972. Ch. 10, #173, p. 260.
Holt, S. B.: 1964. Ch. 14, #30, #33, p. 337.
Holzel, A.: 1956. Ch. 17, #43, p. 390.
Holzinger, K. J.: 1929. Ch. 6, #33, p. 100; Ch. 9, #13, p. 201.
  1937. Ch. 6, #83, p. 131; Ch. 8, #52, p. 174; Ch. 9, #52, p. 203, p. 204, p. 213, p. 216; Ch. 11, #40, p. 294.
Honeyman, M. G.: 1966. Ch. 9, #57, p. 206.
Honeyman, M. S.: 1967. Ch. 17, #133, p. 392.
Honzik, M. P.: 1957. Ch. 8, #56, p. 175.
Hood, A. B.: 1963. Ch. 10, #44, p. 231, p. 251.
Hooker, E.: 1960. Ch. 12, #40, p. 313.
Hooton, E. A.: 1939. Ch. 10, #206, p. 263.

1952. Ch. 10, #70, p. 234, p. 236, p. 239.
Hopkins, H. K.: 1971. Ch. 5, #32, p. 86.
Hopkinson, D. A.: 1963. Ch. 17, #147, p. 392.
　1964. Ch. 17, #138, p. 392; #150, p. 392.
　1968. Ch. 17, #24, p. 390; #139, p. 392; #140, p. 392; #153, p. 392.
Hopkinson, G.: 1969. Ch. 11, #62, p. 297.
Horn, A. S.: 1971. Ch. 16, #70, p. 374.
Horn, D.: 1963. Ch. 18, #36, p. 411.
Horn, J. L.: 1968. Ch. 8, #12, p. 165.
Howard, M.: 1956. Ch. 8, #21, p. 166, p. 172.
Howarth, E.: 1971. Ch. 9, #6, p. 199.
Howells, W. W.: 1951. Ch. 10, #109, p. 241.
　1968. Ch. 10, #120, p. 249.
Hoyler, A.: 1958. Ch. 9, #98, p. 216.
Hrubec, Z.: 1969. Ch. 11, #1, p. 279, p. 284, p. 286; Ch. 13, #26, p. 321.
Hsaio, S.: 1967. Ch. 16, #94, p. 377.
Hsia, D. Y. Y.: 1956. Ch. 17, #8, p. 387.
　1969. Ch. 17, #61, p. 390.
Hsia, Y. E.: 1968. Ch. 17, #119, p. 392.
Hsu, G. L. K.: 1946. Ch. 10, #111, p. 241, p. 259.
　1972. Ch. 10, #97, p. 237, p. 255, p. 259, p. 260, p. 262, p. 263.
Hubby, J. L.: 1966. Ch. 17, #13, p. 388.
Hug, G.: 1966. Ch. 17, #53, p. 390.
Huguley, C. M., Jr.: 1959. Ch. 17, #86, p. 391.
　1961. Ch. 17, #87, p. 391.
Humphreys, L. G.: 1957. Ch. 10, #77, p. 235.
　1971. Ch. 8, #94, p. 191.
Hunt, E. E.: 1953. Ch. 10, #100, p. 239.
Hunt, J. McV.: 1969. Ch. 8, #34, p. 168, p. 177.
Hunt, W. G.: 1969. Ch. 17, #14, p. 388.
Hunter, R.: 1969. Ch. 16, #37, p. 369, p. 371.
Hunter, R. L.: 1957. Ch. 17, #11, p. 388.
Husen, T.: 1953. Ch. 6, #26, p. 99.
Hutchinson, G. E.: 1959. Ch. 12, #36, p. 313.
Hutchinson, J. R.: 1972. Ch. 16, #52, p. 371.
Huter, C.: 1952. Ch. 10, #137, p. 250.
Hutt, M. L.: 1960. Ch. 12, #40, p. 313.
Hutzler, J.: 1963. Ch. 17, #69, p. 390.
　1967. Ch. 17, #68, p. 390, #70, p. 390.

I

Ideström, C.-M.: 1967. Ch. 15, #13, p. 348.
Ihda, S.: 1961. Ch. 11, #23, p. 284.
Ikai, M.: 1961. Ch. 15, #28, p. 350.
Ikeuchi, T.: 1966. Ch. 2, #46, p. 29.
Ikura, Y.: 1965. Ch. 17, #25, p. 390.
Illingworth, B.: 1956. Ch. 17, #49, p. 390.
Im Obersteg, J.: 1952. Ch. 10, #73, p. 234, p. 268.
Inhorn, S. L.: 1971. Ch. 10, #30, p. 230.
Inhoff, G.: In press. Ch. 11, #64, p. 297.
Inouye, E.: 1960. Ch. 15, #2, p. 347.
　1961. Ch. 11, #26, p. 284.

Insel, P. M.: 1971. Ch. 9, #51, p. 202, p. 220.
Irreverre, F.: 1967. Ch. 17, #80, p. 391.
Ismail, A. A. A.: 1970. Ch. 12, #29, p. 311.
　1971. Ch. 12, #30, p. 311.
Isselbacher, K. J.: 1956. Ch. 17, #42, p. 390.
　1966. Ch. 17, #71, p. 391.
Itano, H. A.: 1949. Ch. 17, #9, p. 387.

J

Jablon, S.: 1967. Ch. 6, #16, p. 95.
Jackson, Đ. D.: 1959. Ch. 11, #17, p. 283.
Jackson, L. G.: 1970. Ch. 2, #51, p. 29.
Jacobs, P. A.: 1959. Ch. 12, #13, p. 308.
Jacobson, A.: 1965. Ch. 15, #43, p. 354.
Jacobson, L.: 1966. Ch. 7, #20, p. 162.
Jagiello, G.: 1968. Ch. 14, #26, p. 336.
Jancar, J.: 1968. Ch. 10, #25, p. 230.
Janoff, I. Z.: 1950. Ch. 10, #141, p. 251.
Janovsky, D. S.: 1965. Ch. 15, #19, p. 349.
Jarvik, L. F.: 1963. Ch. 3, #19, p. 40; Ch. 8, #24, p. 166, p. 171.
Janz, D.: 1962. Ch. 10, #179, p. 260.
　1969. Ch. 10, #180, p. 260.
Jaspers, K.: 1971. Ch. 10, #49, p. 232, p. 233, p. 249, p. 250.
Jatzkewitz, H.: 1968. Ch. 17, #91, p. 391.
　1969. Ch. 17, #94, p. 391.
Jenkins, R. L.: 1946. Ch. 6, #113, p. 139.
Jensch, F.: 1941. Ch. 12, #18, p. 310.
Jensen, A. R.: 1967. Ch. 6, #41, p. 104.
　1968. Ch. 8, #49, p. 173.
　1969. Ch. 6, #46, p. 109; Ch. 8, #27, p. 166, p. 171, p. 172, p. 174, p. 181, p. 192.
　1970. Ch. 6, #86, p. 132.
　1971. Ch. 8, #50, p. 173.
　1972. Ch. 8, #54, p. 175.
Jervis, G. A.: 1963. Ch. 17, #59, p. 390.
Jinks, J. L.: 1964. Ch. 9, #108, p. 218.
　1970. Ch. 6, #31, p. 99, p. 127; Ch. 8, #63, p. 179, p. 181, p. 183, p. 192; Ch. 9, #20, p. 201, p. 213, p. 214, p. 215, p. 222, p. 224; Ch. 11, #12, p. 283.
Johnson, A. L.: 1970. Ch. 10, #184, p. 260.
Johnson, R. C.: 1968. Ch. 6, #59, p. 118; Ch. 9, #113, p. 218.
Johnson, W. E.: 1970. Ch. 17, #15, p. 388.
Johnston, H., Jr.: 1969. Ch. 13, #31, p. 324.
Jones, H. E.: 1928. Ch. 9, #31, p. 201.
　1932. Ch. 6, #52, p. 113.
　1940. Ch. 8, #44, p. 172.
　1947. Ch. 10, #158, p. 256.
　1955. Ch. 9, #109, p. 218.
Jones, J. W.: 1959. Ch. 12, #14, p. 308.
Jones, M. B.: 1971. Ch. 5, #12, p. 79.
Jones, M. C.: 1950. Ch. 3, #9, p. 36.
Jones, M. T.: 1967. Ch. 10, #93, p. 237, p. 253.

Jonson, E.: 1961. Ch. 6, #14, p. 94.
Joreskog, K. G.: 1967. Ch. 5, #36, p. 87.
Jori, A.: 1970. Ch. 16, #29, p. 368.
Jost, M.: 1944. Ch. 15, #39, p. 352.
Judd, L. L.: 1969. Ch. 5, #30, p. 86.
Juel-Nielsen, N.: 1965. Ch. 6, #84, p. 131; Ch. 9, #75, p. 209.
   1968. Ch. 6, #17, p. 95.
Julian, T.: 1967. Ch. 10, #95, p. 237, p. 262.
Julou, L.: 1969. Ch. 16, #58, p. 372.

K

Kadushin, A.: 1968. Ch. 7, #13, p. 157.
Kahlbaum, K. L.: 1874. Ch. 13, #13, p. 317.
Kahn, J.: 1969. Ch. 2, #32, p. 20.
Kaij, L.: 1959. Ch. 18, #14, p. 401.
   1960. Ch. 16, #83, p. 376.
   1961. Ch. 6, #14, p. 94.
Kaizer, S.: 1965. Ch. 15, #21, #22, p. 349.
Kajii, T.: 1965. Ch. 2, #45, p. 29.
Kajujama, S.: 1970. Ch. 15, #36, p. 352.
Kakihana, R.: 1966. Ch. 16, #92, p. 377.
Kalckar, H. M.: 1956. Ch. 17, #42, p. 390.
Kales, A.: 1965. Ch. 15, #43, p. 354.
Kallmann, F. J.: 1946. Ch. 9, #45, p. 202; Ch. 14, #1, p. 331, p. 332.
   1950. Ch. 10, #190, p. 262.
   1952. Ch. 12, #26, p. 310, p. 311.
   1953. Ch. 14, #2, p. 331, p. 332.
   1962. Ch. 6, #72, p. 123.
   1966. Ch. 13, #33, p. 325.
Kalmar, Z.: 1966. Ch. 6, #76, p. 124.
Kalmus, H.: 1949. Ch. 18, #18, p. 401, p. 402, p. 403, p. 410.
   1951. Ch. 18, #9, p. 401.
   1957. Ch. 18, #12, p. 401.
   1958. Ch. 18, #19, p. 401, p. 403.
   1962. Ch. 18, #32, p. 410.
Kalow, W.: 1957. Ch. 16, #10, p. 366; Ch. 17, #102, p. 391, #143, p. 392.
   1959. Ch. 17, #103, p. 391.
   1962. Ch. 16, #2, p. 363.
   1969. Ch. 16, #47, p. 371.
   1970. Ch. 16, #45, p. 371, #48, p. 371.
Kamaryt, J.: 1965. Ch. 17, #160, p. 392.
Kamionkowski, M.: 1964. Ch. 18, #43, p. 418.
Kamitake, S.: 1961. Ch. 15, #32, p. 351.
Kanfer, J. N.: 1966. Ch. 17, #89, p. 391, #97, p. 391.
Kantor, R.: 1953. Ch. 5, #24, p. 85.
Kanzler, M.: 1971. Ch. 15, #18, p. 349.
Kaplan, A. R.: 1963. Ch. 18, #24, p. 402, p. 410.
   1964. Ch. 18, #25, p. 402, #43, p. 418.
   1965. Ch. 18, #27, p. 404, #30, p. 407, #34, p. 410, #39, p. 413, p. 417, #40, p. 413, p. 415, p. 417.
   1966. Ch. 2, #37, p. 20.
   1967. Ch. 18, #28, p. 405.
   1968. Ch. 2, #29, p. 20, p. 26, p. 29; #38, p. 20, p. 26; Ch. 14, #18, p. 332, p. 335.
   1969. Ch. 2, #1, p. 7, #24, p. 19, p. 26, p. 29.
   1970. Ch. 14, #19, p. 332, p. 335.
   1972. Ch. 1, #2, #4, p. 3; Ch. 2, #6, p. 12, #25, p. 19, p. 25; Ch. 14, #20, p. 332, p. 335.
Kaplan, W. D.: 1959. Ch. 2, #34, p. 20.
Kappas, A.: 1971. Ch. 16, #39, p. 369.
Karczmar, A. G.: 1967. Ch. 16, #31, p. 369.
Karlsson, J.: 1966. Ch. 14, #5, p. 331, p. 332.
   1972. Ch. 14, #12, p. 332.
Karras, A.: 1967. Ch. 18, #28, p. 405.
Katnovsky, M. L.: 1968. Ch. 17, #79, p. 391.
Katz, I.: 1960. Ch. 9, #40, p. 202.
Kaveggia, E. G.: 1971. Ch. 10, #30, p. 230.
Keitt, A. S.: 1966. Ch. 17, #35, p. 390.
Keller, B.: 1963. Ch. 6, #66, p. 120.
Kelley, T. L.: 1927. Ch. 5, #8, p. 73.
Kelley, W. N.: 1967. Ch. 17, #112, p. 392.
   1972. Ch. 16, #44, p. 370.
Kelsall, M. A.: 1968. Ch. 2, #29, p. 20, p. 26, p. 29.
Kempthorne, O.: 1961. Ch. 6, #43, p. 106.
Kemsley, W. F. F.: 1950. Ch. 10, #101, p. 239.
Kendall, R. E.: 1969. Ch. 9, #77, p. 210.
Kennedy, R. J. R.: 1944. Ch. 9, #24, p. 201.
Kerckhoff, A. C.: 1962. Ch. 9, #41, p. 202.
Kerr, M.: 1936. Ch. 9, #88, p. 216.
Kessel, N.: 1963. Ch. 18, #41, #42, p. 417.
Keys, A.: 1952. Ch. 10, #67, p. 234.
Kety, S. S.: 1959. Ch. 11, #16, p. 283.
   1968. Ch. 13, #28, p. 321; Ch. 14, #6, p. 331, p. 332, #16, p. 332; Ch. 16, #86, p. 376.
   1969. Ch. 16, #58, p. 372.
   1970. Ch. 13, #30, p. 322.
Kiev, M.: 1936. Ch. 9, #87, p. 216.
King, L. J.: 1970. Ch. 16, #61, p. 372.
Kinosita, R.: 1959. Ch. 2, #34, p. 20.
Kinsey, A. C.: 1948. Ch. 12, #1, p. 301.
   1953. Ch. 12, #2, p. 301.
Kirk, R. L.: 1968. Ch. 17, #122, p. 392.
Kirkham, K. E.: 1971. Ch. 12, #30, p. 311.
Kirkman, H. N.: 1959. Ch. 17, #6, p. 387.
   1964. Ch. 17, #37, p. 390.
Kitchin, F. D.: 1964. Ch. 17, #125, p. 392.
Klawans, H. L., Jr.: 1970. Ch. 16, #53, p. 372.
   1972. Ch. 16, #112, p. 372.
Klerman, G. L.: 1959. Ch. 10, #241, p. 268.
Klett, C. J.: 1972. Ch. 10, #60, p. 233, p. 240, p. 249.
Kline, N. S.: 1952. Ch. 10, #167, p. 259.
Klinke, R.: 1937. Ch. 10, #50, p. 232.
Klinenberg, J. R.: 1963. Ch. 17, #78, p. 391.
Klintworth, G. K.: 1962. Ch. 12, #42, p. 314.
Kluckhohn, R.: 1958. Ch. 4, #14, p. 56, p. 61.
Knight, R. Z.: 1959. Ch. 16, #15, p. 366.

Knox, W. E.: 1956. Ch. 17, #8, p. 387.
Knudson, A. G.: 1971. Ch. 16, #80, p. 376.
Knussmann, R.: 1960. Ch. 10, #166, p. 258.
  1968. Ch. 10, #5, p. 230, p. 237, p. 241, p. 242.
Koch, H.: 1966. Ch. 6, #56, p. 115; Ch. 9, #71, p. 209.
Koch-Weser, J.: 1970. Ch. 16, #105, p. 379.
Koegler, S. J.: 1963. Ch. 17, #75, p. 391.
Koeller, D. -M.: 1970. Ch. 10, #133, p. 249, #134, p. 249, p. 262.
Kolb, L. C.: 1960. Ch. 12, #39, p. 313.
Koler, R. D.: 1968. Ch. 17, #4, p. 386.
Koller, S.: 1942. Ch. 12, #19, p. 310.
Kolodny, R. C.: 1971. Ch. 12, #32, p. 311.
Komrower, G. M.: 1956. Ch. 17, #43, p. 390.
Kopin, I. J.: 1971. Ch. 16, #26, p. 368.
Korkhaus, G.: 1936. Ch. 15, #26, p. 350.
Korn, S.: 1963. Ch. 7, #4, p. 152, #15, p. 158.
  1970. Ch. 7, #7, p. 152.
  1971. Ch. 7, #8, p. 152.
Kraepelin, E.: 1899. Ch. 13, #9, p. 317.
Krane, S. M.: In press. Ch. 17, #120, p. 391.
Kraus, A. P.: 1968. Ch. 17, #31, p. 390.
Kraus, R.: 1960. Ch. 9, #40, p. 202.
Krauss, W.: 1972. Ch. 10, #204, p. 263.
Krech, D.: 1968. Ch. 3, #16, p. 38.
Kreitman, N.: 1961. Ch. 13, #3, p. 317.
  1964. Ch. 9, #46, p. 202.
  1968. Ch. 9, #47, p. 202.
Kretschmann, H. J.: 1971. Ch. 10, #247, p. 268.
Kretschmer, E.: 1936. Ch. 10, #125, p. 249.
  1967. Ch. 10, #40, p. 231, p. 232, p. 233, p. 236, p. 249, p. 257, p. 258, p. 260, p. 262.
Kringlen, E.: 1966. Ch. 13, #23, p. 320.
Krisch, H.: 1927. Ch. 10, #119, p. 248.
Krug, R. S.: 1964. Ch. 6, #112, p. 138.
Krut, L. H.: 1961. Ch. 18, #31, p. 409.
Kuder, G. F.: 1939. Ch. 5, #5, p. 72.
Kurland, A. A.: 1962. Ch. 15, #20, p. 349; Ch. 16, #66, p. 373.
Kurland, L. T.: 1962. Ch. 15, #20, p. 349; Ch. 16, #66, p. 373.
Kutt, H.: 1971. Ch. 16, #16, p. 367, p. 375, p. 379.

L

Labhart, A.: 1953. Ch. 17, #29, p. 390.
Lader, M. H.: 1966. Ch. 15, #40, p. 352.
  1971. Ch. 9, #80, p. 211; Ch. 10, #223, p. 264.
LaDu, B. N.: 1958. Ch. 17, #2, p. 386, p. 390.
  1967. Ch. 17, #62, p. 390.
Lafourcale, J.: 1968. Ch. 14, #25, p. 336.
Lake, B. D.: 1969. Ch. 17, #109, p. 392.
Lang, T.: 1960. Ch. 12, #17, p. 310.
Langner, T. S.: 1962. Ch. 11, #6, p. 282.
Langston, M. F.: 1968. Ch. 17, #31, p. 390.
LaPolla, A.: 1969. Ch. 5, #30, p. 86.

LaRochelle, J.: 1967. Ch. 17, #63, p. 390.
Lasker, G. W.: 1936. Ch. 17, #55, p. 390.
  1947. Ch. 10, #217, p. 264.
Lasker, M.: 1936. Ch. 17, #55, p. 390.
  1941. Ch. 17, #45, p. 390.
Lasten, L.: 1958. Ch. 17, #2, p. 386, p. 390.
Laster, L.: 1967. Ch. 17, #80, p. 391, #96, p. 391.
  1970. Ch. 16, #78, p. 376.
Laurell, A. B.: 1956. Ch. 17, #126, p. 392.
Laurell, C. B.: 1967. Ch. 17, #134, p. 392.
Laxova, R.: 1965. Ch. 17, #160, p. 392.
Layzer, R. B.: 1967. Ch. 17, #26, p. 390.
Lee, A.: 1903. Ch. 9, #26, p. 201.
Lee, D.: 1955. Ch. 10, #202, p. 263.
Lefton, M.: 1959. Ch. 13, #5, p. 317.
Leguébe, A.: 1960. Ch. 18, #20, p. 401, p. 403, p. 410.
Lehmann, D.: 1965. Ch. 15, #43, p. 354.
Lehmann, H.: 1952. Ch. 17, #101, p. 391.
  1956. Ch. 16, #9, p. 366.
  1960. Ch. 17, #144, p. 392.
Lehmann, H. E.: 1967. Ch. 10, #168, p. 259, p. 262.
Lehmann, W.: 1964. Ch. 10, #28, p. 230.
Lejeune, J.: 1959. Ch. 2, #40, p. 26.
  1967. Ch. 2, #23, p. 19.
  1968. Ch. 2, #25, p. 19, p. 25; Ch. 14, #25, p. 336.
Lemberger, L.: 1971. Ch. 16, #26, p. 368.
Lennenberg, E. H.: 1964. Ch. 4, #6, p. 51, p. 61.
  1967. Ch. 4, #3, p. 51, p. 61.
Lennox, W. G.: 1945. Ch. 9, #104, p. 217.
Lenoir, J.: 1961. Ch. 17, #128, p. 392.
Lenz, W.: 1967. Ch. 10, #27, p. 230.
Lerner, I. M.: 1950. Ch. 6, #1, p. 90.
  1968. Ch. 8, #98, p. 191.
Lesch, M.: 1964. Ch. 16, #43, p. 370, p. 379; Ch. 17, #111, p. 392.
Lessa, W. A.: 1940. Ch. 10, #4, p. 230, p. 231.
Levan, A.: 1956. Ch. 2, #20, p. 18.
Levin, B.: 1962. Ch. 17, #76, p. 391.
  1967. Ch. 17, #118, p. 392.
Levine, S.: 1965. Ch. 12, #5, p. 304.
Levitz, M.: 1963. Ch. 17, #69, p. 390.
Levy, D. M.: 1943. Ch. 3, #24, p. 43.
Lewis, A. J.: 1963. Ch. 14, #32, p. 337.
Lewis, D. G.: 1957. Ch. 8, #35, p. 168.
Lewis, G. M.: 1963. Ch. 17, #54, p. 390.
Lewis, N. D. C.: 1969. Ch. 10, #81, p. 236.
Lewis, W. H. P.: 1967. Ch. 17, #147, p. 392.
  1968. Ch. 17, #148, p. 392.
  1969. Ch. 17, #161, p. 392.
Lewontin, R. C.: 1966. Ch. 17, #13, p. 388.
  1970. Ch. 17, #15, p. 388.
Ley, P.: 1969. Ch. 11, #62, p. 297.
Li, C. C.: 1954. Ch. 8, #33, p. 166.
  1971. Ch. 8, #29, p. 166.
Li, J. C. R.: 1964. Ch. 5, #2, p. 68, p. 78, p. 87.

Lichtenberger, W.: 1965. Ch. 6, #97, p. 136.
Lichtenstein, E. A.: 1936. Ch. 18, #7, p. 401.
Lieberman, D. M.: 1968. Ch. 14, #26, p. 336.
Lienert, G. A.: 1961. Ch. 11, #51, p. 292.
Lifschitz, F.: 1966. Ch. 17, #58, p. 390.
Lilljeqvist, A. -Ch.: 1968. Ch. 17, #119, p. 392.
Lind, M.: 1967. Ch. 15, #13, p. 348.
Lindegård, B.: 1953. Ch. 10, #61, p. 233, p. 234, p. 237, p. 239.
   1956. Ch. 10, #62, p. 233, p. 239; #104, p. 239, p. 248, p. 252, p. 255, p. 256, p. 266; #114, p. 247; #218, p. 264.
Lindsley, D. B.: 1942. Ch. 16, #101, p. 378.
Lindsten, J.: 1964. Ch. 14, #30, p. 337.
Lindzey, G.: 1957. Ch. 10, #136, p. 250, p. 255; Ch. 11, #61, p. 296.
   1968. Ch. 9, #1, p. 198.
   1971. Ch. 9, #65, p. 208, p. 209.
Lipcon, H. H.: 1956. Ch. 15, #48, p. 354.
Lisboa, B. P.: 1957. Ch. 15, #10, p. 348; Ch. 16, #14, p. 366.
Livingstone, F. B.: 1967. Ch. 17, #121, p. 392.
Lloyd, J. K.: 1969. Ch. 17, #117, p. 392.
Locher, W. G.: 1969. Ch. 16, #47, p. 371.
Lodge Patch, I. C.: 1965. Ch. 10, #169, p. 259, p. 262.
Loeb, P. M.: 1970. Ch. 16, #78, p. 376.
Loehlin, J.: 1971. Ch. 9, #65, p. 208, p. 209.
Loehlin, J. C.: 1965. Ch. 6, #30, p. 99; Ch. 9, #82, p. 216.
   1968. Ch. 6, #117, p. 140.
   1972. Ch. 16, #85, p. 376.
Loh, H. H.: 1968. Ch. 16, #96, p. 377.
Löhr, G. W.: 1962. Ch. 17, #39, p. 390.
Lombroso, C.: 1911. Ch. 10, #138, p. 251.
Long, W. K.: 1967. Ch. 17, #159, p. 392.
Lopez, R. E.: 1965. Ch. 16, #99, p. 378.
Loraine, J. A.: 1970. Ch. 12, #29, p. 311.
   1971. Ch. 12, #30, p. 311.
Lorenz, K.: 1965. Ch. 4, #1, p. 49, p. 61.
   1970. Ch. 4, #10, p. 53, p. 61.
Lorr, M.: 1954. Ch. 10, #78, p. 235.
Lovell, R. R. H.: 1962. Ch. 16, #46, p. 371.
Lowell, F. M.: 1969. Ch. 13, #31, p. 324.
Lubin, A.: 1950. Ch. 10, #139, p. 251.
Lucas, A. R.: 1970. Ch. 17, #19, p. 389.
Luchsinger, R.: 1961. Ch. 15, #3, p. 347.
Luffman, J. E.: 1967. Ch. 17, #151, p. 392.
   1968. Ch. 17, #139, p. 392.
Luke, J. E.: 1959. Ch. 15, #41, p. 352.
Lundstrom, A.: 1941. Ch. 6, #38, p. 102.
Luria, A. R.: 1936. Ch. 6, #94, p. 134.
Lush, J. L.: 1945. Ch. 6, #32, p. 99.
Lutz, F.: 1918. Ch. 9, #22, p. 201.
Lynch, B. L.: 1968. Ch. 17, #31, p. 390.
Lyon, M. F.: 1961. Ch. 2, #35, p. 20.

M

McArthur, N.: 1953. Ch. 6, #22, p. 99.
McCall, R. B.: 1970. Ch. 8, #57, p. 175.
McCandless, B. R.: 1964. Ch. 6, #101, p. 137.
McClearn, G. E.: 1966. Ch. 16, #92, p. 377.
McCloy, C. H.: 1940. Ch. 10, #107, p. 241.
McCurdy, P. R.: 1964. Ch. 17, #37, p. 390.
McDermott, E.: 1949. Ch. 10, #82, p. 236, p. 239, p. 250, p. 263.
   1954. Ch. 10, #76, p. 234, p. 237, p. 242.
McGirr, E. M.: 1965. Ch. 17, #67, p. 390.
McGraw, M. B.: 1935. Ch. 6, #91, p. 134.
   1939. Ch. 6, #92, p. 134.
McKay, R. J.: 1967. Ch. 2, #28, p. 19.
McKeown, T.: 1970. Ch. 6, #79, #80, p. 126.
McKusick, V. A.: 1960. Ch. 3, #14, p. 38.
   1970. Ch. 16, #41, p. 370.
   1972. Ch. 16, #52, p. 371.
McLean, P.: 1967. Ch. 17, #74, p. 391.
McMurray, J.: 1967. Ch. 17, #84, p. 391.
McMurray, W. C.: 1963. Ch. 17, #75, p. 391.
McNeil, T. F.: 1968. Ch. 5, #33, p. 86; Ch. 10, #244, p. 268.
McNemar, Q.: 1942. Ch. 5, #21, p. 83.
   1964. Ch. 8, #9, p. 165.
MacAlpine, I.: 1969. Ch. 16, #37, p. 369, p. 371.
MacCabae, H. J.: 1965. Ch. 17, #67, p. 391.
Maccoby, E. E.: 1956. Ch. 6, #105, p. 138.
MacDonald, E. M.: 1965. Ch. 17, #67, p. 391.
MacDonald, J. W.: 1961. Ch. 17, #105, p. 391.
MacKinnon, D. W.: 1962. Ch. 8, #100, p. 191.
Mack, C.: 1972. Ch. 10, #173, p. 260.
Mack, J. W.: 1971. Ch. 16, #60, p. 372.
MacLean, N.: 1964. Ch. 2, #21, p. 19, p. 25; Ch. 14, #22, p. 335.
MacMahon, B.: 1970. Ch. 10, #245, p. 268.
MacSweeney, D. A.: 1970. Ch. 10, #184, p. 260.
Magaro, P.: 1971. Ch. 5, #31, p. 86.
Makino, S.: 1965. Ch. 2, #45, p. 29.
   1966. Ch. 2, #44, #46, p. 29.
Malamud, I.: 1941. Ch. 10, #198, p. 262.
Malamud, W.: 1941. Ch. 10, #198, p. 262.
Maldonado, N.: 1969. Ch. 17, #83, p. 391.
Malitz, S.: 1971. Ch. 15, #18, p. 349.
Manosevitz, M.: 1971. Ch. 9, #65, p. 208, p. 209.
Mantle, D. J.: 1964. Ch. 2, #21, p. 19, p. 25; Ch. 14, #22, p. 335.
Maplestone, P. A.: 1962. Ch. 16, #46, p. 371.
Marcus, J.: 1969. Ch. 7, #6, p. 152.
Margolese, M. S.: 1970. Ch. 12, #31, p. 311.
Market, C. L.: 1957. Ch. 17, #11, p. 388.
   1959. Ch. 17, #12, p. 388.
Markkanen, T.: 1957. Ch. 15, #23, p. 349.
   1966. Ch. 6, #119, p. 140; Ch. 9, #67, p. 208; Ch. 16, #84, p. 376.

Marks, V.: 1967. Ch. 10, #96, p. 237, p. 262.
Marsters, R.: 1964. Ch. 18, #25, p. 402.
　1967. Ch. 18, #28, p. 405.
Mårtens, S.: 1969. Ch. 15, #14, p. 348.
Martensson, E.: 1967. Ch. 17, #96, p. 391.
Martin, C. E.: 1948. Ch. 12, #1, p. 301.
　1953. Ch. 12, #2, p. 301.
Martin, R.: 1957. Ch. 10, #113, p. 245, p. 249.
Martynova, R. P.: 1936. Ch. 18, #7, p. 401.
Marver, H. S.: 1970. Ch. 16, #38, p. 369.
　1972. Ch. 16, #36, p. 369.
Mason, H. H.: 1951. Ch. 17, #41, p. 390.
Masters, W. H.: 1971. Ch. 12, #32, p. 311.
Matarazzo, J. D.: 1960. Ch. 18, #35, p. 411.
Mathai, C. K.: 1966. Ch. 17, #158, p. 392.
Matheny, A. P.: 1971. Ch. 6, #81, p. 126; Ch. 9, #102, p. 217.
Mather, K.: 1949. Ch. 9, #19, p. 201.
　1964. Ch. 2, #8, p. 12, p. 13.
　1965. Ch. 2, #9, p. 13.
Matsubara, S.: 1968. Ch. 17, #82, p. 391.
Matzilewich, B.: 1972. Ch. 10, #173, p. 260.
Mauz, F.: 1937. Ch. 10, #178, p. 260.
May, J.: 1951. Ch. 11, #13, p. 283.
Mcardle, R. B.: 1951. Ch. 17, #51, p. 390.
Mead, A. P.: 1953. Ch. 4, #13, p. 54, p. 61.
Mednick, B.: 1971. Ch. 6, #78, p. 126.
Mednick, S. A.: 1968. Ch. 5, #33, p. 86; Ch. 10, #227, p. 264, #244, p. 268.
　1971. Ch. 6, #78, p. 126.
　1973. Ch. 10, #228, p. 264, p. 268.
Mehl, E.: 1969. Ch. 17, #94, p. 391.
Mehlman, B.: 1952. Ch. 13, #4, p. 317.
Meier, H.: 1963. Ch. 16, #27, p. 368.
Meili, R.: 1972. Ch. 10, #243, p. 268.
Meili-Dworetzki, G.: 1972. Ch. 10, #243, p. 268.
Meisler, A.: 1964. Ch. 17, #34, p. 390.
Meisner, L.: 1971. Ch. 10, #30, p. 230.
Mellman, W. J.: 1967. Ch. 17, #5, p. 386.
Mellor, C. S.: 1967. Ch. 14, #44, p. 337.
Mendelson, M.: 1960. Ch. 10, #197, p. 262.
Meredith, W.: In press. Ch. 6, #115, p. 139.
Merleau-Ponty, M.: 1962. Ch. 1, #6, p. 3.
Merrill, M. A.: 1960. Ch. 5, #22, p. 84.
Merton, B. B.: 1958. Ch. 18, #13, p. 401.
Mesnikoff, A.: 1960. Ch. 12, #39, p. 313.
Metrakos, J. D.: 1969. Ch. 16, #77, p. 375.
Metrakos, K.: 1969. Ch. 16, #77, p. 375.
Meyer, A.: 1951. Ch. 13, #15, p. 318.
Meyer, A. E.: 1960. Ch. 10, #38, p. 231.
Meyerhoff, J. L.: 1970. Ch. 16, #71, p. 374.
Michael, S. T.: 1962. Ch. 11, #6, p. 282.
Mickey, J. S.: 1946. Ch. 9, #45, p. 202.
Midlo, C.: 1943. Ch. 6, #18, p. 96.
　1966. Ch. 14, #36, p. 337, p. 338.
Miele, F.: 1969. Ch. 6, #49, p. 109.

Miles, C. C.: 1968. Ch. 10, #15, p. 253.
Milgram, S.: 1961. Ch. 6, #63, p. 118.
Miller, A. L.: 1967. Ch. 17, #74, p. 391.
Miller, J. R.: 1966. Ch. 14, #37, p. 337.
Miller, M. C.: 1970. Ch. 10, #211, p. 263.
Miller, R. B.: 1943. Ch. 10, #164, p. 258.
Miller, R. W.: 1966. Ch. 2, #42, p. 29.
Ming, M. P.: 1967. Ch. 9, #25, p. 201.
Mirenova, A. N.: 1935. Ch. 6, #93, p. 134.
　1936. Ch. 6, #94, p. 134.
Mitoma, C.: 1957. Ch. 17, #60, p. 390.
Mitsuda, H.: 1967. Ch. 14, #10, p. 331, p. 332.
　1972. Ch. 14, #11, p. 331, p. 332.
Mittler, P.: 1971. Ch. 6, #6, p. 91.
Mix, L.: 1971. Ch. 16, #90, p. 377.
Mock, M. B.: 1966. Ch. 17, #97, p. 391.
Mohrman, R. K.: 1962. Ch. 3, #5, p. 35.
Mohyuddin, F.: 1963. Ch. 17, #75, p. 391.
Moller, F.: 1959. Ch. 17, #12, p. 388.
Moller, N. B.: 1935. Ch. 14, #51, p. 337, p. 341.
Money, J.: 1955. Ch. 12, #21, #22, p. 310.
　1968. Ch. 10, #21, p. 230, p. 260; #230, p. 264, p. 265, p. 266.
　1969. Ch. 10, #29, p. 230, p. 260, p. 264, p. 266.
　1970. Ch. 12, #4, p. 304.
Monod, J.: 1971. Ch. 4, #7, p. 52, p. 61.
Montagu, A.: 1968. Preface, #1, p. xi.
　1969. Preface, #2, p. xi.
　1970. Ch. 1, #7, p. 3.
Moore, T. V.: 1946. Ch. 10, #111, p. 241, p. 259.
Moran, P. A. P.: 1969. Ch. 12, #28, p. 311.
Morel, B. A.: 1852-1853. Ch. 13, #7, p. 317.
Morganti, C.: 1954. Ch. 2, #43, p. 29.
　1967. Ch. 17, #130, p. 392.
Morikawa, T.: 1967. Ch. 17, #70, p. 390.
Moroz, L.: 1964. Ch. 17, #137, p. 392.
Morrison, J. R.: 1971. Ch. 16, #98a, p. 378.
Morrison, J. R.: 1973. Ch. 16, #98c, p. 378.
Morrow, A. C.: 1968. Ch. 16, #12, p. 366.
Morsing, C.: 1956. Ch. 10, #114, p. 247.
Morton, N. E.: 1959. Ch. 2, #2, p. 7.
　1962. Ch. 2, #3, p. 7.
　1967. Ch. 9, #25, p. 201.
Mosely, E. C.: 1964. Ch. 10, #170, p. 259.
Moser, H. W.: 1971. Ch. 17, #95, p. 391.
　1972. Ch. 10, #173, p. 260.
Mossell, D. A. A.: 1966. Ch. 17, #57, p. 390.
Motulsky, A. G.: 1957. Ch. 16, #1, p. 363.
　1964. Ch. 16, #4, p. 363, p. 367.
　1968. Ch. 16, #12, p. 366.
　1969. Ch. 16, #11, p. 366.
　1971. Ch. 16, #6, p. 363, p. 366.
　1972. Ch. 16, #88, p. 376.
Mudd, S. H.: 1967. Ch. 17, #80, p. 391.
　1970. Ch. 16, #78, p. 376.
Muller, J. C.: 1955. Ch. 16, #63, p. 373.

Mura, E.: 1971. Ch. 6, #78, p. 126.
Murken, J. D.: 1973. Ch. 10, #175, p. 260.
Murray, P.: 1965. Ch. 17, #67, p. 390.
Murphy, J. V.: 1971. Ch. 17, #95, p. 391.
Mussacchio, J. M.: 1969. Ch. 16, #58, p. 372.
Myrianthopoulos, N. C.: 1962. Ch. 15, #20, p. 349; Ch. 16, #66, p. 373.
   1967. Ch. 16, #67, p. 373.
   1970. Ch. 6, #21, p. 96.
Myrtek, M.: 1966. Ch. 10, #157, p. 255.

N

Nadler, H. L.: 1970. Ch. 17, #106, p. 392.
Naeslund, J.: 1956. Ch. 6, #96, p. 135.
Naiman, J. L.: 1964. Ch. 17, #37, p. 390.
   1967. Ch. 17, #22, p. 390.
Naka, S.: 1965. Ch. 15, #50, p. 355.
Nakanishi, M.: 1961. Ch. 15, #28, p. 350.
Necheles, T. F.: 1969. Ch. 17, #83, p. 391.
Nedoma, K.: 1958. Ch. 12, #24, p. 310.
Neel, J. V.: 1954. Ch. 9, #14, p. 201.
   1967. Ch. 6, #16, p. 95.
   1970. Ch. 3, #10, p. 36.
Nelson, D. H.: 1953. Ch. 5, #6, p. 73.
Nesselroade, J. R.: 1967. Ch. 9, #42, p. 202.
Newman, H. H.: 1937. Ch. 6, #83, p. 131; Ch. 8, #52, p. 174; Ch. 9, #52, p. 201; Ch. 11, #40, p. 294.
   1972. Ch. 8, #53, p. 175.
Newman, R. W.: 1952. Ch. 10, #102, p. 239.
Ng, L.: 1971. Ch. 16, #109, p. 379.
Ng, W. G.: 1968. Ch. 17, #156, p. 392.
Nias, D.: 1973. Ch. 9, #50, p. 202.
Nichols, J. R.: 1967. Ch. 16, #94, p. 377.
Nichols, R. C.: 1965. Ch. 6, #40, p. 103, p. 140; Ch. 9, #15, p. 201.
   1966. Ch. 6, #15, p. 95; Ch. 9, #63, p. 208.
Nicholay, E.: 1939. Ch. 9, #94, p. 216.
Nielsen, J.: 1965. Ch. 11, #8, p. 282.
   1967. Ch. 6, #7, p. 92.
   1969. Ch. 10, #176, p. 260.
Nimer, R. A.: 1960. Ch. 5, #7, p. 73.
Nishikawa, M.: 1965. Ch. 17, #25, p. 390.
Nolan, E. G.: 1959. Ch. 9, #91, p. 216.
   1961. Ch. 9, #92, p. 216.
Norring, A.: 1963. Ch. 14, #46, p. 337.
Norum, K. R.: 1967. Ch. 17, #110, p. 392.
Noyes, A.: 1954. Ch. 10, #199, p. 262.
Nugent, C. A.: 1960. Ch. 5, #7, p. 73.
Nunnally, J. C.: 1967. Ch. 5, #1, p. 67, p. 71, p. 78, p. 79.
Nyhan, W. L.: 1964. Ch. 16, #43, p. 370, p. 379; Ch. 17, #111, p. 392.
Nyman, G. E.: 1956. Ch. 10, #104, p. 239, p. 248, p. 252, p. 255, p. 256, p. 266, #114, p. 247, #148, p. 252.

O

Obetholzer, V. G.: 1962. Ch. 17, #76, p. 391.
   1967. Ch. 17, #118, p. 392.
O'Brien, J. S.: 1968. Ch. 17, #98, p. 391.
   1969. Ch. 17, #90, p. 391.
Ödegärd, O.: 1963. Ch. 14, #7, p. 331, p. 332.
   1972. Ch. 14, #13, p. 332.
Ohno, S.: 1959. Ch. 2, #34, p. 20.
   1971. Ch. 12, #16, p. 309.
Okada, S.: 1968. Ch. 17, #98, p. 391.
Oki, T.: 1967. Ch. 11, #32, p. 287.
Okuno, G.: 1965. Ch. 17, #25, p. 390.
Omenn, G. S.: 1972. Ch. 8, #2, p. 164; Ch. 16, #88, p. 376.
Opitz, J. M.: 1971. Ch. 10, #30, p. 230.
Opler, M. K.: 1962. Ch. 11, #6, p. 282.
Oppenheim, A. N.: 1952. Ch. 10, #167, p. 259.
Orgain, E. S.: 1955. Ch. 16, #63, p. 373.
Orlyndgen, E.: 1945. Ch. 9, #96, p. 216.
Orr, H. K.: 1965. Ch. 6, #109, p. 138, p. 139.
Osborne, R. H.: 1959. Ch. 6, #24, p. 97; Ch. 10, #9, p. 230, p. 264.
   1961. Ch. 6, #43, p. 106.
Osborne, R. T.: 1968. Ch. 6, #50, p. 109.
   1969. Ch. 6, #49, p. 109.
Oski, F. A.: 1967. Ch. 17, #22, p. 390.
Osserman, E. F.: 1951. Ch. 10, #68, p. 234.
Ostergaard, L.: 1968. Ch. 14, #16, p. 332.
Ostertag, W.: 1958. Ch. 10, #10, p. 230.
Ottinger, D. R.: 1964. Ch. 7, #10, p. 153.
Ousted, C.: 1972. Ch. 10, #39, p. 231.
Overall, J. E.: 1972. Ch. 10, #60, p. 233, p. 240, p. 249.
Owen, D. R.: 1972. Ch. 10, #177, p. 260.

P

Page, J. G.: 1968. Ch. 15, #15, p. 348; Ch. 16, #107, p. 379.
   1969. Ch. 16, #111, p. 379.
   1971. Ch. 16, #89, p. 377.
Paglia, D. E.: 1966. Ch. 17, #28, p. 390.
   1967. Ch. 17, #22, p. 390.
   1968. Ch. 17, #23, p. 390.
Paine, R. S.: 1968. Ch. 16, #98, p. 378.
Pallister, P. D.: 1971. Ch. 10, #30, p. 230.
Pant, S. S.: 1972. Ch. 10, #173, p. 260.
Pare, C. M. B.: 1956. Ch. 12, #23, p. 310.
   1962. Ch. 15, #6, p. 348; Ch. 16, #59, p. 372.
   1970. Ch. 15, #7, p. 348.
   1971. Ch. 16, #60, p. 372.
Parker, J. B.: 1963. Ch. 10, #129, #132, p. 249.
Parker, N.: 1964. Ch. 12, #43, p. 314.
   1966. Ch. 6, #58, p. 118; Ch. 11, #27, p. 284.
Parnell, R. W.: 1955. Ch. 10, #202, p. 263.
   1958. Ch. 10, #80, p. 236, p. 239, p. 251, p. 256, p. 263.

Parr, C. W.: 1963. Ch. 17, #154, p. 392.
    1968. Ch. 17, #155, p. 392.
Parr, D.: 1960. Ch. 12, #25, p. 310.
Partanen, J.: 1966. Ch. 6, #119, p. 140; Ch. 9, #67, p. 208; Ch. 16, #84, p. 376.
Partington, M. W.: 1967. Ch. 17, #64, p. 390.
Partridge, E.: 1958. Ch. 8, #6, p. 164.
Pasamanick, B.: 1959. Ch. 13, #5, p. 317.
    1961. Ch. 18, #22, p. 401, p. 402.
    1965. Ch. 18, #47, p. 421.
Passananti, G. T.: 1971. Ch. 16, #89, p. 377, #109, p. 379.
Patau, K.: 1962. Ch. 14, #34, p. 337.
Paterson, D. G.: 1930. Ch. 10, #115, p. 248.
Patrick, A. D.: 1965. Ch. 17, #88, p. 391.
    1969. Ch. 17, #109, p. 392.
Pattabiraman, T.: 1963. Ch. 17, #93, p. 391.
Patterson, J. C.: 1970. Ch. 9, #116, p. 220.
Patton, R. G.: 1969. Ch. 10, #23, p. 230.
Pauling, L.: 1949. Ch. 17, #9, p. 387.
    1970. Ch. 16, #82, p. 376.
Paulson, G. W.: 1970. Ch. 16, #53, p. 372.
    1972. Ch. 16, #112, p. 372.
Pearson, K.: 1903. Ch. 9, #26, p. 201.
    1906. Ch. 9, #28, p. 201.
Penrose, L. S.: 1935. Ch. 2, #16, p. 16.
    1955. Ch. 6, #10, p. 94.
    1959. Ch. 17, #66, p. 390.
    1963. Ch. 8, #38, p. 168.
Pereira, G.: 1971. Ch. 14, #41, p. 337.
Perel, J. M.: 1971. Ch. 15, #18, p. 349; Ch. 16, #110, p. 379.
Perkoff, G. T.: 1960. Ch. 5, #7, p. 73.
Perrin, M. J.: 1961. Ch. 18, #31, p. 409.
Perris, C.: 1966. Ch. 10, #131, p. 249, p. 262, #193, #194, p. 262.
Peterson, M. R.: 1973. Ch. 14, #17, p. 332.
Pfeiffer, C. C.: 1956. Ch. 18, #45, p. 420.
Phillips, L.: 1968. Ch. 5, #28, p. 86.
Philpot, G. R.: 1954. Ch. 17, #74, p. 391.
Pickford, J. H.: 1967. Ch. 9, #44, p. 202.
Pickford, R. W.: 1949. Ch. 8, #31, p. 166.
Pierce, C. M.: 1956. Ch. 15, #48, p. 354.
Pihkanen, T. A.: 1957. Ch. 15, #23, p. 349.
Pinel, P.: 1801. Ch. 13, #10, p. 317.
Pinnell, S. R.: In press. Ch. 17, #120, p. 392.
Pitts, G. C.: 1951. Ch. 10, #68, p. 234.
Pivnicki, D.: 1968. Ch. 10, #171, p. 259, p. 262.
Plank, E. N.: 1958. Ch. 6, #120, p. 140.
Plank, R.: 1964. Ch. 6, #67, p. 120.
Plotrowski, Z. A.: 1960. Ch. 12, #40, p. 313.
Point, P.: 1964. Ch. 10, #170, p. 259.
Polani, P. E.: 1959. Ch. 12, #14, p. 308.
    1968. Ch. 14, #26, p. 336.
Polednak, A. P.: 1971. Ch. 10, #185, p. 262.
    1972. Ch. 14, #40, p. 337.

Poll, H.: 1935. Ch. 14, #52, p. 337, p. 341.
Pollin, W.: 1962. Ch. 11, #36, p. 287.
    1968. Ch. 6, #77, p. 125; Ch. 11, #63, p. 297.
    1969. Ch. 11, #1, p. 279, p. 284, p. 286; Ch. 13, #26, p. 321.
    1970. Ch. 11, #28, p. 286.
    1972. Ch. 10, #188, p. 262.
    In press. Ch. 11, #64, p. 297.
Pollock, R. A.: 1971. Ch. 16, #51, p. 371.
Pomeroy, W. B.: 1948. Ch. 12, #1, p. 301.
    1953. Ch. 12, #2, p. 301.
Ponicki, D.: 1968. Ch. 10, #170, p. 259.
Pons, J.: 1959. Ch. 14, #48, p. 337.
Popenoe, P.: 1928. Ch. 3, #7, p. 36.
Porter, I. H.: 1962. Ch. 17, #38, p. 392.
Porter, P. J.: 1968. Ch. 2, #31, p. 20.
Post, F.: 1965. Ch. 10, #169, p. 259.
Postnikova, E. N.: 1936. Ch. 18, #7, p. 401.
Povey, S.: 1968. Ch. 17, #24, p. 390.
Powell, W.: 1964. Ch. 18, #25, p. 402, #43, p. 418.
    1967. Ch. 18, #28, p. 405.
Prader, A.: 1963. Ch. 17, #29, p. 390.
Pratt, R. T. C.: 1956. Ch. 17, #30, p. 390.
    1965. Ch. 15, #12, p. 348; Ch. 16, #17, p. 367, p. 373.
    1967. Ch. 16, #76, p. 375.
Prell, D.: 1951. Ch. 9, #12, p. 200, p. 210, p. 216.
    1952. Ch. 9, #78, p. 210; Ch. 11, #45, p. 292.
Price, J.: 1962. Ch. 10, #191, p. 262.
Price-Evans, D. A.: 1965. Ch. 15, #11, p. 348, #12, p. 348; Ch. 16, #17, p. 367, p. 373.
    1969. Ch. 16, #5, p. 363, p. 366.
    1970. Ch. 15, #16, p. 348.
    1971. Ch. 16, #25, p. 368.
Proctor, A.: 1970. Ch. 14, #42, p. 337.
Propp, R. P.: 1968. Ch. 17, #135, p. 392.
Pryor, W. W.: 1955. Ch. 16, #63, p. 373.
Puck, T. T.: 1967. Ch. 2, #22, p. 19.
Pugh, T. F.: 1970. Ch. 10, #245, p. 268.
Pugliath, C.: 1970. Ch. 16, #29, p. 368.

Q
Quinn, G. P.: 1962. Ch. 15, #17, p. 348.

R
Rabin, A. J.: 1952. Ch. 9, #90, p. 216.
Rabach, J.: 1958. Ch. 12, #24, p. 310.
Rachild, I.: 1968. Ch. 6, #17, p. 95.
Radeker, A. G.: 1970. Ch. 16, #38, p. 369.
Rado, S.: 1949. Ch. 12, #35, p. 313.
Ranier, J. D.: 1960. Ch. 12, #39, p. 313.
    1966. Ch. 13, #33, p. 325.
Raney, E.: 1939. Ch. 9, #105, p. 217.
Ranney, H. M.: 1967. Ch. 17, #26, p. 390.
Raphael, L. G.: 1962. Ch. 14, #47, p. 337.
Raphael, T.: 1962. Ch. 14, #47, p. 337.

Rapley, S.: 1968. Ch. 17, #139, p. 392.
Rappaport, M.: 1971. Ch. 5, #32, p. 86.
Rathbun, J. C.: 1948. Ch. 17, #104, p. 391.
　1961. Ch. 17, #105, p. 391.
　1963. Ch. 17, #75, p. 391.
Rauth, J. E.: 1937. Ch. 18, #21, p. 401, p. 409.
Rechtschaffen, A.: 1962. Ch. 15, #44, p. 354.
　1964. Ch. 15, #45, p. 354.
Record, R. G.: 1970. Ch. 6, #80, p. 126.
Reding, G. R.: 1964. Ch. 15, #45, p. 354.
Redlich, F.: 1966. Ch. 11, #5, p. 281.
Reed, E. W.: 1962. Ch. 8, #108, p. 191.
　1965. Ch. 8, #28, p. 166, p. 182, #112, p. 192.
Reed, S. C.: 1962. Ch. 8, #108, p. 191.
　1965. Ch. 8, #28, p. 166, p. 182, #112, p. 192.
　1968. Ch. 8, #42, p. 172.
Rees, L.: 1945. Ch. 10, #66, p. 233, p. 241, p. 247, p. 248, p. 256, p. 258, p. 259, p. 263.
　1950. Ch. 10, #64, p. 233, p. 255; #65, p. 233, p. 258.
　1962. Ch. 15, #6, p. 348; Ch. 16, #59, p. 372.
　1973. Ch. 10, #2, p. 230, p. 231, p. 233, p. 239, p. 240, p. 241, p. 248, p. 249, p. 250, p. 259, p. 260, p. 262, p. 263, p. 267.
Reich, T.: 1969. Ch. 16, #55, p. 372.
Reichenbach, H.: 1954. Ch. 2, #11, p. 13.
Reiss, D.: 1967. Ch. 11, #58, p. 296.
　1971. Ch. 11, #59, p. 296.
Reisse, H.: 1961. Ch. 11, #51, p. 292.
Rempel, H.: 1967. Ch. 9, #44, p. 202.
Rendle-Short, J.: 1969. Ch. 3, #31, p. 43.
Rennie, T. A. C.: 1962. Ch. 11, #6, p. 282.
Rethioré, M. O.: 1968. Ch. 14, #25, p. 336.
Rey, E. R.: 1970. Ch. 10, #133, p. 249, #134, p. 249, p. 262.
Reynolds, E. L.: 1949. Ch. 10, #88, p. 237.
　1950. Ch. 10, #71, p. 234, p. 236.
Reznikoff, M.: 1966. Ch. 9, #57, p. 206.
Richardson, M. W.: 1939. Ch. 5, #5, p. 72.
Rife, D. C.: 1938. Ch. 18, #8, p. 401.
Ringel, S. A.: 1972. Ch. 16, #112, p. 372.
Rinkel, M.: 1959. Ch. 10, #241, p. 251.
Ritter, H.: 1958. Ch. 6, #19, p. 96.
Rivas, M. L.: 1972. Ch. 16, #52, p. 371.
Rivat, L.: 1961. Ch. 17, #128, p. 392.
Rivers, S.: 1959. Ch. 17, #86, p. 391.
Roberts, J. A. F.: 1941. Ch. 8, #104, p. 191.
　1958. Ch. 8, #40, p. 168.
Roberts, R. C.: 1967. Ch. 8, #73, p. 181.
Robbins, P. W.: 1959. Ch. 17, #52, p. 390.
Robinson, A.: 1967. Ch. 2, #22, p. 19.
Robinson, H. M.: 1961. Ch. 17, #105, p. 391.
Robinson, L. R.: 1966. Ch. 2, #41, p. 29.
Robson, E. B.: 1963. Ch. 17, #145, p. 392.
　1967. Ch. 17, #142, p. 392.
Rodnick, E. H.: 1969. Ch. 5, #30, p. 86.

Röhrborn, G.: 1971. Ch. 2, #50, p. 29.
Rokkones, T.: 1967. Ch. 17, #68, p. 390.
Roman-Goldzieher, K.: 1945. Ch. 9, #97, p. 216.
Roos, D. E.: 1956. Ch. 9, #37, p. 202.
Ropartz, C.: 1961. Ch. 17, #128, p. 392.
Rosen, S.: 1959. Ch. 15, #41, p. 352.
Rosenberg, L. E.: 1968. Ch. 17, #119, p. 392.
　1972. Ch. 16, #79, p. 376.
Rosenbloom, F. M.: 1967. Ch. 17, #112, p. 392.
Rosenthal, D.: 1958. Ch. 11, #9, p. 282.
　1968. Ch. 13, #28, p. 321; Ch. 14, #6, p. 331, p. 332, #16, p. 332; Ch. 16, #86, p. 376.
　1970. Ch. 3, #32, p. 43, p. 44; Ch. 8, #45, p. 172; Ch. 11, #10, p. 282, p. 292, p. 293.
　Ch. 13, #25, p. 321; Ch. 16, #65, p. 373.
　1971. Ch. 8, #46, p. 172.
Rosenthal, R.: 1966. Ch. 7, #20, p. 162.
Rosenzweig, M. R.: 1968. Ch. 3, #16, p. 38.
　1971. Ch. 11, #56, p. 295.
Rosner, F.: 1968. Ch. 14, #43, p. 337.
Rothhammer, F.: 1971. Ch. 14, #41, p. 337.
Rowland, L. P.: 1967. Ch. 17, #26, p. 390.
Royce, J. R.: 1966. Ch. 10, #155, p. 255.
Rubricht, W. C.: 1964. Ch. 15, #45, p. 354.
Rudersdorf, M.: 1968. Ch. 10, #186, p. 262.
Rüdin, E.: 1953. Ch. 11, #31, p. 287.
Russell, A.: 1962. Ch. 17, #76, p. 391.
Russell, J. D.: 1967. Ch. 17, #7, p. 387.
Rutter, M.: 1963. Ch. 7, #15, p. 158.
　1972. Ch. 10, #24, p. 230.
Ryan, E.: 1956. Ch. 16, #9, p. 366.

S

Sainsbury, M. J.: 1962. Ch. 15, #6, p. 348; Ch. 16, #59, p. 372.
Saksi, T.: 1967. Ch. 11, #34, p. 287.
Salmon, M. R.: 1931. Ch. 18, #3, p. 401.
Sammon, J. W.: 1970. Ch. 14, #42, p. 337.
Samuels, I.: 1965. Ch. 15, #29, p. 351, p. 352.
Samuels, L. T.: 1960. Ch. 5, #7, p. 73.
Sanchez, C. A.: 1964. Ch. 14, #38, p. 337.
Sands, S. L.: 1941. Ch. 10, #198, p. 262.
Sandberg, A. A.: 1953. Ch. 5, #6, p. 73.
Sandhoff, K.: 1968. Ch. 17, #91, p. 391.
Sanford, R. N.: 1943. Ch. 10, #164, p. 258.
Sank, D.: 1968. Ch. 14, #53, p. 337, p. 341.
Sarswathi, S.: 1963. Ch. 17, #93, p. 391.
Sartorius, N.: 1969. Ch. 9, #77, p. 210.
Saslow, G.: 1960. Ch. 18, #35, p. 411.
Saudek, R.: 1932. Ch. 9, #93, p. 216.
Saunders, D. R.: 1963. Ch. 6, #116, p. 140.
Saunders, S. J.: 1968. Ch. 16, #49, p. 371.
Scarr, S.: 1964. Ch. 9, #111, p. 218.
　1966. Ch. 9, #68, p. 209, #69, p. 209, p. 218, p. 219.
　1968. Ch. 6, #65, p. 119; Ch. 11, #18, p. 284.

1969. Ch. 6, #75, p. 124; Ch. 9, #70, p. 209.
Scarr-Salapatek, S.: 1971. Ch. 6, #51, p. 111, p. 112; Ch. 8, #62, p. 179, #93, p. 190.
Schachter, M.: 1952. Ch. 9, #85, p. 216.
Schaefer, E. S.: 1958. Ch. 6, #106, p. 138, p. 139.
  1963. Ch. 6, #107, p. 138.
  1964. Ch. 6, #108, p. 138; Ch. 10, #19, p. 230, p. 268.
Schaefer, O.: 1971. Ch. 16, #90, p. 377.
Schaefer, T., Jr.: 1961. Ch. 10, #20, p. 230.
Schaffer, H. R.: 1971. Ch. 3, #26, p. 43.
Schallek, W.: 1952. Ch. 18, #46, p. 420.
Schapira, F.: 1961. Ch. 17, #46, p. 390.
Schapira, G.: 1961. Ch. 17, #46, p. 390.
Scheinfeld, A.: 1943. Ch. 10, #216, p. 263.
Schellenberg, J. A.: 1960. Ch. 9, #39, p. 202.
Schildkraut, J. J.: 1969. Ch. 16, #57, p. 372.
Schlegel, W. S.: 1957. Ch. 10, #41, p. 231, p. 233.
Schlesinger, K.: 1970. Ch. 16, #32, p. 369, p. 375.
Schmid, K.: 1964. Ch. 17, #137, p. 392.
Schmid, R.: 1959. Ch. 17, #52, p. 390.
  1972. Ch. 16, #36, p. 369.
Schneider, A. S.: 1965. Ch. 17, #27, p. 390.
  1966. Ch. 17, #28, p. 390.
  1967. Ch. 17, #22, p. 390.
Schoenfeldt, L. F.: 1968. Ch. 9, #120, p. 208.
Schorer, C. E.: 1968. Ch. 13, #6, p. 317.
Schroter, W.: 1965. Ch. 17, #36, p. 390.
Schubert, W. K.: 1966. Ch. 17, #53, p. 390.
Schuckit, M. A.: 1972. Ch. 16, #87, p. 376.
Schull, W. J.: 1955. Ch. 6, #11, p. 94.
  1965. Ch. 9, #14, p. 201.
Schulsinger, F.: 1968. Ch. 10, #227, p. 264; Ch. 13, #28, p. 321; Ch. 14, #6, p. 331, p. 332, #16, p. 332.
  1971. Ch. 6, #78, p. 126.
  1973. Ch. 10, #228, p. 264, p. 268.
Schultz, A.H.: 1956. Ch. 6, #3, p. 90.
Schumer, F.: 1965. Ch. 9, #10, p. 200, p. 209.
Schwarz, V.: 1956. Ch. 17, #43, p. 390.
Schwidetzky, I.: 1970. Ch. 10, #3, p. 230.
Scoggins, R.: 1959. Ch. 17, #86, p. 391.
Scott, J. P.: 1957. Ch. 3, #25, p. 43.
  1958. Ch. 3, #11, p. 36.
  1965. Ch. 3, #4, p. 35, p. 36, p. 41, p. 42, p. 44, #17, p. 39.
  1968. Ch. 3, #35, p. 44.
  1970. Ch. 3, #12, p. 37.
  1973. Ch. 3, #28, p. 43.
Scottish Council for Research in Education: 1949. Ch. 8, #106, p. 191.
Scowen, E. F.: 1967. Ch. 17, #113, p. 392.
Scriver, C. R.: 1962. Ch. 17, #72, p. 391.
  1967. Ch. 17, #63, p. 390.
Scudder, C. G.: 1967. Ch. 16, #31, p. 369.
Searle, L. V.: 1949. Ch. 3, #22, p. 41.

Seegmiller, J. E.: 1958. Ch. 17, #2, p. 386, p. 390.
  1963. Ch. 17, #78, p. 391.
  1967. Ch. 17, #112, p. 392.
Seeman, E.: 1932. Ch. 9, #93, p. 216.
Segraves, R. T.: 1970. Ch. 10, #96, p. 237, p. 253, p. 255, p. 264, p. 266.
Selander, R.: 1969. Ch. 17, #14, p. 388.
  1970. Ch. 17, #15, p. 388.
Selin, M. J.: 1959. Ch. 16, #15, p. 366.
Sellers, E. M.: 1970. Ch. 16, #105, p. 379.
Sells, S. B.: 1968. Ch. 9, #7, p. 199.
Seltzer, C. C.: 1943. Ch. 10, #159, p. 256.
  1945. Ch. 10, #86, p. 237, p. 248, p. 253, p. 254, p. 258.
  1946. Ch. 10, #83, p. 236, p. 254.
  1964. Ch. 10, #209, p. 263.
Senay, E. C.: 1973. Ch. 3, #28, p. 43.
Shafer, I. A.: 1962. Ch. 17, #72, p. 391.
Shäfers, F.: 1940. Ch. 6, #53, p. 113.
Shapiro, A.: 1962. Ch. 15, #44, p. 354.
Shapiro, D.: 1966. Ch. 17, #89, p. 391.
Shapiro, R.: 1968. Ch. 6, #17, p. 95.
Shaw, C. R.: 1962. Ch. 17, #17, p. 388.
  1965. Ch. 17, #18, p. 388.
  1970. Ch. 17, #16, p. 388, #19, p. 389.
  1971. Ch. 17, #21, p. 389.
Sheard, M. H.: 1971. Ch. 10, #242, p. 268.
Sheldon, W.: 1964. Ch. 17, #108, p. 392.
Sheldon, W. H.: 1940. Ch. 10, #42, p. 231, p. 234, p. 236, p. 237, p. 239, p. 245.
  1941. Ch. 10, #116, p. 248, p. 251, p. 253.
  1942. Ch. 10, #43, p. 231, p. 250, p. 253, p. 257, p. 258.
  1949. Ch. 10, #82, p. 236, p. 239, p. 250, p. 263.
  1954. Ch. 10, #76, p. 234, p. 237, p. 242.
  1969. Ch. 10, #81, p. 236.
Shen, F. H.: 1968. Ch. 16, #96, p. 377.
Sherman, L. J.: 1964. Ch. 10, #170, p. 262.
Shetty, T.: 1971. Ch. 16, #103, p. 378.
Shields, J.: 1954. Ch. 11, #21, p. 284.
  1962. Ch. 6, #60, p. 118, p. 131; Ch. 8, #64, p. 179; Ch. 9, #81, p. 212, p. 213, p. 218.
  Ch. 10, #16, p. 230, p. 231, p. 264; Ch. 11, #41, p. 292; Ch. 16, #8, p. 365.
  1966. Ch. 13, #22, p. 320.
  1967. Ch. 6, #36, p. 100; Ch. 14, #9, p. 331.
  1968. Ch. 12, #41, p. 314.
  1971. Ch. 3, #33, p. 44; Ch. 5, #29, p. 86; Ch. 11, #48, p. 292.
Shimakawa, M.: 1966. Ch. 15, #51, #52, #53, p. 355.
Shipley, P. W.: 1967. Ch. 6, #20, p. 96.
Shopsin, B.: 1971. Ch. 16, #72, p. 374.
Shreffler, D. C.: 1967. Ch. 17, #133, p. 392.
Siegel, B.: 1961. Ch. 10, #235, p. 266.
Siegel, M.: 1966. Ch. 14, #27, p. 337.
Siemens, H.: 1924. Ch. 15, #34, p. 351, p. 352.

Signori, E. I.: 1967. Ch. 9, #44, p. 202.
Silk, E.: 1952. Ch. 17, #101, p. 391.
  1960. Ch. 17, #144, p. 392.
Sills, F. D.: 1950. Ch. 10, #72, p. 234, p. 238, p. 241, p. 242, p. 256.
Silverman, J.: 1971. Ch. 5, #32, p. 86.
Silverberg, M.: 1967. Ch. 17, #63, p. 390.
Simmons, J. E.: 1964. Ch. 7, #10, p. 153.
Simon, E. R.: 1968. Ch. 17, #23, p. 390.
Simon, J. R.: 1962. Ch. 14, #28, p. 337.
Simon, Th.: 1905. Ch. 5, #17, p. 82; Ch. 8, #19, p. 165.
Simpson, N. E.: 1966. Ch. 17, #146, p. 392.
Sinclair, L.: 1962. Ch. 17, #76, p. 391.
Singer, J.: 1949. Ch. 17, #9, p. 387.
Singer, K.: 1972. Ch. 10, #97, p. 237, p. 255, p. 259, p. 260, p. 262, p. 263.
Singer, M. T.: 1967. Ch. 11, #60, p. 296.
Singh, S.: 1967. Ch. 14, #45, p. 337.
Sinnot, J. J.: 1937. Ch. 18, #21, p. 401, p. 409.
Sjöbring, H.: 1973. Ch. 10, #149, p. 252.
Sjöerdsma, A.: 1963. Ch. 17, #78, p. 391.
Sjöqvist, F.: 1967. Ch. 15, #13, p. 348.
  1969. Ch. 15, #14, p. 348.
  1970. Ch. 15, #24, p. 349; Ch. 16, #24, p. 368, p. 373.
  1971. Ch. 16, #22, p. 368.
Skells, H. M.: 1949. Ch. 8, #55, p. 175.
Skerlj, B.: 1953. Ch. 10, #100, p. 239.
Skodak, M.: 1949. Ch. 8, #55, p. 175.
Skude, G.: 1963. Ch. 18, #23, p. 402.
Slater, E.: 1944. Ch. 10, #205, p. 263; Ch. 11, #52, p. 293.
  1953. Ch. 11, #20, p. 284.
  1958. Ch. 14, #4, p. 331, p. 332.
  1961. Ch. 11, #22, p. 284.
  1962. Ch. 12, #27, p. 310, p. 311.
  1965. Ch. 9, #115, p. 220; Ch. 11, #49, p. 292.
  1968. Ch. 13, #19, p. 320, p. 321.
  1971. Ch. 10, #225, p. 264; Ch. 11, #46, p. 292, p. 293.
  1972. Ch. 14, #14, p. 332.
Slater, P.: 1944. Ch. 10, #205, p. 263; Ch. 11, #52, p. 293.
  1965. Ch. 10, #169, p. 262.
Smith, C.: 1970. Ch. 6, #35, p. 100.
  1971. Ch. 12, #44, p. 314.
Smith. C. A. B.: 1947. Ch. 2, #18, p. 16.
Smith, D. R.: 1954. Ch. 12, #12, p. 308.
Smith, D. W.: 1957. Ch. 10, #143, p. 251.
  1962. Ch. 14, #34, p. 337.
Smith, G.: 1949. Ch. 9, #101, p. 217.
Smith, H. C.: 1949. Ch. 10, #117, p. 248, p. 251.
Smith, J. B.: 1967. Ch. 2, #27, p. 19.
Smith, L. H., Jr.: 1961. Ch. 17, #87, p. 391.
  1964. Ch. 17, #114, p. 392.
  1968. Ch. 17, #115, p. 392.
Smith, R. T.: 1965. Ch. 6, #64, p. 119; Ch. 9, #110, p. 218.
Smith, S. M.: 1955. Ch. 6, #10, p. 94.
Smithies, O.: 1955. Ch. 17, #10, p. 388.
  1959. Ch. 17, #123, p. 392.
Snow, R. E.: 1969. Ch. 8, #65, p. 179.
Snyder, L. H.: 1931. Ch. 18, #5, p. 401.
  1932. Ch. 18, #6, p. 401.
Snyder, S. H.: 1970. Ch. 16, #71, p. 374.
  1971. Ch. 16, #70, p. 374.
Sobel, E. H.: 1969. Ch. 10, #18, p. 264.
Soboleva, G. Y.: 1936. Ch. 18, #7, p. 401.
Sohval, A. R.: 1961. Ch. 2, #36, p. 20.
Sontag, R.: 1944. Ch. 15, #39, p. 352.
Spearman, C.: 1904. Ch. 5, #19, p. 82.
Spencer, N.: 1963. Ch. 17, #149, p. 392.
  1964. Ch. 17, #138, p. 392, #150, p. 392.
  1968. Ch. 17, #153, p. 392.
Spencer, S. J. G.: 1955. Ch. 10, #202, p. 239.
Spencer-Booth, Y.: 1971. Ch. 3, #27, p. 43.
Spencer-Peet, J.: 1963. Ch. 17, #54, p. 390.
Spielberger, C. D.: 1963. Ch. 10, #129, p. 249, #132, p. 249.
Squires, R.: 1964. Ch. 17, #107, p. 392.
Srole, L.: 1962. Ch. 11, #6, p. 282.
Stabenau, J.: 1968. Ch. 6, #77, p. 125; Ch. 11, #63, p. 297.
  1969. Ch. 11, #1, p. 279, p. 284, p. 286; Ch. 13, #26, p. 321.
Staffieri, J. R.: 1972. Ch. 10, #234, p. 266.
Stafford, R. E.: 1961. Ch. 6, #100, p. 137.
  1967. Ch. 9, #64, p. 208, #73, p. 209.
  1968. Ch. 6, #27, p. 99.
Staron, N.: 1957. Ch. 17, #143, p. 392.
Stegerhoek, L. J.: 1959. Ch. 18, #15, p. 401.
Steinberg, A. G.: 1967. Ch. 17, #127, p. 392.
Steinberg, F. S.: 1968. Ch. 14, #43, p. 337.
Stenstedt, A.: 1966. Ch. 11, #30, p. 287.
Stern, C.: 1960. Ch. 2, #15, p. 14, p. 15, p. 16.
Stertz, G.: 1920. Ch. 10, #182, p. 260.
Stevens, S. S.: 1940. Ch. 10, #42, p. 231, p. 234, p. 236, p. 237, p. 239, p. 245.
  1942. Ch. 10, #43, p. 231, p. 250, p. 253, p. 257, p. 258.
Stevenson, A. C.: 1961. Ch. 6, #23, p. 97.
Stewart, K. M.: 1963. Ch. 17, #54, p. 390.
Stewart, M. A.: 1971. Ch. 16, #98a, p. 378.
  1973 Ch. 16, #98c, p. 378.
Stewart, N.: 1947. Ch. 8, #80, p. 184.
Stowens, D.: 1970. Ch. 14, #42, p. 337.
Strand, L. J.: 1970. Ch. 16, #38, p. 369.
Strandskov, H. H.: 1953. Ch. 3, #20, p. 40.
Strayer, L. C.: 1930. Ch. 6, #88, p. 134.

Strijbasch, H.: 1969. Ch. 16, #30, p. 368.
Strömgren, E.: 1937. Ch. 10, #53, p. 233.
  1965. Ch. 11, #8, p. 282.
Strong, J. A.: 1959. Ch. 12, #13, p. 308.
Strong, P. S.: 1951. Ch. 17, #41, p. 390.
Stumpfl, F.: 1937. Ch. 11, #19, p. 284.
Sturgeon, P.: 1966. Ch. 17, #157, p. 392.
Suda, M.: 1965. Ch. 17, #25, p. 390.
Sullivan, M.: 1961. Ch. 17, #87, p. 391.
Sulzbacher, S. I.: 1971. Ch. 16, #102, p. 378.
Suter, H.: 1968. Ch. 17, #82, p. 391.
Sutherland, E. W.: 1970. Ch. 16, #95, p. 377.
Sutton, H. E.: 1955. Ch. 6, #11, p. 94.
  1962. Ch. 18, #16, p. 401.
Suzuki, H.: 1965. Ch. 15, #50, p. 355.
Suzuki, K.: 1970. Ch. 17, #99, p. 391.
  1971. Ch. 17, #92, p. 391.
Suzuki, Y.: 1970. Ch. 17, #99, p. 391.
Svancard, J.: 1971. Ch. 9, #53, p. 201.
Swyer, G. I. M.: 1960. Ch. 12, #25, p. 310.
Syner, F. N.: 1962. Ch. 17, #17, p. 388.
Szekely, L.: 1957. Ch. 12, #38, p. 313.

T

Tabershaw, I. R.: 1966. Ch. 16, #92, p. 377.
Tada, K.: 1967. Ch. 17, #70, p. 390.
Takahara, S.: 1952. Ch. 17, #81, p. 391.
Takala, M.: 1957. Ch. 15, #23, p. 349.
Takuma, T.: 1971. Ch. 9, #112, p. 218.
Tanaka, K. R.: 1966. Ch. 17, #71, p. 391.
  1968. Ch. 17, #33, p. 390.
Tanaka, T.: 1965. Ch. 17, #25, p. 390.
Tanner, J. M.: 1949. Ch. 10, #99, p. 239, p. 245.
  1951. Ch. 10, #87, p. 237.
  1952. Ch. 10, #215, p. 263.
  1959. Ch. 10, #229, p. 264.
  1962. Ch. 10, #89, p. 237, p. 263, p. 268.
  1964. Ch. 10, #6, p. 230, p. 241, p. 243, p. 266.
Tarlow, M. J.: 1969. Ch. 17, #117, p. 392.
Tashian, R. E.: 1962. Ch. 17, #17, p. 388.
Tarui, S.: 1965. Ch. 17, #25, p. 390.
Tatum, E. L.: 1941. Ch. 3, #3, p. 34.
Taut, R. R.: 1959. Ch. 17, #52, p. 390.
Taylor, C. C.: 1971. Ch. 6, #55, p. 113.
Taylor, D. C.: 1972. Ch. 10, #39, p. 231.
Taylor, G. G.: 1954. Preface, #3, xi.
Taylor, K. M.: 1970. Ch. 16, #71, p. 374.
Tedesco, T. A.: 1967. Ch. 17, #5, p. 386.
Tellenbach, H.: 1961. Ch. 10, #196, p. 262.
Tellenbach, R.: 1974. Ch. 10, #201, p. 263.
Tenney, A. M.: 1969. Ch. 10, #81, p. 236
Terbalche, J.: 1968. Ch. 16, #49, p. 371.
Terman, L. M.: 1916. Ch. 5, #18, p. 82.
  1954. Ch. 8, #95, p. 191.
  1960. Ch. 5, #22, p. 84.
  1968. Ch. 10, #152, p. 253.
Terreau, M. E.: 1970. Ch. 16, #48, p. 371.
Tharp, R. G.: 1963. Ch. 9, #38, p. 202.
Thiessen, D.: 1971. Ch. 9, #65, p. 208, p. 209.
Thoday, J. M.: 1953. Ch. 1, #3, p. 3.
  1970. Ch. 8, #92, p. 188, p. 190.
Thomas, A.: 1963. Ch. 7, #4, p. 152.
  1967. Ch. 7, #19, p. 162.
  1968. Ch. 7, #3, p. 152, p. 153, p. 161; #17, p. 162.
  1969. Ch. 7, #6, p. 152.
Thomas, D. H. H.: 1959. Ch. 17, #66, p. 390.
  1969. Ch. 16, #66, p. 373.
Thompson, H.: 1929. Ch. 6, #87, p. 134.
  1941. Ch. 6, #90, p. 134.
  1965. Ch. 2, #26, p. 19.
Thompson, W. R.: 1954. Ch. 3, #23, p. 41.
  1960. Ch. 10, #224, p. 264; Ch. 11, #50, p. 292, p. 295.
  1961. Ch. 10, #20, p. 230.
  1966. Ch. 9, #103, p. 217.
  1967. Ch. 5, #35, p. 86.
  1971. Ch. 9, #66, p. 208, p. 217.
Thomson, J. A.: 1965. Ch. 17, #67, p. 390.
Thornton, J. A.: 1963. Ch. 17, #145, p. 392.
Thurston, L. L.: 1935. Ch. 5, #20, p. 82.
  1946. Ch. 10, #108, p. 241.
  1947. Ch. 9, #5, p. 199; Ch. 10, #110, p. 241.
Tienari, P.: 1963. Ch. 11, #25, p. 284; Ch. 13, #24, p. 320.
  1966. Ch. 6, #69, p. 122.
Tiger, L.: 1971. Ch. 4, #15, p. 56, p. 61.
Tilling, T.: 1904. Ch. 10, #124, p. 249.
Titley, W. B.: 1936. Ch. 10, #200, p. 262.
Tjio, J. H.: 1956. Ch. 2, #20, p. 18.
Tokita, K.: 1964. Ch. 17, #137, p. 392.
Toro, G.: 1971. Ch. 12, #3, p. 303.
Torrey, E. F.: 1973. Ch. 14, #17, p. 332.
Townes, P. L.: 1965. Ch. 17, #116, p. 392.
Townsend, E. H.: 1951. Ch. 17, #41, p. 390.
Trap-Jensen, J.: 1964. Ch. 17, #107, p. 392.
Troll, W.: 1956. Ch. 17, #8, p. 387.
Trotter, W. R.: 1962. Ch. 18, #32, p. 410.
Troup. E. A.: 1938. Ch. 9, #86, p. 216.
Trudeau, J. G.: 1965. Ch. 14, #35, p. 337.
Tryon, R. C.: 1930-1941. Ch. 3, #21, p. 40.
Tsuda, J.: 1967. Ch. 11, #35, p. 287.
Tuck, D.: 1967. Ch. 15, #13, p. 348.
  1970. Ch. 16, #24, p. 368, p. 373.
Tucker, W. B.: 1940. Ch. 10, #4, p. 230, p. 231; #42, p. 231, p. 234, p. 236, p. 237, p. 239, p. 245.
Tuddenham, R. D.: 1962. Ch. 8, #16, p. 165.
Turpin, R.: 1959. Ch. 2, #40, p. 26.
Tyler, F. H.: 1960. Ch. 5, #7, p. 73.
Tyler, L.: 1965. Ch. 8, #78, p. 184; Ch. 10, #17, p. 230, p. 231, p. 249, p. 266.

U

Uchida, J. A.: 1962. Ch. 14, #34, p. 337.
　1963. Ch. 14, #32, p. 337.
Udenfriend, S.: 1957. Ch. 17, #60, p. 390.
Undeutsch, U.: 1959. Ch. 10, #47, p. 231.
U.S. Dept. of Health, Education and Welfare: 1967. Ch. 11, #4, p. 280.

V

Vacca, E.: 1949. Ch. 9, #89, p. 216.
Vale, C. A.: 1969. Ch. 8, #71, p. 181.
Vale, J. R.: 1969. Ch. 8, #71, p. 181.
Valentine, W. N.: 1965. Ch. 17, #27, p. 390.
　1966. Ch. 17, #28, p. 390.
　1967. Ch. 17, #22, p. 390.
　1968 Ch. 17, #23, p. 390, #32, #33, p. 390.
Van Abeelen, J. H. F.: 1969. Ch. 16, #30, p. 368.
Van deKamer, J. H.: 1960. Ch. 17, #57, p. 390.
Vandenberg, S. G.: 1962. Ch. 9, #59, p. 207, p. 217.
　1965. Ch. 15, #29, p. 351, p. 352.
　1966. Ch. 9, #60, p. 207, p. 216; Ch. 16, #7, p. 365.
　1967. Ch. 6, #111, p. 138; Ch. 9, #58, p. 207, #64, p. 208, #73, p. 209, Ch. 15, #1, p. 347; Ch. 11, #47, p. 292.
　1968. Ch. 6, #27, p. 99, #59, p. 118, #117, #118, p. 140; Ch. 8, #4, p. 164, p. 165; Ch. 9, #61, p. 207, #113, p. 218.
　1969. Ch. 8, #30, p. 166.
　1970. Ch. 6, #48, p. 109.
　In Press. Ch. 6, #25, p. 98.
Van Hoof, F.: 1968. Ch. 17, #100, p. 391.
Vartanyan, M. E.: 1972. Ch. 14, #21, p. 334, p. 335, p. 342.
Verkade, P. E.: 1959. Ch. 18, #15, p. 401.
Vernon, P. E.: 1933. Ch. 9, #99, p. 217.
　1965. Ch. 8, #10, p. 165.
　1969. Ch. 6, #102, p. 137.
Vesell, E. S.: 1968. Ch. 15, #15, p. 348; Ch. 16, #107, p. 379.
　1969. Ch. 16, #111, p. 379.
　1971. Ch. 16, #89, p. 377, #104, p. 378, #109, p. 379.
Videbech, T.: 1968. Ch. 6, #17, p. 95.
Vierucci, A.: 1967. Ch. 17, #130, p. 392.
Villers, J. D.: 1962. Ch. 16, #46, p. 371.
Vincent, B. L.: 1967. Ch. 16, #67, p. 373.
Vogel, F.: 1956. Ch. 10, #13, p. 264.
　1958. Ch. 9, #106, p. 217.
　1959. Ch. 16, #3, p. 363.
　1970. Ch. 15, #5, p. 347; Ch. 16, #74, p. 374.
　1971. Ch. 2, #50, p. 29.
Vojtisek, J. E.: 1971. Ch. 5, #31, p. 86.
Von Bracken, H.: 1969. Ch. 6, #68, p. 122.
Von Rohden, F.: 1926. Ch. 10, #128, p. 249, p. 263.
　1937. Ch. 10, #51, p. 233.
Von Verschuer, O.: 1952. Ch. 10, #11, p. 230, p. 264.
　1954. Ch. 10, #12, p. 230, p. 264, p. 268.
Von Zerssen, D.: 1960. Ch. 10, #38, p. 231.
　1963. Ch. 10, #7, p. 230, p. 231, p. 232.
　1964. Ch. 10, #63, p. 233, p. 234, p. 236, p. 237, p. 239, p. 242, p. 243.
　1965. Ch. 10, #55, p. 233, p. 249, #75, p. 234, p. 250, p. 251, p. 255, p. 266.
　1966. Ch. 10, #45, p. 231, p. 237, p. 241, p. 242, p. 248, p. 256, p. 258, #103, p. 239, p. 249, p. 258, p. 259, p. 260.
　1968. Ch. 10, #90, p. 237, p. 238, p. 239, p. 242, p. 245.
　1969. Ch. 10, #46, p. 231, p. 242, #79, p. 236, p. 239, p. 249, p. 250, p. 260, p. 267, #133, p. 249.
　1970. Ch. 10, #134, p. 249, p. 262.
　1971. Ch. 10, #163, p. 257, p. 260.
　In press. Ch. 10, #165, p. 258.

W

Wada, Y.: 1967. Ch. 17, #70, p. 390.
Waisman, H. A.: 1964. Ch. 17, #65, p. 390.
Waldenström, J.: 1937. Ch. 16, #34, p. 369.
Waldrop, F. N.: 1967. Ch. 16, #67, p. 373.
Wallace, T. J.: 1965. Ch. 17, #67, p. 390.
Wallach, M. A.: 1971. Ch. 8, #96, p. 191.
Waller, H. D.: 1962. Ch. 17, #39, p. 390.
　1968. Ch. 17, #40, p. 390.
Waller, J. H.: 1971. Ch. 8, #32, p. 166, p. 184, p. 185, p. 186, #110, p. 192.
Waller, M. I.: In press. Ch. 6, #25, p. 98.
Wallin, P.: 1944. Ch. 9, #23, p. 201, p. 202.
Wallner, J.: 1953. Ch. 5, #24, p. 85.
Walsh, M. J.: 1970. Ch. 16, #93, p. 377.
Wanklin, J. M.: 1961. Ch. 17, #105, p. 391.
Ward, I. L.: 1972. Ch. 12, #8, p. 304.
Warren, R.: 1965. Ch. 15, #21, #22, p. 349.
Warshaw, A. L.: 1967. Ch. 17, #96, p. 391.
Watson, R. L.: 1971. Ch. 16, #51, p. 371.
Watts, R. W. E.: 1963. Ch. 17, #78, p. 391.
　1967. Ch. 17, #113, p. 392.
Waxman, S. H.: 1967. Ch. 2, #27, p. 19.
Way, E. L.: 1968. Ch. 16, #96, p. 377.
Ways, P. O.: 1968. Ch. 17, #23, #24, p. 390.
Wechsler, D.: 1955. Ch. 5, #23, p. 84.
　1958. Ch. 8, #7, p. 164.
Weibacher, R. G.: 1962. Ch. 17, #38, p. 390.
Weijers, H. A.: 1960. Ch. 17, #57, p. 390.
Weiner, J. S.: 1949. Ch. 10, #99, p. 239, p. 245.
Weinberg, W.: 1908. Ch. 2, #13, p. 13.
Weinshelboum, R. M.: 1971. Ch. 16, #42, p. 370.
Weir, R. H.: 1962. Ch. 4, #4, p. 51, p. 61.
Weitz, W.: 1924. Ch. 15, #25, p. 350.
Weizmann, F.: 1971. Ch. 8, #58, p. 175.
Welham, W. C.: 1951. Ch. 10, #68, p. 234.

Weller, G. M.: 1965. Ch. 7, #11, p. 153.
Wells, I. C.: 1949. Ch. 17, #9, p. 387.
Wells, W. D.: 1961. Ch. 10, #235, p. 266.
Welner, J.: 1968. Ch. 14, #16, p. 332.
Wender, P.: 1968. Ch. 13, #28, p. 321.
Wender, P. H.: 1968. Ch. 14, #6, p. 331, p. 332; #16, p. 332.
Wendt, G. G.: 1951. Ch. 14, #49, p. 337.
  1956. Ch. 10, #13, p. 230, p. 264.
Wenner, W. H.: 1965. Ch. 15, #43, p. 354.
Wepster, B. M.: 1959. Ch. 18, #15, p. 401.
Werner, M.: 1934. Ch. 15, #31, p. 351.
  1935. Ch. 15, #27, p. 350, #30, p. 351.
  In press. Ch. 11, #64, p. 297.
Wertheimer, F. I.: 1925. Ch. 10, #54, p. 233.
Westphal, K.: 1931. Ch. 10, #127, p. 249, p. 260.
Westland, R.: 1962. Ch. 6, #8, p. 93.
Wharton, R. N.: 1971. Ch. 15, #18, p. 349.
Wheeler, L. R.: 1929. Ch. 10, #174, p. 260.
Whitehouse, R. L.: 1959. Ch. 10, #229, p. 264.
Whiting, J. W. M.: 1958. Ch. 4, #14, p. 56, p. 61.
Whittaker, J. A.: 1970. Ch. 15, #16, p. 348.
Whittaker, M.: 1960. Ch. 17, #144, p. 392.
Widdowson, E. M.: 1964. Ch. 10, #33, p. 230.
Wilcke, H. L.: 1962. Ch. 3, #5, p. 35.
Wilde, G. J. S.: 1964. Ch. 6, #61, p. 118; Ch. 9, #72, p. 209, p. 211; Ch. 11, #44, p. 292.
  1970. Ch. 6, #62, p. 118.
  1971. Ch. 9, #66, p. 208, p. 217.
Wilde, K.: 1964. Ch. 10, #91, p. 237, p. 245, p. 251, p. 253, p. 255, p. 258.
Willerman, L.: 1967. Ch. 6, #74, p. 124.
Will, D. P.: 1968. Ch. 9, #7, p. 199.
Williams, H. E.: 1968. Ch. 17, #115, p. 392.
Williams, M.: 1971. Ch. 17, #95, p. 391.
Williams, R. J.: 1956. Ch. 10, #239, p. 268; Ch. 16, #81, p. 376.
Willoughby, R. R.: 1927. Ch. 9, #29, p. 201.
  1928. Ch. 9, #30, p. 201.
Wilsnack, W.: 1965. Ch. 11, #8, p. 282.
Wilson, D. A.: 1970. Ch. 10, #184, p. 260.
Wilson, G. D.: 1970. Ch. 9, #116, p. 220.
Wilson, P. T.: 1932. Ch. 6, #52, p. 113.
Wilson, R. S.: 1970. Ch. 6, #13, p. 94.
Wilson, W. P.: 1967. Ch. 15, #46, p. 354.
Winch, R. F.: 1958. Ch. 9, #36, p. 202.
Winder, C.: 1953. Ch. 5, #24, p. 85.
Winestine, M. C.: 1969. Ch. 6, #82, p. 131.

Wing, L.: 1966. Ch. 15, #40, p. 352.
Wingert, F.: 1971. Ch. 10, #247, p. 268.
Winokur, G.: 1969. Ch. 16, #55, p. 372.
Winter, E.: 1958-1959. Ch. 10, #203, p. 263.
Wiseman, L.: 1966. Ch. 8, #61, p. 179.
Witte, E. H.: 1967. Ch. 2, #28, p. 19.
Wittgenstein, L.: 1953. Ch. 4, #16, p. 57, p. 61.
Wolf, H. P.: 1963. Ch. 17, #29, p. 390.
Wolf, R. M.: 1964. Ch. 6, #103, p. 137.
Wolf, T. H.: 1969. Ch. 8, #14, #15, p. 165.
Wolfe, O. H.: 1969. Ch. 17, #117, p. 392.
Wolff, P. H.: 1972. Ch. 16, #91, p. 377.
World Health Organization: 1966. Ch. 6, #121, #122, p. 140.
Wray, J. A.: 1967. Ch. 6, #20, p. 96.
Wright, S.: 1931. Ch. 8, #51, p. 174.
  1955. Ch. 2, #14, p. 14.
  1968. Ch. 3, #1, p. 33.
Wurtman, R. J.: 1969. Ch. 3, #6, p. 36.
Wyngaarden, J. B.: 1972. Ch. 16, #44, p. 370.
Wynne, L. C.: 1967. Ch. 11, #60, p. 296.
  1968. Ch. 13, #29, p. 322.

Y

Yalom, I. D.: 1973. Ch. 10, #232, p. 265.
Yamada, K.: 1965. Ch. 2, #45, p. 29.
Yang, S. Y.: 1969. Ch. 17, #14, p. 388.
  1970. Ch. 17, #15, p. 388.
Yarrow, L. J.: 1965. Ch. 7, #12, p. 153.
Yates, F.: 1963. Ch. 2, #4, p. 9.
Yoshida, A.: 1969. Ch. 16, #11, p. 366.
Yoshizaki, H.: 1964. Ch. 17, #137, p. 392.
Young, J. P. R.: 1971. Ch. 9, #80, p. 211, #107, p. 217; Ch. 10, #223, p. 264.
Young, W. F.: 1967. Ch. 17, #118, p. 392.

Z

Zacharias, L.: 1969. Ch. 3, #6, p. 36.
Zannoni, V. G.: 1958. Ch. 17, #2, p. 386, p. 390.
Zarrow, M. X.: 1971. Ch. 11, #57, p. 295.
Zazzo, R.: 1960. Ch. 6, #54, p. 113.
Zeidenberg, P.: 1971. Ch. 15, #18, p. 349.
Zell, W.: 1951. Ch. 14, #49, p. 337.
Zigler, E.: 1967. Ch. 8, #39, p. 168.
  1968. Ch. 5, #28, p. 86.
Zubek, J.: 1958. Ch. 8, #59, p. 177.
Zubin, J.: 1965. Ch. 9, #10, p. 200, p. 209.
Zung, W. W. K.: 1967. Ch. 15, #46, p. 354.

SUBJECT INDEX

A

Abdominal pain, 369, 370
Acalasia, 387, 391
Acentric chromosomal fragments, 28, 335, 342
Acetanilide, 367
Acetylation, 365, 366, 367, 373, 379
Acetylcholine, 369
Achondroplasia, 260
Acidosis, 371
Acknowledgments, xiii
Acrocyanosis, 351
Acroparesthesia, 351
ACTH, 367, 368, 373
Adaptability, 154
Addiction, 377, 380, 418
Adjective Check List, 209
Affective disorder, 319, 372
Age, paternal, associated with homosexuality, 310
Age & taste threshold, 409
Agitated depression, 369
ALA, *see* delta-aminolevulinic acid
ALA-synthetase, 369
Alcaptonuria, 390
Alcohol, 349, 369, 374, 376, 377, 418
Alcoholism, 376, 418
Alkaptonuria, 385, 386
Allopurinol, 370
Ambivalence, 319
Aminoquinoline, 367
Amitriptyline, 347
Amphetamine, 368, 374, 378
Analysis, factor, *see* Factor analysis
Andersen's disease, 390
Androgyny, 237, 238, 245
Anemia, 367, 387, 391, 392
   hemolytic, 367, 390
   progressive, 390
   sickle cell, 387
Anesthetic catastrophes predisposition, 371
Aneuploidies, 19
Anticholinergic, 368
Anticholinesterase, 368
Anticoagulant, 377, 379
Antidepressant, 347, 348, 367, 368, 372
Antimalarial, 367
Antimode, 417
Antipyrine, 367, 377
Apnea, 366
Aspirin, 367
Associational disturbances, 319

Assortative mating, 50
Asthenic, 232
Ataxia, 367
Athletic, 232, 234, 239
Attenuation correction, 77
Audiogenic seizure, 369
Authors, contributing, vii, viii
Autism, 43, 319
Autonomic, 369, 377
Avoidance, 368
Awakening behavior, 354
Azathioprine, 370

B

B12 vitamin, 376
Baboon, 57
Barbiturate, 369
Barr body, *see* Sex chromatin
Basal metabolism, 350
Beagle, 39
Bean, fava, 367
Behavior analysis, 151
Behavior-genetic analysis, 3
Bernreuter Personality Inventory, 292
Bimodal distribution, 363, 401, 403, 404
Binding sites in plasma protein, 379
Biological acuity, 420
Birth order and homosexuality, 310
Bishydroxycoumarin, 377
Bladder control, 355
Blindness, 391
Blinking, 351
Blood glucose, 350, 351
Blood pressure, 350
Blood sugar, 350, 351
Blood types, 93
Bloom's syndrome, 29
Brain size & environment, 38
Brain damage, 378
Brain degeneration, 391
Brain dysfunction syndrome, minimal, 378
Buccal cell, 21, 22, 332
Butyric acid, 375

C

C vitamin, 376
Caffeine, 349, 371, 375
California Psychological Inventory, 207
Cardiac arrest, 371
Cardiotachography, 350

Carrier, heterozygous, 9
Catalase, 387
Catatonia, 317
Cattell High School Personality Questionnaire, 206
Childhood diseases, 355
Chimpanzee, 57
Chloral hydrate, 379
Chlordiazepoxide, 368
Chloroquine, 369
Chlorpromazine, 349, 368
Cholinergic, 369
Choreiform movements, 371
Choreoathetosis, 370, 392
Chromosome, 26, 28, 29, 91, 92, 343
    abnormalities, 26, 29
    differences in MZ twins, 91, 92
    fragments, acentric, 28
    Philadelphia, 343
Citrullinemia, 386
Coffee, 349, see also Caffeine
Cognitive disturbance, 319
Complementarity, 202
Complementary epistasis, 7
Comrey Personality Questionnaire, 136, 207
Concordance, 16, 100, 284, 320, 357
Consanguinity, 9
Consanguinous combination, 8
Conservatism, 222
Construct validity, 78
Contents table, xv
Continuous variation traits, 3, 364, 403, 404, see also Gaussian distribution
Contributing authors, vii, viii
Coproporphyrin, 369
Corticosteroids, 367, 368, 379
Co-twin control studies, 133, 313
Coumarin, see anticoagulant
Cretinism, 230, 390
Cricket, 389
Criminal, 263
Cultural experience, 3, 4
Cumulative gene action, 403
Cyanosis, 350, 391
Cystathionuria, 376, 390
Cystic fibrosis, 386

D

Dapsone, 366
Decaffeinated coffee, 349, see also Caffeine
Dehydrogenase, 379
Delta-aminolevulinic acid, 369
Dementia praecox, 317
Depression, 369, 372, 373, 380
Dermatitis, photosensitive, 369
Dermatoglyphics, 337, 342

Desipramine, 368
Desmethyl imipramine, 348
Determinism, 296
Developmental genetics, 35, 36
Diarrhea, 390
Dibucaine, 366, 373
Diet, 407, 408
Dilantin, see Diphenylhydantoin
Dimethylase, 379
Diphenylhydantoin, 367, 375, 379
Discontinuous variation, 364, 404, see also Bimodal distribution
Diseases of childhood, 355
Distractability, 156, 378
DNA, 18
Dog, 36, 39, 40, 41, 42, 43, 368
Dopa, 367, 371, 372, 373, 379, see also Dopamine
Dopamine, 370, 374, see also Dopa
Down's syndrome, 26, 29, 92, 230, 310, see also G_1-trisomy syndrome
Dreaming, 354
Drift, genetic, 14
Drooling, 353, 358
Drosophila, 188, 306, 307, 388
Drowsiness, 367
Drug-drug interactions, 378, 379
Drug-genotype interactions, 378
Drug-induced Parkinsonism, 349
Duodenal ulcer, 418
Dysautonomia, 369, 379
Dysplasia, 237, 260
Dystonic, 373
Dystrophy, myotonic, 371

E

ECG, see EKG
Ectomorph, 234, 235, 239, 252
Ectopenia, 239
EEG, 374, 375, 378
EKG, 350
Electrocardiogram, see EKG
Electroencephalogram, see EEG
Emotional lability, 370
Emotionality, 199
Endomorph, 234, 235, 251, 252
Environment and brain size, 38
Environment-genetic interactions, 3, 13, 157, 171, 177, 230, 295
Environmental manipulation, 326
Enzyme deficiency, 386, 390, 391, 392
Enzyme polymorphism, 392
Enzyme replacement therapy, 366
Epilepsy, 260, 375, 380
Equilibration, 54, 55
Equipotentiality, 303

Eserine, 368
Ethanol, see Alcohol
Ether, 371
Eugenics, 325
Eurymorphy, 258
Extrachromosomal inheritance, 218
Extrapyramidal, 373
Extraversion, 199, 206, 214, 221, 252

F

Fabry's disease, 391
Factor analysis, 86, 87, 200, 240, 255
Facts vs. theories, xi
Falling asleep, 356, 359
Family studies, 286, 290
Fanconi's syndrome, 29
Fava bean, 367
Favism, 367, 390
Femininity, 256
Fibrosis, cystic, 386
Fingerprints, see Dermatoglyphics
Flaccid paralysis, 369
Fly, 188, 306, 307
Food dislikes, 407
Food preferences, 407
Forbes' disease, 390
Foster children studies and intelligence, 172
Frostbite, 351, 352, 355
Fruit fly, 188, 306, 307

G

G_1-trisomy syndrome, 26, 29, 92, 230, 310, see also Down's syndrome
G6PD, see Glucose-6-phosphate dehydrogenase
Galactosemia, 387
Galvanic skin reflex, 352
Gangliosidosis, 391
Garrod, 385
Gastric ulcer, 418
Gaucher's disease, 391
Gaussian distribution, 401, 403, 404
Gene action, 34, 35, 385
Genetic, 7, 10, 14, 33, 306, 325, 330, see also Genotype
　analysis, 7
　counseling, 325
　drift, 14
　engineering, 325
　linkage, 10
　manipulation, 325
　theories of homosexuality, 306
　theories of schizophrenia, 330
Genotype, 5, 6, see also Genetic
Genotype-environment, 3, 13, 157, 171, 177, 230, 295
Glaucoma, open-angle, 367, 368
Glucose, blood, see Blood sugar

Glucuronidation, 379
Glycogen storage disease, 387, 390
Gout, 370
Griseofulvin, 369
Growth retardation, 391
Guinea pig, 368
Gynandromorphy, 237
Gynandrophrenia, 253

H

Hallucinations, 369
Halothane, 371, 377
Handedness, left, 116
Hardy-Weinberg principle, 13, 14
HCL taste, 402, 403
Heartbeat interval, 350
Hebephrenia, 317
Hemizygous, 9, 10, 20
Hemoglobin, 387
Hemolytic anemia, 367, 390, 391
Heritability, 11, 40, 99, 100, 101, 108, 201, 208, 377
　categorical data, 100
　continuous traits, 101
　twins, 99
Heroin addiction, 418
Heterosexual-homosexual rating scale, 302
Heterozygous carrier, 9
HGPRT, see Hypoxanthine-guanine phosphoribosyl
Histamine, 370
Histidinemia, 390
Holandric, 10
Homocystinuria, 376, 390
Homogamy, 202
Homogentisic acid, 385
Homosexual sibships, 309, 310
Homosexuality, 301, 302, 304, 306, 310, 311, 313
　& birth order, 310
　genetic theories, 306
　& hormonal influence, 304, 310
　& paternal age association, 310
　psychoanalytic theories, 304
　rating scale, 302
　twin studies, 311, 313
Hormones & homosexuality, 304, 310
Human individual differences, xi
　intraspecies variation, xi, see also Human species polymorphism
　species polymorphism, 4, see also Polymorphism
Humanization, 3
Huntington's chorea, 35, 371, 379
Hurler's syndrome, 376
Hydralazine, 366
Hydrochloric acid, 402, 403
Hydrocortisone, 367
Hydroxylating enzymes, 348

Hydroxylation, 348, 368, 372, 375, 379
Hydroxytryptamine, 375
Hyperactive child syndrome, 378
Hyperactivity, 378
Hyperhidrosis, 352
Hyperkinetic child, 378
Hyperkalemia, 371
Hyperpyrexia, 370, 371, 379
Hypertension, 379
Hyperthermia, 370, 371, 379
Hyperuricemia, 370
Hypervalinemia, 390
Hypnogogic movements, 354
Hypocalcemia, 371
Hypochondriasis, 202
Hypoglycemia, 387, 390
Hypomanic, 372
Hypothermia, 368, 371
Hypotonia, 392
Hypoxanthine, 370
Hypoxanthine-guanine phosphoribosyl transferase, 370
Hypoxia, 371

I

Imipramine, 347, 348, 368, 372
Impulsiveness, 200, 253, 263, 378
INH, see Isoniazid
Insensitivity, taste, 401
Intelligence, 40, 81, 164, 165, 166, 168, 172, 175, 181, 183, 191, 193, 248, 376, see also IQ
  distribution, 168
  & family size, 193
  & foster children studies, 172
  & future, 191
  genetic model, 165, 175
  genetics, 164
  heritability, 181
  nature, 164
  & physique, 164
  quotient, see IQ
  regression, 166
  stratification, 193
  test score variance, 183
  testing, 40, 81
Intraocular pressure, 367
Introversion, 256
IQ, 40, 41, 81, 82, 83, 84, 135, 173, 174, 185, 222, 317, see also Intelligence
  correlations, 185, 222
  & home environment, 174
  measurement, 81
  tests, 317
  & twins, 135
  variance, 173, 185

Isocarboxazid, 347
Isochromosome, 21
Isoniazid, 348, 365, 366, 375, 379
Itinerant antimode, 417

K

Karyotype, 23, 24, 332
Klinefelter's syndrome, 25, 29, 265, 310
Krabbe's disease, 391

L

Landrace pig, 371
Language, 58, 59
Learning, 39, 378
Left-handedness, 116
Leptosome, 232, 239
Lethargy, 392
Leucodystrophy, 391
Leukemia, chronic myelogenous, 343
Lesch-Nyhan syndrome, 370, 379, 392
Lie scale, 221
Light sleep, 356, 358
Limb weakness, 369
Linkage, genetic, 10, 15
Lockjaw, 373
Love capacity, 52, 53
LSD, 374
Lymantria, 307, 308
Lysergic acid diethylamide, see LSD

M

Maladaptive behavior, 43
Mania, 372
Manic-depressive psychosis, 372
MAO, see Monamine oxidase
MAOI, see Monamine oxidase inhibitor
Maple-syrup urine disease, 376, 390
Maudsley Personality Inventory, 220, 292
MAVA, 99
Maze tests, 39, 40, 41
MBD, see Minimal brain dysfunction syndrome
McArdle's syndrome, 390
Melanoma, cutaneous malignant, 343
Memory, 377
Mendel, 385
Menstrual cycle & taste, 412
Mental retardation, 370, 375, 380, 390, 391, 392
Meprobamate, 374
Mercaptoimidazole, 401
Mercaptopurine, 370
Meritocracies, 193
Mesomorph, 234, 235
Metabolism, 350
Methacholine, 370
Methemoglobinemia, 391

Methoxyflurane, 371
Methylmalonic aciduria, 376
Methylphenidate, 378, 379
Middlesex Hospital Questionnaire, 211, 212
Minimal brain dysfunction syndrome, 378, 380
Minnesota Multiphasic Personality Inventory, *see* MMPI
MMPI, 199, 202, 205, 206, 209, 402
Mobility, 188
Monamine oxidase, 372
Monamine oxidase inhibitor, 347, 348, 372, 373, 379
Monkey, 43
Monofactorial, 403
Mood, 156
Morbidity risk, 7
Mosaicism, 19, 332, 342
Mother, and emotional development, 52, 53
Motion sickness, 352, 355
Mouse, 368, 369, 375, 377
Movements, hypnogogic, 354
MS, *see* Multiple sclerosis
Mucopolysaccharides, 376
Multifactorial inheritance, 16
Multiple Abstract Variance Analysis (MAVA), 99
Multiple etiologies, 7
Multiple sclerosis, 367
Multivariate, 86
Muscle, 371, 390
  cramps, 390
  relaxant, 371
  rigidity, 371
  weakness, 390
Mutation, 14
Mutilation, self, 370
Myers-Briggs Type Indicator, 207
Myotonic dystrophy, 371

N

Natural selection, 49
NE, *see* Norepinephrine
Neonate, 153
Nephropathy, 370
Neurological, 390, 391
Neuropathy, autonomic, 369
Neurosis, 280, *see* Psychoneurosis, *see also* Neuroticism
  family studies, 286
  polygenic model, 293
  twin studies, 284
Neuroticism, 199, 206, 221, *see* also Neurosis twins, 290, 291, 292
Neurotransmitter, 369
Nicotine, 375
Niemann-Pick disease, 391
Night terror, 355

Nitrofuran, 367
Norepinephrine, 370, 372, 373, 375
Norm of reaction, 11, 175, 324
Normal distribution, *see* Gaussian, *see* Continuous distribution
Normative development, 79
Norms, 79
Nortriptyline, 348, 368, 379
Nystagmus, 367
Nutrition & twinning, 96

O

Oculogyric crisis, 373
Odor of phenylthiourea, 402
Oral mucosa ulceration, 387
Oxidants, 367, 379
Oxidative phosphorylation, 375
Oxyphenylbutazone, 348

P

Pain sensitivity, 370
Pair concordance analysis, 16, 17, 18, 357
Palsy, 392
Para-amino-salicylic acid, 367
Paralysis, flaccid, 369
Parameters, 67
Paranoid, 369
Parkinsonism, 349, 371, 372, 373
PAS, *see* Para-amino-salicylic acid
Paternal age and homosexuality, 310
Pea, garden, 385
Pedigree analysis, 7
Pentose-phosphate shunt, 367
Pentylenetetrazol, 375
Peptic ulcer, 417
Persistence, 156
Personality, 84, 85, 198, 409, 421
Pharmacogenetics, 347, 363
Phenacetin, 367
Phenelzine, 347, 348, 366, 367, 373
Phenobarbitol, 369, 374, 375, 378, 379
Phenothiazine, 349, 373, 374
Phenylalanine, 376, 385, 387
Phenylbutazone, 377
Phenylketonuria, 376, 387, 390
Phenylthiourea, 401
  odor, 402
Philadelphia chromosome, 343
Phobias, 355
Phosphorylation, 375
Physiological genetics, 350
Physique, 248
  & intelligence, 248
  & temperament, 248
Pig, guinea, 368

Pig, Landrace, 371
Plasma protein binding sites, 379
Polygenic inheritance, 16
Polymorphism, 388, 392
Polymyalgia rheumatica, 367
Pompe's disease, 390
Population genetics, 50
Population taste threshold distribution, 403
Porphobilinogen, 369
Porphyria, 369, 379
   intermittent acute, 369
   Swedish type, 369
   variegata, 369
   South African type, 369
Posture, sleep, 354
Pressure, blood, 350
Pressure, intraocular, 367
Primate social systems, 53, 54
Primaquine, 367, 390
Probenecid, 367
Procaine, 371
Progressive anemia, 390
Projective tests, 209
Proof of genetic etiology, 330
PROP, 402
Propylthiouracil, 401
Protein binding sites, 379
Proteinuria, 392
Protoporphyrin, 369
Pseudocholinesterase, 365, 366, 371, 373
Psilocybin, 374
Psychoanalytic theories of homosexuality, 304
Psychogalvanic reflex, 352
Psychological family studies, 290
Psychoneurosis, 282, see also Neurosis
Psychopathology, 86
Psychopharmacogenetics, 363, 379, see also Psychotropic drugs
Psychophysiology, 350
Psychosis, 368, 372, see also Schizophrenia
   manic-depressive, 372
   schizophrenic, see Schizophrenia
   steroid, 368
Psychoticism, 202, 221
Psychotropic drugs, 348, 349, see also Psychopharmacogenetics
PTC, 401
Pulse rate, 350
Purine, 370
Pyknic, 232, 234, 239
Pyknomorphy, 258
Pyridoxine, 367, 376

Q
Quinine, 402

R
Rabbit, 368
Rat, 38, 40, 41, 368, 377, 379
Relaxant, muscle, 371
Reliability, 69
REM sleep, 354
Reserpine, 368, 372, 373
Respiratory rate, 350
Rheumatica, polymyalgia, 367
Ribonucleotide, 370
Rigidity, muscle, 371
Riley-Day syndrome, 369
Risk, 7
RNA, 18

S
Salivary secretion, 352
Sampling, 68
Sandhoff's disease, 391
Schiff bases, 377
Schizophrenia, 43, 44, 125, 126, 202, 260, 317, 320, 322, 323, 325, 326, 330, 331, 332, 335, 337, 340, 341, 343, 349, 373, 374, 376, 380
   acentric chromosomal fragments, 335, 342
   Bleuler, Manfred, theory, 331
   Böök theory, 331
   chronicity index, 341
   dermatoglyphics, 337
   dominant gene theory, 331
   environmental etiology, 322
   environmental manipulation, 326
   genetic counseling, 325
   genetic-environment interaction, 319, 320, 330
   Gottesman theory, 331
   infectious factor, 343
   infinite etiologies theory, 331
   Kallmann theory, 331
   Karlsson theory, 331
   karyotype, 332
   Kety theory, 331
   length of first hospitalization, 340
   Mitsuda theory, 331
   mosaics, 332
   multifactorial theory, 331
   multiple etiologies, 331, 343
   multiple-gene theory, 331
   multiple-locus theory, 331
   Ödegärd theory, 331
   parent-to-offspring transmission evidence, 332
   polygenic theory, 331
   prevalence, 325
   recessive genes theory, 331
   Shields theory, 331
   single-locus genetic theories, 331
   Slater theory, 331

slow-acting virus theory, 332
twin studies, 320, 323
two-locus genetic theory, 331
X-chromosome-aneuploidy mosaicisms, 332, 342
Schizophrenic behavior, 369
Scoliosis, 392
Scopolamine, 368
Secretor locus, 371
Selection, 41, 44, 49, 50
Self-mutilation, 370
Sensory threshold, 155, 402
Serotonin, 372, 373
Sex, 20, 35, 36, 37, 97, 308, 309, 310, 409
  & body size, 35, 36, 37
  chromatin, 20, 308
  chromosome, 20
  determination, 308
  deviants, 308
  differences, 35, 36, 37
  ratio & homosexual sibships, 309, 310
  & taste threshold, 409
  & twinning, 97
Siblings & taste thresholds, 404
Sickle cell anemia, 387
Seizure, audiogenic, 369
Seizure, epileptic, 260, 375, 380
Sleep, 354, 355, 356, 357, 358, 359
  behavior genetics, 354
  falling into, 356, 359
  light, 356, 358
  posture, 354
  REM, 354
  talking, 354, 355, 356, 357, 358, 359
  tryptophan effect, 358
  walking, 354, 355, 356, 359
Smoking & taste threshold, 409
Social behavior selection, 44
Sociopathy, 294
Somatotonia, 250, 253
Somatotype, 232, 234, 235, 244, 245, 246
Somnambulism, 354, 355, 356, 359
Spasticity, 370
Speech defect, 116
Spouses & taste thresholds, 407, 409
Standard deviation, 68
Standard error, 68
Standardization, 80, 81
Startle response, 352
Statistical significance, 12, 13
Steroid, 367, 368
Steroid psychosis, 368
Stimulation, 155
Stratification in sampling, 68, 69
Strychnine, 375
Stuttering, 116

Succinyl choline, 366, 371
Sugar, blood, 350, 351
Sulfa, 366, 367, *see* also Sulfonamides
Sulfonamides, 369, *see* also Sulfa
Suxamethonium, 366, 371, 373, 379, 391
Sweat glands, 352
Sweating, 352
Sympathomimetic amines, 379

T
Table of contents, xv
Tachypnea, 371
Tachycardia, 371
Taste, 370, 401, 402, 403, 404, 407, 408, 409, 412, 417, 418, 420, 421
  acuity, 401
  & age, 409
  & alcoholism, 418
  antithyroid compounds, 401, 403
  biological acuity, 420
  & dietary preferences, 407, 408
  & duodenal ulcer, 418
  & gastric ulcer, 418
  HCL, 402
  & heroin addiction, 418
  hydrochloric acid, 402
  insensitivity, 401
  & menstrual cycle, 412
  & MMPI, 402
  & peptic ulcer, 417
  & personality, 421
  phenylthiourea, 401
  PROP, 402
  PTC, 401
  quinine, 402
  & sex, 409
  & siblings, 404
  & smoking, 409
  test solutions molarity, 402
  thresholds, 402
  & twins, 404
  & ulcer, 417
Tay-Sachs disease, 391
Teeth-grinding, 354, 355, 359
Temperament, 152, 153, 154, 156, 158, 248, 256, 257
  categories, 153, 154
  & neonate, 153
  origins, 156
  & physique, 248
  twin studies, 158
Terror, night, 355
Theories *vs.* facts, xi
Thought disorder, 319
Threshold, 155, 402
  taste, 402

Thurstone Temperament Schedule, 207
Topaceous, 370
Torticollis, 373
Trichloracetic acid, 379
Tricyclic antidepressants, 347, 348, 368, 372, 379
Triptyline, 379
Trunk index, 236
Tryptophan affecting sleep, 358
Turner's syndrome, 21, 23, 92, 310
Twin method, 91, 283
Twinning, 91, 96
   frequency, 96
   rates, 96
Twins, 91, 96, 99, 116, 117, 118, 119, 120, 122, 123, 125, 127, 131, 135, 140, 146, 158, 203, 210, 218, 283, 284, 310, 320, 349, 350, 365, 366
   activities, 119
   & alcohol, 349, 377
   associations, 140
   attitudes, 116, 127
   birth weights, 123
   & coffee, 349
   dominance, 117, 125
   dominance-submissiveness, 125
   & drug reactions, 365
   dyadic relationship, 122
   habits, 119
   & heritability, 99
   & homosexuality, 310
   & insomnia, 349
   & isoniazid, 366
   IQ, 135, 203
   & left-handedness, 116
   mental abilities, 116
   names, 120
   neuroses, 284
   & minimal brain dysfunction syndrome, 378
   mothers' organizations, 140
   parental reactions, 218
   personality, 116, 210
   pharmacology, 365
   physiology, 350
   questionnaires, 146
   reared apart, 131
   & schizophrenia, 320
   separation, 118, 131
   size, 116
   stuttering, 116
   & taste thresholds, 403, 404
   temperament, 158
Tyramine, 379
Tyrosine, 385, 386
Tyrosine hydroxylase, 372
Tyrosinemia, 390
Tyrosinuria, 386

U
Ulcer, 417
   duodenal, 417
   gastric, 417
   peptic, 417
Universals, 59
Uremia, 370
Uric acid, 370
Urine, 385, 386

V
Validity, 74, 78
   construct, 78
Variance, 11, 200
Variation, continuous, 364
Variation, discontinuous, 364
Vascular reactivity to alcohol, 377
Vicia fava, 367
Viscerotonia, 252, 253
Vitamin B12, 376
Vitamin C, 376
Vomiting, 370
Von Gierke's disease, 387, 390

W
WAIS, 421
Warfarin, *see* Anticoagulant
Weakness, limb, 369
Wechsler Adult Intelligence Scale, 421
Withdrawal, 374
Wolman's disease, 392

X
Xanthine, 370
X chromosome, 10, 20, 334
X-chromosome aneuploidy, 332, 342
X-chromosome-aneuploidy mosaicism, 332, 342
X-linked, 10, 15
Xanthomatosis, 392
XXX female, 22, 24, 25, 334
XXY male, 25
XYY male, 25

Y
Y chromosome, 10
Y-linked, 10

Z
Zygosity determination, 93, 94, 95, 96